Adrenergic Receptor Protocols

METHODS IN MOLECULAR BIOLOGY™

John M. Walker, Series Editor

METHODS IN MOLECULAR BIOLOGY™

Adrenergic Receptor Protocols

Edited by

Curtis A. Machida

Oregon Regional Primate Research Center
Oregon Health Sciences University
Beaverton, OR

Humana Press ✳ Totowa, New Jersey

This publication is printed on acid-free paper. ∞
ANSI Z39.48-1984 (American Standards Institute)
Permanence of Paper for Printed Library Materials.

Background illustration courtesy of Amy Lee, Amy E. Wissekerke, Diane L. Rosin, and Kevin R. Lynch. Illustration depicts cells ventrolateral to locus ceruleus in the subceruleus region.

Cover design by Patricia F. Cleary.

For additional copies, pricing for bulk purchases, and/or information about other Humana titles, contact Humana at the above address or at any of the following numbers: Tel.: 973-256-1699; Fax: 973-256-8341; E-mail: humana@humanapr.com; or visit our Website: http://humanapress.com

Printed in the United States of America. 10 9 8 7 6 5 4 3 2 1

Library of Congress Cataloging in Publication Data

Main entry under title:

Methods in molecular biology™.

Adrenergic receptor protocols/edited by Curtis A. Machida.
    p.   cm.—(Methods in molecular biology; v. 136)
    Includes index.
    ISBN 0-89603-602-2 (alk. paper)
    1. Adrenaline—Receptors Laboratory manuals. I. Machida, Curtis A. II. Series: Methods in molecular biology (Totowa, NJ); v. 126.
    [DNLM: 1. Receptors, Adrenergic—analysis. 2. Gene Expression Regulation. 3. Receptors, Adrenergic—genetics. 4. Sequence Analysis, RNA—methods. W1 ME9616J v. 126 2000/WL 102.8 A2417 2000]
    QP572.A35A   2000
    612.8'9—dc21
    DNLM/DLC
    for Library of Congress   99-34211
                                                                CIP

Preface

Adrenergic receptors are important modulators in the sympathetic control of various metabolic processes in the central and peripheral nervous systems. These receptors are localized at multiple sites throughout the central nervous system (CNS) and serve as important regulators of CNS-mediated behavior and neural functions, including mood, memory, neuroendocrine control, and stimulation of autonomic function.

Adrenergic Receptor Protocols consists of 35 chapters dealing with various aspects of adrenergic receptor analyses, including the use of genetic, RNA, protein expression, transactivator, second messenger, immunocytochemical, electrophysiological, transgenic, and *in situ* hybridization approaches. This volume details the use of various methods to examine the adrenergic receptor system, using aspects of the genetic flow of information as a guide (DNA→ RNA → transactivator → protein expression → second messenger analyses → cellular analyses → transgenic whole animal approaches).

Adrenergic Receptor Protocols displays step-by-step methods for successful replication of experimental procedures, and would be useful for both experienced investigators and newcomers in the field, including those beginning graduate study or undergoing postdoctoral training. The Notes section contained in each chapter provides valuable troubleshooting guides to help develop working protocols for your laboratory. With *Adrenergic Receptor Protocols*, it has been my intent to develop a comprehensive collection of modern molecular methods for analyzing adrenergic receptors.

I would like to thank the many chapter authors for their contributions. They are all experts in various aspects of adrenergic receptors, and I appreciate their efforts and hard work in developing comprehensive chapters. As Volume Editor, it was a privilege to preview the development of *Adrenergic Receptor Protocols*, and to acquire insight on the various methodological approaches from different contributors. I would like to thank Professor John Walker, Series Editor for *Methods in Molecular Biology*, for his guidance and help in the development of this volume, and Thomas Lanigan, President of Humana Press. I would also like to thank Carol Houser for her administrative assistance in the preparation of manuscripts, and for members of my laboratory group, who have been helpful in the early stages of identification of potential chapter topics and contributors. Special thanks are extended to my wife, Dr. Cindy Machida, and my daughter, Cerina, for their support during the long hours involved in the compilation and editing of this volume.

Curtis A. Machida

Contents

Contributors

CHIYE AOKI • *Center for Neural Science, New York University, New York, NY*

SULEIMAN W. BAHOUTH • *Department of Pharmacology, University of Tennessee, Memphis, TN*

BURNS C. BLAXALL • *Department of Pharmacology, University of Colorado Health Sciences Center, Denver, CO*

DAVID B. BYLUND • *Department of Pharmacology, University of Nebraska Medical Center, Omaha, NE*

PATRICIA K. CURRAN • *Membrane Biochemistry Section, Laboratory of Molecular and Cellular Neurobiology, National Institute of Neurological Disorders and Stroke, National Institutes of Health, Bethesda, MD*

JEAN D. DEUPREE • *Department of Pharmacology, University of Nebraska Medical Center, Omaha, NE*

JEAN-CHRISTOPHE DEVEDJIAN • *Laboratoire d'Ingénierie des Systèmes Macromoléculaires, Marseille, France*

JANE C. DEWAR • *Division of Therapeutics, University Hospital, Nottingham, UK*

CHERYL D. DUNIGAN • *Membrane Biochemistry Section, Laboratory of Molecular and Cellular Neurobiology, National Institute of Neurological Disorders and Stroke, National Institutes of Health, Bethesda, MD*

LINCOLN EDWARDS • *Departments of Nutrition and Pharmacology, Case Western Reserve University School of Medicine, Cleveland, OH*

STEFAN ENGELHARDT • *Institut für Pharmakologie, Universität Würzburg, Germany*

PAUL ERNSBERGER • *Departments of Nutrition and Pharmacology, Case Western Reserve University School of Medicine, Cleveland, OH*

PETER H. FISHMAN • *Membrane Biochemistry Section, Laboratory of Molecular and Cellular Neurobiology, National Institute of Neurological Disorders and Stroke, National Institutes of Health, Bethesda, MD*

BIN GAO • *Department of Pharmacology and Toxicology, Medical College of Virginia, Richmond, VA*

ULRIK GETHER • *Division of Molecular and Cellular Physiology, Department of Medical Physiology, The Panum Institute, University of Copenhagen, Denmark*

JEAN-LUC GUILLAUME • *Immuno-Pharmacologie Moléculaire, Institut Cochin de Génétique Moléculaire, Paris, France*

IAN P. HALL • *Division of Therapeutics, University Hospital, Nottingham, UK*

BRIAN B. HOFFMAN • *Division of Endocrinology, Gerontology, and Metabolism, Department of Medicine, VA Palo Alto Health Care Systems, Palo Alto, CA; Department of Medicine, Stanford University School of Medicine, Stanford, CA*

ZHOU-WEI HU • *Division of Endocrinology, Gerontology, and Metabolism, Department of Medicine, VA Palo Alto Health Care Systems, Palo Alto, CA; Department of Medicine, Stanford University School of Medicine, Stanford, CA*

PAUL A. INSEL • *Department of Pharmacology, University of California, San Diego, CA*

ANNE DAM JENSEN • *Division of Molecular and Cellular Physiology, Department of Medical Physiology, The Panum Institute, University of Copenhagen, Denmark*

RALF JOCKERS • *Immuno-Pharmacologie Moléculaire, Institut Cochin de Génétique Moléculaire, Paris, France*

MELANIE E. M. KELLY • *Departments of Pharmacology and Ophthalmology, Dalhousie University, Halifax, NS, Canada*

PHILBERT KIRIGITI • *Division of Neuroscience, Oregon Regional Primate Research Center, Oregon Health Sciences University, Beaverton, OR*

GEORGE KUNOS • *Department of Pharmacology and Toxicology, Medical College of Virginia, Richmond, VA*

HITOSHI KUROSE • *Laboratory of Pharmacology and Toxicology, Graduate School of Pharmaceutical Sciences, University of Tokyo, Japan*

FRANCES M. LESLIE • *Department of Pharmacology, College of Medicine, University of California, Irvine, CA*

FUBAO LIN • *Department of Molecular Pharmacology-HSC, Diabetes and Metabolic Diseases Research Center, School of Medicine, State University of New York, Stony Brook, NY*

MARTIN J. LOHSE • *Institut für Pharmakologie, Universität Würzburg, Germany*

CURTIS A. MACHIDA • *Division of Neuroscience, Oregon Regional Primate Research Center, Oregon Health Sciences University, Beaverton, OR; Department of Biochemistry and Molecular Biology and Program in Neuroscience, Oregon Health Sciences University, Portland, OR*

JUDITH C. W. MAK • *Department of Thoracic Medicine, National Heart and Lung Institute, Imperial College, London, UK*

CRAIG C. MALBON • *Department of Molecular Pharmacology-HSC, Diabetes and Metabolic Diseases Research Center, School of Medicine, State University of New York, Stony Brook, NY*

KENNETH P. MINNEMAN • *Department of Pharmacology, Emory University, Atlanta, GA*

RENNOLDS S. OSTROM • *Department of Pharmacology, University of California, San Diego, CA*

JAMES F. PADBURY • *Department of Pediatrics, Women and Infants' Hospital of Rhode Island, Brown University School of Medicine. Providence, RI*

HERVÉ PARIS • *Institut National de la Santé et de la Recherche Médicale, Institut Louis Bugnard, Toulouse, France*

DIANNE M. PEREZ • *Department of Molecular Cardiology, Lerner Research Institute, Cleveland Clinic Foundation, Cleveland, OH*

J. DAVID PORT • *Department of Pharmacology and Division of Cardiology, Department of Medicine, University of Colorado Health Sciences Center, Denver, CO*

STEVEN R. POST • *Department of Pharmacology, University of California, San Diego, CA*

JOHN F. RESEK • *Department of Pharmacology, University of Wisconsin, Madison, WI*

SARINA RODRIGUES • *Center for Neural Science, New York University, New York, NY*

DANIEL K. ROHRER • *Department of Molecular Pharmacology, Roche Bioscience, Palo Alto, CA*

YAJING RONG • *Department of Pharmacology, University of Wisconsin, Madison, WI*

DIANE L. ROSIN • *Department of Pharmacology, University of Virginia Health Sciences Center, Charlottesville, VA*

ARNOLD E. RUOHO • *Department of Pharmacology, University of Wisconsin, Madison, WI*

JENNIFER S. RYAN • *Department of Pharmacology, Dalhousie University, Halifax, NS, Canada*

STÉPHANE SCHAAK • *Institut National de la Santé et de la Recherche Médicale, Institut Louis Bugnard, Toulouse, France*

DEBRA A. SCHWINN • *Department of Anesthesiology, Duke University Medical Center, Durham, NC*

MARGARET A. SCOFIELD • *Department of Pharmacology, Creighton University, Omaha, NE*

CHANJUAN SHI • *Department of Pharmacology, Dalhousie University, Halifax, NS, Canada*

MICHAEL K. SIEVERT • *Department of Pharmacology, University of Wisconsin, Madison, WI*

MARK STAFFORD SMITH • *Department of Anesthesiology, Duke University Medical Center, Durham, NC*

RUTH L. STORNETTA • *Department of Pharmacology, University of Virginia, Charlottesville, VA*

A. DONNY STROSBERG • *Immuno-Pharmacologie Moléculaire, Institut Cochin de Génétique Moléculaire, Paris, France*

YI-TANG TSENG • *Department of Pediatrics, Women and Infants' Hospital of Rhode Island, Brown University School of Medicine, Providence, RI*

HSIEN-YU WANG • *Department of Physiology and Biophysics, Diabetes and Metabolic Diseases Research Center, School of Medicine, State University of New York, Stony Brook, NY*

AMANDA P. WHEATLEY • *Division of Therapeutics, University Hospital, Nottingham, UK*

URSULA WINZER-SERHAN • *Department of Pharmacology and Toxicology, Medical College of Virginia, Virginia Commonwealth University, Richmond, VA*

ZHONGREN WU • *Department of Molecular and Cellular Engineering, University of Pennsylvania, Philadelphia, PA*

YONG-FENG YANG • *Division of Neuroscience, Oregon Regional Primate Research Center, Oregon Health Sciences University, Beaverton, OR*

HONGYING ZHONG • *Department of Pharmacology, Emory University, Atlanta, GA*

MICHAEL J. ZUSCIK • *Department of Molecular Cardiology, Lerner Research Institute, Cleveland Clinic Foundation, Cleveland, OH*

I

GENETIC ANALYSIS

1

Construction of Libraries for Isolation of Adrenergic Receptor Genes

Margaret A. Scofield, Jean D. Deupree, and David B. Bylund

1. Introduction

1.1. Adrenergic Receptors

Adrenergic receptors mediate the central and peripheral actions of norepinephrine and epinephrine. Both of these catecholamine messengers play important roles in the regulation of diverse physiological systems and are widely distributed throughout the body. Agonists and antagonists interacting with adrenergic receptors have proven useful in the treatment of a variety of cardiovascular, respiratory, and mental disorders *(1,2)*.

Adrenergic receptors were originally divided into two major types, α-adrenergic receptor (α-AR) and β-adrenergic receptor (β-AR), based on their pharmacological characteristics (i.e., rank order potency of agonists) *(3)*. Subsequently, the α-AR and β-AR types were further subdivided into α_1-AR, α_2-AR, β_1-AR, and β_2-AR subtypes (for a more complete historical perspective *see* **refs. 4,5**). Based on both pharmacological and molecular evidence, it is now clear that a more useful classification scheme is based on three major types—α_1-AR, α_2-AR, and β-AR—each of which is further divided into three or four subtypes (**Fig. 1**) *(4)*.

1.1.1. α_1-AR Subtypes

α_{1A}-AR and α_{1B}-AR subtypes were defined pharmacologically based on the differential affinities of WB 4101 and phentolamine *(6–8)*, and on selective receptor inactivation by the alkylating agent chlorethylclonidine. Three α_1-AR subtypes have been identified by molecular cloning (**Table 1**). The α_{1B}-AR from hamster was cloned first *(9)*, followed by the bovine α_{1A}-AR, which

From: *Methods in Molecular Biology, vol. 126: Adrenergic Receptor Protocols*
Edited by: C. A. Machida © Humana Press Inc., Totowa, NJ

Adrenergic Receptors

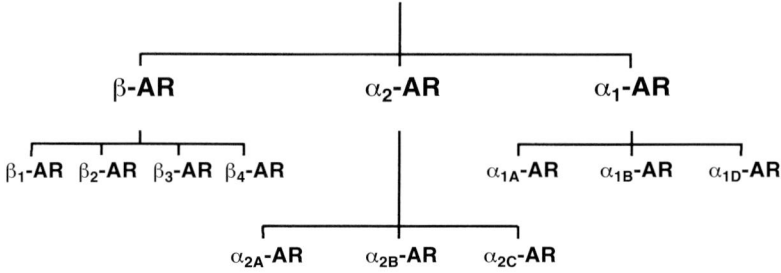

Fig. 1. The current classification scheme for adrenergic receptors.

unfortunately was prematurely identified as the α_{1C}-AR subtype *(10,11)*. The α_{1D}-AR was cloned from the rat *(12,13)*, although this receptor was prematurely called the α_{1A}-AR. A fourth α_1-AR subtype, called the α_{1L}-(based on its low affinity for prazosin), has been suggested *(14,15)*, although its existence is debatable *(16)*.

1.1.2. α_2-AR Subtypes

The evidence for α_2-AR subtypes initially came from binding and functional studies in various tissues and cell lines *(17)*. On the basis of these studies, three genetic and four pharmacological α_2-AR subtypes have been defined. The α_{2A}-AR and α_{2B}-AR subtypes were initially defined based on differential affinity for adrenergic agents, such as prazosin and oxymetazoline *(4)*. These subtypes were subsequently cloned from a variety of species (**Table 2**). A third subtype, α_{2C}-AR, was identified originally in an opossum kidney cell line *(18,19)* and has also been cloned from several species (**Table 2**). A fourth pharmacological subtype, the α_{2D}-AR, has been identified in the rat and cow *(20,21)*. This pharmacological subtype is a species ortholog of the human α_{2A}-AR subtype and, thus, is not considered to be a separate genetic subtype.

1.1.3. β-AR Subtypes

The β_1-AR and β_2-AR subtypes were identified as early as 1967 based on a comparison of the rank orders of potency of a variety of agonists *(22)*. Highly selective antagonists for both β_1-AR and β_2-AR have been subsequently developed. More recently, it became apparent that not all of the β-AR-mediated responses can be classified as either β_1-AR or β_2-AR, and thus, the β_3-AR was identified *(23,24)*. This receptor has low affinity for the commonly used β antagonists and has often been referred to as the "atypical" β-AR. These

(text continued on page 13)

Table 1

α₁-Adrenergic Receptors

Adrenergic receptor subtype[a]	Species, common name	Definition in GenBank database, subtype identification in text	Accession number	cDNA or genomic[b] tissue type	5' Flanking region	Number of nucleotides, coding sequence
α_{1A}-AR	*Oryctolagus cuniculus* rabbit	α_{1a}-adrenoceptor	U81982	cDNA (c) liver	No	1401 bp (1..1401)
	O. cuniculus rabbit	α_{1c}-adrenoceptor subtype	S75999	cDNA (p) brain	No	432 bp (1..430)
	Homo sapiens human	adrenergic α_{1c} receptor	U03866	cDNA (c) prostate	No	1500 bp (66..1466)
	H. sapiens human	α_{1C}-AR	L31774	cDNA (c) prostate	No	1401 bp (1..1401)
	H. sapiens human	α_{1C}-AR	D25235	cDNA (c) prostate	No	2290 bp (437..1837)
	H. sapiens human	α_{1C}-AR	U02569	cDNA (c) prostate	No	1902 bp (425..1825)
	H. sapiens human	α_{1A}-AR isoform 4	AF013261	cDNA (c) prostate	No	1765 bp (201..1568)
	H. sapiens human	α_{1C}-adrenoceptor receptor isoform 3	D32201	cDNA (c) prostate	No	2089 bp (437..1726)
	H. sapiens human	α_{1C}-AR isoform 2	D32202	cDNA (c) prostate	No	2306 bp (437..1936)
	H. sapiens human	α_{1c}-adrenoceptor subtype	S76001	cDNA (p) brain, saphenous vein	No	432 bp (1..431)

(continued)

Table 1 (continued)
α_1-Adrenergic Receptors

Adrenergic receptor subtype[a]	Species, common name	Definition in GenBank database, subtype identification in text	Accession number	cDNA or genomic[b] tissue type	5' Flanking region	Number of nucleotides, coding sequence
	H. sapiens human	α_{1a}-AR	U72653	Genomic	Yes	6195 bp (no coding region)
	Bos taurus cow	α_{1C}-AR	J05426	cDNA (c) adult brain cortex	No	2461 bp (97..1497)
	Mus musculus house mouse	α_{1a}-AR	S80220	cDNA (p) brain	No	252 bp (1..252)
	M. musculus house mouse	α_{1A}-AR	AF031431	cDNA (c) brain, liver, kidney	No	1401 bp (1..1401)
	Rattus norvegicus Norway rat	α_{1C}-AR	U13368	cDNA (c) cardiac myocytes	No	1862 bp (36..1436)
	R. norvegicus Norway rat	α_{1c}-AR	U07126	cDNA (c) brain	No	1428 bp (16..1416)
	Meriones unguiculatus Mongolian gerbil	α_{1a}-AR	AF047188	cDNA (p) spiral modiolar artery	No	175 bp (1..175)
	Oryzias latipes Japanese medaka fish	α_{1A} adrenoceptor	D63859	Genomic (c)	No	1770 bp (228..1640)
α_{1B}-AR	*M. musculus* house mouse	α_{1B}-AR	Y12738	cDNA (c) brain	No	2268 bp (724..2268)

Species / common name	Gene	Accession	Type / tissue	Coding region	Length
Mus. musculus house mouse	α_{1b}-AR	S80219	cDNA (p) brain	No	120 bp (3..19)
R. norvegicus Norway rat	α_{1B}-AR	X51585	cDNA (c) brain	No	2086 bp (241..1788)
R. norvegicus Norway rat	α_{1B}-AR	M60655	cDNA (c) brain	No	2108 bp (15..1562)
R. norvegicus Norway rat	α_{1B}-AR	D32045	Genomic	Yes	2387 bp (1628..2387)
R. norvegicus Norway rat	α_{1B}-AR	U83985	Genomic	Yes	513 bp (no coding region)
R. norvegicus Norway rat	α_{1B}-AR exon 1	L08609	Genomic (p)	Yes	2207 bp (1209..2157, L08610:51..649)
R. norvegicus Norway rat	α_{1B}-AR exon 2	L08610	Genomic (p)	No	1365 bp (L08609: 1209..2157, 51..649)
M. unguiculatus Mongolian gerbil	α_{1b}-AR	AF047189	cDNA (p) spiral modiolar artery	No	258 bp (1..258)
Mesocricetus auratus Syrian golden hamster	α_{1B}-AR	J04084	cDNA (c) smooth muscle	No	2089 bp (15..1562)
H. sapiens human	adrenergic α_{1b} receptor	U03865	cDNA (c) brainstem	No	1738 bp (124..1686)
H. sapiens human	α_{1B}-AR	L31773	cDNA (c) heart	No	1560 bp (1..1560)
H. sapiens human	α_{1B}-AR	M99589	Genomic	Yes	2669 bp (no coding region)

(continued)

Table 1 (continued)
α_1-Adrenergic Receptors

Adrenergic receptor subtype[a]	Species, common name	Definition in GenBank database, subtype identification in text	Accession number	cDNA or genomic[b] tissue type	5′ Flanking region	Number of nucleotides, coding sequence
	Canis familiaris dog	RDC5 mRNA for G protein-coupled receptor	X14050	cDNA (p) thyroid	No	1695 bp (1..1256)
α_{1D}-AR	*R. norvegicus* Norway rat	α_{1A}-AR	M60654	cDNA (c) brain	No	2936 bp (480..2162)
	R. norvegicus Norway rat	$\alpha_{1a/d}$-AR	L31771	cDNA (c) cerebral cortex	No	2939 bp (480..2165)
	R. norvegicus Norway rat	α_{1D}-AR	AF071014	Genomic (p)	Yes	1783 bp (1597..1783)
	M. musculus house mouse	α_{1d}-AR	S80044	cDNA (c) brain	No	1902 bp (118..1806)
	M. musculus house mouse	α_{1A}-AR homolog	L20333	cDNA (p) testis	No	483 bp (1..483)
	H. sapiens human	$\alpha_{1a/d}$-AR	L31772	Genomic and cDNA (c) prostate	No	1831 bp (1..1719)
	H. sapiens human	adrenergic α_{1a} receptor	U03864	cDNA (c) hippocampus	No	1860 bp (58..1776)
	H. sapiens human	$\alpha_{1A/D}$-AR	D29952	Genomic and cDNA (c) prostate	No	2077 bp (5..1723)
	H. sapiens human	α_{A1}-AR	M76446	cDNA (c) brain	No	2002 bp (56..1561)
	O. cuniculus rabbit	α_{1d} adrenoceptor	U64032	cDNA (c) liver	No	1731 bp (1..1731)

[a]Assignment of subtype is based on the alignment and groupings of coding sequences generated by the denogram of the program Pileup, GCG.

[b] G = complete coding sequence; (p) partial coding sequence

Table 2

α₂-Adrenergic Receptors

Adrenergic receptor subtype[a]	Species, common name	Definition in GenBank database, subtype identification in text	Accession number	cDNA or genomic[b] tissue type	5′ Flanking region	Number of nucleotides, coding sequence
α₂A-AR	*Rattus rattus* black rat	α₂D-AR	U79031	cDNA (c) brain	No	1552 bp (1..1353)
	R. norvegicus Norway rat	α₂-AR-RG20	M62372	Genomic (c)	No	1380 bp (1..1353)
	R. norvegicus Norway rat	α₂D-AR	U49747	Genomic (p)	Yes	2836 bp (2831..2836)
	M. musculus house mouse	α₂-AR (α₂-AR-C10 homolog)	M99377	Genomic (c)	No	1454 bp (51..1403)
	M. musculus house mouse	α₂A-AR	U29693	Genomic	Yes	2828 bp (no coding region)
	M. auratus golden hamster	α₂ receptor adrenergic (α₂A-AR)	L28124	cDNA (p) adipocytes	No	313 bp (1..313)
	Sus scrofa pig	α₂A-AR (α₂-AR-C10 homolog)	J05652	Genomic (c)	No	1728 bp (130..1482)
	Bos taurus cow	α₂D-AR	U79030	Genomic (c)	Yes	2923 bp (1509..2867)
	B. taurus cow	adrenergic receptor subtype α₂D-AR	S66295	cDNA (p) pineal gland	No	120 bp (1..120)
	H. sapiens human	α₂-AR	M18415	Genomic (c)	No	1521 bp (59..1411)
	H. sapiens human	α₂-AR (α-2A)	M23533	Genomic (c)	Yes	3604 bp (2078..3430)
	Gallus gallus chicken	adrenergic receptor subtype α₂A-AR	S66185	Genomic (p)	No	117 bp (1..117)

(continued)

Table 2 (continued)
α₂-Adrenergic Receptors

Adrenergic receptor subtype[a]	Species, common name	Definition in GenBank database, subtype identification in text	Accession number	cDNA or genomic[b] tissue type	5' Flanking region	Number of nucleotides, coding sequence
α_{2B}-AR	*Cavia porcellus* guinea pig	α_{2A}-AR adrenoceptor	U25722	Genomic (c)	No	2291 bp (49..1401)
	R. norvegicus Norway rat	α_{2B}-AR (RNG-α_2-AR)	M32061	cDNA (c) kidney	No	2319 bp (366..1727)
	R. norvegicus Norway rat	α_{2B}-AR	X74400	Genomic (c)	No	1639 bp (178..1524)
	M. musculus house mouse	α_2-AR (α_{2C2}-AR homolog)	L00979	Genomic (c)	No	1650 bp (227..1573)
	M. musculus house mouse	α_2-AR (α_{2C2}-AR homolog)	M94583	Genomic (c)	Yes	5265 bp (1146..2513)
	Elephas maximus Indian elephant	α_{2B}-AR	Y12525	Genomic (p)	No	1153 bp (1..1153)
	Dugong dugon sea cow	α-AR subtype 2B	Y15947	Genomic (p)	No	1171 bp (1..1171)
	Procavia capensis cape rock hyrax (shrewmouse)	α_{2B}-AR	Y12523	Genomic (p)	No	1168 bp (1..1168)
	Orycteropus afer aardvark	α_{2B}-AR	Y12522	Genomic (p)	No	1165 bp (1..1165)
	Amblysomus hottentotus golden moles	α_{2B}-AR	Y12526	Genomic (p)	No	1159 bp (1..1159)
	Echinops telfairi Madagascar hedgehog	α-AR subtype 2B	Y17692	Genomic (p)	No	1153 bp (1..1153)
	Macroscelides proboscideus short-eared elephant shrew	α_{2B}-AR	Y12524	Genomic (p)	No	1162 bp (1..1162)

Species	Receptor	Accession	Source	Intron	Length
B. taurus cow	α-AR subtype 2B	Y15944	Genomic (p)	No	1177 bp (1..1177)
Equus caballus horse	α-AR subtype 2B	Y15945	Genomic (p)	No	1168 bp (1..1168)
Erinaceus europaeus western European hedgehog	α$_{2B}$-AR	Y12521	Genomic (p)	No	1174 bp (1..1174)
Talpa europaea European mole	α$_{2B}$-AR	Y12520	Genomic (p)	No	1192 bp (1..1192)
H. sapiens human	α$_{2B}$-AR (α$_{2C2}$-AR)	AF005900	Genomic (c)	Yes	9842 bp (5398..6750)
H. sapiens human	α$_2$-AR (α$_2$-AR c2)	M34041	Genomic (c)	No	2072 bp (413..1765)
H. sapiens human	α$_2$-AR-ADRA2C	M38742	Genomic (p)	No	885 bp (1..885)
O. cuniculus rabbit	α-AR subtype 2B	Y15946	Genomic (p)	No	1183 bp (1..1183)
C. porcellus guinea pig	α$_{2B}$ adrenoceptor gene	U25723	Genomic (c)	No	1987 bp (328..1674)
Didelphis marsupialis opossum	α-AR subtype 2B	Y15943	Genomic (p)	No	1147 bp (1..1147)
α$_{2C}$-AR *M. musculus* house mouse	α$_2$-AR (α$_2$-C4 homolog)	M97516	Genomic (c)	No	2409 bp (415..1791)
M. musculus house mouse	α$_2$-AR (α$_2$-C4 homolog)	M99376	Genomic (c)	No	1503 bp (51..1427)
R. norvegicus Norway rat	α$_2$-AR-RG10	M62371	Genomic (c)	No	1380 bp (1..1377)
R. norvegicus Norway rat	α$_2$-C4-AR	X57659	Genomic and cDNA (c) brain	?	2991 bp (907..2283)

(continued)

11

Table 2 (continued)
α₂-Adrenergic Receptors

Adrenergic receptor subtype[a]	Species, common name	Definition in GenBank database, subtype identification in text	Accession number	cDNA or genomic[b] tissue type	5' Flanking region	Number of nucleotides, coding sequence
	R. norvegicus Norway rat	α₂-AR	D00819	Genomic (c)	No	1745 bp (62..1438)
	R. norvegicus Norway rat	α₂B-AR	M58316	cDNA (c) brain	No	1704 bp (91..1467)
	H. sapiens human	α₂-AR (α₂-C4)	J03853	cDNA (c) kidney	No	1491 bp (39..1424)
	H. sapiens human	α₂-C4-AR	U72648	Genomic (c)	Yes	4850 bp (2807..4192)
	H. sapiens human	gDNA encoding adrenaline α₂-AR CII receptor	E07358	unknown (c) (patent)	No	1382 bp (3..1379)
	H. sapiens human	α₂CII-AR	D13538	Genomic (c)	No	1382 bp (3..1379)
	H. sapiens human	α₂-AR 1.8 (α₂-C4)	X59684	Genomic (p)	No	387 bp (1..387)
	C. porcellus guinea pig	α₂C-adrenoceptor	U25724	Genomic (c)	No	1995 bp (205..1572)
	Didelphis virginiana opossum	α₂c-AR	U04310	cDNA (c) kidney	No	1410 bp (1..1410)
α₂C-AR like	*L. ossifagus* cuckoo wrasse-teleost fish	α₂-adrenoceptor	U07743	Genomic (c)	Yes	2898 bp (1115..2413)
unde-fined	*C. auratus* goldfish	α₂-AR	L09064	Genomic (c)	Yes	2764 bp (1188..2498)

[a]Assignment of subtype is based on the alignment and groupings of coding sequences generated by the denogram of the program Pileup, GCG (*see* Chapter 2, **Fig. 1**).
[b](c) Complete coding sequence; (p) partial coding sequence.

three β-AR subtypes have been cloned from a variety of species (**Table 3**). There is strong evidence for a β_4-AR, although this subtype has not yet been cloned *(25–28)*.

1.2. Library Construction

The information obtained from gene-specific cDNA or genomic DNA is important for determining the coding region, promoter region, regulatory elements, or introns of a gene. For the intronless genes, such as the α_2-AR, β_1-AR, and β_2-AR, the genes are relatively small. The nucleotide sequence of the cDNA or intronless genome can be determined from a λ library that can contain inserts of 0–12 kb in size, or if using replacement vectors, 9–23 kb in size. λ Phage libraries are useful in that they are easily duplicated and are not susceptible to selective amplification. For the well-studied species, such as the human, mouse, or rat, it may be far less time-consuming to purchase a commercial library. However, for other species (i.e., gerbil) or for specific microsized tissues (i.e., inner ear tissues) for which commercial libraries are not available, it may be necessary to synthesize your own genomic or cDNA library. Further, for genomic libraries of genes, such as the α_{1A}-AR, α_{1B}-AR, and α_{1D}-AR, which contain 15- to 30-kb introns, it will also be necessary to construct a library with large genomic inserts as found in cosmid libraries. For a genomic DNA library, we describe the synthesis of a cosmid library that results in a plasmid- (bacteria) based library, which is easy to manipulate after isolation of the clone. For making a tissue-specific library from a small amount of RNA we describe a polymerase chain reaction- (PCR) based synthesis of a cDNA library.

2. Materials

All aqueous solutions should be made from distilled, deionized water. Extreme care should be taken not to contaminate any of the working solutions with DNA or RNA and any of the RNase enzymes that are inherently present on skin. This will require wearing gloves and frequently changing gloves. For RNA work, plasticware that has only been touched with a gloved hand should be used. If it is necessary to use glassware, it should be rinsed with RNase-free water (*see* **step 1** in **Subheading 2.2.1.**) and baked at 300°C for 4 h. In addition, solutions should be sterilized by autoclaving or filtration, and should be stored in small frozen aliquots that are then discarded after having been opened several times. Listed below are stock solutions that are used in many molecular biology protocols.

(text continued on page 20)

Table 3
β-Adrenergic Receptors

Adrenergic receptor subtype[a]	Species, common name	Definition in GenBank database, subtype identification in text	Accession number	cDNA or genomic[b] tissue type	5' Flanking region	Number of nucleotides, coding sequence
β₁-AR	*H. sapiens* human	β₁-AR	J03019	cDNA (c) placenta	No	1723 bp (87..1520)
	H. sapiens human	ADRB1R gene	X69168	Genomic	Yes	3100 bp (no coding region)
	M. unguiculatus Mongolian gerbil	β₁-AR	AF055349	cDNA (c) vestibular labyrinth, stria vascularis, brain	No	221 bp (1..221)
	Macaca mulatta rhesus monkey	β₁-AR	X75540	Genomic (c)	Yes	4401 bp (1425..2867)
	C. familiaris dog	β₁-AR	U73207	Genomic (c)	No	1845 bp (330..1751)
	S. scrofa pig	β₁-AR	AF042454	Genomic (c)	No	2284 bp (456..1862)
	Ovis aries sheep	β₁-AR	S81783	Genomic (p)	No	1040 bp (1..1040)
	O. aries sheep	β₁-AR	AF072433	Genomic (c)	Yes	4749 bp (2289..3692)
	R. norvegicus Norway rat	β₁-AR	J05561	Genomic (c)	No	1645 bp (69..1469)
	R. norvegicus Norway rat	β₁-AR	D00634	Genomic (c)	Yes	3686 bp (1257..2657)

Subtype	Species / common name	Accession	Source	Coding region	Length
β_1-AR	*R. norvegicus* Norway rat	X75538	Genomic	Yes	1365 bp (no coding region)
β_1-AR	*R. norvegicus* Norway rat	X75539	Genomic	3' Flanking region	1525 bp (no coding region)
β_1-AR	*M. musculus* house mouse	L10084	Genomic (c)	No	1525 bp (100..1500)
β_1-AR	*Cervus dama* fallow deer	AF041457	cDNA (p)	No	144 bp (1..144)
β-AR	*Meleagris gallopavo* turkey	M14379	cDNA (c) fetal red blood cells	No	1806 bp (70..1521)
β_1-AR like	*Xenopus laevis* African clawed frog	Y09213	cDNA (c) embryo	No	1584 bp (301..1458)
β_1-AR like	*Meleagris gallopavo* turkey adrenergic β_{4c} receptor	U13977	cDNA (c) fetal red blood cells	No	1533 bp (89..1375)
β_1-AR like	*M. gallopavo* turkey adrenergic β_{4c} receptor	U13978	Genomic (c)	Yes	3445 bp (892..2138, 2548..2587)
β_2-AR	*B. taurus* cow	Z86037	cDNA (c) oviduct	No	2032 bp (224..1480)
β_2-AR	*B. taurus* cow adrenergic β_2 receptor	X67213	cDNA (p)	No	376 bp (1..376)
β_2-AR	*S. scrofa* pig	U53185	cDNA (p)	No	329 bp (1..329)
β_2-AR	*S. scrofa* pig	AF000134	Genomic (c)	Yes	5288 bp (1758..3014)

(continued)

Table 3 (continued)
β-Adrenergic Receptors

Adrenergic receptor subtype[a] / Species, common name	Definition in GenBank database, subtype identification in text	Accession number	cDNA or genomic[b] tissue type	5' Flanking region	Number of nucleotides, coding sequence
Rattus sp. rat	β₂-AR	X17607	Genomic (c)	Yes	4190 bp (2252..3508)
R. norvegicus Norway rat	β₂-AR	L39264	Genomic (c)	Yes	4197 bp (2309..3565)
R. norvegicus Norway rat	β₂-AR	J03024	cDNA (c) heart	No	1959 bp (102..1358)
R. norvegicus Norway rat	β₂-AR	U35448	Genomic (p)	Yes	3780 bp (3712..3780)
M. musculus house house	β₂-AR	X15643	Genomic (c)	Yes	4928 bp (2212..3468)
M. auratus golden hamster	β-AR	X03804 J02728 M16107	cDNA (c)	No	2018 bp (211..1461)
M. unguiculatus Mongolian gerbil	β₂-AR	AF055350	cDNA (p) Vestibular labyrinth; stria vascularis; lung	No	756 bp (1..756)
H. sapiens human	β₂-AR	J02960	Genomic (c)	Yes	3458 bp (1264..2505)
H. sapiens human	β-AR	X04827	cDNA (c) neonatal human brain stem	No	1970 bp (178..1419)

3. Liquid nitrogen.
4. Digestion buffer: Combine 20 mL 0.5 *M* EDTA, pH 8.0, 1 mL 1 *M* Tris-HCl, pH 8.0, 10 mL 5% SDS, 2 mg of pancreatic RNase (DNase-free), and 100 mL of distilled water. The final concentration of the digestion buffer is 0.1 *M* EDTA, 10 m*M* Tris-HCl, 20 µg/mL pancreatic RNase (DNase-free), 0.5% SDS.
5. 20 mg/mL proteinase K (in water).
6. Phenol buffered to pH 8:chloroform:isoamyl alcohol (25:24:1). One volume of phenol, pH 8.0, is mixed with 1 vol of chloroform:isoamyl alcohol (24:1). Make only the amount needed, and discard the rest.
7. TE buffer (*see* **item 4** in **Subheading 2.**).
8. 0.3 (w/v) and 1% (w/v) Electrophoresis-grade agarose gel with 0.5 µg/mL ethidium bromide.
9. TAE electrophoresis buffer: Dilute 10X TAE buffer 10-fold with water.
10. λ DNA (Life Technologies, Gaithersburg, MD).
11. Dialysis bag: 12,000–14,000 mol-wt cutoff. Boil for 10 min in 1 L of 2% (w/v) sodium bicarbonate and 1 m*M* EDTA, pH 8.0. Rinse with distilled water, and autoclave in water for 10 min. Store at 4°C, and handle with gloves.

2.1.2. Preparation of the Cosmid Library

1. Genomic DNA, 200 µg, ≥200 kb (100–500 µg/mL in TE buffer).
2. Restriction endonucleases: Any of the isoschizomers *Nde*II, *Sau*3A, or *Mbo*I, and the respective 10X restriction digestion buffers supplied by the manufacturer.
3. Gel-loading buffer: 50 g glycerol, 40 mL 10X TAE buffer, 1 g sodium dodecyl sulfate (SDS), and 0.1 g bromophenol blue in 100 mL water.
4. 0.3% Agarose gel.
5. 0.5 *M* EDTA, pH 8.0.
6. Phenol, pH 8.0:chloroform:isoamyl alcohol (25:24:1).
7. Chloroform.
8. 3 *M* sodium acetate, pH 5.5.
9. 100% Ethanol.
10. 70% Ethanol.
11. Calf intestinal alkaline phosphatase (CIP), 20–80 U/µL (Clontech, Palo Alto, CA).
12. 1X CIP buffer: 50 m*M* Tris-HCl, pH 8.0. Dilute the 10X buffer supplied with enzyme 1 to 10 with water.
13. TE buffer.
14. SuperCos I (Stratagene, La Jolla, CA).
15. Gigapack® III XL packaging extract (Stratagene).

2.2. Synthesis of a cDNA Library Using Small Amounts of RNA

2.2.1. cDNA Synthesis

1. RNase-free water (diethyl pyrocarbonate [DEPC] water): Treat 1 L of water with 1 mL DEPC by stirring for 24 h at 37°C, and then autoclaving one to three times or until all residual DEPC is gone (*see* **Notes 1–3**).

2. Total RNA (1 μg) or Poly(A)$^+$RNA (100 ng to 1 μg) is treated with RNase-free DNase I and dissolved in DEPC water (*see* **Notes 4–10**).
3. Degenerate anchored oligo dT primer TTCCGGAATTCAGCGGCCGC(T)$_{17}$ MN (10 μM) where M represents G, A, or C and N represents G, A, T, and C (*see* **Note 11**).
4. Reverse transcriptase: SUPERSCRIPT® II RNase H-Reverse Transcriptase (200 U/μL) (*see* **Note 12**).
5. 5X First-strand buffer: 250 nM Tris-HCl, pH 8.3, 375 mM KCl, 15 mM MgCl$_2$ (supplied with enzyme).
6. 10 mM dNTP mix (10 mM each dATP, dGTP, dCTP, dTTP) from Life Technologies (Gaithersburg, MD).
7. 100 mM Dithiothreitol (DTT): Dissolve 0.309 g of DTT in 20 mL of 0.01 M sodium acetate, pH 5.2, and sterilize by filtration. Store at –20°C.
8. *Escherichia coli* RNase H (2 U/μL).
9. Phenol (saturated with water):chloroform:isoamyl alcohol (25:24:1).
10. Chloroform.
11. 7.5 M ammonium acetate.
12. 100% Ethanol.
13. 70% Ethanol.
14. Terminal deoxynucleotidyl transferase (TdT), 10 U/μL (Stratagene).
15. 5X buffer TdT: 500 mM potassium cacodylate, pH 7.2, 10 mM CoCl$_2$, 1 mM DTT (Stratagene).
16. 50 μM dGTP (dilute 10 mM dGTP 200-fold).
17. 5′ Sense primer: GACTCGAGTCGACATCGA(C)$_{13}$, 10 μM in water (*see* **Note 13**).
18. Advantage™ KlenTaq polymerase (Clontech, Palo Alto, CA).
19. Advantage™ KlenTaq polymerase buffer 10X (Clontech): 400 mM Tricine-KOH, pH 9.2, 150 mM potassium acetate, 35 mM magnesium acetate, 750 μg/mL bovine serum albumin.
20. 3′ Antisense adapter primer: TTCCGGAATTCAGCGGCCGC, 10 μM water.
21. Perkin Elmer 480 Thermal Cycler (PE Applied Biosystems, Foster City, CA).

2.2.2. Cloning of the cDNA

1. 100 ng of amplified cDNA.
2. 10X universal buffer (Stratagene): 1 M KOAc, 250 mM Tris-acetate, pH 7.6, 100 mM magnesium acetate, 5 mM β-mercaptoethanol, 100 μg/mL bovine serum albumin (BSA).
3. *Sal*I 10 U/μL.
4. *Not*I 10 U/μL.
5. Phenol, pH 8.0 chloroform:isoamyl alcohol (25:24:1).
6. Chloroform.
7. 100% Ethanol.
8. 70% Ethanol.
9. Zap Express® vector (Stratagene).
10. 1 μL (400 U) T4 DNA ligase (Life Technologies, Gaithersburg, MD).

11. 10X ligase buffer (Life Technologies): The final concentration is 0.5 M Tris-HCl, pH 7.5, 50 mM MgCl$_2$, 50 mM DTT, 0.5 mg/mL BSA, or gelatin.
12. 10 mM ATP.
13. CIP (Stratagene).
14. Gigapack III Plus® or Gigapack III Gold® packaging extracts (Stratagene).

3. Methods

3.1. Genomic Library Synthesis

3.1.1. Isolation of Genomic DNA

1. Immediately freeze the freshly isolated tissue in liquid nitrogen (*see* **Note 14**).
2. Grind the tissue in a mortar and pestle in the presence of liquid nitrogen until the tissue is a fine powder (*see* **Note 15**).
3. Gradually add 100 mg of powdered tissue to the surface of 1.2 mL digestion buffer, and gently shake until the tissue is suspended. Incubate at 37°C for 1 h.
4. Using a glass rod, gently stir in 5 µL of proteinase K for every 1 mL of solution. The DNA solution is gently swirled at 50°C overnight (12–18 h).
5. In a hood, add equal volumes of phenol/chloroform/isoamyl alcohol, and extract the proteins by slowly inverting the tube for 10 min until an emulsion has been formed.
6. Centrifuge the emulsion for 15 min at 5000g, and remove the DNA by slowly pipeting the top aqueous layer with a wide-bore pipet tip (pipet tip cut off with a sterile razor blade) without removing the white protein from the interface. Repeat this procedure two more times or until no more protein is visible at the interface (*see* **Note 16**).
7. Use wide-bore pipet tips to transfer the DNA into a dialysis bag, and dialyze against 4 L of TE buffer with four changes of solution. Store the DNA at 4°C, and avoid any repeated freezing and thawing of the DNA solution *(29)* (*see* **Note 17**).
8. Measure the DNA concentration at an A_{260} and A_{280}, where 1 absorption unit at 260 nm equals 50 µg DNA/mL. Further, the 260/280 absorption ratio should be about 1.8 (*see* **Note 18**).
9. Analyze the integrity of 0.5 µg of DNA on a 0.3% (w/v) agarose gel electrophoresed under low voltage (1 V/cm) at 4°C, and visualize by ethidium bromide staining. Pour a 1% agarose layer to a thickness of 1/3 the normal capacity of the gel form, and solidify. On the top of this layer, pour a 0.3% agarose up to the final thickness of the gel, insert the comb in the 0.3% layer, and solidify in the refrigerator. The wells are formed in the 0.3% agarose layer, and the 1% agarose will act as a solid support. The genomic DNA should migrate much more slowly than the λ DNA (48.5-kb) size marker *(29)* (*see* **Notes 19–21**).

3.1.2. Preparation of the Cosmid Library

Prior to ligation into cosmid vectors, the genomic DNA must be cut into random 45-kb fragments. The generation of random fragments of DNA is best accomplished by using restriction endonucleases that cleave DNA frequently and recognize four base sequences (i.e., *Sau*3A). The restriction sites should also have ends that are cohesive with the *Bam*HI site present in the SuperCos 1 cosmid vector (Stratagene) used in the procedure below. A test reaction is run on 10 μg of DNA to determine the ideal conditions for producing ~30–45-kb genomic DNA fragments. Then these exact conditions are used to digest 100 μg of DNA.

1. Prepare six tubes with gel-loading buffer to stop the timed reactions. Add 10 μL of gel-loading buffer to six microcentrifuge tubes, and label as 0-, 5-, 10-, 20-, 30-, and 45-min time-points.
2. In another empty microcentrifuge tube, combine 10 μL of 10X restriction enzyme buffer, water, and 10 μg of genomic DNA, such that the final volume is 100 μL. Gently mix the solution such that the DNA is evenly dispersed, and keep the tube at 4°C (*see* **Notes 22** and **23**).
3. Remove 15 μL from the genomic DNA solution in **step 2** above, and add to the zero-time-point tube in **step 1** above.
4. Add 0.5 U of *Nde*II, *Sau*3A, or *Mbo*I to the genomic DNA mixture in **step 2** above, and gently mix at 4°C. Incubate the tube at 37°C, and rotate the tubes gently during the incubation (*see* **Note 24**).
5. At 5, 10, 20, 30, and 45 min after the start of incubation at 37°C, stop the reaction by removing 15 μL from the digesting genomic DNA. Add the aliquot to the respective tubes containing the loading buffer prepared in **step 1** above, and gently mix.
6. Electrophorese the samples on a 0.3% agarose gel at 1 V/cm at 4°C using λ DNA as a marker. Observe the time of digestion that results in DNA that comigrates as one large band with monomeric λ DNA (48.5 kb). These exact conditions for the digestion will be duplicated in **step 7** below (*see* **Note 25**).
7. Repeat the above digestion on 10 aliquots each containing 10 μg of genomic DNA using the same ideal digestion reaction conditions determined in **steps 2–6** above. These next reactions are performed in 10 separate reactions of 100-μL vol for identical time periods. The reactions are stopped by the addition of 1.5 μL of 0.5 *M* EDTA, pH 8.0. Electrophorese a 5-μL sample to determine the size of the digested DNA (*see* **Note 26**).
8. Pool the digested DNA, and extract with an equal volume of phenol/chloroform/ isoamyl alcohol as in **steps 5** and **6** in **Subheading 3.1.1.**
9. Extract the aqueous phase with one equal volume of chloroform (*see* **Note 27**).
10. Add 0.1 vol of 3 *M* sodium acetate, pH 5.5, to the extracted DNA solution, and mix gently to ensure a homogenous solution.
11. Carefully mix in 2.5 vol of 100% ethanol, stir, and place on ice for 30 min. Centrifuge for 20 min at 15,000*g* at 4°C.
12. Wash the pellet with 500 μL of 70% ethanol, recentrifuge at 15,000*g*, and air-dry the DNA pellet (*see* **Note 28**).
13. Gently mix the pellet in 50 μL of 1X CIP buffer to resuspend the DNA. Allow adequate time for complete resuspension.

14. Dephosphorylate the genomic DNA to prevent ligation of the ends by adding 2 µL of CIP 20–80 U/µL and increasing the final volume to 100 µL with 1X CIP buffer. Mix well, and incubate for 1 h at 37°C.
15. Add 3 µL 0.5 *M* EDTA, pH 8.0, and heat-denature at 68°C for 10 min to denature CIP.
16. Repeat **steps 8–12** above to extract protein and precipitate the DNA.
17. Resuspend the DNA to a concentration of 1 µg/µL in TE buffer, and analyze 1 µg on a 0.3% gel to ensure the integrity of the DNA has been maintained. The DNA is now ready for ligation and packaging into the SuperCos I cosmid vector (*see* **Note 29**).
18. The cosmid vector SuperCos I (Stratagene) is prepared for ligation to the genomic DNA according to the manufacturer's instructions (*see* **Note 30**).
19. One microgram of SuperCos I vector is ligated to 2.5 µg of chromosomal DNA, and the ligation products are packaged into the Gigapack III XL packaging extract (Stratagene) according to manufacturer's directions (*see* **Notes 31** and **32**).

3.2. Synthesis of a cDNA Library Using Small Amounts of RNA

The following procedure can be used to synthesize a directionally cloned cDNA library from small amounts of RNA by taking advantage of the ability of the PCR to amplify the cDNA. The restriction enzyme sites are generated at the 5′ and 3′ end using enzymes that rarely cut mammalian DNA. This allows the cDNA to be readily cloned into vectors. This procedure is based on the method of Domec et al. *(31)*. In addition, the cDNA synthesized in this procedure can be used to determine the sequences of 5′ and 3′ specific ends of mRNAs (*see* Chapter 2, **Subheading 3.6.**). It should also be noted that kits for synthesis of cDNA libraries are also available from a number of commercial sources as reviewed by DeFrancesco *(32)*.

3.2.1. cDNA Synthesis

In order to ensure that there is no DNA in the RNA sample used for reverse transcriptase (RT) PCR in the following procedure, a second control tube for cDNA synthesis should be prepared as described below, except that RT is absent from the reaction mix. The absence of any PCR product in this tube indicates that there is no genomic DNA contamination.

1. Combine 1 µg of total RNA (or 200 ng of poly A$^+$ RNA) with 2 µL of degenerate anchored oligo dT primer, and add DEPC-treated water to 12 µL. Heat to 70°C for 10 min, and cool on ice (*see* **Note 33**).
2. Centrifuge the mixture for 10 s (pulse centrifuge) to spin down any condensed liquid, and add 4 µL 5X first-strand buffer, 2 µL 0.1 *M* DTT, and 1 µL 10 m*M* dNTP mix. Mix and incubate at 42°C for 2 min (*see* **Note 34**).
3. Add 1 µL of SUPERSCRIPT II, and gently mix by pipeting. Incubate at 42°C for 50 min. Terminate the reaction by heating at 70°C for 15 min.

4. Pulse centrifuge the tube, add 1 µL of *E. coli* RNase H, and incubate at 37°C for 20 min to remove RNA.
5. Add enough DEPC-treated water to increase the volume to 50 µL. Extract the diluted mixture once with 50 µL of phenol, pH 8.0: chloroform: isoamyl alcohol (25:24:1) and once with chloroform (*see* **Note 35**).
6. Add 0.5 vol of 7.5 *M* ammonium acetate to the DNA solution, mix, and precipitate with 2 vol of 100% ethanol, centrifuge and wash the pellet with 1 vol of 70% ethanol, and air-dry as before in **steps 10–12** of **Subheading 3.1.2.** (*see* **Note 36**).
7. Dissolve the cDNA pellet in 9 µL water, add 4 µL of 5X buffer TdT, 2 µL dGTP, and 5 µL terminal TdT. Incubate the mixture for 30 s at 37°C, and terminate the reaction by immediately placing on ice (*see* **Note 37**).
8. Immediately add 10 µL (0.5 vol) of 7.5 *M* ammonium acetate mix to the DNA pellet and mix. Add 2 vol or 60 µL of 100% ethanol, mix, centrifuge the pellet, and air-dry the pellet.
9. Dissolve the single-strand G-tailed cDNA in 41 µL of water, 5 µL 5X KlenTaq polymerase buffer, 1 µL dNTP mix, 1 µL 5′ sense primer, 1 µL 3′ antisense adapter primer, and 1 µL Advantage™ KlenTaq polymerase, final volume of 50 µL. Incubate at 95°C for 5 min, followed by 25 cycles of 1 min at 95°C (denaturation), 1 min at 50°C (annealing), and 4 min at 70°C (extension). The final extension cycle conditions are for 5 min at 70°C. The cDNA is now double-stranded and amplified (*see* **Note 38**).
10. Remove 5 µL of the amplified cDNA, and analyze the products by electrophoresis on a 1.0% ethidium bromide-stained agarose gel. Most of the double-stranded cDNA should be > 1.5 kb in size.
11. Increase the volume of the amplified cDNA mixture to 150 µL, and add 0.1 vol of 3 *M* sodium acetate, pH 5.5. Extract the mixture with 1 vol of phenol, pH 8:0 chloroform:isoamyl alcohol (25 : 24:1). Remove the aqueous layer, and extract it again with 1 vol of chloroform. Precipitate the aqueous layer with 2 vol of 100% ethanol, wash the precipitate with 70% ethanol, and dry. Add sterile distilled water to an approximate concentration of 1 to 0.1 µg/µL (*see* **Note 39**).

3.2.2. Cloning of the cDNA

1. Digest 1 µg of the amplified cDNA with 2 µL of 10X universal buffer (final concentration of 2X universal buffer), 2 U of *Sal*I, and 2 U of *Not*I in a final volume of 10 µL for 60 min at 37°C (*see* **Note 40**).
2. Increase the volume of the digested amplified cDNA mixture to 50 µL, add 0.1 vol of 3 *M* sodium acetate, pH 5.5, and extract with 1 vol of phenol-chloroform-isoamyl alcohol. Extract the aqueous phase again with 1 vol of chloroform. Precipitate the aqueous layer with 2 vol of 100% ethanol, wash the precipitate with 70% ethanol, and dry. Add sterile distilled water to an approximate concentration of 100 ng/µL.
3. The Zap Express® vector is prepared for ligation by double-digesting the vector with *Not*I and *Sal*I, and dephosphorylating the vector with CIP according to the manufacturer's instructions (Stratagene) (*see* **Note 41**).

4. Mix 50–100 ng cDNA with *Not*I- and *Sal*I-digested ends with 1 µg of Zap Express® arms with 1 µL of 10X ligase buffer, 1 µL 10 m*M* ATP, and water up to a final volume of 9 µL. Add 1 µL (400 U) T4 DNA ligase, mix, and incubate at 4°C overnight (*see* **Note 42**).

5. Package the ligation mix into Gigapack III Plus or Gigapack III Gold packaging extracts (Stratagene) according to manufacturer's directions (*see* **Note 43**).

6. The library should be amplified only once so that slower-growing clones will not be underrepresented. Plaques are plated for library screening, and the isolated plaque excised with a helper phage to form recombinant plasmids (phagemid vector) for transformation of *E. coli* as described in the Stratagene manual.

4. Notes

1. Materials should be free of RNase. Gloves should be worn at all times, and contamination should be diligently avoided by using sterile microbiological techniques. Autoclaving does not denature RNases.

2. With the exception of Tris solutions, all aqueous solutions should be made RNase-free by treating with DEPC. Tris inactivates DEPC and, thus, should be autoclaved and filter-sterilized (0.2-µm filter).

3. DEPC is a carcinogen and absorbs water. Thus, the DEPC container should be stored in a desiccator at 4°C. The bottle of DEPC should be opened underneath the hood with full eye, face, and skin protection in the event of the buildup of pressure inside the bottle. When DEPC absorbs water from the air, it reacts with water, and decomposes to form carbon dioxide and ethanol. The carbon dioxide builds up pressure in a tightly capped bottle, and can explode the bottle and spray the remaining DEPC everywhere *(33)*.

4. Total RNA isolation procedures based on the acid guanidinium thiocyanate-phenol–chloroform extraction of Chomczynski and Sacchi *(34)* have been described previously *(29,35)*. Several kits, such as TRIzol (Life Technologies) are also available and are very convenient for isolation of intact total RNA.

5. The A_{260}/A_{280} ratio of clean RNA is 2.0, and samples will range from 1.7 to 2.0. An additional absorption reading can be taken at 230 nm to identify polysaccharide levels where an A_{260}/A_{230} ratio of 2.0 is ideal. Scans from 200 to 300 nm will also be useful in revealing the presence of contaminants *(36)*. One absorbance unit at 260 nm is the equivalent of 44.19 µg/mL of RNA *(37)*.

6. Poly (A)$^+$ RNA, which accounts for about 1–2% of total RNA, can enrich the synthesis of cDNA from messenger RNA. If there is enough tissue, the poly(A)$^+$ RNA can be isolated via oligo dT cellulose, which binds the poly (A)$^+$ tail *(29,37,38)*. Kits, such as The FastTrack 2.0 mRNA Isolation Kit from Invitrogen Corp. (San Diego, CA), can also be conveniently used to isolate Poly(A)$^+$RNA either directly from tissue or from total RNA.

7. Any remaining DNA must be removed by treating the RNA sample with RNase-free DNase I according to manufacturer protocols, followed by phenol/chloroform extraction and ethanol precipitation.

8. The integrity of the total RNA must also be determined. The 18S and 28S riboso-

mal RNA bands can be observed on a regular (nondenaturing) 1.0% agarose gel after electrophoresis of 2 µg of total RNA, which is not degraded *(37)*. It is not absolutely necessary to run a formaldehyde-denaturing agarose gel to ascertain the integrity of the RNA unless you are planning to analyze message content by Northern blot analysis. Undegraded poly A$^+$ RNA should be apparent after electrophoresis as a smear of 200 bp to 10 kb RNA stained with ethidium bromide.

9. The isolated RNA should be stored at –70°C in formamide rather than water for long-term storage *(39)*. Four volumes of 100% ethanol can be used to precipitate the RNA from the formamide. Alternatively, RNA can be stored as a precipitate in ethanol at –70°C.

10. There are several kits available for RNA isolation, and they are described by Lewis *(40)* and DeFrancesco *(41,42)*.

11. The degenerate primer is designed with an adapter containing *Eco*RI-*Not*I sites at the 5' end, a center stretch of Ts for annealing/priming to the poly A$^+$ RNA, and two degenerate bases at the 3' end for anchoring the primer specifically to the 5' end of the poly A tail of mRNA. This eliminates heterogeneity in the initial priming position. These are similar to primers used in the differential display technique *(43)*.

12. This enzyme is a Moloney Murine Leukemia Virus (M-MLV) with a deleted RNase H activity for the reverse transcription of full-length products. It can be incubated at 50°C for the synthesis of cDNA from RNA containing extensive secondary structures. Other RT enzymes can also be used, such as Avian Myeloblastosis Virus (AMV) or M-MLV.

13. This primer will bind to the oligo dG-tailed cDNA (from **step 9** in **Subheading 3.2.1.**) and will provide multiple cloning sites of *Xho*I, *Sal*I, and *Cla*I.

14. For cells in a monolayer, first trypsinize the cells, and collect the cell pellet by centrifugation (5 min at 500*g*). Then wash the cells twice with 1–10 mL ice-cold phosphate-buffered saline (PBS), resuspend 10^8 cells in 1 mL digestion buffer cells, and proceed to **step 3** in **Subheading 3.2.1.** for digestion.

15. It is critical that the tissue be pulverized, and any "chunks" should be removed from the powder. The tissue can be minced before freezing to aid in the pulverization. Alternatively, a stainless-steel Waring Blendor can be used to blend the tissue into a powder in the presence of liquid nitrogen. Finally, a prechilled hammer can be used to pulverize the frozen tissue (placed between two right-side-up plastic weighing boards) on a block of dry ice.

16. If the pH of the buffered phenol is < 8.0, DNA will be trapped at the interface. In addition, phenol is very caustic, so gloves, goggles, and proper laboratory attire should be worn. If skin contact is made with the phenol, wash with copious amounts of water, and apply a sodium bicarbonate paste. Do not wash with ethanol, which will act as a vehicle for absorption of phenol into the skin!

17. Plan on leaving sufficient room in the dialysis bag for the solution to double in volume. Ethanol precipitation of the DNA will result in the isolation of DNA of smaller fragment sizes (100–150 kb). Since the initial digestion buffer included RNase, there is no need to remove RNA at this step, thus saving additional organic

extractions and dialysis. DNA should be ready for use. Expect about 2 mg of DNA/g of tissue or 10^9 cells *(44)*.

18. Lower ratios indicate protein contamination, and the DNA should be deproteinized again by adding SDS to a concentration of 0.5% and following **steps 4–7** in **Subheading 3.2.1.**

19. Pulsed-field gel electrophoresis can also be used to separate high-mol-wt DNA and will do so with better resolution than standard gel electrophoresis.

20. Concatemers of λ DNA in agarose blocks can be ordered from Pharmacia Biotech (Piscataway, NJ) for more accurate size markers. Alternatively, λ DNA molecules can be ligated together at 16°C in a T4 DNA ligation reaction. The λ DNA cohesive ends are disrupted by heating the DNA for 5 min at 56°C followed by ligation at 37°C. This results in the formation of a λ 50-kb concatemer ladder *(29)*.

21. After pipeting the viscous DNA into the well, the pipet tip should be gently drawn across the back of the well to break off the DNA from the tip. If this is not done, the sample may "wick" out of the well as the tip is drawn out. Alternatively, the sample may be loaded in a "dry" well, and then the electrophoresis buffer carefully added to immerse the gel. After loading the wells, allow a few minutes for the DNA to distribute evenly throughout the well.

22. Use a wide-mouth pipet tip for aliquoting DNA to avoid shearing.

23. Occasionally, restriction enzyme preparations, such as *Sau*3A or *Mbo*I, do not produce cohesive ends that can be ligated to *Bam*HI overhangs. Thus, the 4-base cutter restriction enzyme should be tested prior to cutting genomic DNA to make sure that the overhangs will ligate into the *Bam*HI cut vector. This is accomplished by digesting DNA of a known sequence with the 4-base cutter enzyme and ligating the expected fragments into a vector cut with *Bam*HI. The size of the cloned recombinant vector can then be determined by electrophoresis, and the presence of ligated products evaluated.

24. Incompletely mixing the enzyme with the viscous DNA can interfere with the digestion. An alternative method involves incubating the genomic DNA, the restriction enzyme, and the restriction enzyme buffer in the absence of Mg^{2+} overnight at 4°C while gently mixing. The reaction is initiated the following day at 37°C by the addition of Mg^{2+} and gentle stirring *(45)*.

25. Alternatively, this partial digestion test could be performed by using different amounts of restriction endonuclease for identical time periods. The integrity of the DNA can also be visually ascertained by observing the viscosity of the DNA, where the loss of viscosity indicates that the fragments are too small.

26. The same DNA stock solution used previously for the test reaction should also be used for the scaled-up reaction. Digestion of the DNA in 10 identical reactions is preferred, since scaling up the reaction to 1 mL does not always successfully duplicate the previous smaller reaction volume conditions.

27. This step removes excess phenol and remaining protein from the aqueous layer or top layer. If it is difficult to remove the last portion of the chloroform because of the inversion of the chloroform layer as a bubble, recentrifuge and place the

pipet tip in the lower organic phase, and remove as much chloroform as possible. Recentrifuge and remove the remaining top aqueous layer.

28. Do not dry completely, or the DNA will be difficult to dissolve.

29. As an added precaution, the genomic DNA can be further fractionated by size using either a sucrose gradient *(29)* or agarose gel *(46)*. Size fractionation will ensure that short DNA fragments (<30 kb) will not compete for the SuperCos I vector. However, even if these fragments are incorporated, the cosmid will be too small to be packaged into the library. Thus, size fraction is not considered to be absolutely necessary, especially when chromosomal DNA is in short supply.

30. This involves linearizing the vector by digestion with *Xba*I, dephosphorylating the ends with CIP, and creating two cosmid vector arms (1.1 and 6.5 kb) by digestion with *Bam*HI. SuperCos I vector also has two tandem cos sites that are separated by the *Xba*I restriction site. The cos sites on each arm will prevent the packaging of cosmid concatemers without inserts and will efficiently package DNA that has not been size fractionated *(47)*. *Bam*HI provides a site with phosphorylated ends within the polylinker for ligation of the dephosphorylated genomic DNA.

31. λ/Cosmid packaging extracts can also be made, but for convenience and quality assurance, it is recommended that these packaging extracts be purchased *(29,48)*.

32. When small quantities of tissue are involved and it is difficult to isolate 200-kb size DNA using the above standard procedures, an alternative technique can be used. The 200-kb genomic DNA is isolated in agarose blocks and is separated by pulsed-field gel electrophoresis. The methodology involves the immobilization of the pulverized tissue or other cellular samples in agarose blocks using 1-mL transfer pipets as a form for the agarose block. Deproteinization and restriction enzyme digestion and dephosphorylation are all done within the agarose block to maintain the high molecular weight of the genomic DNA. The DNA is then phenol/chloroform-extracted from the agarose and ligated into the SuperCos I vector *(29,49)*.

33. Heating the RNA to a high temperature will melt secondary structures, and quick cooling will keep the RNA denatured.

34. At this point, an RNase inhibitor, such as RNasin® Ribonuclease Inhibitor from Promega Corporation (Madison, WI), can be added to a concentration of 1 U/μL to prevent any extraneous RNA degradation. However, the water from the first step should be reduced in anticipation of the addition of the RNase inhibitor.

35. The yield of DNA will be somewhat improved if the lower phenol layer is back-extracted. That is, the phenol layer is saved, and an equal volume of TE buffer (pH 8.0) is again added to the phenol. The sample is vortexed, centrifuged, and the aqueous layers from each extraction are combined.

36. It is important to remove all the dNTPs in preparation for the following step. Precipitation with ammonium acetate removes dNTPs more efficiently than does sodium acetate.

37. This procedure adds approx 15 dGp residues to the 3′ terminus of the cDNA. Addition of more than 15 dG can create a long, GC-rich region with the second-

ary structure of DNA that can interfere with transcription of cDNA species. The appropriate conditions for the tailing reaction can be checked by testing the conditions with a linear plasmid (pBluescript® II) cut with *Pst*I and then tailing with dGTP and terminal deoxynucleotidyl transferase. The length of the tailed DNA can be determined after digestion of the DNA with *Hin*fI and electrophoresing the sample on a 6% denaturing polyacrylamide gel stained with ethidium bromide.

38. The Advantage™ KlenTaq polymerase uses a mix of two enzymes to perform long-distance PCR while proofreading the DNA synthesis. *Taq* DNA polymerase is the major enzyme and is deficient in 5′ exonuclease activity. *Tth* DNA polymerase is present in small amounts, and has 5′ exonuclease activity for removing mismatches between a target and primer, thus providing proofreading and efficient primer extension capabilities. In addition, the enzyme mix includes a TaqStart antibody that mediates a "hot-start" reaction after the temperature is raised to 70°C. This will activate the DNA polymerases at elevated temperatures and eliminate nonspecific priming to the template.

39. At this point, the amplified cDNA can either be used for rapid amplification of cDNA ends (RACE or one-sided anchored PCR) (*see* **Subheading 3.6.** in Chapter 2), or it can be cloned into a vector to synthesize a cDNA library for screening (*see* **Subheading 3.2.2.**).

40. Restriction digest of the *Sal*I and *Not*I restriction endonuclease sites at the 5′ and 3′ ends of the cDNA, respectively, will allow directional cloning into the Zap Express® vector. The frequency with which these enzymes cleave mammalian DNA is extremely low, and as a result, the cDNA does not need to be protected *(29)*.

41. The Zap Express® vector has 12 unique cloning sites for directional cloning, and it has the high cloning efficiency typical of λ vectors. Recombinant DNA can be selected by blue/white screening. In addition, the DNA insert within the λ phage can be excised and converted into a Bluescript plasmid. The library can also serve as an expression library where a cytomegalovirus (CMV) promoter initiates transcription. A neomycin-resistant gene enables selection of stable transfected eukaryotic cells.

42. Ligate using an equal molar ratio of vector to insert to prevent multiple inserts from ligating together.

43. Extracts can also be prepared, but those supplied by the manufacturers are more efficient and easier to use *(29,50)*.

References

1. Emilien, G. and Maloteaux, J. M. (1998) Current therapeutic uses and potential of β-adrenoceptor agonists and antagonists. *Eur. J. Clin. Pharmacol.* **53,** 389–404.
2. Ruffolo, R. R., Bondinell, W., and Hieble, J. P. (1995) α- and β-adrenoceptors: From the gene to the clinic. 2. Structure–activity relationships and therapeutic applications. *J. Med. Chem.* **38,** 3681–3716.

3. Ahlquist, R. P. (1948) A study of adrenotropic receptors. *Am. J. Physiol.* **153,** 586–600.
4. Bylund, D. B. (1988) Subtypes of α_2-adrenoceptors: pharmacological and molecular biological evidence converge. *Trends Pharmacol. Sci.* **9,** 356–361.
5. Bylund, D. B., Eikenberg, D. C., Hieble, J. P., Langer, S. Z., Lefkowitz, R. J., Minneman, K. P., et al. (1994) IV. International Union of Pharmacology Nomenclature of Adrenoceptors. *Pharmacol. Rev.* **46,** 121–136.
6. Johnson, R. D. and Minneman, K. P. (1987) Differentiation of α_1-adrenergic receptors linked to phosphatidylinositol turnover and cyclic AMP accumulation in rat brain. *Mol. Pharmacol.* **31,** 239–246.
7. Minneman, K. P., Han, C., and Abel, P. W. (1988) Comparison of α_1-adrenergic receptor subtypes distinguished by chloroethylclonidine and WB4101. *Mol. Pharmacol.* **33,** 509–514.
8. Morrow, A. L. and Creese, I. (1986) Characterization of α_1-adrenergic receptor subtypes in rat brain: a reevaluation of ^{3}H-WB 4101 and ^{3}H-prazosin binding. *Mol. Pharmacol.* **29,** 321–330.
9. Cotecchia, S., Schwinn, D. A., Randall, R. R., Lefkowitz, R. J., Caron, M. G., and Kobilka, B. K. (1988) Molecular cloning and expression of the cDNA for the hamster α_1-adrenergic receptor. *Proc. Natl. Acad. Sci. USA* **85,** 7159–7163.
10. Ford, A. P. D. W., Williams, T. J., Blue, D. R., and Clarke, D. E. (1994) α_1-Adrenoceptor classification: sharpening Occam's razor. *Trends Pharmacol. Sci.* **15,** 167–170.
11. Schwinn, D. A., Lomasney, J. W., Lorenz, W., Szklut, P. J., Fremeau, R. T., Jr., Yang-Feng, T. L., et al. (1990) Molecular cloning and expression of the cDNA for a novel α_1-adrenergic receptor subtype. *J. Biol. Chem.* **265,** 8183–8189.
12. Lomasney, J., Cotecchia, S., Lorenz, W., Leung, W. Y., Schwinn, D. A., Yang-Feng, T. L., et al. (1991) Molecular cloning and expression of the cDNA for the α_{1A}-adrenergic receptor: the gene for which is located on human chromosome 5. *J. Biol. Chem.* **266,** 6365–6369.
13. Perez, D. M., Piascik, M. T., and Graham, R. M. (1991) Solution-phase library screening for the identification of rare clones: isolation of an α_{1D}-adrenergic receptor cDNA. *Mol. Pharmacol.* **40,** 876–883.
14. Muramatsu, I., Ohmura, T., Hashimoto, S., and Oshita, M. (1995) Functional subclassification of vascular α_1-adrenoceptors. *Pharmacol. Commun.* **6,** 23–28.
15. Oshita, M., Kigoshi, S., and Muramatsu, I. (1991) Three distinct binding sites for [^{3}H]-prazosin in the rat cerebral cortex. *Br. J. Pharmacol.* **104,** 961–965.
16. Ford, A. P. D. W., Daniels, D. V., Chang, D. J., Gever, J. R., Jasper, J. R., Lesnick, J. D., et al. (1997) Pharmacological pleiotropism of the human recombinant α_{1A}-adrenoceptor: implications for α_1-adrenoceptor classification. *Br. J. Pharmacol.* **121,** 1127–1135.
17. Bylund, D. B. (1992) Subtypes of α_1- and α_2-adrenergic receptors. *FASEB J.* **6,** 832–839.
18. Blaxall, H. S., Murphy, T. J., Baker, J. C., Ray, C., and Bylund, D. B. (1991)

Characterization of the alpha-2C adrenergic receptor subtype in the opossum kidney and in the OK cell line. *J. Pharmacol. Exp. Ther.* **259**, 323–329.

19. Murphy, T. J. and Bylund, D. B. (1988) Characterization of alpha-2 adrenergic receptors in the OK cell, an opossum kidney cell line. *J. Pharmacol. Exp. Ther.* **244**, 571–578.

20. Michel, A. D., Loury, D. N., and Whiting, R. L. (1989) Differences between α_2 adrenoceptor in rat submaxillary gland and the α_{2A}- and α_{2B}-adrenoceptor subtypes. *Br. J. Pharmacol.* **98**, 890–897.

21. Simonneaux, V., Ebadi, M., and Bylund, D. B. (1991) Identification and characterization of α_{2D}-adrenergic receptors in bovine pineal gland. *Mol. Pharmacol.* **40**, 235–241.

22. Lands, A. M., Arnold, A., McAuliff, J. P., Luduena, F. P., and Brown, T. G. (1967) Differentiation of receptor systems activated by sympathomimetic amines. *Nature* **214**, 597,598.

23. Arch, J. R. S., Ainsworth, M. A., Cawthorne, M. A., Piercy, V., Sennitt, M. V., Thody, V. E., et al. (1984) Atypical β-adrenoceptors on brown adipocytes as a target for anti-obesity drugs. *Nature* **309**, 163–165.

24. Bond, R. A. and Clarke, D. E. (1988) Agonist and antagonist characterization of a putative adrenoceptor with distinct pharmacological properties from the alpha- and beta-subtypes. *Br. J. Pharmacol.* **95**, 723–734.

25. Galitzky, J., Langin, D., Verwaerde, P., Montastruc, J. L., Lafontan, M., and Berlan, M. (1997) Lipolytic effects of conventional β_3-adrenoceptor agonists and of CGP 12,177 in rat and human fat cells: preliminary pharmacological evidence for a putative β_4-adrenoceptor. *Br. J. Pharmacol.* **122**, 1244–1250.

26. Kaumann, A. J. (1997) Four β-adrenoceptor subtypes in the mammalian heart. *Trends Pharmacol. Sci.* **18**, 70–76.

27. Kaumann, A. J., Preitner, F., Sarsero, D., Molenaar, P., Revelli, J. P., and Giacobino, J.-P. (1998) (–)-CGP 12177 causes cardiostimulation and binds to cardiac putative β4-adrenoceptors in both wild-type and β_3-adrenoceptor knockout mice. *Mol. Pharmacol.* **53**, 670–675.

28. Sarsero, D., Molenaar, P., and Kaumann, A. J. (1998) Validity of (–)-[3H]-CGP 12177A as a radioligand for the 'putative beta4-adrenoceptor' in rat atrium. *Br. J. Pharmacol.* **123**, 371–380.

29. Sambrook, J., Fritsch, E. F., and Maniatis, T. (1989) *Molecular Cloning: A Laboratory Manual,* 2nd ed., Cold Spring Harbor Laboratory Press, Cold Spring Harbor, NY.

30. Delidow, B. C., Lynch, J. P., Peluso, J. J., and White, B. A. (1996) Polymerase chain reaction: basic protocols, in *Methods in Molecular Biology: Basic DNA and RNA Protocols,* vol. 58 (Harwood, A. J., ed.), Humana, Totowa, NJ, pp. 275–292.

31. Domec, C., Garbay, B., Fournier, M., and Bonnet, J. (1990) cDNA library construction from small amounts of unfractionated RNA: Association of cDNA synthesis with polymerase chain reaction amplification. *Anal. Biochem.* **188**, 422–426.

32. DeFrancesco, L. (1997) Don't clone alone: A profile of cDNA libraries and kits. *The Scientist* **11**, 16, http://www.the-scientist.library.upenn.edu/yr1997/sept/profile1_970915.html.

33. Scofield, M. A., Sun, L., and Pettinger, W. A. (1992) Spontaneous hazardous explosion of unopened bottles of diethyl pyrocarbonate. *Biotechniques* **12**, 820,821.

34. Chomczynski, P. and Sacchi, N. (1987) Methods of RNA isolation by acid guanidinium thiocyanate-phenol-chloroform extraction. *Anal. Biochem.* **162**, 156–159.

35. Mukhopadhyay, T. and Roth, J. A. (1998) Isolation of total RNA from tissues or cell lines, in *Methods in Molecular Biology, vol. 86: RNA Isolation and Characterization Protocols* (Rapley, R. and Manning, D. L., eds.), Humana, Totowa, NJ, pp. 55–59.

36. Rapley, R. and Heptinstall, J. (1998) UV Spectrophotometric analysis of ribonucleic acids, in *Methods in Molecular Biology, vol. 86: RNA Isolation and Characterization Protocols* (Rapley, R. and Manning, D. L., eds.), Humana, Totowa, NJ, pp. 65–68.

37. Farrell, R. E. (1993) A laboratory guide for isolation and characterization, in *RNA Methodologies* Academic, San Diego, CA, pp. 76–82.

38. Bryant, S. and Manning, D. L. (1998) Isolation of messenger RNA, in *Methods in Molecular Biology, vol. 86: RNA Isolation and Characterization Protocols* (Rapley, R. and Manning, D. L., eds.), Humana, Totowa, NJ, pp. 61–64.

39. Chomczynski, P. (1992) Solubilization in formamide protects RNA from degradation. *Nucleic Acids Res.* **20**, 3791–3792.

40. Lewis, R. (1997) Kits take the trickiness out of RNA isolation, purification. *The Scientist* **11**, 16,17, http://www.the-scientist.library.upenn.edu/yr1997/mar/tools_970331.html.

41. DeFrancesco, L. (1998) Getting RNA out of cells quickly is the name of the game with total RNA purification kits. *The Scientist* **12**, 20–23, http:www.the-scientist.library.upen.edu/yr1998/sept/profile2_980914.html.

42. DeFrancesco, L. (1998) Oligo(dT) takes on a variety of faces in kits for the purification of mRNA. *The Scientist* **12**, 21, http://www.the-scientist.library.upenn.edu/yr1998/may/profile2_980525.html.

43. Ausubel, F., Brent, R., Kingston, R. E., Moore, D. D., Seidman, J. G., Smith, J. A., et al. (1994) Differential display of mRNA by PCR, in *Current Protocols in Molecular Biology,* vol. 2, Wiley, Interscience, New York, pp. 15.8.1.–15.8.8.

44. Ausubel, F., Brent, R., Kingston, R. E., Moore, D. D., Seidman, J. G., Smith, J. A., et al. (1998) Preparation of genomic DNA from mammalian tissue, in *Current Protocols in Molecular Biology,* vol. 1, Wiley, Interscience, New York, pp. 2.2.1–2.2.3.

45. Pierce, J. C. and Sternberg, N. L. (1992) Using Bacteriophage P1 system to clone high molecular weight genomic DNA, in *Methods in Enzymology,* vol. 216 (Wu, R., ed.), Academic, San Diego, CA, pp. 549–574.

46. Ausubel, F., Brent, R., Kingston, R. E., Moore, D. D., Seidman, J. G., Smith, J. A., and Struhl, K. (1987) Size fractionation using agarose gels, in *Current Protocols in Molecular Biology,* vol. 1, Wiley, Interscience, New York, pp. 5.4.1.–5.4.4.
47. Evans, G. A., Lewis, K., and Rothenberg, B. E. (1989) High efficiency vectors for cosmid microcloning and genomic analysis. *Gene* **79,** 9–20.
48. Dale, J. W. and Greenaway, P. J. (1996) In vitro packaging of DNA, in *Methods in Molecular Biology, vol. 58: Basic DNA and RNA Techniques* (Harwood, A. J., ed.), Humana, Totowa, NJ, pp. 171–175.
49. Briley, G. P. and Bidwell, C. A. (1994) Use of agarose block DNA to make cosmid libraries. *Biotechniques* **17,** 278,279.
50. Ausubel, F., Brent, R., Kingston, R. E., Moore, D. D., Seidman, J. G., Smith, J. A., et al. (1988) Plating lambda phage to generate plaques, in *Current Protocols in Molecular Biology,* vol. 1, Wiley, Interscience, New York, pp. 1.11.1.–1.11.4.

2

Isolation of Adrenergic Receptor Genes

Margaret A. Scofield, Jean D. Deupree, and David B. Bylund

1. Introduction

In order to isolate a single gene, phage or cosmid libraries can be screened by the conventional technique of hybridization as described by Sambrook et al. *(1)* using end-labeled oligonucleotide probes or gene-specific probes. The probes are labeled either by nick translation, end labeling, or random priming using radioactive or nonradioactive techniques. Newer methods that use polymerase chain reaction (PCR) to screen phage libraries have been described by Yu and Bloem *(2)*. Below, we describe a method for screening a cosmid library using PCR rather than a conventional colony hybridization technique. The methodology is based on a report by Takumi and Lodish *(3)*. The cosmid libraries produce transformed bacterial colonies containing large cosmid vectors that behave as plasmids. These plasmids can be extracted from the bacteria by standard techniques for plasmid isolation once the appropriate clone is selected.

2. Materials
2.1. Cosmid Genomic Library Screening

1. Cosmid genomic library (*see* **Subheading 3.1.2.** in Chapter 1).
2. Luria-Bertani (LB) media: Dissolve 10 g Bacto-tryptone, 5 g Bacto-yeast extract, and 10 g NaCl in 1 L distilled water. Final solution is adjusted to pH 7.0 with approx 200 µL of 5 N NaOH and is sterilized by autoclaving.
3. LB agar plates: 15 g Bacto-agar in 1 L of LB media, sterilized by autoclaving, cooled to 55°C, and 30–35 mL poured into an 85-mm Petri dish and cooled.
4. 10 mg/mL kanamycin in water, sterilized by filtration through a 0.22-µm filter and stored at –20°C.

From: *Methods in Molecular Biology, vol. 126: Adrenergic Receptor Protocols*
Edited by: C. A. Machida © Humana Press Inc., Totowa, NJ

5. LB/kanamycin plates: sterile kanamycin is added to 50–55ºC molten LB agar (**item 3** above) to a concentration of 25 μg/mL, mixed, and 30–35 mL of the molten agar poured into an 85-mm Petri dish and cooled.
6. Gene-specific sense primer (25 pmol/μL water) and gene-specific antisense primer (25 pmol/μL water): synthetic oligonucleotide primers designed by the investigator (*see* **Subheading 3.3.**).
7. 10X PCR buffer: 500 mM KCl and 100 mM Tris-HCl, pH 8.3 (supplied with *Taq* DNA polymerase).
8. 25 mM MgCl$_2$ in water (supplied with *Taq* DNA polymerase).
9. 10 mM dNTP mix: dATP, dCTP, dTTP, dGTP, each dissolved in water at a concentration of 10 mM.
10. *Taq* DNA polymerase (5 U/μL) (Perkin-Elmer Foster City, CA or other supplier of licensed *Taq* DNA polymerase).
11. PCR mix for one 50-μL reaction: 5 μL 10X PCR buffer, 4 μL 25 mM MgCl$_2$ (final 2 mM), 0.5 μL 25 pmol/μL gene-specific sense primer (final 0.25 pmol/μL), 0.5 μL 25 pmol/μL gene-specific antisense primer (final 0.25 pmol/μL), 1 μL dNTP mix, 0.25 μL *Taq* DNA polymerase (5 U/μL), and 38.75 μL sterile water.
12. 2% Agarose gel.
13. 10X TAE electrophoresis buffer: 0.40 M Tris-acetate and 10 mM disodium EDTA (TAE), 48.4 g Tris base, and 3.72 g disodium EDTA in 850 mL water are adjusted to pH 8.0 with 10.6 mL glacial acetic acid, and the final volume is increased to 1 L with water.
14. Sterile 96-well microtiter dishes.
15. Glycerol: sterilized.

2.2. Analysis and Sequencing

1. Restriction endonuclease enzymes.
2. Hybrid phage/plasmid vector: Bluescript II (Stratagene, La Jolla, CA), pUC 18/19 (Life Technologies, Gaithersburg, MD).
3. Sequencing primers.
4. DNA sequencing facility or DNA sequencing kit: Sequenase DNA polymerase Version 2.0 kit (Amersham Life Sciences, Arlington Heights, IL) or Thermo Sequenase cycle sequencing kit (Amersham Life Sciences).
5. Oligonucleotide synthesis facility.

2.3. Selection of Primers for PCR

1. Oligonucleotide synthesis facility.
2. Sequences of adrenergic receptor subtype genes.
3. Computer programs for analyzing primers: Oligo 5.0 (National Biosciences, Plymouth, MN) or Prime (Genetics Computer Group [GCG], Madison, WI).
4. Computer programs for comparing DNA sequences: Pileup program (GCG).
5. PCR enzymes and reagents.
6. Thermal cycler.

2.4. Reverse Transcription and Selection of Primers

1. Oligonucleotide synthesis facility.
2. Sequences of adrenergic receptor gene subtypes.
3. Computer programs for analyzing primers.
4. Computer programs for comparing DNA sequences.
5. Random hexamers, oligo dT_{17} primers or gene-specific antisense primers.
6. Reverse transcriptase enzymes.
7. PCR enzymes and reagents.
8. Thermal cycler.

2.5. Cloning of PCR Products

1. Cloning vectors.
2. TA cloning kits: TA Cloning or TOPO-Cloning kit (Invitrogen, Carlsbad, CA).
3. Restriction endonucleases.
4. Supplies for bacterial plating.
5. Procedures and kits for plasmid isolation.

2.6. Rapid Amplification of cDNA Ends (RACE)

1. cDNA with 3' antisense adapter primer (*see* **item 2**) and 5' sense adapter primer (*see* **item 3**) on respective ends (*see* Chapter 1, **step 11** in **Subheading 3.2.1.**).
2. 3' antisense adapter primer: TTCCGGAATTCAGCGGCCGC 25 μM (*see* Chapter 1, **item 20** in **Subheading 2.2.1.**).
3. 5' Sense adapter primer: GACTCGAGTCGACATCGAC 25 μM: primer derived from 5' sense primer without the oligo dC tail (*see* Chapter 1, **item 17** in **Subheading 2.2.1.**).
4. Gene-specific sense primer (25 pmol/μL in water) and gene-specific antisense primer (25 pmol/μL in water): synthetic oligonucleotide primers designed by the investigator based on the sequence of the gene-specific PCR product isolated in **step 4** of **Subheading 3.5.**
5. 10 m*M* dNTP mix.
6. 10X PCR buffer: 500 m*M* KCl and 100 m*M* Tris-HCl, pH 8.3 (Perkin-Elmer).
7. 25 m*M* $MgCl_2$ (supplied with *Taq* DNA polymerase).
8. *Taq* DNA polymerase (5 U/μL) Perkin-Elmer or other supplier of licensed *Taq* DNA polymerase.
9. 100-Fold dilution of amplified cDNA from above (10–100 ng/μL).
10. Second internal gene-specific sense primer (25 pmol/μL in water) and second internal gene-specific antisense primer (25 pmol/μL in water): nested PCR synthetic oligonucleotide primers designed by the investigator based on the sequence of the gene-specific PCR product isolated in **step 4** of **Subheading 3.5.**, and internal to the gene-specific sense and antisense primers in **item 4** above.

3. Methods

3.1. Cosmid Genomic Library Screening

1. Titer the cosmid library by making the appropriate serial dilutions in cold LB/ kanamycin (25 µg/mL) media and plating the dilutions on LB agar/kanamycin (25 µg/mL) plates. The plates are incubated overnight at 37°C and the colony-forming units/mL (CFU/mL) are determined (*see* **Note 1**).
2. Mix the cosmid library, and remove 16 aliquots such that each aliquot contains 1×10^5 colonies/1 mL. Place the aliquots in separate microcentrifuge tubes.
3. Prepare a PCR master mix for a fraction more than the number of reactions (16 reactions and 2 controls) actually required (i.e., 18.5 reactions), or 92.5 µL 10X PCR buffer, 74 µL MgCl$_2$, 9.25 µL gene-specific sense primer, 9.25 µL gene-specific antisense primer, 18.5 µL dNTP mix, 4.6 µL *Taq* DNA polymerase, and 716.9 µL sterile water (*see* **Note 2**).
4. After mixing the master mix, distribute 49 µL into 18 tubes. Place 1 µL from each of the 16 aliquots of the library into 16 of the tubes containing the master mix. Add 1 µL of water to 49 µL of master mix of the 17th tube, thus providing a negative control or no DNA template. Use the 18th tube as the positive control by adding the appropriate nanogram amount of genomic DNA in 1–49 µL of master mix.
5. Amplify the 1 µL aliquots of the library and the two controls according to previously determined PCR conditions for approx 35 cycles (*see* **Notes 3** and **4**).
6. Electrophorese the PCR products through a 2% agarose gel containing 0.5 µg/mL ethidium bromide, and visualize with UV light.
7. Dilute the specific library pools or aliquots of cosmid that give a PCR product of the appropriate size to a concentration of about 30,000 clones/mL with LB/ kanamycin media, or increase the volume of the aliquot by 3.3 vol. Aliquot 100 µL of each dilution (3000 colonies/well) into wells of a 96-well microtiter dish (*see* **Note 5**).
8. Pool the rows and columns of the above dilutions in the following way: combine 10 µL of each well in the column into a microcentrifuge tube to give a final volume of 120 µL from the columns; combine 10 µL from each well in the rows into a microcentrifuge tube to give a final volume of 80 µL for the rows.
9. Remove 2.5 µL from each of the tubes containing pools of the respective columns and rows, and amplify the diluted cosmids using PCR master mixes and positive and negative controls as in **steps 3** and **4**.
10. Analyze the amplified products by gel electrophoresis to identify the tube from the pooled columns and the tube from the pooled rows that demonstrate the presence of the appropriate PCR product. The well that intersects between a positive pooled column and a positive pooled row is indicative of a positive clone being present in the 30,000 clones/mL diluted wells.
11. Dilute this positive aliquot to 300 colonies/mL by increasing the volume 100-fold with the addition of approx 10 mL with LB/kanamycin media.

12. Aliquot 100 μL (30 colonies) of the diluted mixture into additional 96-well microtiter dishes.
13. Repeat **steps 7–9** above to identify a positive clone.
14. Plate the entire volume of the well with the positively identified clone on LB agar/kanamycin plates.
15. Toothpick each colony into individual wells of the 96-well microtiter dishes containing 100 μL of LB/kanamycin media, and grow for 6–8 h.
16. Pool and PCR 1-μL aliquots to identify positive clones as in **steps 7–9** above.
17. Grow any individual colonies that amplify the apparent gene-specific PCR product in 5 mL LB/kanamycin cultures for 6–8 h at 37°C with continuous shaking (*see* **Note 6**).
18. Add 15% (v/v) glycerol to 250 μL of the culture. Mix and freeze at –70°C for long-term storage.
19. Take the remainder of the culture and extract the cosmid from the bacteria using a standard alkaline lysis/phenol-chloroform extraction, small-scale plasmid DNA procedure (*see* **Note 7**).
20. Standard restriction digestion techniques are then used to digest 1 μg of the cosmid DNA with *Not*I, and analyze the digestion products by electrophoresis on a 0.8–1.0% agarose gel (*see* **Note 8**).

3.2. Analysis and Sequencing

1. The genomic insert should be restriction-mapped *(4)*, and the genomic DNA fragment that contains the coding sequence should be determined by Southern analysis using a gene-specific cDNA probe *(1)*. The relevant restriction-digested genomic fragments should be subcloned into a hybrid phage/plasmid vector such as Bluescript II (Stratagene) or pUC18/19 (Life Technologies).
2. Both strands of the recombinant plasmid inserts should be sequenced. A final sequence of the contiguous cosmid insert will be determined based on the restriction map and sequence of the restriction fragments.
3. If a sequencing facility is available, the plasmids can be readily sequenced using cycle sequencing techniques and fluorescence-based dideoxynucleotides. About 500 bp of sequence are generated in a sequencing run from four separate fluorescent dideoxynucleotide tags in one lane of an acrylamide sequencing gel.
4. Sequencing primers may be either vector-specific sequences adjacent to the cloning site or homologous to the gene-specific sequence. These gene-specific sequences would be determined after sequencing runs using vector-specific sequences near the cloning site.
5. The complete cosmid insert can also be directly sequenced using internal primers from the original genomic PCR product (*see* **item 6** in **Subheading 2.1.**) to sequence the cosmid in both directions (5′ and 3′). The resultant sequence is then used to develop new primers for the next sequencing run.
6. If a sequencing facility is not readily available, then the single-stranded or double-

stranded dideoxy chain-termination methods can be performed *(5)* using a Sequenase DNA polymerase Version 2.0 kit (Amersham Life Sciences) or a Thermo Sequenase cycle sequencing kit (Amersham Life Sciences) (*see* **Note 9**).

3.3. Selection of Primers for PCR

The sequence of the oligonucleotides used for the PCR should be selected based on the guidelines listed below.

1. Primers should span a region of DNA with less than a 60% average GC content (*see* **Note 10**).
2. The sense (upstream) and antisense (downstream) primers should not be complementary to one another especially at the 3′ end. In addition, they should not be complementary internally (palindromes), such that the primer can fold back on itself (*see* **Note 11**).
3. Oligonucleotides can range from 18 to 40 nucleotides in length, but for most applications 18–24 bp are sufficient.
4. The sense and antisense primers should have approximately the same G + C content (40–60%). The melting temperatures, T_ms, for each primer should be within 1–2°C of each other (*see* **Note 12**).
5. The primer annealing temperature for PCR is approx 5°C lower than the T_m of the oligonucleotides (*see* **Note 13**).
6. The selection of primers from known sequences can be determined visually or with computer programs. Two such programs are Oligo 5.0 (National Biosciences, Plymouth, MN) or Prime from GCG (*see* **Note 14**).
7. Primers for amplifying DNA of more than 2 kb in length (long-distance PCR, LDPCR) are designed to have higher annealing temperatures to provide greater specificity. When amplifying with these primers, cosolvents are added to lower the DNA melting temperature. It is also important to make sure that the selected primers do not contain repetitive sequences (Alu sequences) (*see* **Note 15**).
8. Noncomplementary bases (extensions) can be added at the 5′ end of primers. These extensions may code for restriction sites or promoter sequences or other sequences that are useful for cloning the amplified product into a vector or in vitro synthesis of RNA. When adding extensions for restriction endonuclease recognition sequences two to three extra bases (G or C) should be added on the 5′ end, so that the enzyme has enough room to recognize the restriction site. These extensions will not hinder the PCR unless these sequences are present within the DNA region to be amplified.
9. The nonspecific binding and extension of primers prior to the initial denaturation of the template during the first step of PCR can be significantly reduced by keeping the reaction mixes at 0°C before thermal cycling and using "hot-start" techniques (*see* **Note 16**).
10. New receptor subtypes or multiple subtypes from species, which have not been rigorously studied, can be determined by designing primers based on two consensus regions from all the members of the same families. However, these

consensus regions should distinguish the family of interest from other families (i.e., other G-protein receptor gene families). Thus, the sense primer is identical to the upstream consensus region, and the antisense primer is identical to the downstream consensus region. We have successfully used this approach to clone the gerbil α-AR and β-AR *(6,7)*.

11. Consideration should also be given to the development of primers that span introns if a cDNA preparation is used as a template. The genomic database should be used to locate introns and then consensus primers designed on either side of the intron. Both the β_3-AR and α_1-AR contain introns. The α_1-AR in particular has at least one large 20- to 30-kb intron within the sixth transmembrane domain. However, the α_2-AR, β_1-AR, and β_2-AR are intronless (*see* **Note 17**).

12. Primers for cloning subtype-specific adrenergic sequences should be designed based on the consensus sequences that distinguish one subtype from another and are poorly conserved between subtypes, such as the sequences within the third intracellular loop of adrenergic receptors. The sense primer then is on the 5′ end of the loop and the antisense primer is on the 3′ end of the loop.

13. Database resources and computer programs can be utilized to help in the design of consensus sequences. The GenBank database can be accessed through the NCBI program Entrez (www.ncbi.nlm.nih.gov/Entrez/nucleotide.html), and keywords can be entered under the nucleotide search program. This search will identify all the sequences and their accession numbers, including expressed sequence tags (EST) sequences (or randomly transcribed cDNA) that have been submitted to GenBank. One should be aware that more than one keyword might be necessary to find all the sequences (i.e., adrenergic vs adrenoceptor). Occasionally, sequences published before sequence submission to GenBank was common may not have been entered into GenBank.

14. The investigator should understand that the name of the subtype for α_1-AR, α_2-AR, and β_3-AR that is described with the accession number is not necessarily the correct or current subtype nomenclature (*see* **Tables 1–3** in Chapter 1). The subtype specificity of the sequences can be determined by performing a multiple alignment comparison of the cDNA sequences of only the coding regions. We have successfully used the Pileup program from GCG to distinguish between the subtypes for the α_2-AR. This program will line up all the input sequences (nucleotide as well as amino acid) and output these data as a file. It will also produce a denogram that will group the sequences according to similar homology (**Fig. 1**). However, this is not an evolutionary tree and the grouping is only based on the similarities of the respective sequences (*see* Chapter 3, **Subheading 3.2.3.**) (*see* **Note 18**).

15. Once consensus sequence regions have been defined based on both nucleotide comparisons and amino acid homologies, it is possible that even though there may be an exact match in the amino acid consensus sequences, the nucleotide sequence may vary, especially at the third base of the codon. Thus, it may be useful to synthesize degenerate primers to encompass all the possible primer variations. Degenerate bases may be chosen based on the bases specific for each

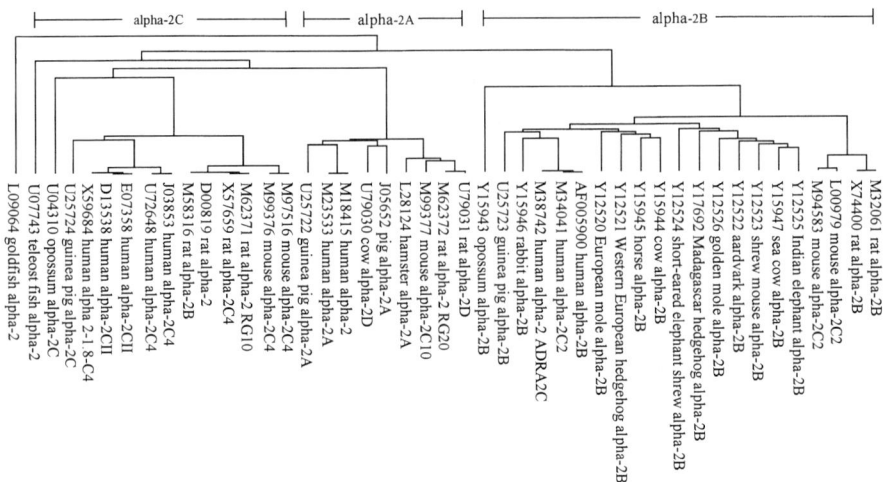

Fig. 1. Denogram of α_2-AR database sequences: Sequences were entered into the multiple sequence alignment Pileup program of GCG. Accession numbers and description of the sequences from GenBank are indicated. The three α_2-AR subtypes, α_{2A}-AR, α_{2B}-AR, and α_{2C}-AR, are indicated above the sequences. These assignments were made on the basis of the three major groupings of sequences as a result of the high identity of the nucleotide sequences within each group.

species or on all the possible nucleotide sequences for the amino acids at the primer site. However, the number of degenerate bases should be kept to a minimum. In particular, the base composition at the 3′ end should be conserved among all the species or subtypes (*see* **Note 19**).

3.4. Reverse Transcription and the Selection of Primers

Several different types of primers and reverse transcriptase (RT) enzymes can be used to form the first-strand or cDNA from either total or poly A$^+$ RNA templates (*see* **Notes 20** and **21**).

1. In general, random hexamers are used as primers when a transcript is very long, has a short poly A tail, and/or contains regions with a high % GC content, thus creating areas with significant secondary structures. Random hexamers are initially incubated at room temperature with RT to extend the primers such that they can anneal to the RNA at the higher RT incubation temperatures. Thus, these primers are not specific for any RNA species, either rRNA, tRNA, or mRNA, but the specificity for the amplified fragments can be achieved by using gene-specific primers at the higher annealing temperatures of PCR.
2. Oligo dT provides the next level of specificity by priming the synthesis of cDNA

predominantly from the poly A tail of mRNA. However, oligo dT may not transcribe efficiently if the RNA is partially degraded or the amplification target is significantly upstream of the poly A tail. The oligo dT primers are reverse-transcribed in the same manner as random hexamers, and the gene-specific products are amplified at the more stringent conditions of PCR using gene-specific primers (*see* **Note 22**).

3. Finally, gene-specific primers that represent sequences from the antisense strand (or downstream primers) can be used for cDNA synthesis and can be designed based on consensus sequences. They are particularly useful for transcribing regions of mRNA that are distant from the poly A tail. We have successfully used this approach for amplifying adrenergic receptor subtypes from species, such as mouse or gerbil, and for quantifying RNA by competitive RT-PCR *(8)* (*see* **Notes 23** and **24**).

3.5. Cloning of PCR Products

PCR products have been historically difficult to clone because of the terminal transferase activity of *Taq* polymerase that adds single deoxynucleotides, in particular deoxyadenosine residues, to the 3′ end of the amplification products. This creates an overhang rather than a blunt end and cannot be cloned into a blunt-ended vector.

1. PCR products can be cloned by engineering a restriction endonuclease recognition sequence at the 5′ ends of PCR primers to create a compatible overhanging end with a cloning vector. Primers of this type are used in Chapter 1, **Subheading 2.2.1., items 3** and **17**.
2. In other procedures, the products are blunt-ended, and cloned by filling the recessed ends with Klenow polymerase or by removing the extended bases with *Pyrococcus furiosus* (*Pfu*) DNA polymerase.
3. Finally, vectors with T overhangs can provide a sticky end for the A overhang in the PCR product in a procedure called TA cloning. This is one of the more popular methods for cloning PCR products and is described by Trower *(9)* (*see* **Notes 25** and **26**).
4. The identity of the PCR products can be confirmed by sequencing, Southern hybridization with known probes, or restriction endonuclease analysis *(1)*.

3.6. Rapid Amplification of cDNA Ends (RACE)

After a product from RT-PCR of cDNA has been cloned and identified, that fragment can be used to screen a cDNA library in a similar manner to that described for screening a genomic library (*see* **Subheading 3.1.**). Alternatively, the small amount of sequence information derived from the RT-PCR fragment can be used to amplify sequences from both the 3′ end (3′ RACE) and 5′ end (5′ RACE) of the target message. The cDNA used above to construct the library (*see* Chapter 1, **step 11** in **Subheading 3.2.1.**) can be used as a template to amplify the respective ends of a specific message. The adapter primers on the library cDNA provide a known primer sequence, or tag, on the 3' or 5' ends of all the cDNA species. Internal primers can be synthesized based on the sequence of the previously identified RT-PCR product and will serve as the gene-specific primers for the amplification reaction. Thus, the terminal adapter primers and gene-specific primers will be used during PCR to amplify two overlapping fragments of the target cDNA to complete the cDNA sequence. This methodology also has the potential to isolate products that are a result of alternative splicing of the same gene.

1. 3' Ends are amplified by performing 30–35 cycles of PCR on the amplified cDNA in 50 µL as follows: 10–100 ng of amplified cDNA, 5 µL 10X PCR buffer, 4 µL MgCl$_2$ (final 2 mM), 0.5 µL gene-specific sense primer (final 0.25 pM/µL), 0.5 µL 3' adapter primer (final 0.25 pM/µL), 1 µL dNTP mix (final 20 nM each), 0.25 µL *Taq* DNA polymerase, and sterile water to 50 µL. The PCR annealing temperature should be between 50 and 55°C, but these conditions will vary depending on the T_m of the gene-specific and adapter primers (*see* **Note 27**).
2. Analyze the amplified products by gel electrophoresis (*see* **Note 28**).
3. Perform a nested PCR reaction on the 3'-amplified product as follows: PCR 1 µL of the 3'-amplified product above (*see* **step 1**) with 5 µL 10X PCR buffer, 4 µL MgCl$_2$, 0.5 µL second internal gene-specific sense primer, 0.5 µL 3' adapter primer, 1 µL dNTP mix, 0.25 µL *Taq* DNA polymerase, and sterile water to 50 µL (*see* **Note 29**).
4. Analyze the amplified products by gel electrophoresis and clone, and/or sequence the appropriate distinct band(s).
5. Amplify the 5' end by repeating **steps 1** and **2** above using the gene-specific antisense primer in place of the gene-specific sense primer, and the 5' sense adapter primer in place of the 3' antisense adapter primer.
6. In a reaction similar to that in **step 3** above, perform a nested PCR on the 5'-amplified product from **step 5** using the 5' sense adapter primer and the second internal gene-specific antisense primer.
7. Analyze the amplified products by gel electrophoresis and clone, and/or sequence any distinct band(s) (*see* **Note 30**).

4. Notes

1. Mix the library so that all the bacteria are suspended, and make three 10^{-2} serial dilutions and a final 10^{-1} dilution. Plate 100 µL from the final dilution (10^{-7} dilution) and the previous dilution (10^{-6} dilution) by spreading the aliquot with a bent glass rod that has been sterilized by immersion in 70% ethanol and flamed in a Bunsen burner. The rod is allowed to cool before spreading the bacteria. Ideally, dilutions that plate 100–300 CFU will allow a reasonably accurate determination of the titer.
2. The volume for the master mix should always be slightly larger than that required

for distribution to the tubes to eliminate the possibility of not having enough master mix for all the tubes.

3. The PCR reagent and cycling conditions for these reactions and the gene-specific primers (*see* **Subheading 3.3.**) should be predetermined using genomic DNA (100–500 ng/PCR reaction) as a template to amplify a specific portion of the gene. The ability of the primers to anneal to genomic DNA can be enhanced by denaturing the DNA (i.e., boiling followed by quick cooling on ice) before addition to the PCR mix. The identity of the PCR product should be confirmed (*see* **step 4** in **Subheading 3.5.**). Mg^{2+} concentrations for PCR may vary from 1 to 5 m*M*. Primers that amplify the 5′ end of the gene will select for clones with DNA near the transcription start site, whereas primers from the 3′ end of the gene will select clones nearer the 3′ untranslated region. Ideally, a cosmid library will allow the selection of the complete gene irrespective of the primers used. These reactions can also be performed in a 25-μL vol depending on the thermal cycler and test tube requirements.

4. There is no need to purify the DNA from the library, because the denaturing conditions at 95°C during PCR will lyse the cells to release the plasmid for amplification.

5. These plates can be stored for a day at 4°C after covering the dishes with parafilm. Adhesive acetate plate sealing material (Linbro, Flow Laboratories, McLean, VA) can also be used to cover the plates. The colonies should not be allowed to grow within the liquid media, since some clones will multiply at a faster rate than other clones and thus will potentially select against the target clone. For permanent storage, 15 μL of glycerol are added to each well and mixed. The plate is then stored at −70 or −80°C.

6. The yield of the cosmids will decrease with longer incubation times.

7. A typical procedure can be found in Sambrook et al. *(1)*. In addition, many plasmid isolation kits are also available from manufacturers and have been reviewed by Mack *(10)*.

8. Typical restriction digestion procedures can be found in Sambrook et al. *(1)*. A 7.9-kb vector and large (20–40 kb) insert should be observed. Smaller gene fragments might be observed if *Not*I cleaves within the gene insert.

9. Further information on cycle sequencing can be found in Brush *(11)*.

10. If the GC content of the PCR product is >60%, then cosolvents, such as 5–10% solutions of glycerol, dimethyl sulfoxide (DMSO), or formamide, may be required in the PCR reaction.

11. Primers that are complementary at the 3′ end will tend to form primer-dimers, which will compete with the genomic template.

12. The T_ms of the oligonucleotides can be approximated according to the following formula: $T_m = 4(G + C) + 2(A + T)$. Alternatively, the T_ms can be calculated using computer programs, such as Oligo 5.0 (National Biosciences) or Prime (GCG).

13. The exact annealing temperature for an optimum PCR can be experimentally modified by varying the annealing temperature in 2–5°C increments in either direction to establish primer conditions that are specific to the DNA. Magnesium

concentrations for PCR can also be varied between 1.5 and 5 mM to determine the optimum concentration.

14. The Prime program will design primers that span a selected region of the DNA sequence and will allow you to select the primer length, guanine-cytosine content, and T_m difference between your primers. In addition, the program will eliminate primers with internal and external complementarity.

15. The conditions for LDPCR can also be modified by lengthening the PCR extension times. In addition, small amounts of proofreading thermostable DNA polymerase enzymes (*Pfu*) are added to the *Taq* polymerase mix *(12)*.

16. "Hot start" can be carried out manually by heating all the PCR reagents, but *Taq* polymerase to 70–80°C, and then adding *Taq* polymerase at these higher temperatures. Immediately after the enzyme addition, the temperature is increased further to 95°C to commence denaturation and the thermal cycling procedure. Other techniques employ waxes that melt at 70–80°C and, thus, mix the aqueous enzyme mixture above the wax with the reagents below. An easier and more efficient procedure uses a *Taq* antibody, such as TaqStart (Clontech, Palo Alto, CA), which denatures and releases from the enzyme at 70–80°C as the PCR reagents are heated up to 95°C during the first denaturation step.

17. When cDNA is used as a template for intronless genes, untranscribed RNA templates should be included as controls to demonstrate the absence of genomic DNA.

18. The investigator can utilize the Pileup denogram to determine the subtypes of the sequences in the database based on the grouping of the sequences as a result of homology comparisons. This multisequence lineup is invaluable for determining consensus sequences between species by utilizing EST sequences and complete cDNA sequences.

19. This region of the primer will serve as an anchor for initiation of replication and extension by DNA polymerase, and heterogeneity in the primer at this position will result in too much mispriming. It is important to note that the annealing temperature and magnesium concentration may also need to be varied to account for the differences in bases within the degenerate primer subset. Conditions that might work for a known rat sequence may not be appropriate for the DNA of other species, depending on the degeneracy of the sequences. Further information on the synthesis of degenerate primers is reported by Preston *(13)*.

20. The reverse transcriptase enzymes most often used in these reactions are derived from Avian Myeloblastosis Virus (AMV) or Moloney Murine Leukemia Virus (M-MLV). AMV can reverse-transcribe at higher temperatures (48–55°C), thus eliminating secondary structure problems, and M-MLV can reverse-transcribe more efficiently and with less RNase H activity, but at lower temperatures (37–42°C). Other protocols are also available that utilize the same enzyme (*Thermus thermophilus, Tth*) for reverse transcription (70°C) as for PCR by changing the reaction conditions. A more complete description of reverse transcriptase enzymes and RT-PCR kits has been written by Wilkinson *(14)*.

21. In our laboratory, a typical reverse transcriptase reaction is performed in a 10-μL

vol with 5 mM MgCl$_2$, 1X PCR buffer, 1 mM dNTP, 1 U/µL RNase inhibitor, 2.5 U/µL M-MLV-RT, 2.5 µM antisense primers (or 2.5 µM random hexamer or 2.5 µM oligo dT primers) and ~1 µg total RNA. After incubating at 42°C for 50 min and then at 95°C for 5 min to denature the enzyme, the entire mixture is amplified by PCR in a 50-µL reaction with the appropriate sense or upstream primer and PCR reagents.

22. In comparison to random primers, oligo dT will tend to copy a greater proportion of the 3' end of the mRNA, and random primers will transcribe more efficiently at the 5' region of the mRNA.

23. The success of this approach is dependent on how specific the antisense primers are at the lower-temperature conditions of 42°C for cDNA synthesis by reverse transcriptase. Occasionally, we have found that the antisense primers are too non-specific at 42°C and will generate a RT-PCR product that appears to be of the expected size. However, when the product is sequenced, it is apparent that the antisense primer has served both as a sense and antisense primer in the PCR. Where the antisense cDNA primers give nonspecific products, then either more stringent PCR conditions could be employed or oligo dT and random hexamers could be used during reverse transcription to aid in the successful amplification of the gene-specific product.

24. In general, we have found that it is necessary to try each of the above primer approaches for cDNA synthesis. If the amplified product is of the expected size, it is useful to run PCR controls on this product in the presence of only the antisense primer and only the sense primer in PCR assays. Sequence analysis of the RT-PCR products should be confirmed by analyzing the products from three individual RT-PCR reactions to ensure that *Taq* polymerase has not introduced errors into the sequence.

25. In our laboratory, we routinely use the Original TA Cloning® kit or TOPO-Cloning® Kit from Invitrogen (Carlsbad, CA). These kits provide the materials for cloning the PCR product into a supplied linear vector. The insert can be conveniently excised from the vector with *Eco*RI or other convenient restriction endonuclease enzymes. Further, the vector contains M13 forward and reverse primers for sequencing. These kits are also designed to use PCR products directly from the PCR mix, from products isolated from low-melt gels, or extracted by other methods. Competent cells are also provided for transformation. Blue/white screening can be used to identify the recombinant plasmids. The colonies can also be picked with a toothpick, and a PCR reaction directly performed on the minute amount of bacteria using primers that flank the insert or gene-specific primers. The sizes of the products are then identified by gel electrophoreses *(15)*. DeFrancesco *(16)* gives a complete discussion of the PCR-based cloning kits that are available for cloning these products.

26. The investigator should keep in mind that PCR products generated by polymerases with 3' to 5' exonuclease activity, such as *Pfu*, will automatically create blunt ends that must be blunt-end-cloned or must be changed to single nucleotide overhangs in an extension reaction with *Taq* polymerase for TA cloning.

27. The DNA polymerase can also be mixed with proofreading enzymes to amplify potentially longer cDNA products with fewer sequencing errors.
28. Many times a smear will be observed for this initial reaction, and a second internal sense primer is then used to amplify the gene-specific product in a nested PCR reaction.
29. The amplified product from **Subheading 3.6., step 1** that is used as a template in this reaction can be removed directly from the PCR. Alternatively, the product can be used as the template if the DNA has been isolated from the gel. This can be done by either using a toothpick to add the PCR product directly from the agarose into the PCR mixture, or by cutting out the band containing the product isolation from a low-melt agarose gel and then melting the gel.
30. Variations of this procedure have also been described by Ausubel et al. *(17)* and Frohman *(18)*. In addition, several companies also supply kits for these procedures, such as the Marathon™ cDNA amplification kit from Clontech or 3′ RACE and 5′ RACE kits from Life Technologies.

References

1. Sambrook, J., Fritsch, E. F., and Maniatis, T. (1989) *Molecular Cloning: A Laboratory Manual*, 2nd ed., Cold Spring Harbor Laboratory, Cold Spring Harbor, NY.
2. Yu, L. and Bloem, J. J. (1996) Use of polymerase chain reaction to screen phage libraries in *Methods in Molecular Biology, vol. 58: Basic DNA and RNA Protocols*, (Harwood, A. J., ed.), Humana, Totowa, NJ, pp. 335–339.
3. Takumi, T. and Lodish, H. F. (1994) Rapid cDNA cloning by PCR screening. *Biotechniques* **17**, 443–444.
4. Ausubel, F., Brent, R., Kingston, R. E., Moore, D. D., Seidman, J. G., Smith, J. A., et al. (1995) Mapping by multiple endonuclease digestions, in *Current Protocols in Molecular Biology*, vol. 1. Wiley, Interscience, New York.
5. Sanger, F., Nicklen, S., and Coulson, A. R. (1977) DNA sequencing with chain-terminating inhibitors. *Proc. Natl. Acad. Sci. USA* **74**, 5463–5467.
6. Gruber, D. D., Dang, H., Shimozono, M., Scofield, M. A., and Wangemann, P. (1998) α_1-Adrenergic receptor mediate vasoconstriction of the isolated spiral modiolar artery *in vitro*. *Hear. Res.* **119**, 113–124.
7. Wangemann, P., Liu, J., Shimozono, M., Bruchas, M., and Scofield, M. A. (1997) Pharmacological and molecular biological demonstration of β_1- and β_2-adrenergic receptors in inner ear tissues. *Assoc. Res. Otolaryngol.* **21**, 55.
8. Xiao, L., Scofield, M. A., and Jeffries, W. B. (1998) Cloning, expression and pharmacological characterization of the mouse α_{1A}-adrenoceptor. *Br. J. Pharmacol.* **124**, 213–221.
9. Trower, M. K. and Elgar, G. S. (1996) Cloning PCR products using T-vectors, in *Methods in Molecular Biology, vol. 58: Basic DNA and RNA Protocols*, (Harwood, A. J., ed.), Humana, Totowa, NJ, pp. 313–324.

10. Mack, A. (1996) Kits and other new developments streamline DNA purification process. *The Scientist* **10,** 17,18, http://www.the-scientist.library.upenn.edu/yr1996/apr/tools_960429.html.
11. Brush, M. (1997) Cycle sequencing kits. *The Scientist* **11,** 16–18, http:www.the-scientist.library.upen.edu/yr1997/july/profile1_970721.html.
12. Cheng, S. (1995) Longer PCR amplifications, in *PCR Strategies* (Innis, M. A., Gelfand, D. H., and Sninsky, J. J., eds.), Academic, San Diego, CA, pp. 313–324.
13. Preston, G. (1996) Polymerase chain reaction with degenerate oligonucleotide primers to clone gene family members, in *Methods in Molecular Biology, vol. 58: Basic DNA and RNA Protocols*, (Harwood, A. J., ed.), Humana, Totowa, NJ, pp. 303–312.
14. Wilkinson, D. A. (1998) Getting the message with RT-PCR. *The Scientist* **12,** http://www.the-scientist.library.upenn.edu/yr1998/august/profile2_980817.html.
15. Trower, M. K. (1996) A rapid PCR-based colony screening protocol for cloned inserts, in *Methods in Molecular Biology, vol. 58: Basic DNA and RNA Protocols*, (Harwood, A. J., ed.), Humana, Totowa, NJ, pp. 329–333.
16. DeFrancesco, L. (1998) PCR based cloning kits: something for everybody. *The Scientist* **12,** 22, http://www.the-scientist.library.upenn.edu/yr1998/apr/profile2_980413.html.
17. Ausubel, F., Brent, R., Kingston, R. E., Moore, D. D., Seidman, J. G., Smith, J. A., et al. (1992) Production of a complete cDNA library, in *Current Protocols in Molecular Biology*, vol. 1. Wiley Interscience, New York, pp. 5.8.1.–5.8.8.
18. Frohman, M. A. (1990) RACE: rapid amplificiation of cDNA ends, in *PCR Protocols—A Guide to Methods and Applications* (Innis, M. A., Gelfand, D. H., Sninsky, J. J., and White, T. J., eds.), Academic, San Diego, CA, pp. 28–38.

3

Analyses of Adrenergic Receptor Sequences

Jean D. Deupree, Margaret A. Scofield, and David B. Bylund

1. Introduction

Once a cDNA sequence for one of the adrenergic receptors (ARs) is obtained, it is useful to determine the restriction endonuclease sites in the sequence, to translate the nucleotide sequence into the protein sequence, to compare the sequence with known adrenergic receptor sequences in order to verify the receptor and subtype, to submit the sequence to GenBank, and to analyze the predicted secondary structure of the receptor.

2. Materials

The methods reported here require the use of one of the programs in the Genetic Computer Group (GCG) program package or computer programs found at various web sites. The GCG program package is published by the Computer Group of Madison, WI (*see* **Note 1**). The most current version is volume 10. In addition, the URLs for various web sites that have programs that will perform functions similar to the GCG program are listed in this section.

2.1. Protein Translation and Identification of Restriction Endonuclease Sites

1. GCG programs: TRANSLATE, MAP, and FRAMES.
2. Translate and protein machine at ExPASy Proteomics tools: http://expasy.hcuge.ch/www/tools.html#translate.

2.2. Sequence Alignments

2.2.1. Search for Known Adrenergic Receptor Sequences

URLs for commonly used data bases are:

From: *Methods in Molecular Biology, vol. 126: Adrenergic Receptor Protocols*
Edited by: C. A. Machida © Humana Press Inc., Totowa, NJ

1. Entrez Browser from NCBI: http://www3.ncbi.nlm.nih.gov/Entrez/.
2. European Bioinformatics Institute (EBI), SRS: http://srs.ebi.ac.uk/.
3. GCRDb: http://www.gcrdb.uthscsa.edu/ (*see* **Note 2**).
4. GPCRDB: http://swift.embl-heidelberg.de/7tm/.
5. GenBank: http://www.ncbi.nlm.nih.gov/Web/Genbank/index.html.
6. GenomeNet (Kyoto University and University of Tokyo), DBGET: http://www.genome.ad.jp/dbget/dbget2.html.
7. Prosite: http://expasy.ch/swissmod/SWISS-MODEL.html.

2.2.2. Comparison of Sequences with One Other Sequence

2.2.2.1. DOT MATRIX COMPARISON

GCG programs are: COMPARE and DOTPLOT.

2.2.2.2. GLOBAL SEQUENCE ALIGNMENT

1. GCG program: GAP.
2. ROBUST: Biology Workbench: http://biology.ncsa.uiuc.edu/.

2.2.2.3. LOCAL SEQUENCE ALIGNMENT

The GCG program is BESTFIT.

2.2.3. Multiple Sequence Alignment (MSA)

1. GCG programs: PILEUP and PRETTY.
2. MSA: (Sum of Pairs Criterion) Biology Workbench: http://biology.ncsa.uiuc.edu/.

2.2.4. Homology Searching

2.2.4.1. FASTA

1. GCG program: FASTA.
2. Biology Workbench: http://biology.ncsa.uiuc.edu/.
3. G-protein Coupled Receptor Data Base (GCRDb): http://www.gcrdb.uthscsa.edu/.
4. EMBL, EBI: http://www.ebi.ac.uk.

2.2.4.2. BASIC LOCAL ALIGNMENT SEARCH TOOL (BLAST)

1. GCG program: BLAST.
2. http://www.ncbi.nlm.nih.gov/BLAST/.
3. Biology Workbench: http://biology.ncsa.uiuc.edu/.
4. EMBL, EBI: http://www.ebi.ac.uk.
5. Other programs that are specific for comparison of protein sequence alignments can be found at Biology Workbench: http://biology.ncsa.uiuc.edu/.
   These programs include:
   LALIGN: calculates N-best local protein sequence alignments (part of FASTA).
   CLUSTALW: multiple sequence alignment.

ALIGN: optimal alignment of two protein sequence.
SSEARCH: Smith-Waterman local alignment of proteins (part of FASTA).
SIM: N-best local Similarities using affine weights.
BESTSCORE: calculates the best comparison score.
PRSS: compares a protein sequence to a shuffled protein sequence.
SAPS: statistical analysis of protein sequence.

2.3. Submission of Nucleotide Sequence to GenBank Database

1. BANKIT: NCBI: http://www.ncbi.nlm.nih.gov/BankIt/.
2. SEQUIN: NCBI: http://www.ncbi.nlm.nih.gov/Sequin/.

2.4. Methods for Analyzing Protein Structure

2.4.1. Primary Structure of Adrenergic Receptor

2.4.1.1. AMINO ACID COMPOSITION

1. ExPASy Proteomics tools: http://expasy.hcuge.ch/www/tools.html#translate.
2. GCG program: COMPOSITION.
3. AASTATS: Statistics based on amino acid abundance, The Biology WorkBench: http://biology.ncsa.uiuc.edu/.

2.4.1.2. SEARCH FOR KNOWN PROTEIN MOTIFS

1. BLOCKS: http://www.blocks.fhcrc.org/.
2. GCG programs: MOTIFS, PROFILESCAN and SPSCAN.
3. PFAM: http://www.sanger.ac.uk/Software/Pfam/.
4. PFSCAN: http://biology.ncsa.uiuc.edu/.
5. PPSEARCH: PROSITE pattern search, EMBL, EBI: http://www.ebi.ac.uk/.
6. PRINTS: http://www.biochem.ucl.ac.uk/bsm/dbbrowser/PRINTS.
7. PROSEARCH: The Biology WorkBench: http://biology.ncsa.uiuc.edu/.
8. SMART: http://coot.embl-heidelberg.de/SMART.

2.4.1.3. PEPTIDE FRAGMENTS

The GCG program is PEPTIDESORT and PEPTIDEMAP.

2.4.1.4. ISOELECTRIC POINT

1. GCG program: ISOELECTRIC.
2. PEPIDENT, TAGIDENT, COMPUT pI/Mw: ExPASy Proteomics tools: http://expasy.hcuge.ch/www/tools.html#translate.

2.4.2. Secondary Structure of Adrenergic Receptor

2.4.2.1. α AND β HELIXES

The GCG programs are PEPPLOT and PEPSTRUCTURE.

2.4.2.2. HYDROPHOBICITY

1. GCG programs: PEPTIDESTRUCTURE and PEPPLOT.
2. GREASE: The Biology WorkBench: http://biology.ncsa.uiuc.edu/.
3. Physicochemical result, Pôle Bio-Informatique Lyonnais (PBIL), http://pbil.univ-lyon1.fr/.

2.4.2.3. HELICAL HYDROPHOBIC MOMENT

The GCG programs are HELICALWHEEL, MOMENT, and PEPPLOT.

2.4.2.4. SURFACE PROBABILITY

The GCG program is PEPTIDESTRUCTURE.

2.4.2.5. FLEXIBILITY

The GCG program is PEPTIDESTRUCTURE.

2.4.2.6. GLYCOSYLATION SITES

The GCG program is PEPTIDESTRUCTURE.

2.4.2.7. ANTIGENIC INDEX

1. GCG program: PEPTIDESTRUCTURE.
2. Physicochemical result, Pôle Bio-Informatique Lyonnais, http://pbil.univ-lyon1.fr/.

2.4.2.8. COILED-COIL SEGMENTS

The GCG program is COILSCAN.

3. Methods

3.1. Protein Translation and Identification of Restriction Endonuclease Sites

There are several translation programs available for determining the amino acid sequence. The program TRANSLATE from GCG will translate the sequence starting and stopping where you designate. It does not recognize the start and stop codons or exons. The MAP program will translate the nucleotide sequence into an amino acid sequence as well as identify restriction endonuclease sites in the DNA. The protein translation portion of the MAP program will allow you to see open reading frames in any of the six possible reading frames based on traditional translation coding or other user-supplied nonconventional translation schemes. Identification of restriction endonuclease sites is useful for subcloning and confirming the identify of the sequence. Any subset of restriction endonucleases may be selected to determine which sites are present in the sequence. The FRAMES program (GCG) will identify the start and stop codons in all six reading frames, but cannot identify the location of the exons or which start codon translation normally starts from. There are two methods for identifying the correct start and stop codons and the location of exons. One is to translate the cDNA. Alternatively, the start codons from genomic DNA can be determined by comparing the sequence to known adrenergic receptor sequences using the programs COMPARE, GAP, and PRETTY discussed below. More detailed information on how to locate start and stop codons, splice sites, branch points, promoters and terminators of transcription, and introns and exons is described by Haussler *(1)*.

3.2. Methods for Conducting Sequence Alignments

Once a DNA sequence for one of the adrenergic receptor subtypes is obtained, comparisons to one or more known sequences are useful. There are several programs available for conducting various types of sequence alignments. Factors to consider in choosing a method include: the availability of software, the number of sequences to be compared, and the relative lengths of the sequences. In addition, you need to decide whether you want to compare DNA sequences or protein sequences. DNA level searches are best for locating nearly identical regions of sequences. For example, it is helpful to compare your sequence with the sequence for the same gene from other species. Protein-level searches with other adrenergic receptors are helpful in locating the transmembrane regions and for detecting evolutionarily related genes. The description for each of the methods listed below provides examples of computer programs, World Wide Web sites that are available, and a brief description of the capabilities of the methods. Examples using these methods for comparing human α_2-AR subtypes are provided. Extensive details on how to use the GCG computer programs can be found in volume 24 of *Methods in Molecular Biology (2)*. More extensive searches for sequence similarities with genes that are not adrenergic receptors may also be done. Two excellent articles that discuss some of these search strategies in more detail are Altschul *(3)* and Brenner *(4)*.

3.2.1. Searches for Known Adrenergic Receptor Sequences

A list of known sequences for adrenergic receptors can be obtained by searching one of the databases using search engines provided by Entrez, Sequence Retrieval System (SRS), DBGET, or one of the G-protein-coupled receptor databases (GCRDb or GPCRDB). The Entrez database and Browser are provided by the National Center for Biotechnology Information (NCBI). NCBI also builds, maintains, and distributes the GenBank Sequence Database. GenBank is the NIH genetic sequence database that contains all publicly available DNA sequences. GenBank is provided by the National Center for

Biotechnology Information. It is part of the International Nucleotide Sequence Database Collaboration, which also includes the DNA DataBank of Japan (DDBJ) and the European Molecular Biology Laboratory (EMBL). SRS retrieval system has been developed by the EBI. The DBGET search engine has been maintained by the Institute for Chemical Research, Kyoto University and the Human Genome Center of the University of Tokyo. GCRDb is a database of all G-protein-coupled receptors. However, this site is not as current as the GPCRDB site. The GCRDb data can be searched by looking at families, ligands, or species. Family A, Group II contains the data for the biogenic amine receptors. Receptors are listed by accession number, name, and species. (Note: α_{1A}-AR are currently listed as α_{1C}-AR). GPCRDB is an excellent protein database for G-protein-coupled receptors. From this site, you can link to nucleotide sequences or obtain a phylogenic tree for the different families of G-protein-coupled receptors. Links to snake-like models for known G-protein-coupled receptors are also available. PROSITE on the ExPASy Molecular Biology Server is a database of protein families and domains. It consists of biologically significant sites, patterns, and profiles that help to identify reliably to which known protein family a new sequence belongs.

3.2.2. Comparison of Sequence with One Other Sequence

3.2.2.1. Dot Matrix Comparisons to Determine Locally Matching Areas

A dot matrix analysis compares two sequences and generates a table of coordinates where two sequences are similar. There are two methods for comparing the sequence: the stringency comparison method or the word comparison method. Stringency comparison uses the method of Maizel and Lenk *(5)* to compare two sequences in every register, searching for all the places where a given number of matches (stringency) occur within a given range (window). The word comparison method looks for short, perfectly matched words of a set length. Either the window and stringency are defined or the word length is defined, depending on which method is used. The data generated by the comparison program of the two sequences are then graphically plotted on the *x*- and *y*-axes. A dot is placed where there is a similarity between the two sequences as determined in the comparison program. **Figure 1** compares the sequences for the human α_{2A}- and α_{2B}-ARs adrenergic receptors. The diagonal line shows areas of sequence similarities between these two sequences. These areas include the seven-transmembrane domains as well as other areas of sequence similarity. The gap in the line is caused by the dissimilarity between the third intracellular loops of α_{2A}-AR and α_{2B}-AR.

The computer programs used to do this analysis are COMPARE and

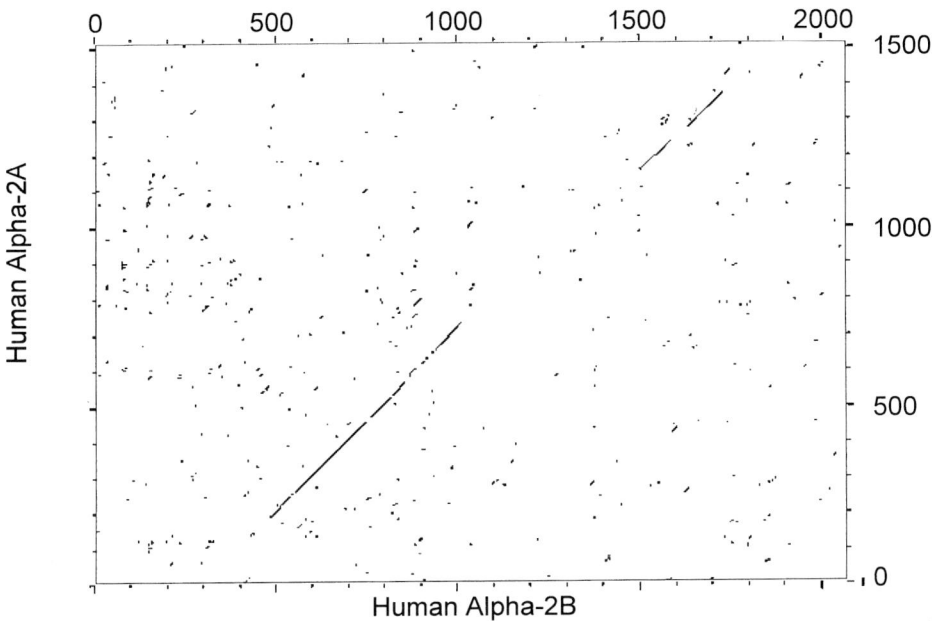

Fig. 1. DOTPLOT figure comparing human α_{2A}-AR and human α_{2B}-AR. The DNA sequences for human α_{2A}-AR (M18415) and human α_{2B}-AR (M34041) were compared using the COMPARE program in GCG. The DOTPLOT shows results of the COMPARE fit in graphical form. The lines on the graph show the regions of identity. The large gap in the line is the location of the third intracellular loop.

DOTPLOT, which are part of the GCG computer package. Detailed explanations of these programs can be found at http://molbio.unmc.edu/Class/ under class notes.

3.2.2.2. Global Sequence Alignment

Global sequence alignment allows for the complete alignment of two DNA or protein sequences. This requires putting in gaps in the sequence so that regions of similarity will be aligned. One computer program that does this is GAP of the GCG computer package (*see* **Note 1**). GAP uses the algorithm of Needleman and Wunsch *(6)*. **Figure 2** provides an example of GAP using a comparison of the human α_{2A}-AR and α_{2B}-AR. The vertical bars between the sequences indicate identity between the two sequences. The gaps inserted in the sequence are noted by dots. The GAP figure gives the four figures of merit for the alignments: Quality, Ratio, Identity, and Similarity. GCG defines these values as follows from next page (*see* **Note 1**):

```
        Gap Weight:        50      Average Match:  10.000
     Length Weight:         3   Average Mismatch:   0.000

          Quality:      8625             Length:    2084
            Ratio:     5.671               Gaps:       9
Percent Similarity: 61.829   Percent Identity: 61.829

 607 GGCCGAGCCGCGCTGCGAGATCAACGACCAGAAGTGGTACGTCATCTCGTCGTGCATCGGCTCCTTCTTCGCTCCCTGCCTCATCATGATCCTGGTCTAC 706
     |     ||  |   ||| || || |||||| | ||   |||||| ||| | | ||  |||||||| || ||||| ||||| ||||||||||||| ||||||
 889 CGGGCGCCCCCAGTGCAAGCTCAACCAGGAGGCCTGGTACATCCTGGCCTCCAGCATCGGATCTTTCTTTGCTCCTTGCCTCATCATGATCCTTGTCTAC 988

 707 GTGCGCATCTACCAGATCGCCAAGCGTCGCACCCGCGTGCCACCCA................GCCGCCGGGGTCGGACGCCGTCGCCGCGCCGCCG 787
     |||||||||| | ||||||||| |||||||| ||  ||| |||||             ||  |   ||||| |  |||| |   ||| | |
 989 CTGCGCATCTACCTGATCGCCAAACGCAGCAACCGCAGAGGTCCCAGGGCCAAGGGGGGGCCTGGGCAGGGTGAGTCCAAGCAGCCCCGACCCGACCATG 1088

 788 GGGGG.................CACCGAGCGCAGGCCCAACGGTCTGGGCCCCGAGCGCAGCGCGGGCCCGGG.......GGGCGCAGAGGCCGAACCG 862
     | |||                  | ||| | |||| | ||  ||| |    | |  | |  | ||    |  |  | |||| | | |
1089 GTGGGGCTTTGGCCTCAGCCAAACTGCCAGCCCTGGCCTCTGTGGCTTCTGCCAGAGAGGTCAACGGACACTCGAAGTCCACTGGGGAGAAGGAGGAGGG 1188

 863 CTGCCCACCCAGCTCAACGGCGCCCCTGGCGAGCCCGCGCCGGCCGGGCCGCGCGCGACACCGACGCGCTGGACCTGGAGGAGAGCTCGTCTTCCGACCACG 962
     | |||                ||  | | ||| ||  |   ||  ||  ||  |  |  |   | ||  | |  |||  |  |  |  | |  || ||
1189 GGAGACCCCTGAAGATACTGGGACCCGGGCCTTGCCACCCAGTTGGGCTGCCCTTCCCAACTCAGGCCAGGGCCAGAAGGAGGGTGTTTGTGGGGCATCT 1288

 963 CCGAGCGGCCTCCAGGGCCCCGCAGACCCGAGCGCGGTCCCCGGGGCAAAGGCAAGGCCCGAGCGAGCCAGGTGAAGCCGGGCGACAGCCTGCCGCGGCG 1062
     | | ||    |      |  |  | || | |   | |   |  | || |  | |   ||  || ||| |   |||      | ||  || |   | |  | |
1289 CCAGAGGATGAAGCTGAAGAGGAGGAAGAGGAGGAGGAGGAGGAAGAGTGTGAACCCCAGGCAGTGCCAGTGTCTCCGGCCTCAGCTTGCAGCCCCC 1388

1063 CGGGCCGGGGGCGACGGGGATCGGG...............ACGCCGGCTGCAGGGCCGGGGGAGGAGCGCGTCGG....GGCTGCCAAGGCGTCGCGCTG 1143
     || | || | ||| |  |||          |||  |  | | || |||  | ||  | |   ||||  | || |  |      |  | |  |  | |
1389 CGCTGCAGCAGCCACAGGGCTCCCGGGTGCTGGCCACCCTACGTGGCCAGGTGCTCCTGGGCAGGGGCGTGGGTGCTATAGGTGGGCAGTGGTGGCGTCG 1488

1144 GCGCGGGCGGCAGAACCTCGAGAAGCGCTTCACGTTCGTGCTGGCCGTGGTCATCGGAGTGTTCGTGGTGTGCTGGTTCCCCTTCTTCTTCACCTACACG 1243
     |  | |   ||| ||   ||||||||||| ||| |||||||||||| |||||||||| || |||||||||||||||||||||| ||||| |||||| |||
1489 AAGGGGCGCACGTGACCCGGGAGAAGCGCTTCACCTTCGTGCTGGCTGTGGTCATTGGCGTTTTTGTGCTGCTGGTTCCCCTTCTTCTTCAGCTACAGC 1588
```

Fig. 2. GAP of human α_{2A}-AR and human α_{2B}-AR. GAP program in GCG was run on the full, known sequences of human α_{2A}-AR (M18415) and human α_{2B}-AR (M34041). GAP showed a good correlation for all except upstream of the translation codon and the third intracellular loop. Shown in the figure are the third intracellular loops (indicated by a solid black line above the sequence) for the α_{2A}-AR and the areas on either side of the third intracellular loop. The default optional parameters for GAP weight, length weight, average match, and average mismatch were used. The top sequence is human α_{2A}-AR, and the bottom sequence is human α_{2B}-AR.

The Quality is the metric maximized in order to align the sequences. Ratio is the quality divided by the number of bases in the shorter segment. Percent Identity is the percent of the symbols that actually match. Percent Similarity is the percent of symbols that are similar. Symbols across from gaps are ignored. A similarity is scored when the scoring matrix value for a pair of symbols is greater than or equal to the average positive non-identical comparison value in the matrix, the similarity threshold.

The GAP program does not work well if there are few regions of similarity separated by divergent regions of variable length. It also does not work well if there is a large difference between the lengths of the two sequences. One way to overcome the latter problem is first to compare the two sequences using COMPARE and DOTPLOT. This will show where the regions of similarity are located. GAP then can be run using only those portions of the sequences where there is similarity. If you compare sequence alignment of the DOTPLOT (**Fig. 1**) with that of the sequence alignment produced by the GAP program

(**Fig. 2**), you can see that the lines shown in the dot plot agree with the areas of greatest sequence similarity between α_{2A}-AR and α_{2B}-AR.

An alternate approach to solving the problem of alignment of long sequences with shorter regions of high similarity or identity is to use the implementation of FASTA within the GCG package. By selecting one of the sequences to be compared as the search sequence and by selecting the other one as the database, you will in turn obtain from the FASTA program the alignments of all areas of high similarity. When one or more of the sequences you want to compare is longer than 32,000 bases, you must use the FASTA approach, because GAP (and BESTFIT) is limited to 32,000-base sequences. Another global sequence alignment program that allows for the alignment of two protein sequences is ROBUST.

3.2.2.3. LOCAL SEQUENCE ALIGNMENT

Another way to compare two DNA, RNA, or protein sequences is by local sequence alignment. This allows for local alignment in areas of similarity that are found in large areas of divergence (gaps). One program that will do this is BESTFIT (GCG). BESTFIT uses the local homology algorithm of Smith and Waterman *(7)*. BESTFIT uses the same figure of merit for the alignments as GAP. BESTFIT works well for aligning sequences that have only a small overlap in sequences, for example, merging two separate, but overlapping sequences from the same gene. Local sequence alignments are designed to align the most similar segments and do not do as well as global alignments in finding the third, fourth, and fifth areas of similarity. It tends to leave off part of the beginning or the end of the sequence. FASTA will find these areas of similarity much better, but will not attempt to align the intervening areas with much lower levels of similarity. The BESTFIT match for the comparison of human α_{2A}-AR and α_{2B}-AR is similar to that of GAP and COMPARE for the regions of great homology. However, the alignment of the third intracellular loop is quite different in BESTFIT than in GAP (**Fig. 3**). Note that you cannot determine whether the fit is better using GAP or BESTFIT using the similarity or identity matrices, since these are not optimized by either BESTFIT or GAP.

3.2.3. Multiple Sequence Alignments (MSA)

There are several methods for doing MSA. The GCG program PILEUP produces multiple sequence alignments from a group of related sequences using progressive, pairwise alignments. The method used is similar to the method described by Higgins and Sharp *(8)*. This program also plots a dendrogram showing the clustering relationships used to create the alignment. Although the dendrogram looks like a phylogenetic tree, it is only a measure of the similarity

```
       Gap Weight:        50     Average Match:  10.000
    Length Weight:         3  Average Mismatch:  -9.000

          Quality:      4217            Length:    1343
            Ratio:     3.371              Gaps:      16
Percent Similarity:   70.540  Percent Identity:  70.540


 636 AGAAGTGGTACGTCATCTCGTCGTGCATCGGCTCCTTCTTCGCTCCCTGCCTCATCATGATCCTGGTCTACGTGCGCATCTACCAGATCGCCAAGCGTCG 735
     || ||||| | | || ||||||| | ||||| ||||| |||||| |||||||||||||||| | |||||||| ||||||||| |||||||| | | |
 918 AGGCCTGGTACATCCTGGCCTCCAGCATCGGATCTTTCTTTGCTCCTTGCCTCATCATGATCCTTGTCTACCTGCGCATCTACCTGATCGCCAAACGCAG 1017

 736 CACCCGCGTGCCACCCAGCCGCCGGGGTCGGACGCCGTCGCCGCGCCGCCGGGGGGCACCGAGCGCAGGCCCAAC......GGTCTGGGCCCCGAGC. 827
     ||||| |||||||      ||||| | ||   |    | ||| | || ||  | ||  || || || | ||| |       ||| || | || |||
1018 CAACCGCAGAGGTCCCAGGGCCAAGGG.........GGGGCCTGGGCAGGGTGAGTCCAAGCAGCCCCGACCCGACCATGGTGGGGCTTTGGCCTCAGCC 1108

 828 ......GCAGCGCGGGCCCGGGGGGC.....GCAGAG........GCCGAACCGCTGCCCACCCAGCTCAACGGCG.........CCCCTGGCGAGCCCG 899
            |||| | |||| | |||      |||||        |  | |  || | ||||  |  || || |         ||||||| || | |
1109 AAACTGCCAGCCCTGGCCTCTGTGGCTTCTGCCAGAGAGGTCAACGGACACTCGAAGTCCACTGGGGAGAAGGAGGAGGGGGGGGAGACCCCTGAAGATACTG 1208

 900 CGCCGGCCGGGCCGCGCGACACCGACGCGCT..................GGACCTGGAGGAGAGCTCGTCTTCCGACCACGCCGAGCGGCCTCCAGGG 979
     | | ||||||| || |||| | | | | | |                  | || |||| ||||| ||  |  |||| |||| | || | | |||
1209 GGAC..CCGGGCCTTGCCACCCAGTTGGGCTGCCCTTCCCAACTCAGGCCAGGGCCAGAAGGAGGG..TGTTTGTGGGGCATCTCCAGAGG.....ATGA 1299

 980 CCCCGCAGACCCGAGCGCGGTCCCCGGGGCAAAGGCAAGGCCCGAGCGAGCCAGGTGAAGC........CGGGCGACAGCCTGCCGCGGCGCGGGCCGGG 1071
     | | | ||  |||||||| || || |||| ||||||| || | |||| ||  ||  ||         || || |||| ||| ||| ||| ||| ||
1300 AGCTGAAGAGGAGGAAGAGGAGGAGGAGGAGGAAGAGTGTGAAC..CCCAGGCAGTGCCAGTGTCTCCGGCCTCAGCTTGCAGCCCCCCGCTGCAGC 1397

1072 GGCGACGGGGATCGGG..............ACGCCGGCTGCAGGGCCGGGGGGAGGAGCGCGTCGG....GGCTGCCAAGGCGTCGCGCTGGCGCGGGCG 1152
     || || ||| || |                 ||| | || | |  || ||| |||| ||| |  |        | ||  || | || || || || ||
1398 AGCCACAGGGCTCCCGGGTGCTGGCCACCCTACGTGGCCAGGTGCTCCTGGGCAGGGGCGTGGGTGCTATAGGTGGGCAGTGGTGGCGTCGAAGGGCGCA 1497

1153 GCAGAACCTCGAGAAGCGCTTCACGTTCGTGCTGGCCGTGGTCATCGGAGTGTTCGTGGTGTGCTGGTTCCCCTTCTTCTTCACCTACACGCTCACGGCC 1252
     || || ||||||| ||||||||||| ||||||||||||||| ||  |  |||| |||||||||||| |||||||||||||||||||||| || ||  |||
1498 CGTGACCCGGGAGAAGCGCTTCACCTTCGTGCTGGCTGTGGTCATTGGCGTTTTTGTGCTCGTGGTTCCCCTTCTTCTTCAGCTACAGCCTGGGCGCC 1597
```

Fig. 3. BESTFIT of human α_{2A}-AR and human α_{2B}-AR. BESTFIT program in GCG was run on the full, known sequences of human α_{2A}-AR (M18415) and human α_{2B}-AR (M34041). BESTFIT did not fit the beginning or the end of the sequence and did not fit the third intracellular loop as well as GAP. Shown in the figure are the third intracellular loop (indicated by solid black line above the sequence) for the α_{2A}-AR adrenergic receptor and the areas on either side of the third intracellular loop. The default optional parameters were used for gap weight, length weight, average match, and average mismatch. The top sequence is human α_{2A}-AR and bottom sequence is human α_{2B}-AR.

among sequences. The example in **Fig. 4** is part of the comparison of the human α_{2A}-AR, α_{2B}-AR, and α_{2C}-AR. PILEUP aligns the sequences, inserting gaps as needed; however, it does not give a consensus sequence. A consensus sequence can be obtained by running PRETTY, which prints multiple sequence alignments and can determine the consensus of these sequences. Note that a sequence alignment program, such as PILEUP, needs to be run before PRETTY

Fig. 4. *(facing page)* Results of PILEUP and PRETTY analysis of human α_{2A}-AR, α_{2B}-AR and α_{2C}-AR. The sequences for human α_{2A}-AR (M18415), α_{2B}-AR (M34041), and α_{2C}-AR (J03853) were compared using the PILEUP and PRETTY programs in GCG. Consensus was defined as three bases out of three being the same. Shown is the demogram indicating that α_{2C}-AR and α_{2A}-AR were matched first and then α_{2B}-AR was compared to the match for α_{2C}-AR and α_{2A}-AR and part of the results of the PRETTY program.

Dendrogram labels (left to right): Alpha-2B, Alpha-2C, Alpha-2A

Plurality: 3.00   Threshold: 1   AveWeight 1.00   AveMatch 1.00   AvMisMatch 0.00

```
          101                                                                       200
Alpha-2A  ~~~~~~~~~~ ~CCCGCCTTC ATCTTCCCGCC AGGAGGCCAA GGCCGTTGGC CGAGGGCAGC TTTGCGCCCA TGGCTCCCT GCAGCCGGAC GCGGGCAACG
Alpha-2C  GCTCGCGGGA GGACCATGGC GTCCCCGGCCG CTGGCGGCGGT CTGGCGCCGGT GGCGGCAGCG GCGGGCCCCA ATGGCAGCGG CCGGGGCGGAG AGGGGCAGCG
Alpha-2B  CCAGGCCATG GGGCTCCAGC GCCCTCGCGG CGCCCGAGGG GCGACGCTCT TGTCTAGCCG AGCCGGGCAG CGCTGTCGTC CACGGTGCGC ACTGGGCGGG
Consensus ---------- --C-----C -----C--- -----G---- ---------- -G------- -----C--- --------- --G-----G ---GG---G

          201                                                                       300
Alpha-2A  CGAGCTGGAA CGGGACCGAG GCGCCGGGGG GCGGCGCCCG GGCCACCCCT TACTCCCTGC AGGTGACGCT GACGCTGGTG TGCCTGGCCG GCCTGCTCAT
Alpha-2C  GCGGGGGTTGC CAATGCCTCG GGGGCTTCCT GGGGGCCGGC TACTCGGGCGG TACTCGGGCGG GCGCGGTGGC AGGGCTGGCT GCCGTGGTGG GCTTCCTCAT
Alpha-2B  CAGCGCTCCC TCTGCCCACC TCCCGCCCCG TCA?GGACCA CCAGGACCCC TACTCGGTGC TACTCGGTGC AGGCCACAGC GGCCATAGCG CCTTCCTCAT
Consensus --------- ----CC--- --------- -----C--- -----CC-- TACTC---G- ------G-- ----T-G-- ----T---- -C-T-CTCAT

          301                                                                       400
Alpha-2A  GCTGCTCACC GTGTTCGGCA ACGTGCTCGT CATCATCGCT GTGTTCACGA GCCGCGCGCT CAAGGCGCCC CAAAACCTCT TCCTGGTGTC TCTGGCCTCG
Alpha-2C  CGTCTTCACC GTGGTGGGCA ACGTGCTCGT GGTGATCGCC GTGCTGACCA GCCGGGCGCT GCGCGCGCCA CAGAACCTCT TCCTGGTGTC GCTGGCCTCG
Alpha-2B  TCTCTTTACC ATCTTCGGCA ACGCTCTGGT CATCCTGGCT GTGTTGACCA GCCGTCGCT GCGCGTCGCT CAGAACCTGT TCCTGGTGTC GCTGGCCGCC
Consensus --T-T-ACC -T--T-GGCA ACG--CT-GT --T--T-GC- GTG-T-AC-A GCCG--CGCT ---GC--CC- CA-AACT-T TCCTGGTGTC -CTGGCC-C-

          401                                                                       500
Alpha-2A  GCCGACATCC TGGTGGCCAC GCTCGTCATC CCTTTCTGC TGGCCAACGA GGTCATGGGC TACTGGTACT TCGGCCAAGAC TTGGTGCGAG ATCTACCTGG
Alpha-2C  GCCGACATCC TGGTGGCCAC GCTGGTCATG CCCTTCTCGT TGGCCAACGA GCTCATGGGC TACTGGTACT TCGGGCCAGGI GTGGTGCGGC GTGTACCTGG
Alpha-2B  GCCGACATCC TGGTGGCCAC GCTCATCATC CCTTTCTGC TGGCCAACGA GCTGCTGGGC TACTGGTACT TCCGGCGCAC GTGGTGCCGAG GTGTACCTGG
Consensus GCCGACATCC TGGTGGCCAC GCT--TCAT- CC-TTCTGC TGGCCAAGGA G-T--TGG-C TACTGGTACT TC-G--G--- -TGGTGCG-- -T-TACCTGG

          501                                                             600
Alpha-2A  CGCTCGAGGT GCTCTTCTGC ACGTCGTCCA TCGTGCACCT GTGCGCCATC AGCCTGGACC GCTACACACAG GCCATCGAGT ACAACCTGAA
Alpha-2C  CGCTCGATGT GCTGTTTTGC ACCTCGTCGA TCGTGCATCT GTGTGCCATC AGCCTGGACC GGTGACGCAG GCCGTCGAGT ACAACCTGAA
Alpha-2B  CGCTCGAGGT GCTCTTCTGC ACCTCGTCCA TCGTGCACCT GTGCGCCATC AGCCTGGACC CGTGAGCCGC GCGCTGGAGT ACAACTCCAA
Consensus CGCTCGA-GT GCT-TT-TGC AC-TCGTC-A TCGTGCA-CT GTG-GCCATC AGCCTGGACC --T-A--C-- GC--T-GAGT ACAAC---AA
```

can be run. Alternatively, sequences containing appropriate gaps can be used. The user can define what constitutes a consensus, for example, three matches out of five or five matches out of five. The consensus sequence in **Fig. 4** was generated using PILEUP followed by PRETTY. Another program that will conduct multiple sequence alignments is MSA.

3.2.4. Homology Searching

3.2.4.1. FASTA

FASTA is one of two programs that are generally used to find other sequences homologous with the search sequence. BLAST is the other primary homology search program in use. FASTA is a global sequence alignment tool that works on similar principles to that of GAP using the method of Pearson and Lipman *(9)*. FASTA is used to search for similarities between one sequence (the query) and any group of sequences of the same type (nucleic acid or protein) as the query sequence. FASTA can be used to search an entire database for sequences or areas of similarities within the query sequence. Some variations of FASTA are:

Library search programs: FASTA, TFASTA, SSEARCH
Local homology programs: LFASTA, PLFASTA, LALIGN, PLALIGN
Statistical significance: PRDF, RELATE, PRSS
Global alignment: ALIGN.

FASTA, as implemented in GCG, is only available in a single version, which performs both FASTA and TFASTA, and which can perform Smith-Waterman alignments *(7)* for the final alignment of similar regions.

3.2.4.2. BASIC LOCAL ALIGNMENT SEARCH TOOL (BLAST)

BLAST is another tool designed to allow one to determine the presence of sequence homology between a query sequence and all the sequences in a database. BLAST is a local alignment search tool that uses the method of Altschul et al. *(10)*. It can search databases on your own computer or databases maintained at the NCBI in Bethesda, MD. The query and databases search can be nucleotide or protein or a combination thereof. BLAST will not report all of the occurrences of matches between two sequences. For example, if you want to determine whether a 20-mer that might be used as a primer occurs more than once in a sequence, BLAST will find the first match, but will not find multiple matches in the same sequence. The primary difference between BLAST and FASTA is that BLAST looks for local areas of similarities and FASTA looks for global similarities. The BLAST server at NCBI now has a family of BLAST programs. A new version of BLAST (BLAST 2.0) uses the method of Altschul et al. *(11)*. The new version provides significant performance enhancements, the addition of "gapping" routines, and position-specific iterated BLAST (PSI-Blast). There are various versions of BLAST that allow for the comparison of:

Nucleotide vs nucleotide sequence (BLASTn).

Amino acid query sequence against a protein sequence database (BLASTp).

Six-frame conceptual translation products of a nucleotide query sequence (both strands) against a protein sequence database (BLASTx).

Comparison of a protein query sequence against a nucleotide sequence database dynamically translated in all six reading frames (both strands) (tBLASTn).

The six-frame translations of a nucleotide query sequence against the six-frame translations of a nucleotide sequence database (tBLASTx).

PSI-BLAST allowing for more sensitivity to weak, but biologically meaningful sequence similarities.

3.3. Submission of Nucleotide Sequence to GenBank Database

Once the sequence has been confirmed, it can be submitted to the GenBank database. Submission programs, such as BANKIT, are available from NCBI, and provide for a quick and convenient submission of simple sequence data. Submission can also be done on the Web using your browser. The submission sequence requires information regarding the source of the DNA (organism and cDNA or DNA library), the sequence, name of the gene, authors, journal references, transcription start sites, translation start and stop site, putative amino acid sequence, location of introns, and promoter regions if known. The nucleotide and amino acid sequence data can be easily copied and pasted into the program. Approximately 24 h after submission, the accession number is returned to the investigator via e-mail. The accession number should then be included in the paper reporting the sequence.

Sequin version 2.70 is a stand-alone software tool that is available by anonymous FTP. Although it is capable of handling simple submissions, it has been designed to simplify the submission of complex sequence data containing long sequences, multiple annotations, segmented sets of DNA, or phylogenetic and population studies. It will allow complex annotations, and provide for submission of multiple sequences. The output files for submission are either mailed on a floppy disk or sent by e-mail. Accession numbers are again assigned by the NCBI staff.

3.4. Methods for Analyzing Protein Structure

There are several computer programs and Web sites that provide information about protein structure. Comparison of protein sequences with known

Table 1
Amino Acid Composition of Human α_{2A}-AR[a]

A (Ala): 43	C (Cys): 12	D (Asp): 13	E (Glu): 19
F (Phe): 24	G (Gly): 44	H (His): 3	I (Ile): 30
K (Lys): 17	L (Leu): 38	M (Met): 4	N (Asn): 13
P (Pro): 39	Q (Gln): 10	R (Arg): 36	S (Ser): 29
T (Thr): 24	V (Val): 31	W (Trp): 9	Y (Tyr): 12
Other: 0			
Total: 451			

[a]The GCG program COMPOSITION of the human α_{2A}-AR produced these results.

structures of adrenergic receptors or other proteins is explained in **Subheading 3.2.** Information that can be obtained about protein sequences includes the amino acid composition, presence of sequence motifs, identification of sites for cleavage by known proteases, peptide fragments that would be formed by amino acid digestion, and the isoelectric point of the protein. In addition, information on secondary structure of the protein, including the hydrophobic and hydrophilic domains, location of α helixes and β sheets, and antigenic sites is also available (**Note 3**).

3.4.1. Primary Structure of the Adrenergic Receptor

3.4.1.1. AMINO ACID COMPOSITION

The amino acid composition of the receptor can be determined using COMPOSITION, which is part of the GCG package. The analysis of the human α_{2A}-AR is shown in **Table 1**.

3.4.1.2. SEARCH FOR KNOWN PROTEIN MOTIFS

The protein sequence can be searched for known protein motifs (a short conserved region in a protein sequence), which are found in protein families. There are several profile and motif databases available. The GCG program MOTIFS can recognize the patterns with some of the symbols mismatched, but not with gaps. Analysis of the human α_{2A}-AR using MOTIFS (GCG) identified a region of 45 amino acids, which has a consensus pattern identical to the consensus pattern in the third transmembrane helix of G-protein-coupled receptors.

3.4.1.3. PEPTIDE FRAGMENTS

Peptide fragments that would be formed from an amino acid digest can be determined. PEPTIDESORT (GCG) sorts the peptides by weight, position, and high-performance liquid chromatography (HPLC) retention at pH 2.1, and

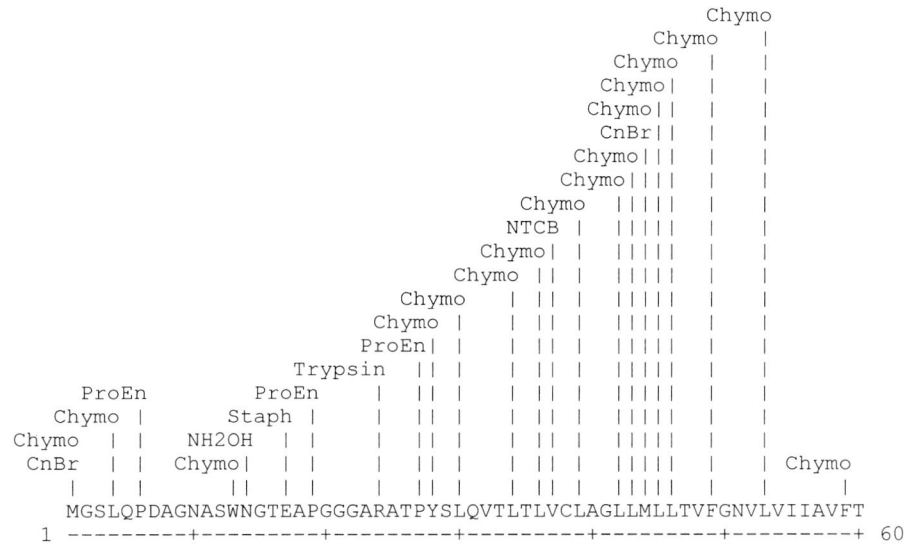

Fig. 5. PEPTIDE MAP of the amino acid sequence for human α_{2A}-AR. The DNA sequence for human α_{2A}-AR (M18415) was translated using the TRANSLATE program in GCG. The amino acid sequence was then analyzed to determine where the sequence could be cut by known proteolytic enzymes or reagents. The results of the analysis for the first 60 amino acids are shown.

shows the composition of each peptide. It also prints a summary of the composition of the whole protein.

Sites where the peptide can be cut by known proteolytic enzymes or reagents can be determined using PEPTIDEMAP (GCG). An example of the analysis of the human α_{2A}-AR is shown in **Fig. 5**.

3.4.1.4. ISOELECTRIC POINT

The isoelectric point of the protein can be determined using the ISOELECTRIC program in GCG. ISOELECTRIC produces plots of the total positive, total negative charges and the net charge of a protein as a function of pH. The isoelectric point (pH at which the net charge is zero) is indicated on the plot.

3.4.2. Secondary Structure of the Adrenergic Receptor

Various aspects of the secondary structure of adrenergic receptors can be determined from the location of α and β helixes, hydrophobicity, helical hydrophobic movement, surface probability, flexibility, glycosylation sites, antigenic index, and coiled-coiled segments. The results of protein structural analysis using GCG program PLOTSTRUCTURE of the human α_{2A}-AR are

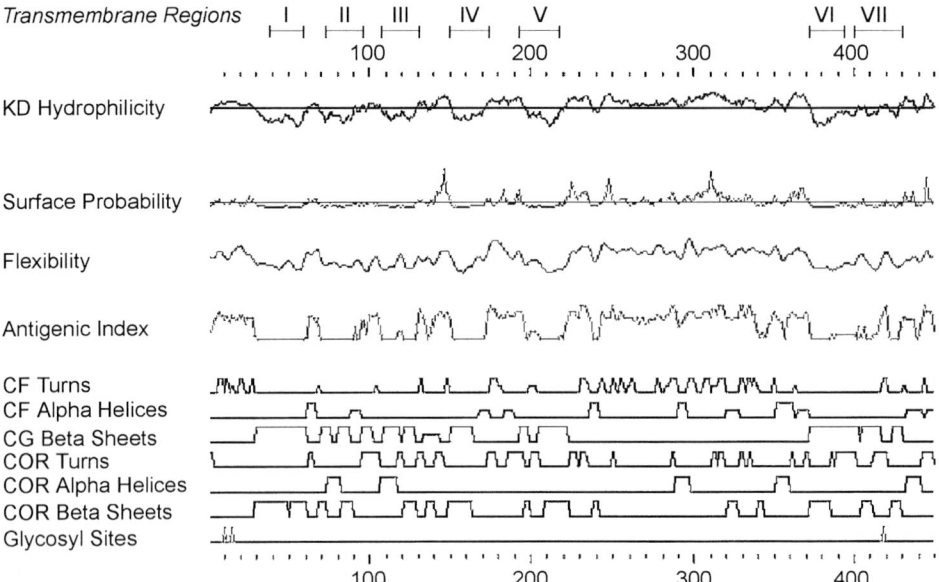

Fig. 6. Graphics from PLOTSTRUCTURE for human α_{2A}-AR. The DNA sequence for human α_{2A}-AR (M18415) was translated using the TRANSLATE program in GCG. The amino acid sequence was then analyzed using PEPTIDESTRUCTURE in GCG. The figure is then obtained using PLOTSTRUCTURE. Markings showing the location of the seven-transmembrane domains have been superimposed on the image.

shown in **Fig. 6**. There are also various programs for determining the tertiary structure of the protein (*see* **Note 4**).

3.4.2.1. α AND β HELIXES

Several methods can be used to obtain information about α helixes and β sheets. PEPPLOT in GCG does the following analysis and provides the results in graphical form:

The Residue Schematic: Each residue is represented by a line at the position where it occurs. The length of the line, color of line, and position of the line are used to indicate the chemically similar groups of amino acids.

Chou and Fasman β Sheet Forming and Breaking Residues: This shows residues that are β sheet-forming and breaking as defined by Chou and Fasman *(12)*.

Chou and Fasman α and β Propensities: This shows the propensity measures for α helix and β sheet according to Chou and Fasman *(12)*.

Chou and Fasman α Helix-Forming and Breaking Residues: The residues that are α helix-forming and breaking, as defined by Chou and Fasman *(12)*, are shown.

Chou and Fasman Amino Ends: Regions of the sequence that resemble sequences typically found at the amino end of α helices and β structures *(12)* are shown here.

Chou and Fasman Carboxyl Ends: This shows the regions of the sequence typically found at the carboxyl end of α helices and β structures *(12)*.

Chou and Fasman Turns: Regions of the sequence typically found in turns *(12)* are presented in this analysis.

Goldman, Engelman, and Steitz Transbilayer Helices: This curve identifies the nonpolar transbilayer helices *(13)*.

PEPTIDESTRUCTURE (GCG) does the following analysis providing the results in graphical form:

Secondary structure according to the method: This method predicts helices, sheets, and turns *(12)*.

Secondary structure according to the Garnier-Osguthorpe-Robson method *(14)*.

3.4.2.2. HYDROPHOBICITY

Hydrophobicity or hydrophilicity can be determined by looking at the Kyte Doolittle Hydropathy plot. This curve is the average of a residue-specific hydrophobicity index over a window of nine residues *(15)*.

3.4.2.3. HELICAL HYDROPHOBIC MOMENT

The helical hydrophobic moment at each position of the sequence is determined by this analysis. The moment statistic is the probability that the sequence at each position is amphiphilic, meaning the position has hydrophobic residues on one side and hydrophilic residues on the other. The hydrophobic moment is calculated as described by Eisenberg et al. *(16)*, except that the program normalizes the hydrophobic moment for the local hydrophobicity of the amino acids in the window where the moment is determined as described by Finer-Moore and Stroud *(17)*. This can be determined using PEPPLOT, GCG. The program MOMENT (GCG) makes a contour plot of the helical hydrophobic moment of a peptide sequence. This program identifies amphiphilic structures by identifying when the residues on one side of the structure are more hydrophobic than on the other. The program HELICALWHEEL (GCG) plots a peptide sequence as a helical wheel to help you recognize amphiphilic regions.

Each residue is offset from the preceding one by 100°, the typical angle of rotation for an α helix.

3.4.2.4. SURFACE PROBABILITY

The surface probability can be done according to the Emini et al. method *(18)*, using PEPTIDESTRUCTURE, GCG.

3.4.2.5. FLEXIBILITY

The flexibility can be determined using the method of Karplus-Schulz *(19)* with PEPTIDESTRUCTURE, GCG.

3.4.2.6. GLYCOSYLATION SITES

Glycosylation sites are predicted for sites where the residues have the composition NXT or NXS. When X is D, W, or P, the site is taken to be a weak glycosylation site. Otherwise, it is a strong glycosylation site. PEPTIDE-STRUCTURE in GCG will accomplish this analysis.

3.4.2.7. ANTIGENIC INDEX (AI)

The AI is a measure of the probability that a region is antigenic. It is calculated by summing several weighted measures of secondary structure as described by Jameson and Wolf *(20)*. AI can be determined using the PEPTIDE-STRUCTURE program in GCG and physicochemical result, Pôle Bio-Informatique Lyonnais.

3.4.2.8. COILED-COIL SEGMENTS

Coiled-coil segments in protein sequences can be located with COILSCAN (GCG).

4. Notes

1. Information on how to purchase or run the different programs of GCG package can be obtained at http://www.gcg.com. The Genetics Computer Group is located at 575 Science Drive, Madison, WI 53711.
2. GCRDb contains the following information for some of the receptors:
   GenBank information.
   Reference: via Medline to the actual reference for the sequence.
   FASTA DNA Format for each sequence.
   FASTA protein format.
   Graphic report.
   2D snake model.
   Swiss protein.
   Mouse genome database.
   OMIM.
   tGRAP mutant.

3. Additional information on how to do protein analysis using GCG can be found at http://molbio.unmc.edu/class/ under class notes, Chapter 9.

4. Web sites that will provide programs for two- and three-dimensional structural analysis of proteins are:
   • Brookhaven National Laboratories, protein data base: http://www.pdb.bnl.gov/. This site provides an archive of experimentally determined three-dimensional structures of biological macromolecules.
   • Entrez:http://www3.ncbi.nlm.nih.gov/Entrez/. The site is provided by the NCBI.
   • The Biology WorkBench: http://biology.ncsa.uiuc.edu/. The Biology Work-Bench is a Web-based tool that allows biologists to search protein and nucleic acid sequence databases. This site has many nucleic acid alignment and protein alignment and modeling tools.
   PELE: Protein structure prediction.
   GOR4: Predict secondary structure of proteins.
   CHOFAS: Predict secondary structure of protein sequence *(12)*.
   MPSSP: Protein secondary structure prediction.
   • ExPASy Molecular Biology Server: http://expasy.hcuge.ch/. The ExPASy is a server of the Swiss Institute of Bioinformatics (SIB). This server is dedicated to the analysis of protein sequences and structures. There are three basic divisions of this site:
   SWISS-PROT: Annotated protein sequence database: http://expasy.hcuge.ch/sprot/sprot-top.html.
   PROSITE: Database of protein families and domains: http://expasy.hcuge.ch/sprot/prosite.html.
   ExPASy Proteomics tools: http://expasy.hcuge.ch/www/tools.html.
   • PBIL: http://pbil.univ-lyon1.fr/.
   The PBIL World Wide Web server has been developed at the Laboratory of Biometry, Genetics and Population Biology and the Institute of Biology and Chemistry of Proteins. This site is dedicated to molecular biology and ecology.

References

1. Haussler, D. (1998) Computational genefinding, in *Trends Guide in Bioinformatics* (Brenner, S. and Lewitter, F., eds.), Elsevier Science, Cambridge, UK, pp. 12–15.
2. Griffin, H. G. and Griffin, A. M. (1994) Computer analysis of sequence data, in *Methods in Molecular Biology, vol. 24: Computer Analysis of Sequence Data* (Griffin, A. M. and Griffin, H. G., eds), Humana, Totowa, NJ, pp. 1–8.
3. Altschul, S. F. (1998) Fundamentals of database searching, in *Trends Guide in Bioinformatics* (Brenner, S. and Lewitter, F., eds.), Elsevier Science, Cambridge, UK, pp. 7–9.
4. Brenner, S. (1998) Practical database searching, in *Trends Guide in Bioinformatics* (Brenner, S. and Lewitter, F., eds.), Elsevier Science, Cambridge, UK, pp. 9–12.

5. Maizel, J. V. J. and Lenk, R. P. (1981) Enhanced graphic matrix analysis of nucleic acid and protein sequences. *Proc. Natl. Acad. Sci. USA* **78,** 7665–7669.

6. Needleman, S. B. and Wunsch, C. D. (1970) A general method applicable to the search for similarities in the amino acid sequence of two proteins. *J. Mol. Biol.* **48,** 443–453.

7. Smith, T. F. and Waterman, M. S. (1981) Comparison of biosequences. *Annu. Rev. Biophys. Biophys. Chem.* **2,** 482–489.

8. Higgins, D. G. and Sharp, P. M. (1989) Fast and sensitive multiple sequence alignments on a microcomputer. *Comput. Appl. Biosci.* **5,** 151–153.

9. Pearson, W. R. and Lipman, D. J. (1988) Improved tools for biological sequence comparison. *Proc. Natl. Acad. Sci. USA* **85,** 2444–2448.

10. Altschul, S. F., Gish, W., Miller, W., Myers, E. W., and Lipman, D. J. (1990) Basic local alignment search tool. *J. Mol. Biol.* **215,** 403–410.

11. Altschul, S. F., Madden, T. L., Schaffer, A. A., Zhang, J., Zhang, Z., Miller, W., et al. (1997) Gapped BLAST and PSI-BLAST: a new generation of protein database search programs. *Nucleic Acids Res.* **25,** 3389–3402.

12. Chou, P. Y. and Fasman, G. D. (1978) Prediction of secondary structure of proteins from their amino acid sequence. *Adv. Enzymol.* **47,** 45–147.

13. Engelman, D. M., Steitz, T. A., and Goldman, A. (1986) Identifying nonpolar transbilayer helices in amino acid sequences of membrane proteins. *Annu. Rev. Biophys. Biophys. Chem.* **15,** 321–353.

14. Garnier, J., Osguthorpe, D. J., and Robson, B. (1978) Analysis of the accuracy and implications of simple methods for predicting the secondary structure of globular proteins. *J. Mol. Biol.* **120,** 97–120.

15. Kyte, J. and Doolittle, R. F. (1982) A simple method for displaying the hydropathic character of a protein. *J. Mol. Biol.* **157,** 105–132.

16. Eisenberg, D., Weiss, R. M., and Terwilliger, T. C. (1984) The hydrophobic moment detects periodicity in protein hydrophobicity. *Proc. Natl. Acad. Sci. USA* **81,** 140–144.

17. Finer-Moore, J. and Stroud, R. M. (1984) Amphipathic analysis and possible formation of the ion channel in an acetylcholine receptor. *Proc. Natl. Acad. Sci. USA* **81,** 155–159.

18. Emini, E. A., Hughes, J. V., Perlow, D. S., and Boger, J. (1985) Induction of hepatitis A virus-neutralizing antibody by a virus-specific synthetic peptide. *J. Virol.* **55,** 836–839.

19. Karplus, P. A. and Schulz, G. E. (1985) Predicition of chain flexibility in proteins. *Naturwissen* **72,** 212,213.

20. Jameson, B. A. and Wolf, H. (1988) The antigenic index: a novel algorithm for predicting antigenic determinants. *Comput. Appl. Biosci.* **4,** 181–186.

4

Polymerase Chain Reaction Screening of Genomic Libraries for Adrenergic Receptor Genes

Dianne M. Perez and Michael J. Zuscik

1. Introduction

In the search for novel G-protein-coupled receptor genes, two common approaches have worked fairly well and are relatively easy to perform. One method is homology-based screening approaches, which utilize low-stringency screening of genomic or cDNA libraries with a known cDNA probe. The other method uses polymerase chain reaction-(PCR) based approaches on the same genomic or cDNA libraries. The latter approach is more sensitive based on the inherent amplification in the PCR process, and the strategic design of the PCR primers can ultimately lead to more novel sequences being obtained. The method can also be applied directly on mRNA after it is transcribed into first-strand cDNA. However, libraries offer the assurance that once a PCR product is obtained, the gene must be present in the library because it generated the template needed in the PCR process. This approach presents a quicker means of obtaining a full-length clone.

This chapter emphasizes the PCR-based screening of genomic libraries. The method can be applied to cDNA libraries as well. However, if one is familiar with the intron/exon structure of the receptor family, then primers are designed inside an exon and the possible intron interference in the PCR process is negated (i.e., some introns are quite large, >20 kb, making complete PCR amplification difficult). The genomic approach offers a higher chance of identifying a novel gene, since all genes must be encoded in the genomic DNA and the construction of a genomic library does not favor one gene over another; thus, there is even representation of genes in the library. On the other hand, cDNA libraries are constructed from mRNA, and low-abundance genes would be underrepresented, a common finding for the adrenergic receptors.

From: *Methods in Molecular Biology, vol. 126: Adrenergic Receptor Protocols*
Edited by: C. A. Machida © Humana Press Inc., Totowa, NJ

A general procedural flowchart is illustrated in **Table 1**. The overall strategy in this methodology is first to identify and generate a small PCR product for novel subtypes, which is then used to probe for the full-length gene. This approach takes advantage of the highly conserved regions within a receptor protein to generate the PCR product and, therefore, will not encode a full-length gene. The ability to PCR-amplify a full-length gene of a novel subtype has never been successful, because the amino- and carboxy-terminis of adrenergic receptors are unique and share very low, if any, homology between family members.

2. Materials

2.1. Equipment

This consists of a PCR machine, 3000-V power supply, agarose-gel apparatus, UV transilluminator, and DNA sequencing apparatus.

2.2. Special Reagents

1. Genomic library (either made or bought from Clontech [Palo Alto, CA] or Stratagene [La Jolla, CA]).
2. Deoxynucleotide triphosphates; dATP, dCTP, dGTP, TTP (100 mM stocks from Pharmacia, Piscataway, NJ).
3. GTG-agarose; low melting or agarose; low EEO (FMC).
4. Random Primed Labeling Kit (Boehringer Mannheim, Indianapolis, IN).
5. *Taq* Polymerase or similar version such as *pfu* (Boehringer Mannheim).
6. 50X TAE buffer (per L: 242 g of Tris base, 57.1 mL of glacial acetic acid, and 100 mL of 0.5 M ethylenediamine tetra-acetic acid [EDTA], pH 8.0).
7. 6X Agarose loading buffer: 0.25% bromophenol blue, 0.25% xylene cyanol, 30% sterilized glycerol in sterile water.
8. TE buffer (10 mM Tris-HCl, pH 8.0, 1 mM EDTA, pH 8.0); sterilize by autoclaving.
9. 3 M sodium acetate, pH 5.2 (408.1 g of sodium acetate in 800 mL water). Adjust pH with glacial acetic acid. Add water to 1 L. Sterilize by autoclaving.
10. Calf intestine phosphatase (CIP), Amersham.
11. pBluescript vector, Stratagene.
12. DH5α competent cells (Gibco-BRL, Gaithersburg, MD).
13. NZCYM media (Gibco-BRL; 22 g/L of water, sterilize by autoclaving).
14. Ampicillin plates (NZCYM media plus 16 g of Bacto-agar/L of media, sterilize by autoclaving, and cool to 55°C. Add 100 mg of ampicillin (Sigma, St. Louis, MO)/L. Mix and pour into Petri dishes.
15. RNase A (Boehringer Mannheim): Make stock solution 10 mg/mL in ddH$_2$O. Boil for 20 min, and then slowly cool to room temperature. Aliquot.
16. Geneclean kit (Bio101, Vista, CA).
17. Solution 1: 50 mM glucose, 10 mM EDTA, 25 mM Tris-Cl (pH 8.0), sterilize or prepare from sterile components.

Table 1
PCR-Screening of Genomic Libraries

Strategic-design of PCR primers
Preparation of library for DNA templates
PCR and optimization of conditions
Gel analysis and subcloning of PCR products
Screening of PCR products by sequencing
Computer analysis of DNA and translated protein sequence
Traditional screening of the genomic or cDNA library for the full-length gene

18. Solution 2: 0.2 N NaOH, 1% sodium dodecyl sulfate (SDS). No need to sterilize.
19. Solution 3: 60 mL of 5 M potassium acetate, 11.5 mL glacial acetic acid, and 28.5 mL H$_2$O. No need to sterilize.
20. 20% Polyethylene glycol (PEG) + 2 M NaCl. Weigh out 20 g/100 mL of PEG [8000–10,000 mol wt], appropriate NaCl, and water. Autoclave for 20 min. The solution need not be sterile, but autoclaving helps to dissolve the PEG.
21. Stop solution: 95% formamide, 20 mM EDTA, 0.05% bromophenol blue, 0.05% xylene cyanol.
22. 10X TBE (per L: 108 g of Tris base, 55 g of boric acid, 40 mL of 0.5 M EDTA, pH 8.0).
23. Stock acrylamide: *bis*-acrylamide, 40% (380 g acrylamide, 20 g *N,N'*-methylene *bis*-acrylamide, ddH$_2$O to 1 L). Store protected from light at 4°C. Acrylamide is toxic. Wear gloves.
24. Acrylamide/urea gel stock solution (150 mL stock acrylamide:*bis*-acrylamide; 500 g urea; 100 mL 10X TBE buffer; ddH$_2$O to 1 L). Warm to dissolve. Filter through grade 361 qualitative filter paper. Store protected from light at room temperature.
25. Gel wash solution (200 mL methanol; 200 mL glacial acetic acid; 1.6 L ddH$_2$O).
26. Ethidium bromide solution, Sigma (10 mg/mL in ddH$_2$O).

3. Methods
3.1. Strategic Design of Primers

The starting point in the design of PCR primers is to find two regions in the receptor family that are highly conserved between the members. This is typically in the transmembrane (TM) spanning domains of the G-protein-coupled receptor superfamily, which contain seven of these regions. The two regions should not bridge any potential introns. Most adrenergic receptor genes are intronless, but members of the α_1-AR gene family do contain a large intron (>20 kb) between TM 6 and TM 7 *(1)*. Intron positions can be conserved among receptor gene families. The two regions should also be spaced so that the predicted length of the PCR product is <1 kb but >200 bp. The size of the PCR product can be increased, but the efficiency of the PCR process is reduced with longer extensions, and there is a greater chance of PCR-induced errors in the products. Conversely, PCR products smaller than 200 bp will generate more false positives and be inefficiently labeled with ^{32}P-dCTP when used as probes in the eventual screening for the full-length clone. The adrenergic receptors share an especially high homology in TM domains 3, 5, and 6, which collectively forms the basis of the binding pocket for the endogenous agonists epinephrine and norepinephrine *(2–4)*. As an example of the strategy, the primers described below will use TMs 3 and 6 in the α_1-AR, predicted to generate a PCR product of 500–600 bp. This strategy was the same procedure used to clone the novel α_1-AR subtype, the α_{1D}-AR *(5)*.

3.1.1. Design of Primers

1. Align the amino acid sequence between family members in the two transmembrane domains as shown below. Find the region that has the highest homology (bolded). Determine the consensus sequence.

	TM 3
α_{1A}-AR	F C N V **W A A V D V L C C T A S I** M G
α_{1B}-AR	F C D I **W A A V D V L C C T A S I** L S
Consensus	**W A A V D V L C C T A S I**

	TM 6
α_{1A}-AR	L G I V V G C F V L C W L P F F L V M
α_{1B}-AR	L G I V V G M F I L C W L P F F I A L
Consensus	**I V V G M F I L C W L P F F**

2. Write the amino acid sequence of the consensus sequence between family members, and underneath the amino acid, write the degenerate codons for that amino acid.
   W   A   A   V   D   V   L   C   C   T   A   S   I,
   5′- tgg, gc(a/t/g/c), gc(a/t/g/c), gt(a/t/g/c), ga(t/c), gt(a/t/g/c), (c/t)t(a/t/g/c), tg(c/t), tg(c/t), ac(a/g/c/t), gc(a/t/g/c), (a/t)(g/c)(a/t/g/c), at(t/c/a) -3′. This oligo represents a length of 42 bases with a degeneracy of 12,582,912 (*see* **Note 1**).

3. To reduce degeneracy, the following docking criteria can be employed (*see* **Note 2**). The 3′ end mixed base is eliminated, since oligonucleotides are made from the 3′ end and resin columns used in the process are composed from a single base. Inosine can be incorporated in places with high degeneracy (*see* **Note 3**). Some codons for amino acids with high degeneracy (such as Ser and Arg which have 6) can be eliminated based on the most common codon used in the family members (*see* **Note 4**). Finally, degeneracy is reduced by eliminating the basepair in the degeneracy that is already encoded in existing family members (*see* **Note 5**). Thus, the oligonucleotide is rewritten below to include the above dock-

ing. I is for inosine. 5'-tgg, gc(c/t), gcI, gtI, ga(c/t), gt(g/a), ctI, tgt, tgc, ac(a/c), gc(a/t), tcI at(a/t) ct-3'. This represents an oligonucleotide that is 41 bases long with a degeneracy of 128. This primer is also the same "sense" as the mRNA or the codons as written; it is referred to as a "sense" primer. This primer is also only 83% identical to the known α_{1B}-AR DNA sequence; thus, the PCR reaction does not saturate with a known receptor subtype.

4. To facilitate cloning of the PCR products, you can incorporate restriction sites at the end of the oligonucleotide (*see* **Note 6**) to match the restriction site in the vector used for subcloning. The oligonucleotide is now written with two or three of the same restriction sites at the 5' end. Here, *Eco*RI is used. 5'-gaa, ttc, gaa, ttc, tgg, gc(c/t), gcI, gtI, ga(c/t), gt(g/a), ctI, tgt, tgc, ac(a/c), gc(a/t), tcI at(a/t) ct-3'.

5. This methodology is repeated for the TM 6 consensus sequence. The sense orientation of the primer is: 5'- atc, gtI, gtI, gg(t/g/c), atg, tt(t/c), at(a/t) tt(a/g), tgc, tgg, ctI, cc(a/g/t), tt(t/c), tt-3' This primer is 41 bases long with a degeneracy of 144. However, for the PCR process, one "sense" primer is needed along with an "antisense" primer. Therefore, the above oligonucleotide is written in the "antisense" orientation as below, which means writing its basepaired partner. We have also included two of the *Eco*RI restriction sites, which must be at the 5' end.

   3'-tag, caI, caI, cc(a/c/g), tac, aa(a/g), ta(t/a), aa(t/c), acg, acc, gaI, gg(t/c/a), aa(a/g), aac, tta, agc, tta, ag-5'. When ordering this oligo from a company, and in established standard practice, oligonucleotides are always written 5' to 3'. Therefore, rewrite the oligo as below.

   5'- ga, att, cga, att, caa, (a/g)aa, (t/c/a)gg, Iag, cca, gca, (t/c)aa, (t/a)at, (a/g)aa, cat, (a/c/g)cc, Iac, Iac, gat-3'. Therefore, this oligo is designated the "antisense" primer. Primer design is now complete.

3.2. The PCR Reaction

3.2.1. Preparing the Library as the DNA Template

Place 1, 5, and 10 µL of the genomic library into separate microfuge tubes containing 200 µL of sterile water. Heat to 95°C for 10–15 min, and snap-cool in ice. This process breaks open the phage heads and releases the λ and genomic DNA (*see* **Note 7**). This is your DNA template stocks. Use the stock that generates the best PCR products.

3.2.2. PCR Reaction

1. Set up the PCR reaction as follows: 1 µL of DNA template, 200 pmole each primer, 10 µL of 10X PCR buffer (supplied by manufacturer), deoxynucleotide triphosphates (dNTPs) at 1.25 m*M* each final, 0.5 µL of *Taq* polymerase, and sterile water to 100 µL final volume. Add the *Taq* polymerase last. Overlay reaction with 50 µL of sterile mineral oil. Therefore, you will have four reactions. Three reactions will contain 1 µL of DNA from each stock DNA template vial and one control PCR that contains everything except the DNA template (*see* **Note 8**).

2. PCR conditions (*see* **Note 9**): One cycle: Denature at 95°C for 5 min, anneal at 45°C for 5 min, extend at 72°C for 45 min, then PCR cycle for 35 cycles at 94°C for 1 min, 45°C for 2 min, and 72°C for 3 min.

3.3. Gel Analysis of PCR Products

Electrophorese 20 µL of product in a 1.2% agarose gel. A band at the predicted size based on the primers should be apparent (*see* **Notes 10** and **11**).

3.3.1. Agarose-Gel Electrophoresis

1. Weigh out 1.2 g of Agarose (GTG or low EEO; *see* **Note 12**), and place into 100 mL of 1X TAE buffer. Microwave or heat sample until all of the gel fragments have solubilized. Add 5 µL of an ethidium bromide solution (10 mg/mL in water) for visualization of the DNA, mix, and pour into a casting tray with comb. Allow to solidify for about 1 h. Remove comb. Place gel in running chamber, and submerge with 1X TAE. Place hardened gel in running chamber, and barely cover gel with 1X TAE. Add loading buffer to each sample. Electrophorese mol-wt marker and 20 µL of each PCR sample. Attach positive current cable to end of gel box where the DNA (negatively charged) will migrate.
2. Electrophorese samples at an amperage that will not overheat the gel. Conduct electrophoresis until the bromophenol blue in the loading buffer has migrated to the end of the gel. Localize PCR products by placing gel on a UV transilluminator, and photograph gel using a Polaroid camera.

3.4. Subcloning of PCR Products

The subcloning of the PCR products is required to isolate and identify individual DNA sequences within mixed DNA populations that could exist in a single band. Direct PCR cycle sequencing is not recommended for this reason. The authors also recommend that each band be treated as an independent sample and processed separately, as opposed to pooling the fragments and screening a mastermix.

3.4.1. Isolation of Gel-Extracted DNA

1. If the PCR products are larger than 500 bp, you can gel-extract using a Geneclean (BIO101) or similar product according to manufacturer's specifications.
2. For fragments under 500 bp, use the GTG agarose in the electrophoresis. Excise the individual bands with minimal excess agarose (visualize DNA on the UV transluminator). Place gel slice in a microfuge tube, and incubate at 65°C until the gel is melted.
3. Add 300 µL of TE buffer to the melted gel, mix, and extract out the liquified agarose by adding an equal volume of heated equilibrated phenol (*see* **Note 13**). Vortex and microfuge for 2 min to separate the phases. A heavy white precipitate is present, which is the agarose. Remove and retain top phase, and repeat extractions until no white interface is obtained.

4. Do a final chloroform extraction to get rid of any excess phenol that may inter-fere with restriction digestion. Add an equal volume of chloroform to the top phase, mix, and centrifuge. The DNA is present in the aqueous top phase.
5. Add 1/10 vol of 3 *M* sodium acetate to recovered top phase, and mix. Add 2.5 vol of 95% ice-cold ethanol, mix, and incubate at –20°C for 20 min. Microfuge at top speed for 15 min to pellet DNA.
6. Wash pellet with 70% ice-cold ethanol, and centrifuge if pellet dislodges. Decant ethanol and dry pellet. Resuspend in 20 μL of TE buffer.

3.4.2. Restriction Digestion of PCR Products

1. Prepare the digestion reaction using the restriction enzyme that you encoded in the PCR primers (*see* **Note 14**): 20 μL of the DNA fragment in TE, 2.5 μL of the 10X digestion buffer (supplied by manufacturer of the enzyme), and 2.5 μL of the restriction enzyme (*Eco*RI). Incubate at 37°C (or required temperature) over-night in a heat block (16–18 h).
2. Add an equal volume of equilibrated phenol and chloroform. Mix and microfuge for 2 min to separate phases. Remove and retain top phase.
3. Add 1/10 vol of 3 *M* sodium acetate to recovered top phase, and mix. Add 2.5 vol of 95% ice-cold ethanol, mix, and incubate at –20°C for 20 min. Microfuge at top speed for 15 minutes to pellet DNA.
4. Wash pellet with 70% ice-cold ethanol, and centrifuge if pellet dislodges. Decant ethanol and dry pellet. Resuspend in 10 μL of TE buffer. You can estimate the amount of DNA isolated by comparing the band intensities of the mol-wt mark-ers to the band of interest in the agarose gel and assuming an 80% recovery. This amount of DNA is usually too small to estimate spectrophotometrically.

3.4.3. Preparation of Vector for Subcloning

1. Use 10 μg of vector stock such as pBluescript (*see* **Note 15**). Add 2.5 μL of the 10X digestion buffer (supplied by manufacturer of the enzyme) and 2.5 μL of the restriction enzyme (*Eco*RI). Incubate at 37°C (or required temperature) for 2 h.
2. Purify cut vector DNA using the Geneclean kit or alternately by phenol/chloro-form extraction as above (**Subheading 3.4.2., step 2**) followed by precipitation of the DNA (**Subheading 3.4.2., steps 3** and **4**). Resuspend DNA in 10 μL TE.
3. To the entire preparation of the *Eco*RI-cut vector, add 2.5 μL of CIP 10X buffer (supplied by manufacturer) and 2.5 μL of the enzyme CIP and sterile water to 25 μL total volume. Incubate at 37°C for 2–4 h (*see* **Note 16**).
4. Clean-up the *Eco*RI/CIP-treated vector as in **step 2**. Resuspend the DNA in 30 μL of TE. This is your cut-vector stock.

3.4.4. Ligation Reaction

For information regarding ligation theory, *see* the Maniatis cloning manual (*6*). The basic recipe is to use vector and insert DNA at a molar (not μg) ratio of

1:3. Therefore, there are more moles of the insert than vector, which permits a higher efficiency of ligation. Start with 0.1 pmol of vector or approx 0.2 μg of the *Eco*RI/CIP-treated pBluescript, but this is usually established empirically depending upon how well the vector was digested with both the restriction enzyme and the CIP (*see* **Note 17**).

3.4.5. Ligation of PCR Fragments into pBluescript

1. Prepare a ligation reaction for each isolated and prepared PCR band, in addition to one control that contains everything except insert DNA; this control will determine the number of background colonies following transformation.
2. To 1 μL of cut-vector stock (or amount that yields satisfactory background; *see* **Note 17**), add 5–10 μL of digested PCR DNA (to approx 1:3 molar ratio), 2 μL of 10X ligation buffer (supplied with enzyme), 2 μL of ligase, and sterile water to 20 μL final volume.
3. Mix and incubate overnight at 14°C (obtained by placing a water bath in a cold room) (*see* **Note 18**).

3.4.6. Transformation into Bacteria

1. For simplicity, you can buy premade competent cells from Gibco-BRL. We recommend the DH5α strain. For each ligation, use 100 μL of competent cells. Thaw a vial on ice, and then aliquot 100 μL into prechilled 15-mL sterile polypropylene centrifuge tubes used for each individual ligation reaction. Retain cap to maintain sterility.
2. Add no more than 10 μL of the ligation mixture to centrifuge tube. To use the entire mixture, perform multiple separate transformations. Mix slowly by moving the pipet tip back and forth. Do not vortex or pipet up and down. Incubate on ice for 30 min (*see* **Note 19**).
3. Heat-shock competent cells by placing the centrifuge tubes in a 42°C water bath for 1 min (timed). Do not use a heat block since heat transfer is compromised. Return tubes to ice for another 2 min.
4. Add 2 mL of sterile NZCYM media (no antibiotic). Incubate with shaking for 1 h at 37°C to allow the cells to develop antibiotic resistance.
5. Concentrate cells by centrifugation for 10 min at top speed in a tabletop centrifuge. Pipet and discard most of the supernatant, except for 200 μL. Resuspend the cell pellet in the 200 μL media. Plate and spread on 100-mm ampicillin plates until moisture is absorbed. Invert plates and place in 37°C incubator overnight (*see* **Note 20** if using pBluescript).
6. Colonies should be apparent, with greater number than control plates (2–3×). If using the blue/white selection, recombinants will appear pure white. If not using pBluescript or other vector that discriminates recombinants, the plasmids should be screened for insertion of DNA by restriction digestion before sequencing.

3.5. Sequencing of PCR Products

Screen at least 20–50 recombinant colonies/PCR band to determine if the PCR worked efficiently or was contaminated. Hundreds of colonies may need to be screened to establish any new receptor subtypes.

3.5.1. Plasmid Preps for Double-Stranded DNA Sequencing

1. Pick individual colonies with a sterile toothpick. Touch the toothpick onto an NZCYM agarose master plate (gridded and labeled for cataloging of numerous colonies), and drop the toothpick into a culture tube containing 2.5 mL sterile NZCYM. The NZCYM should contain 100 mg/mL ampicillin.
2. Shake tubes vigorously in a 37°C incubator overnight. Also, incubate the master plate in a 37°C incubator overnight.
3. The next morning, centrifuge culture tubes at 2000 g for 10 min to pellet bacteria. Also, store the master plate at 4°C for later reference.
4. Aspirate medium, and resuspend bacterial cell pellets in 100 μL of solution 1. Transfer resuspended pellets to 1.5-mL Eppendorf tubes.
5. Incubate at room temperature for 5 min, and then add 200 μL solution 2. Mix by inversion—do not vortex.
6. Incubate on ice for 5 min, and then add 300 μL of ice-cold solution 3. Vortex gently for 10 s.
7. Incubate on ice for 5 min, and spin for 5 min at maximum speed in a microfuge. Transfer supernatants to fresh Eppendorf tubes.
8. Phenol/chloroform-extract by adding 100 μL each of equilibrated phenol and chloroform. Vortex samples to mix and centrifuge completely for 3 min at maximum speed in a microfuge. Transfer aqueous phase (upper phase) to fresh tubes.
9. Add 2 vol of 100% ETOH (ethanol; equilibrated at room temperature) to recovered aqueous phase, vortex, incubate at room temperature for 10 min, and centrifuge at maximum speed for 10 min in a microfuge.
10. Decant supernatants, and wash pellets with 70% ETOH (200 μL). Drain all fluid from tubes, and air-dry DNA pellets for 15–30 min at room temperature. Resuspend pellets in 40 μL TE buffer.

    DNA prepared using this alkaline lysis method is of sufficient quality for restriction analysis. However, for use in sequencing reactions, it is suggested that the DNA be further purified. Two common methods used for further purification are outlined below.

1. RNase A Method
   a. Add 1 μL of RNase A (10 mg/mL stock) to each sample. Incubate at 37°C for 30 min.
   b. Repeat **steps 8–10** above to obtain DNA.

2. PEG Method
   a. Add 50 μL of an ice-cold solution of 20% PEG + 2 *M* NaCl. Incubate on ice for 1 h or at 4°C overnight.
   b. Centrifuge at maximum speed for 15 min in a microfuge. Decant.
   c. Wash pellets with 70% ETOH (200 μL). Prior to washing, pellets may not be visible; the 70% ETOH wash may cause pellets to become visible.
   d. Centrifuge at maximum speed in a microfuge for 10 min. Drain all fluid, let air-dry, and resuspend in 35 μL TE.

3.5.2. The Sequencing Reaction

The sequencing reaction may be conducted using any one of a number of commercially available kits, but the protocol described below is optimized for the Sequenase version 2.0 DNA sequencing kit (Amersham). This kit, which employs the chain-termination method, includes a detailed instruction manual that is a valuable troubleshooting guide. Use 1–2 μg of the DNA for sequencing. The yield from a single miniprep is sufficient for two sequencing reactions. Calculate the concentration of DNA before conducting the reaction (*see* **Note 21**).

1. Denature double-stranded template DNA. To 15-μL aliquots of stock plasmid DNA (prepared above), add 2 μL of 4 m*M* EDTA (pH 8.0) and 4 μL of 1 *N* NaOH. Incubate at room temperature for 10 min. Place a 2-μL drop of 2 *M* ammonium acetate on the wall of each tube and wash with 100 μL ice-cold 100% ETOH. Mix by inversion, and incubate in a –20°C freezer for 10 min. Centrifuge tubes at maximum speed for 10 min in a microfuge, and dry pellet in a Speed Vac for 10 min.
2. Anneal primer to single-stranded template DNA (if using pBluescript, use T7 or T3 primer). To the single-stranded DNA samples prepared in **step 1** above, add a cocktail containing 7 μL sterile H₂O, 2 μL of the 5X reaction buffer supplied with the kit, and 1 μL of a 1 pmol/μL stock of primer. Anneal primer to template by heating to 37°C for 10 min.
3. Label 4 tubes/clone and/or per primer with A, C, T, or G (termination tubes). Deposit 3.5 μL of stock dideoxy (dd) ATP, ddCTP, ddTTP, and ddGTP into appropriate tubes. Dideoxy nucleotide stock solutions are supplied with the kit. To annealed DNA, add a cocktail containing 1 μL of 0.1 *M* dithiothreitol (DTT), 2 μL of diluted labeling mix (supplied with kit), and 0.5 μL [^{35}S]deoxy ATP (1 μCi/μL). Dilute polymerase 1:8 in dilution buffer (both supplied with kit), and add 2 μL of diluted enzyme to reaction mixture. Incubate at room temperature for 2 min, and immediately transfer 3.5 μL of reaction mixture to each termination tube. Incubate at room temperature for 10 min, add 4 μL stop solution, and store in a –20°C freezer until needed.

3.5.3. Preparation and Electrophoresis of Sequencing Gel

A number of sequencing apparatus are commercially available. The method described below for pouring and conducting electrophoresis of a sequencing gel is optimized for the Bio-Rad Sequi-Gen sequencing apparatus, which includes a detailed set of instructions.

1. Thoroughly wash gel plates with glass cleaner and then with 100% ETOH. To ease removal of the outside plate after electrophoresis of the gel, coat the gel-facing side of this plate with a solution of 2% silane (in chloroform) (**toxic; wear gloves**). Assemble the gel plate sandwich, and pour a plug consisting of 50 mL of a stock acrylamide/urea solution, 200 μL of a 20% ammonium persulfate stock solution, and 200 μL *N',N',N',N'*-tetramethylethylene diamine (TEMED). Set the bottom edge of the assembled plates into the plug, and allow to polymerize for 45–60 min. After plug is set, pour the gel solution, which consists of 100 mL of a stock acrylamide/urea solution, 100 μL of a 20% ammonium persulfate stock solution (in H_2O) and 40 μL TEMED. Set comb to make wells, and allow gel to polymerize >2 h (preferably overnight).
2. Loading and electrophoresis of gel: Assemble gel to running chamber, and add 1X TBE buffer. Preheat gel to 50°C (temperature of the glass plate) by applying a current of 100 mA for about 30 min. Gel electrophoresis at high temperature (>60°C) may crack glass. Bio-Rad sells glass-attachable temperature gauges for sequencing. Heat sequencing reaction samples to 100°C for 3 min to ensure DNA is denatured, and then immediately place the samples on ice. Set comb. Wash (blow) out lanes just before loading, since urea diffuses out of the gel and disturbs uniform loading. Load 4–7 μL per well. Conduct electrophoresis at a constant current of 75 mA (or to maintain 50°C temperature) for desired length of time based on primer used.
3. Autoradiography: After electrophoresis is complete, disassemble gel plates by removing the outside plate. Gently rinse gel, which is still adhered to the inner gel plate, with 2 L of gel wash solution (10% MeOH, 10% HOAc) to remove urea. Rinse one more time with 1 L H_2O. Transfer (peel) gel to thick chromatography paper, and cover the exposed side of the blotted gel with plastic wrap. Trim edges to fit and dry in a gel dryer for 1 h. Remove plastic wrap, and apply to X-ray film ensuring that the dried gel is touching the film. Expose the film for 1–3 d.

3.6. Computer Analysis of Sequence Data

1. The nucleotide sequence can be directly entered into a BLAST data bank for analysis of homology with known sequences. There is an internet site for The National Center of Biotechnology Information, which contain BLAST search engines. There is no need to sequence the PCR product fully. Sequence runs of 200–300 bases is sufficient to generate a "candidate" clone for further consideration and analysis.

2. Alternately, you can use common DNA manipulation programs, such as Intelligenetics or DNA Star, and translate the DNA sequence obtained in both the sense and complementary strand to obtain a predicted protein sequence (*see* **Note 22**). This should be done using all three reading frames. If you have found an exon that encodes a protein, one of the three reading frames should be open from primer to primer. However, there may be sequencing errors and the open reading frame may have to be verified by sequencing in the opposite direction.

3. The protein sequence is then compared to other members of the known receptor class. Since the transmembrane regions used in the primers are defined, the sequence should be compared within that region. If you have found a new family member, it is generally 60–90% identical in amino acid sequence. However, novel receptor subtypes that do not have high homology with known receptors (orphan receptors) can have identity as low as 20–30% (*see* **Note 23**).

4. If the PCR is successful, you should retrieve members of the receptor family from which the primers were designed and other potential clones. If you identify a sequence that is 95–99% identical to a known sequence, it is likely the same known gene (i.e., that is contained in the lab), is caused by sequencing errors or is a species ortholog. You may have to screen hundreds of the colonies to find a new member. If after screening 20–50 colonies and all candidates are known receptor subtypes, you may have to redesign the PCR primers to deselect better the known subtypes, because the PCR became saturated with these sequences.

5. Once you have identified a new receptor subtype or a potential orphan receptor (i.e, "candidate clone"), use the PCR product to screen a library for the full-length clone. The easiest way is to screen a cDNA library since this will not have introns. However, Northern blot analysis using the PCR product as a probe should be initially performed on a panel of tissues to determine target mRNA abundance. A library prepared from a tissue source exhibiting the highest level of target mRNA should then be probed. If unsuccessful, genomic library screening can be performed and any potential introns can be spliced out manually to form a cDNA for expression studies.

3.7. Traditional Screening of the Genomic or cDNA Library for the Full-Length Gene

Methods for the screening of libraries with a known probe are found in this volume in Chapter 5.

4. Notes

1. Degeneracy is determined by multiplying the amount of individual degeneracies of the codons (i.e., $4 \times 4 \times 4 \times 2 \times 4 \times 2 \times 4 \times 2 \times 2 \times 4 \times 4 \times 2 \times 2 \times 4 \times 3 = 12{,}582{,}912$). This is too much degeneracy, since each unique sequence is represented only once in 12,582,912 copies. It is rare for this oligonucleotide to lead to a PCR product, and the degeneracy must be reduced to a level under 256 for the PCR reaction to contain sufficient copies of each primer for successful amplification. This was found by personal experience in which primers made at 384 and

512 degeneracy did not work, but when reduced in degeneracy to 256, produced the expected PCR products *(5)*. There are many companies available to synthesize oligonucleotides that will incorporate mixed bases and inosines. Order them purified; if not, you will need to purify them in the lab (via cartridge, HPLC, or gel purification). These purification methods are detailed in molecular biology handbooks, but not in this chapter owing to space limitations.

2. The authors recommend starting with a very long oligonucleotide, such as 35–45 bases. Although this will lead to higher degeneracy, the longer oligonucleotide becomes more tolerant of mismatches that will occur if you are trying to find a novel receptor subtype. The longer the oligonucleotide, the higher the likelihood of finding something very similar to known receptor subtypes, and the specificity of the PCR process is increased. The shorter the oligonucleotide, the likelihood increases for finding novel receptors outside the particular family. It is recommended that the range 20–55 bases be used. Visually inspect the oligonucleotide for apparent hairpin structures (i.e, self-annealing), and avoid using more than four runs/repeats of the same nucleotide (i.e., aaaa).

3. Inosine can hydrogen bond with any nucleotide (although it prefers cytosine) and is used in places where there is high degeneracy (three or more). It is not placed within 5 bases from the 3′ end of the oligonucleotide, since this region is critical for specificity and the start of the amplification process. It is generally placed in the middle or 5′ end of the oligonucleotide. The longer you synthesize an oligonucleotide, the more inosine it can incorporate. For a 40-base primer, we generally use about four or five inosines. If you incorporate too many inosines, the specificity of the primer is reduced, since inosine can basepair with any nucleotide.

4. To reduce degeneracy, some of the codons are simply dropped from consideration. This is plausible if the oligonucleotide is sufficiently long to be tolerant of mismatches (*see* **Note 2**). If you have sequences of known receptor subtypes, check which codons are preferred in those species. This offers an educated guess at which codons to dock.

5. This is perhaps the most important docking criteria. If you dock bases in the codons from known receptor subtypes, this increases the chances of finding novel subtypes since they have a different DNA sequence (i.e., α_{1A}-AR and α_{1B}-AR contain a gca and gcg for alanine; you use gc[c/t] for the primer to deselect or mismatch the known receptors). Second, you also reduced the likely possibility of saturating the PCR product with the known receptor subtypes which can reduce your chances of finding novel members. If the PCR process is working well, you will retrieve known family members; however, you do not want this process to be too efficient, since all of the primers will be soaked up, amplifying known members and not available for finding new members.

6. Depending on the *Taq* polymerase utilized, the 5′ and 3′ ends can or cannot end with an "a," which is incorporated into the sequence. To subclone these products, either enzymatically blunt the ends or use a t/a cloning kit, which is commercially available. All other polymerases generate blunt ends that are more difficult to subclone than overhangs. Since a high subcloning efficiency is

desirable to screen several PCR products, it is better to incorporate the restriction sites into the primers. However, use a restriction site that is well characterized and retains properties of cutting near the ends of DNA. This information is generally found in the reference section of catalogs of molecular biology companies, such as Stratagene or New England Biolabs. Some enzymes cut better than others when the restriction site is located near the end of the DNA sequence (within 10–12 bases). Multiple copies of the same restriction site are also used to ensure that the sequence will be cut efficiently. Since we are merely screening sequences and not expressing the protein, it does not make any difference if the reading frame is altered in the PCR product owing to inclusion of the restriction sites.

7. In most cases, this DNA preparation is sufficient to generate PCR products. However, if no products are obtained, one alternate method of DNA preparation is to restriction-digest the released DNA and, after cutting, extract the restriction enzyme from the mixture with phenol/chloroform and then a final chloroform extraction before preceding with the PCR reaction. Alternately, shearing the DNA can be performed using an 18-gage needle. The idea here is to make the DNA smaller (genomic libraries have λ DNA that is 45 kb) and easier to denature during the PCR process.

8. If the control tube yields a PCR product (any size), you have a contamination problem, and it is best to stop here and determine the source of the contaminant. Likely suspects are the pipets, water supply, buffers, hands, aerosols, or primers. The problem becomes much larger if you have pieces of cDNA from other receptors in the lab. It is best to start over and make fresh buffers, water solutions, change pipets to ones not used in PCR or in handling the cDNAs, or conduct the PCR setup in another lab.

9. You can experiment with different types of *Taq* polymerase, but since this is for identification purposes only, any type will suffice. Some of the other polymerases have a lower error rate in incorporation as well as higher stability, but have a slower polymerization rate. The first PCR condition consists of a longer initial denaturing step because the DNA template is large. The longer extension time for one cycle allows a greater amount of first strand to be synthesized, increasing the likelihood of producing a product. The annealing temperature is empirical. You want the lowest temperature possible that provides the expected product without too much background noise and smearing. The lower annealing temperature will also increase the chances of finding new products.

10. If there is no product, the conditions and amounts of primers and template need to be adjusted empirically. You can increase the amount of DNA template and/or amount of primers. If this is not successful, try adding 20% glycerol or formamide to the PCR mixture. This will promote the dissociation of the double-stranded DNA, which may be impaired owing to high GC content in the gene of interest. Alternately, the DNA template can be restriction-digested with a 6-base cutter to reduce the sizes of the DNA template stock (*see* **Note 7**). If after exhausting these possibilities, there is still no PCR product, the primers must be suspected. Test the primers on a known receptor subtype clone to determine if it amplifies a

product. If this does not work, redesign the primers in a different region of the receptor, since there may be too much secondary structure in the DNA to allow efficient PCR conditions. If a smear is obtained in the gel, the concentration of DNA, primers, or nucleotides may be too great. Reduce these amounts and also try increasing the annealing temperature and reducing the time of elongation. Smearing always means too much of something. A common problem in smearing is that the *Taq* polymerase from the company has been contaminated with DNA, but you will see this in your control tube if this is the case. Try another source.

11. Since family members of receptors share a similar length in both the DNA and protein sequence, additional family members are likely to have similar-sized PCR products. It is difficult to discriminate similar-sized mixed products on an agarose gel, so sequencing will need to be conducted to discriminate clones. If a different-sized band(s) is obtained in addition to the known subtypes, it is wise to subclone these different products and screen by DNA sequencing. If you obtain many bands (five or more), your PCR conditions are probably not stringent enough.

12. GTG low-melting agarose is used when the predicted PCR products are small (<500 bp). This special agarose has higher resolution for smaller DNA fragments and offers a higher efficiency of extracting the fragments from the gel for subcloning. Many commercially available kits for extracting DNA from gels (i.e., Geneclean) have a low efficiency for fragments below 500 bp.

13. Premade equilibrated phenol can be purchased from Amersham. Alternately, solid phenol can be purchased, melted in a water bath at 60°C and 1 *M* Tris base is added, mixed, and then settled. Repeat adding the Tris until a phase separation is achieved. The phenol will soak up the water and increase its volume until saturated, at which point the phase separation will appear. The lower phase is the equilibrated phenol. Test a small aliquot with pH paper. It should be ≥pH 8.0. If not, add more 1 *M* Tris base until this pH is achieved. Store in refrigerator with the Tris base still on top. Do not use old phenol, which has acquired a color. If the pH is below 8.0, the phenol may not cause a phase separation when used in extraction.

14. As mentioned in **Note 6**, different enzymes cut better than others (some do not even cut at all) when the restriction site is located near the end of a DNA fragment. Therefore, use an enzyme that has been identified as an efficient cutter, such as *Eco*RI. The reference section of molecular biology catalogs, under cutting DNA near the end of fragments, identifies the particular enzyme you want to use and the conditions for the digestion. Most cut efficiently only after an overnight digestion.

15. There are several choices of the types of vectors used for subcloning. M13 vectors offers single-stranded DNA which gives superior sequencing results and is easier to screen by nitrocellulose lifts if so desired. Plasmid DNA is more universal, less technically difficult, and sequencing can be performed in both directions, but it is harder to screen by nitrocellulose lifts. Since this procedure recommends screening by sequencing, we recommend plasmid DNA using a

vector, such as pBluescript or similar versions that contain two sequencing primer sites (i.e., T7 and T3) on either side of the cloning site for sequencing in both directions. Another advantage of pBluescript is the blue/white selection (*see* **Note 20**).

16. Phosphatase treatment of the 5′ ends of the restriction-cut vector removes the phosphates that are needed for the vector to religate to itself. Therefore, removing the phosphates prevents self-ligation, which promotes recombination with inserts and gives little background of noninserted plasmids in the transformation. It is not recommended to remove this step in the subcloning.

17. Take 1, 3, and 5 µL of the *Eco*RI/CIP-treated vector stock and transform into bacteria according to **Subheading 3.4.6.** Use the same aliquot in the ligation reaction that gives you some colonies, but under 10 colonies as background. If there are too many colonies, 30 or more with the lowest aliquot, your restriction digest and/or CIP treatment was inefficient, and it is best to restart the preparation and digest with longer incubation times. If no colonies were obtained, the DNA was probably lost in the manipulations. The cutting process is never 100% complete to generate no background colonies. Supercoiled DNA (uncut vector) transforms with 100× greater efficiency than linearized DNA. Once a suitable vector stock is prepared, isolate the PCR fragment to achieve a sufficient stock for ligation at a 1:3 molar ratio. A small sample of the original PCR reaction can be retained to allow reamplification if needed. Calculation of moles of either plasmid or insert is achieved by measuring the OD 260_{NM} of the sample and using the conversion 1 OD = 50 µg double-stranded DNA; 660 g/mol = 1 bp or approximating the concentration from the DNA markers run on an agarose gel that is at a known concentration.

18. The temperature of ligation is a trade-off between allowing a low enough temperature to promote a 4-base overhang to anneal, but not too low to inhibit fully the reaction rate of the ligase whose optimum temperature is 37°C. The ligation can be performed at room temperature for 4 h, followed by the transformation. However, higher ligation efficiency can be achieved with the overnight incubation at 4°C.

19. The ligation mixture contains proteins and other molecules that can inhibit transformation. Therefore, do not add the ligation mixture at more than 1/10 the volume of the competent cells, but separate the reaction into two aliquots if needed. The 30-min incubation is a minimum. Longer incubations (hours) can increase efficiency.

20. Another advantage of using pBluescript is blue/white selection. Recombinants disrupt the β-galactosidase gene in the vector, and therefore disrupts the ability of the vector to produce the enzyme and its ability to cleave a substrate (turns blue). Therefore, if the inserts were incorporated, the colonies turn white. This selection reduces the amount of subsequent work needed to identify recombinants. To achieve this selection, add 50 µL of a 2% 5-bromo-4-chloro-3-indolyl-β-D-galactoside (X-Gal)(Amersham) solution in dimethylformamide and 50 µL of a 0.1 *M* isopropyl-β-D-thiogalactopyranoside (IPTG) (Amersham) solution in

water, and spread on the ampicillin plates before the cells are added. The next day, if blue (wild-type) colonies are not seen among the white colonies, incubate the plate for a day in the refrigerator to bring out the blue color.

21. Calculation of moles of either plasmid or insert is achieved by measuring the OD 260_{NM} of the sample and using the conversion 1 OD = 50 μg double-stranded DNA and 660 g/mol = 1 bp of DNA, or approximating the concentration from the DNA markers (with established concentrations) electrophoresed in agarose gels.

22. Translate both strands (one is translated as read; the sequence is inverted and complemented to translate the other strand) since the orientation of the insert in the vector is not known. Correct translation of the sequence is achieved when you can identify the amino acids encoded in your primer as being those contained in the predicted transmembrane domains of the known receptor subtypes. Continue reading the open reading frame from the correct translated primer sequence to see if you have a novel sequence.

23. The first step to establish is that the PCR was not contaminated from cDNAs that are already present in the lab. If the library is derived from human tissue, then finding the rat cDNA sequence for a clone indicates contamination of the PCR reaction. Always compare your sequences to known sequences in the lab. Species homologs generally are 85% or greater in identity in DNA sequence, but never 100%. If you have identified a recombinant that is 98% identical and the species is not close in phylogenetics, it is probably a clone containing sequencing errors.

References

1. Ramarao, C. S., Kincade Denker, J. M., Perez, D. M., Gaivin, R. J., Riek, R. P., and Graham, R. M. (1992) Genomic organization and expression of the human α_{1B}-adrenergic receptor. *J. Biol. Chem.* **267,** 21,936–21,945.

2. Hwa, J., Graham, R. M., and Perez, D. M. (1995) Identification of critical determinants α_1-adrenergic receptor subtype selective agonist binding. *J. Biol. Chem.* **270,** 23,189–23,195.

3. Hwa, J. and Perez, D. M. (1996) The unique nature of the serine residues involved in α_1-adrenergic receptor binding and activation. *J. Biol. Chem.* **271,** 6322–6327.

4. Hwa, J., Graham, R. M., and Perez, D. M. (1996) Chimeras of α_1-adrenergic receptor subtypes identify critical residues that modulate active-state isomerization. *J. Biol. Chem.* **271,** 7956–7964.

5. Perez, D. M., Piascik, M. T., and Graham, R. M. (1991) Solution-phase library screening for the identification of rare clones: Isolation of an α_{1D}-adrenergic receptor cDNA. *Mol. Pharmacol.* **40,** 876–883.

6. Sambrook, J., Fritsch, E. F., and Maniatis, T. (1989) *Molecular Cloning: A Laboratory Manual.* Cold Spring Harbor Laboratory, Cold Spring Harbor, NY.

5

Solution-Phase Library Screening for Identification of Rare Adrenergic Receptor Clones

Dianne M. Perez and Michael J. Zuscik

1. Introduction

Receptors, in general, are not highly expressed on endogenous cell surfaces. Adrenergic receptors (ARs) are commonly expressed in the femtomolar range. Correspondingly, their mRNA levels are also present in limited quantities. Northern analysis of the adrenergic receptors has confirmed this observation by requiring a high amount of poly A^+ RNA (10–20 µg/lane) for the detection of the signal. To overcome this limitation, many investigators now use RNase Protection Assays, which are of higher sensitivity but also require much less poly A^+ RNA for analysis. Thus, the lack of abundant mRNA makes receptor cloning studies difficult because of their underrepresentation in a given cDNA library.

The solution-phase technique as published *(1)* was designed to overcome time and labor difficulties in the cloning of adrenergic receptor genes. The technique overcomes the underrepresentation problem by taking advantage of the amplification ability of phage combined with the quickness and efficiency of probing via solution-phase instead of solid-phase approaches. The technique is pertinent when a specific probe (at least 200 bp) is generated or available for the receptor cDNA, and it is known that a cDNA species exists in the library in question. The methodology is analogous to performing a Southern blot analysis on the library, but the key feature is diluting and reamplifying positive cultures to concentrate the underrepresented phage. A flow diagram of the methodology is summarized in **Table 1**. In addition to the solution-phase technique, this chapter also details methods in the traditional plaque hybridization protocol, which one needs after a positive aliquot is identified.

From: *Methods in Molecular Biology, vol. 126: Adrenergic Receptor Protocols*
Edited by: C. A. Machida © Humana Press Inc., Totowa, NJ

Table 1
Methodology of Solution-Phase Library Screening

Southern blot or PCR screen library cDNA
Make nine 1-μL aliquots
Add 100X plating cells
Amplify aliquots in culture overnight
Isolate DNA from 1 mL; retain an aliquot of culture for future use
Digest DNA with restriction enzyme screen to release inserts
Label probe and conduct solution-phase hybridization with isolated phage DNA
Run hybridization on 4% nondenaturing acrylamide gel
Autoradiography on gel
Realiquot positive cultures
Reamplify and rescreen until concentration of phage stock increases and all cultures
  are positive
Plate out positive culture and by traditional plaque-hybridization technique

The solution-phase technique should be used when traditional plaque hybridization methods (nitrocellulose lifts) come up empty-handed. It is not recommended as an alternative to traditional screening methods, since the time required is longer and the amount of technical expertise needed is greater owing to problems in solution amplification of phage. In the cloning of the α_{1D}-AR cDNA *(1)*, a PCR product was first generated based on degenerate primers to known subtypes. Once identified as novel, the PCR product was used to screen a total of 1.8×10^6 recombinant plaques of a rat hippocampus cDNA library, but failed to identify a positive plaque even though Northern analysis identified a transcript in rat hippocampus mRNA. However, the presence of at least one copy of the cDNA in the library was verified by performing a Southern blot analysis on the library itself. The extremely low abundance of the cDNA in the library prompted us to develop the solution-phase library screening method. Accounting for the low abundance of the cDNA in the library, one would have to screen the equivalent of 2250 Petri dishes of 150 mm (or 90×10^6 plaques) to identify this clone by traditional plaque hybridization. Solution-phase screening enhanced the concentration of the phage containing the cDNA by 300-fold and resulted in the screening of only 10- to 150-mm Petri dishes to obtain finally the full-length clone.

2. Materials
2.1 Equipment
This consists of a PCR machine, autoradiography cassettes with intensifying screens, darkroom and film developer, incubator ovens, tabletop shaker,

polyacrylamide gel electrophoresis apparatus, power supply, shaking incubator for bacteria cultures, β-counter, water bath, agarose-gel apparatus, UV transilluminator, and Stratalinker (UV crosslinker).

2.2. Reagents

1. cDNA library (either made or bought from Clontech [Palo Alto, CA] or Stratagene, La Jolla, CA).
2. RNase-free DNase 1 (Boehringer Mannheim).
3. Deoxynucleotide triphosphates; dATP, dCTP, dGTP, TTP (100 mM stocks from Pharmacia, Piscataway, NJ).
4. Agarose, low EEO (FMC).
5. Nitrocellulose filters (Hybond C; Amersham, Piscataway, NJ), nitrocellulose sheets, Bio-Trace NT (Gelman).
6. C600Hfl cells or other lawn cells (Gibco-BRL, Gaithersburg, MD).
7. Random Prime Labeling Kit (Boehringer Mannheim, Indianapolis, IN).
8. NZCYM powdered media, Gibco-BRL; 22/L, sterilize by autoclaving.
9. 50X TAE buffer (per L): 242 g of Tris base, 57.1 mL of glacial acetic acid, and 100 mL of 0.5 M ethylene diamine tetra-acetic acid (EDTA) pH 8.0.
10. 6X Agarose-loading buffer: 0.05% bromophenol blue, 0.05% xylene cyanol, 30% sterilized glycerol in sterile water.
11. 20X SSC: 175.3 g of NaCl and 88.2 g of sodium citrate/L of water. Sterilize by autoclaving.
12. 3 M Sodium acetate, pH 5.2 (408.1 g of sodium acetate in 800 mL water). Adjust pH with glacial acetic acid. Add water to 1 L. Sterilize by autoclaving.
13. 10X TBE (per L): 108 g of Tris base, 55 g of boric acid, 40 mL of 0.5 M EDTA, pH 8.0.
14. Ampicillin plates (NZCYM media plus 16 g of Bacto-agar/L of media, sterilize by autoclaving, and let cool to 55°C. Add 100 mg of ampicillin (Sigma, St. Louis, MO)/L. Mix and pour into Petri dishes.
15. Agar plates for library screening (NZCYM media at 22 g/L, agarose [low EEO] at 16 g/L, sterilize by autoclaving, and pour into 150-mm Petri dishes).
16. Top agar (NZCYM media at 22 g/L and add agarose [low EEO] at 0.7%. Prepare smaller aliquots at 100 mL, and sterilize by autoclaving).
17. Acrylamide gel-loading buffer (same as agarose-loading buffer).
18. Denaturing solution: 87.75 g NaCl, 20 g NaOH, ddH$_2$O to 1 L.
19. Neutralization solution: 87.75 g NaCl, 6.7 g Tris base, 70.2 g Tris-HCl, ddH$_2$O to 1 L.
20. Hybridization solution (5X SSC), 5X Denhardt's solution, 25% formamide, 1% sodium dodecyl sulfate (SDS); 10% dextran sulfate.
21. 50X Denhardt's stock solution (for 500 mL): 5 g ficoll, 5 g polyvinylpyrrolidine, 5 g bovine serum albumin (BSA) (fraction V), ddH$_2$O to 500 mL. Warm to 50°C until dissolved.
22. TM buffer (10X): 12.1 g Tris base (100 mM final), 20.3 g MgCl$_2$ (100 mM final), ddH$_2$O to 1 L, autoclave.

23. Whatman DE 52 resin: Suspend resin in equal volume of 1 M Tris-HCl, pH 7.6, to swell. Using Whatman #1 filter paper and funnel, remove Tris and wash resin with TM buffer. Transfer resin to bottle, and add TM buffer to 3X vol. Mix and resuspend before using.
24. Depurination solution : 0.2 N HCl or 6.5 mL of 12.1 N HCl in 1 L ddH$_2$O.
25. Geneclean kit (Bio101, Vista, CA).
26. Solution 1: 50 mM glucose, 10 mM EDTA, 25 mM Tris-HCl, pH 8.0.
27. Solution 2: 0.2 N NaOH, 1% SDS.
28. Solution 3: 60 mL of 5 M potassium acetate, 11.5 mL glacial acetic acid, 28.5 mL ddH$_2$O.
29. Sequenase sequencing kit (Amersham).
30. 40% Acrylamide stock solution: 380 g acrylamide, 20 g N,N'-methylene *bis*-acrylamide, ddH$_2$O to 1 L. Filter.
31. 6% Sequencing gel stock solution: 150 mL of acrylamide stock solution, 500 g urea, 100 mL of 10X TBE buffer, ddH$_2$O to 1 L. Warm to dissolve. Filter through grade 361 qualitative filter paper. Store protected from light at room temperature.
32. T$_4$ DNA ligase, T$_4$ DNA kinase, calf intestinal phosphatase (IP), restriction enzymes (Amersham).
33. RNase A (Boehringer Mannheim): Make stock solution in ddH$_2$O; boil for 20 min, let slowly cool at room temperature, and aliquot.
34. 20% Polyethylene glycol (PEG)/2 M NaCl: Weigh out components in ddH$_2$O. Autoclave to dissolve.

It should be noted that all work described below must be carried out in sterile vessels using sterile solutions and media.

3. Methods

3.1. Probe Isolation

Probes can be generated via PCR using primers based on conserved sequences in the transmembrane domains of the receptor. If one is looking for unknown subtypes of a published receptor cDNA, transmembrane domains three, five, and six are generally the most highly conscrved in adrenergic receptors, because they contain key amino acids in the binding of agonist. A detailed methodology for isolation of probes is described Chapter 4 of this volume. Theoretically, one can also develop a probe from a known receptor subtype, and use the probe under low stringency screening in the solution-phase hybridization step. However, you would never be certain of the identity of the amplified phage, until the clone is eventually isolated and sequenced. The same argument holds for the use of oligonucleotides, which will generate a greater number of false positives owing to the even lower specificity. It is recommended that a specific probe of at least 200 bp be used to allow high stringency screening and high specific activity labeling with ^{32}P-dCTP. The longer the probe, the greater the specificity and likelihood of identifying the correct clone.

3.2. Testing for the Presence of the cDNA in the Library

The library in question (*see* **Note 1**) needs to be screened to assure that the probe recognizes a cDNA that is present in the library. This can be accomplished either by Southern blot or by PCR as described below (*see* **Note 2**).

3.2.1. Southern Blot of Library

1. Plating cells are grown by picking an individual colony from a plate and growing in sterile NZCYM media containing 0.2% maltose and 10 m*M* MgSO$_4$. Like all bacterial cultures, there should be good aeration for healthy growth with the ratio of media to container volume never exceeding 1:4. Prepare about 20–50 mL (*see* **Note 3**).
2. Take 5 µL of the library and add to 5 × 10^8 of the appropriate plating cells or the amount of cells that yields the correct multiplicity of infection (MOI). *See* **Subheading 3.3.1.** for the concept of MOI.
3. Grow culture in shaking incubator at 37°C overnight until cells lyse.
4. Add 200 µL of chloroform to lyse remaining cells. Mix. Centrifuge (low speed in tabletop centrifuge, such as T6000, *see* **Note 10**) to remove debris. Remove and retain supernatant.
5. Pellet phage by adding an equal volume of ice-cold 20% PEG/2 *M* NaCl. Mix thoroughly and incubate in ice bucket for 1–2 h (*see* **Note 4**).
6. Centrifuge at least 20 min at 14,000*g*. Decant liquid off carefully; phage pellet is soft. Store tube upside down for a few minutes to drain off excess PEG. Wipe tube of excess moisture with Kimwipe.
7. Resuspend pellet in 300 µL of sterile TE buffer (10 m*M* Tris-HCl, 1 m*M* EDTA, pH 8.0). Transfer to small Eppendorf microfuge tube. Lyse phage by adding equal volume of equilibrated phenol (*see* **Note 5**). Mix gently, but thoroughly. Microfuge for 2 min to separate phases. A large white layer should be apparent, which represents denatured protein. Remove and retain top phase without disturbing the white protein interface. Repeat phenol extraction until no white phase is observed. Add equal volume of chloroform to removed top phase. Mix and centrifuge for 2 min. This step removes the excess phenol. Remove and retain aqueous top phase.
8. To top phase, add 1/10 vol of sterile 3 *M* sodium acetate, and mix. Then add 2.5 vol of 95–100% ice-cold ethanol. Mix gently. Sometimes the DNA is readily visible at this point, forming a cotton ball-like precipitate. Incubate at –20°C for 20 min. Centrifuge in microfuge for 10 min at 14,000*g*. Decant liquid. DNA is white pellet. Let dry thoroughly by inverting tube.
9. Resuspend DNA pellet in 20 µL of TE. Restriction-digest all of the DNA with the appropriate enzyme to release the insert (*see* **Note 6**). λgt10 uses *Eco*RI. To 20 µL of DNA, add 3 µL of 10X restriction enzyme buffer (supplied by

manufacturer), 3 µL of *Eco*RI (high concentration), and sterile water to 30 µL total. Incubate at 37°C for 4 h.

10. Electrophorese digested DNA in a 1.2% agarose gel. Weigh out 1.2 g of agarose (low EEO) and place into 100 mL of 1X TAE buffer. Microwave or heat sample until all of the gel fragments have solubilized. Add 5 µL of an ethidium bromide solution (10 mg/mL in water) for visualization of the DNA, mix, and pour into a casting tray with comb. Let solidify about 1 h, then remove comb. Place gel in running chamber and submerge gel with 1X TAE buffer. Add loading buffer to each sample. Run a mol-wt marker and all of digested DNA sample. Apply positive current to end of box where DNA (negatively charged) will migrate.

11. Conduct electrophoresis at an amperage that will not overheat the gel. Electrophorese gel until the bromophenol blue in the loading buffer has migrated to the end of the gel. Localize DNA by placing gel on a UV transilluminator and photograph gel using a Polaroid camera for documentation. A fluorescent ruler can be used to size bands identified in Southern analysis to DNA markers in ethidium-stained gel.

12. Transfer DNA in gel onto nitrocellulose via the Southern technique.

3.2.1.1. SOUTHERN TECHNIQUE

1. Place gel in acid depurination bath (0.2 *N* HCl) with gentle rocking for 10–15 min or until the bromophenol blue tracking dye has turned from blue to yellow.

2. Decant acid solution and rinse gel three times in ddH$_2$O. Bathe gel in denaturation solution with gentle rocking for 15 min.

3. Decant denaturation solution, and repeat denaturation.

4. Decant denaturation solution. Bromophenol blue tracking dye should be blue again.

5. Bathe gel in neutralization solution with gentle rocking for 30 min.

6. During neutralization, prepare transfer stack:
   a. Cut a piece of nitrocellulose filter paper to be roughly equal to the dimensions of the gel. Wet filter in a bath of ddH$_2$O for 1 min, and place in a 2X SSC bath until ready to use.
   b. Cut 8 pieces of Whatman 3MM paper just smaller than the dimensions of the nitrocellulose.
   c. Cut a stack of paper towels (4–6 cm high) just smaller than the dimensions of the Whatman 3MM papers prepared in the previous step.
   d. To prepare a wick, cut 3 or 4 strips of Whatman 3MM paper that are just wider than the gel and up to 18 or 20 in. long.
   e. Prewet wick in 10X SSC, and construct a platform bridge in a large baking dish that will permit the wick to hang into the dish filled with 10X SSC. A large agarose-gel casting tray or a plate of glass can be used to form a bridge.

7. Assemble the stack.
   a. Place the gel face down on the center of the wick, removing all air bubbles between gel and wick.

b. Set nitrocellulose squarely on gel. Once filter touches gel, do not lift or move since DNA transfer has likely already begun. Again, be sure there are no air bubbles.

c. Prewet one of the eight pieces of Whatman 3MM in 10X SSC, and set it squarely on the nitrocellulose. Avoid short circuits by not allowing 3MM paper to overlap nitrocellulose and touch gel. Remove any bubbles.

d. Stack the remaining Whatman paper and paper towels, and weight stack with 300–500 g to compress sandwich lightly. Again, avoid short circuits; paper towels must not overlap onto Whatman 3MM or nitrocellulose. Allow transfer to proceed overnight.

8. On the next morning, disassemble stack down to nitrocellulose paper.

9. Lift dehydrated gel with adhered nitrocellulose, and place on a piece of Whatman 3MM paper with the nitrocellulose side down. Carefully remove gel and discard. If desired, mark lane locations and label blot with soft pencil or ballpoint pen.

10. Bathe nitrocellulose in 5X SSC for 5 min.

11. Air-dry blot at room temperature (~15 min). Crosslink DNA to filter using automatic UV crosslinker, DNA side up to UV lights.

12. Further dry blot at 80°C for 30 min to 1 h.

13. Store blot until needed at room temperature. Keep blot dry and free from dust by sandwiching it between sheets of Whatman 3MM paper or between paper towels.

3.2.1.2. PREHYBRIDIZATION, HYBRIDIZATION, WASH

Label DNA probe, hybridize, and wash according to **Subheading 3.4.3.**

3.2.2. PCR Screening of Library

1. Dilute 10 µL of library stock and into 200 µL of sterile water. Heat to 95°C for 10 min, and snap-cool on ice. This will disrupt phage heads and release the DNA. This is the DNA template stock for the PCR reaction.

2. Design a "sense" and "antisense" primer from the double-stranded DNA sequence of your probe. If you designed the correct PCR primers in the proper orientation, the 3′ ends of each primer will be pointed toward each other when annealed to their respective regions in the double-stranded probe. The primers should be about 20–25 bases long and generate a PCR product of at least 200 bp.

3. Set up the PCR reaction as follows: 1 µL of DNA template, 200 pmol each primer, 10 µL of 10X PCR buffer (supplied by manufacturer), dNTPs at 1.25 mM each final, 0.5 µL of *Taq* polymerase, and sterile water to 100 µL final volume. Overlay with 50 µL of mineral oil. In addition, set up a control PCR that contains everything except the DNA template (*see* **Note 7**).

4. PCR conditions (*see* **Note 8**): One cycle: Denature at 95°C for 5 min, anneal at 45°C for 5 min, extend at 72°C for 45 min, then PCR regular cycle for 35 cycles at 94°C for 1 min, 45°C for 2 min, and 72°C for 3 min. Electrophorese 20 µL of product in a 1.2% agarose gel. A band of the predicted size determined by the primers should be apparent (*see* **Note 9**).

3.3. Solution-Phase Screening Protocol

3.3.1. Testing for Correct MOI

One of the most difficult steps in the solution-phase screening procedure is to find the correct MOI. Phage has a generally narrow MOI range in which lysis will occur and generate sufficient phage heads for subsequent DNA preparation. A critical point is that the bacterial cells need to be in an appropriate phase of growth for phage infection to occur. When the bacterial cells stop growing and dividing, the phage infection will also stop. You must be within a fivefold dilution of the optimal MOI to obtain sufficient lysate for phage isolation. Too high of an MOI will result in all of the cells being immediately infected and lysed. In this case, the amplification process will stop prematurely and limit subsequent rounds of infection. Too small an MOI will result in limited infection and lysis, resulting in an early saturation of the bacterial culture and cessation of bacterial growth. The testing for the correct MOI is empirical and depends on the total volume used as well as the particular phage type.

1. Determine the volume of media used for phage infection (i.e., 2 mL). To a series of sterile culture tubes, add prepared plating cells (a culture of plating cells provides 1×10^8 cells/mL at 1 OD_{600NM}) to various amounts of phage stock (i.e., library titer is typically around 10^6PFU/μL) starting with a ratio of 100 cells: 1 PFU. Add additional NZCYM media to achieve the 2 mL total volume. It is recommended to set up a series of tubes with half-log changes in ratio above and below the starting 100:1 ratio. Use a large enough tube to achieve sufficient aeration (1:4–5).
2. Incubate with shaking at 37°C overnight. Add one drop of chloroform to each tube and prepare miniprep phage DNA according to **Subheading 3.2.1.**, **steps 4–8**. Measure phage content by analysis on a 1% agarose gel or OD_{260NM} (1 OD_{260NM} = 50 μg double-stranded DNA) and select the MOI ratio that produces the highest yield of DNA for experiments.

3.3.2. Amplification of Library Aliquots

1. Into nine separate sterile culture tubes, add 1-μL aliquots of the library into plating cells used at the correct MOI in a total volume of 2 mL (*see* **Subheading 3.3.1.**).
2. After overnight incubation with shaking at 37°C, one drop of chloroform is added and mixed to lyse any remaining cells. The cultures are centrifuged at a low speed (tabletop centrifuge, such as T6000) to pellet cellular debris (*see* **Note 10**), and the supernatant fraction (phage stock) is retained for future analysis.

3.3.3. λ DNA Minipreparation

1. 1 mL of each supernatant (the rest is retained for subsequent amplification) is used to prepare λ DNA, by adding 27 U of RNase-free DNase 1 with incubation for 20 min at 37°C. This step destroys any bacterial DNA that results from the

phage-induced cell lysis. Three hundred microliters of ice-cold 20%PEG/2 *M* NaCl solution are added to each supernatant, mixed, and incubated on ice for 1 h.

2. Centrifuge in the microfuge at full speed for 20 min. A pinhead sized pellet should be visible at this point (*see* **Note 4**). The pellet is resuspended in 300 μL of TE and DNA is extracted with phenol/chloroform several times as noted in **step 7** of **Subheading 3.2.1.**

3. Sodium acetate is added to 0.3 *M* final concentration, and then 2.5 vol of 100% ice-cold ethanol are added, mixed, and incubated at –20°C for 20 min to precipitate the DNA. The sample is centrifuged in the microfuge at full speed for 10 min, and the resulting DNA pellet dried.

4. Resuspend DNA in 17 μL of TE, and digest all of the resulting DNA for 1–2 h with the appropriate restriction enzyme to release the inserts (*see* **Note 6**). *See* **Subheading 3.2.1.**, **step 9** for digestion protocol.

3.3.4. Labeling of Probe and Solution-Phase Hybridization

1. Label probe according to manufacturer's directions in the random-primed labeling kit (Boehringer Manneheim) using α^{32}P-dCTP (specific activity at least 3000 Ci/mmol). Twenty-five to 50 ng of probe DNA should generate at least 1×10^8 total cpm of purified probe. Probe should be purified from unincorporated ^{32}P-dCTP with the use of a spin-column made from Sephadex G-50. Care is taken to handle and dispose of radioactive materials properly.

2. A spin column is prepared by using a 1 mL syringe, plugging the bottom with glass wool, and adding Sephadex G-50 that is preswollen in sterile TE. Lightly pack the column by placing syringe in a 15-mL polypropylene centrifuge tube to catch eluted liquid. Centrifuge in a tabletop centrifuge at low speed (1*g*) for 1 min. Decant off spin-off liquid. Pack column to a final volume of 0.9 mL. The probe is then added to the top of the prepacked column, washed with 100 μL of 0.1 *M* EDTA, pH 8.0, and centrifuged at 3*g* for 2 min. The spin-off liquid now contains the purified probe; unincorporated nucleotides are held up in the Sephadex beads.

3. 1×10^6 cpm of the labeled probe is added directly to the digestion mixture, and the salt concentration is adjusted to 3X SSC from a concentrated stock (20X SSC). The mixture is boiled for 5 min and transferred immediately to a 55°C water bath for 15 min. Loading buffer is added.

3.3.5. Gel Electrophoresis and Autoradiography

1. The above samples are analyzed on a 4% nondenaturing polyacrylamide minigel (*see* **Note 11**). The entire sample is loaded directly. To determine molecular weights visually, a lane containing ^{32}P-end-labeled DNA markers can be used (*see* **Note 12**). This is prepared by taking 1 μL of DNA marker (i.e., 1-kb ladder usually at 1 μg/μL), adding 1 μL of 10X kinase buffer (supplied by manufacturer), 5 μL of γ^{32}P-ATP (specific activity at least 3000 Ci/mmol), 2 μL of sterile water, and 1 μL of T_4 kinase. Incubate for 40 min at 37°C. Separate marker from unincorporated labeled nucleotides in a 1-mL spin-column of Sephadex G-50.

The amount of marker to electrophorese is determined empirically to limit film overexposure. Only a small amount of the stock is required.

2. A 4% polyacrylamide gel is prepared by adding 13 mL of 30% acrylamide stock (29% acrylamide and 1% *N,N'*-methylene *bis*-acrylamide), 74.9 mL of water, 2.1 mL of 3% ammonium persulfate, and 10 mL of 10X TBE. Add 30 μL of (*N',N',N',N'*-tetramethylethylene diamine (TEMED), swirl, and pour between glass plates before polymerization. Add comb. There is no need to pour a stacker gel. After polymerization, electrophorese gel (8 V/cm) in 1X TBE. Prerun gel for 15 min before loading sample.

3. After electrophoresis, the gel is separated from the glass plates and placed in Saran wrap. The entire gel is placed inside a film cassette with a piece of Kodax X-AR film or its equivalent. The gel does not need to be dried, since it is thin enough for placement within the cassette. Exposure time varies depending on the specific activity of the probe, but usually takes about 4 h at room temperature.

4. At this point, a band is usually observed at a molecular weight corresponding to the size of the cDNA insert; it may be full-length or it may represent a partial clone (*see* **Note 13**). Multiple bands may appear depending on the different sizes of cDNAs represented in the library. Since the cDNA is not abundant, it is likely that similar bands will not be present in all aliquots. At this point, take the aliquot that provides the most intense bands and longest insert and proceed to **Subheading 3.3.6.** for reamplification.

3.3.6. Subdividing Aliquots and Reamplification

1. Once the positive aliquot is identified, the previously saved portion of the phage stock is realiquoted into nine 1 μL samples and plating cells added at the correct MOI as before. **Subheading 3.3.2.** and onward are performed again.

2. Repeat **Subheading 3.3.2.** on the positive aliquot until all realiquoted samples show an intense positive band(s). A solution-phase screening should show similarity to **Fig. 1**, which has undergone two rounds of subdividing and reamplification. The most intense aliquot is then plated and screened by the traditional plaque hybridization technique.

3.4. Traditional Plaque Screening of Positive Aliquot

The traditional approach for screening a genomic or cDNA library involves use of a specific probe to identify DNA transferred to nitrocellulose paper from individual phage-generated plaques. Phage DNA from multiple plaques, which is immobilized onto nitrocellulose filters, is probed much like a Southern blot, and positive plaques are isolated and purified via several rounds of screening. Described below are the experimental procedures required to screen a library.

3.4.1. Plating Procedure

Host bacteria are prepared. The library of interest is titered to determine the appropriate phage-to-bacteria ratio for the generation of dense lawn of plaques

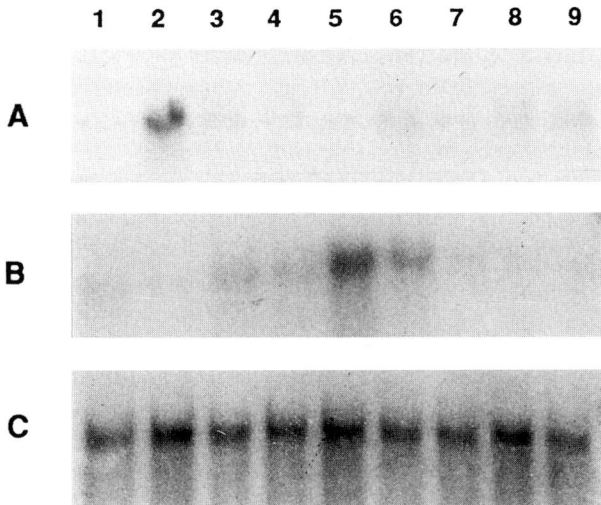

Fig. 1. Solution-phase library screening. Row **A**: One positive culture was found on screening nine 1-µL aliquots of a cDNA library. The band was 2–3 kb in length. Row **B**: Nine 1-µL aliquots of sample 2 in row A were reamplified and rescreened. Two positive cultures were identified. Row **C**: Nine 1-µL aliquots of sample 5 in row B were reamplified and rescreened. All cultures are now positive. Culture 5 stock in row C was screened by traditional plaque hybridization and permitted a cDNA to be isolated.

on 150-mm plates. The goal of the titering experiment is to determine the optimum conditions necessary for generating as many distinct plaques as possible on 150-mm plates. Fixed aliquots of bacterial cells are infected with varying amounts of phage, and a top layer of agarose containing the infected bacteria is poured onto the 150-mm agarose plates. The dilution of phage stock that forms the greatest number of distinct plaques is used to generate many plates (6–20) of crowded plaques for transfer to nitrocellulose.

1. Preparation of host bacteria: Streak an NZCYM plate with the appropriate bacterial strain for the library, and incubate overnight in a 37°C incubator. Using a sterile toothpick, pick a colony from the plate and inoculate a 25-mL vol of sterile NZCYM containing 0.2% maltose and 10 mM MgSO$_4$. Provide maximum aeration by using a 100-mL flask. Grow to log phase in 37°C shaking incubator (4–6 h). When log phase is reached, transfer culture to a 50-mL sterile conical tube and centrifuge at 2000g for 10 min. Resuspend bacterial pellet in 12.5 mL of 10 mM MgSO$_4$. These plating cells can be stored at 4°C for up to 1 wk.
2. Titering of phage stock: Prepare top agarose, which is usually premade and stored in bottles, by melting with an autoclave or microwave oven. During melting, be sure that the cap on the bottle is loose to prevent pressure build-up and explosion.

If using a microwave, frequently check to prevent boil-over. When top agarose is completely melted, place bottle in a 50°C water bath, and allow to equilibrate to 50°C. Temperatures above 60°C will kill the host bacteria.

3. Prepare serial dilutions of phage stock (1:10 down to 1:100,000, if necessary) in TM buffer. Label each dilution to avoid confusion.
4. Infect separate 200-μL aliquots of plating cells (at the bottom of labeled 15-mL sterile conical tubes) with 1–10 μL of each phage dilution. Incubate tubes first at room temperature for 10 min and then at 37°C for an additional 10 min.
5. Label fresh NZCYM plates for each dilution. (To improve spreading of top agarose, it helps to prewarm the plates to 37°C.) Add 9 mL of melted top agarose to a single tube containing the cells, and phage, cap, and invert tube several times to mix. Pour top agarose mixture onto appropriate plate, and rock plate to distribute solution evenly. A uniform layer of top agarose is essential to maximize transfer during the blotting step (**Subheading 3.4.2.**). Mixing and pouring must be performed quickly before agarose hardens (about 10–20 s).
6. Incubate plates inverted at 37°C for 6–16 h (or overnight) or until individual plaques are clearly visible. For future reference, identify the phage dilution that generates the densest lawn of individual plaques (*see* **Note 14**).

3.4.2. Nitrocellulose Lifts

Once the optimum titer is determined for the generation of a densely populated, but not confluent lawn of plaques, multiple plates are poured for the purpose of screening the library. Depending on the type of library (cDNA vs genomic vs cosmid), 300,000–1,000,000 plaques are typically screened. This would require 6–20 (150-mm) plates, each of which can accommodate 20,000–50,000 individual plaques. Once the lawn is grown, phage DNA is transferred from the plaques to nitrocellulose filters by simply blotting filters to the surface of the top agarose. To prevent damage to the top agarose during the lifting procedure, chill plates for at least 1 h at 4°C before attempting lifts. After transfer, DNA fixed to the blots is denatured using basic solutions. Blots are then neutralized, air-dried, and DNA is crosslinked to the matrix either by exposure to UV light or by baking at 80°C.

1. Label two new filters (commercially precut to exactly fit 150-mm round dishes) for each plate using a soft lead pencil. Hybond-C Extra filters from Amersham are recommended. At room temperature, place a filter onto the top agarose of precooled plates being careful that the filter fits exactly into the dish. This is accomplished most easily by holding the round filter lightly on two opposite edges, and letting the sagging center touch the center of the plate. Allow the wetting action to attract the edges of the filter onto the plate. Be sure to handle the filter with gloves to prevent smearing of oil and protein from your fingers. Avoid trapping air bubbles under the filters, and do not attempt to lift and reset a filter once it has touched the agarose. Air bubbles can sometimes be pushed out

if trapped under filter. Mark the orientation by puncturing the filter and the underlying agarose in three or more asymmetrical locations with an 18-gage needle. Mark the location of these puncture marks on the outside of the plastic plate with a felt-tipped marker. Allow the filter to rest on the agarose for 1 min.

2. Using blunt forceps, carefully peel the filter away from the agarose, and immerse it (plaque-side up) in a tray of denaturing solution. After 2 min, transfer filter to a tray of neutralization solution.

3. Place the second numbered filter onto the same plate for 3 min. As before, mark its orientation with respect to the dish by puncturing the filter and the underlying agarose in the same three asymmetrical locations with a needle as the first plate (you can see previous puncture marks under the filter).

4. As in **step 2**, peel the filter away from the agarose, immerse the filter in denaturing solution for 2 min, and transfer to neutralization solution. Filters may be stored in the neutralization solution until all lifts are complete.

5. After lifts are complete, transfer the filters to 5X SSC for 1–2 min, and place on Whatman 3MM paper to air-dry. Crosslink DNA to the filters using either a UV crosslinker (plaque-side up) or by baking in an 80°C oven (place filters between blotting paper) for 1–2 h. Filters should be stored dry at room temperature until hybridization.

3.4.3. Hybridization

One of the challenges encountered when performing hybridization experiments is the determination of the appropriate conditions to maximize and preserve binding of a probe to its desired target sequence. In addition to blocking nonspecific binding of the probe with fish sperm DNA (or some other appropriate blocking DNA), signal-to-noise is affected by the stringency of the hybridization and wash conditions. In general, medium- to low-stringency conditions are used during hybridization to ensure interaction of the probe with its target. Mis-matched interactions that develop between the probe and related, but nonhomologous sequences are eliminated by performing high-stringency washes. Stringency, which is affected by salt and formamide concentration and temperature, must be empirically tailored to fit each situation. Described below is a protocol that has been used successfully for the identification of adrenergic receptor clones using a 600-bp cDNA fragment as probe template.

1. Boil salmon sperm DNA (10 mg/mL stock solution, Boehringer Mannheim) and any other blocking DNA for 10 min. During that time, wet the dried filters in 5X SSC for at least 2 min. Place enough pre-warmed (55°C) hybridization solution to cover the bottom of a plastic container that is large enough to allow the blots to lie flat (Tupperware-like bowls with plastic lids are great). Prehybridization solution is added to cover filters sufficiently and to allow motion of the filters while shaking. Add salmon sperm DNA to a final concentration of 0.1 mg/mL.

Place all of the filters (same container) in this prehybridization solution and incubate at 55°C for >4 h on a rocker platform (platform in an oven works fine).

2. Prepare the probe using the commercially available Random Prime DNA Labeling Kit (Boehringer Mannheim). The steps listed below are optimized for the use of this kit.

a. Denature 25 ng template DNA by boiling for 10 min.

b. After boiling, place DNA on ice and add 3 μL of dATP + dGTP + dTTP mixture, 2 μL of reaction mixture, 1 μL Klenow enzyme (included with the kit), and 5 μL [α-^{32}P]-dCTP (6000 Ci/mmol, New England Nuclear). Incubate at 37°C for 30 min.

c. During incubation of the labeling reaction, prepare spin columns (**Subheading 3.3.4., step 2**) to separate probe from unincorporated [^{32}P]-dCTP.

d. When the labeling reaction is complete, add 100 μL of 0.5 M EDTA, pH 8.0, to reaction mix, and load entire volume onto the top of the spin column. Centrifuge at 3000 g for 2 min. The probe will be located in the eluate. Measure the specific activity by diluting 2 μL of the probe stock into 18 mL scintillation fluid and counting in a scintillation counter.

3. When prehybridization is complete, boil probe for 10 min (*see* **Note 15**), and add probe to the prehybridization mixture to achieve a final concentration of 1×10^6 cpm/mL. Hybridize overnight (>12 h) at 55°C with gentle rocking.

4. When hybridization is complete, recover and retain the hybridization solution. Wash filters at 55°C with gentle rocking using the protocol listed below (*see* **Note 16**). Between each wash, check filters with a survey meter; terminate washes when radioactivity detected from the filters drops to between 1000 and 2000 cpm. You can process many filters at one time.

a. 5X SSC + 0.1% SDS for 10 min.

b. 2X SSC + 0.1% SDS for 10 min.

c. 1X SSC + 0.1% SDS for 10 min.

d. 0.5X SSC + 0.1% SDS for 10 min.

e. 0.1X SSC + 0.1% SDS for 10 min.

f. If necessary, 0.01X SSC + 1% SDS for 10 min (or longer).

5. Mount filters on a piece of old exposed X-ray film that is covered with Saran wrap, cover mounted filters with plastic wrap, and expose to X-ray film using an intensifying screen (*see* **Note 17**). If necessary, mark the filters with radioactive or phosphorescent ink to assist in alignment of the film with individual filters following exposure. You can place about 6 filters/film, usually three sets of duplicates. Try to align the duplicate filters in the same orientation on the backing to make identification of positives easier.

3.4.4. Purification of Positives

After transferring the location of the filter punctures onto the developed autoradiographs with a felt-tipped pen, the actual corresponding plates can be correctly oriented directly over the film. When positioned correctly, positive spots on the autoradiograph correspond to individual positive plaques directly above in the plate. The process of identifying positive plaques is easier if the film and aligned plate are viewed over a light box. Since two filters were lifted from each plate, truly positive plaques will be represented on both filters.

Positive plaques are recovered from the plate by excising an agarose plug over the aligned positive, with the use of the large end of a glass Pasteur pipet or the large end of a 1-ml or 200-µL pipet tip. The plug is blown out by mouth into a microfuge tube containing 1 mL of TM buffer. Phage are eluted from recovered plugs by incubation for >2 h in TM buffer. Eluted phage is titered, replated, and a second round of screening is initiated for subsequent plaque purification. A titer should be chosen that permits a lower plaque density than the initial screen. If positives are not recovered, you may have missed picking the correct plaque. Lifts are probed using the recovered hybridization solution saved in step **Subheading 3.4.3., step 4** above. Usually there is no need to reprehybridized the filters. Just place the dried lifts directly in the used hybridization after reheating (*see* **Note 15**). After the secondary screen is complete, two additional rounds of plaque purification are performed until all plaques on the lift become positive. Once a phage has been purified to homogeneity from a single plaque, it can be amplified for large-scale preparation of λ DNA.

3.5. Analysis of Library Positives

After identifying positive plaques and eluting phage from single recovered plugs, large-scale preparation of phage can proceed to provide DNA for further analysis. There are two common methods used to amplify phage stock for large-scale preps: (1) the lysis of a log-phase bacterial suspension (liquid lysis), and (2) the lysis of bacteria poured as an overlay onto an agarose plate (plate lysis). Because liquid lysis provides a lower yield and because achieving complete lysis of a log-phase suspension of bacteria is typically difficult, the less technical plate lysis method is described below. In short, plate lysis involves mixing phage recovered from a single plaque with bacterial cells and top agarose. This mixture is poured onto an agarose plate to create a barely subconfluent distribution of plaques from which DNA can be harvested.

3.5.1. Large-Scale Preparation of λ DNA

1. Determine dilution of purified phage stock needed to obtain >90% lysis (but under 100%) of the plating cells according to **Subheading 3.4.1.** This dilution will be used for the large-scale preparation of phage.
2. Using the dilution identified to be optimal for producing complete plate lysis, infect fresh, plating cells and pour overlays onto two (or more if desired) NZCYM plates for each purified phage clone.
3. After lysis is complete, carefully scrape the overlay from each plate into a 50-mL conical tube (2 plates/tube), and add sufficient TM buffer to adjust the final volume to 30 mL.

4. Lyse remaining bacteria by adding 200 µL chloroform to each tube. Mix well by inversion, and incubate on ice for 1 h.

5. Centrifuge at 1000g for 10 min to pellet agarose and cell debris. Transfer the supernatants to fresh tubes. Add 1/2 vol of resuspended DE 52 resin and incubate at room temperature with mild shaking for 5–10 min (*see* **Note 19**). Centrifuge at 1000g for 10 min to pellet resin. Transfer supernatant to a 40-mL Sorvall centrifuge tube.

6. Precipitate phage by adding an equal volume of ice-cold 20% PEG/2 *M* NaCl. Mix well by inversion, and incubate on ice for 1 h.

7. Centrifuge at 14,000g for 30 min to pellet phage. Discard supernatant, and invert tubes to drain for 10 min. Wipe excess PEG/NaCl solution out of the tubes using a cotton-tipped swab. Resuspend pellets in 200–400 µL TM buffer. Transfer resuspended phage to microfuge tubes.

8. Break open phage by adding 1/2 vol phenol and 1/2 vol chloroform. Mix well by inversion and centrifuge at maximum speed (2 min) in a tabletop microfuge. Retain top phase. Repeat extraction procedure until white interface is no longer visible between the phases. Perform a final extraction with 1 vol of chloroform to remove all phenol.

9. Add sodium acetate to 0.3 *M* final concentration, and precipitate DNA by adding 2.5 vol of ice-cold 100% ethanol (ETOH). Gently invert tubes until white cotton-like precipitate forms. Immediately centrifuge at maximum speed in a microfuge for 10 s, and carefully remove and discard supernatant. Centrifuge again for 3 min to form a tight pellet, and remove remaining liquid. Wash pellet once with 70% ETOH, and allow pellets to air-dry for 30 min with tubes uncapped.

10. Add 100–300 µL of TE buffer to each pellet, and allow pellet to dissolve into solution overnight. Mix gently. Add more volume if necessary. Do not attempt to rush resuspension of the pellets by any mechanical means (e.g., vortexing, pipeting). High-mol-wt phage DNA will shear easily, and future cloning steps will be compromised.

11. If necessary, RNase-digest phage DNA to remove any extraneous RNA. Perform a second phenol/chloroform extraction and ETOH precipitation to recover purified DNA. As a measure of the quality of the phage DNA, a 1% agarose gel can be used to analyze the DNA. Intact phage DNA will electrophorese as a single, but fuzzy band at around 45 kb near the beginning of the DNA markers. There should be little or no smearing present in the lane; smearing may represent bacterial DNA contamination or degraded phage DNA.

3.5.2. Subcloning of Insert

The insert contained in the phage DNA needs to be sequenced to verify and identify the positive clone. Optional techniques include direct PCR cycle sequencing of the phage DNA itself or subjecting the insert DNA to PCR using primers from the two flanking phage arms (these can be purchased) and then subcloning the fragment into an appropriate vector. PCR can lead to mutations, and eventually the insert must be directly sequenced. Therefore, we recommend releasing the insert from the phage by restriction digestion and subcloning into a general vector for sequencing, such as pBluescript.

3.5.2.1. RESTRICTION DIGESTION OF PHAGE DNA

1. Set up the digestion reaction according to the restriction enzyme needed (usually *Eco*RI; *see* library specification sheets) to release the insert: 20 µL of the phage DNA in TE, 2.5 µL of the 10X digestion buffer (supplied by manufacturer of the enzyme) and 2.5 µL of the restriction enzyme (*Eco*RI). Incubate at 37°C (or required temperature) overnight in a heat block (16–18 h).
2. Electrophorese digested fragment in a 1% agarose gel. Weigh out 1.0 g of agarose (low EEO) and place into 100 mL of 1X TAE buffer. Microwave or heat sample until all of the gel fragments have solubilized. Add 5 µL of an ethidium bromide solution (10 mg/mL in water) for visualization of the DNA, mix, and pour into a casting tray with comb. Allow gel to solidify for about 1 h, and then remove comb. Place hardened gel in running chamber and barely cover with 1X TAE. Add loading buffer to each sample. Electrophorese each digestion sample with additional mol-wt marker.
3. Electrophorese gel at an amperage that will not overheat the gel. Conduct electrophoresis until the bromophenol blue in the loading buffer has migrated to the end of the gel. Localize DNA products by placing gel on a UV transilluminator, and photograph gel with Polaroid camera for documentation. Phage arms will usually be at about 11 and 9 kb, depending on the phage type. The insert will be lower in molecular weight, generally from 500 bp to 4 kb. If the insert contained an internal restriction site, there will be multiple DNA fragments. If the insert is larger than 500 bp, you can isolate the DNA insert by gel extraction using a Geneclean (BIO101) or similar product following the manufacturer's specifications. Resuspend recovered DNA insert in 20 µL of TE buffer.

3.5.2.2. PREPARATION OF VECTOR FOR SUBCLONING

1. Using 10 µg of vector stock, such as pBluescript, add 2.5 µL of the 10X digestion buffer (supplied by manufacturer of the enzyme) and 2.5 µL of the restriction enzyme (*Eco*RI). Incubate at 37°C (or required temperature) for 2 h.
2. Purify cut vector DNA by the Geneclean kit or alternatively by phenol/chloroform extraction as above (**Subheading 3.2.1., step 7**) followed by precipitation of the DNA (**Subheading 3.2.1., step 8**). Resuspend DNA in TE.
3. To the entire *Eco*RI-cut vector preparation, add 2.5 µL of calf intestine phosphatase (CIP) 10X buffer (supplied by manufacturer) and 2.5 µL of the enzyme CIP, and sterile water to 25 µL final volume. Incubate at 37°C for 2–4 h (*see* **Note 20**).
4. Purify the *Eco*RI/CIP-treated vector as in **step 2**. Resuspend the DNA in 30 µL of TE. This is your cut-vector stock.

3.5.2.3. Ligation Reaction

For information regarding ligation theory, *see* the Maniatis cloning manual *(3)*. The basic recipe is to use vector and insert DNA at a molar (not μg) ratio of 1:3. Therefore, there are more moles of the insert than vector, allowing a higher efficiency of ligation. You usually start with 0.1 pmol of vector or approx 0.2 μg of the *Eco*RI/CIP-treated pBluescript, but this is usually established empirically depending on the efficiency of vector digestion and CIP dephosphorylation (*see* **Note 21**).

3.5.2.3.1. Ligation of PCR Fragments into pBluescript

1. Establish a ligation reaction for each isolated DNA insert band. In addition, conduct a control reaction that contains everything except insert DNA to determine the background colonies after transformation.
2. To 1 μL of cut-vector stock (or amount that provides minimal background; *see* **Note 21**), add 5–10 μL of *Eco*RI-digested DNA insert (to approximate 1:3 molar ratio) (*see* **Note 22**), 2 μL of 10X ligation buffer (supplied with enzyme), 2 μL of T_4 DNA ligase, and sterile water to 20 μL final volume.
3. Mix and incubate overnight at 14°C (obtained by placing a water bath in a cold room) (*see* **Note 23**).

3.5.2.4. Transformation into Bacteria

1. For simplicity, you can purchase premade competent cells from Gibco-BRL. We recommend the DH5α strain. For each ligation, use 100 μL of competent cells. Thaw a vial on ice, and then aliquot 100 μL into prechilled 15-mL sterile polypropylene centrifuge tubes on ice for each separate ligation reaction. Retain cap to maintain sterility.
2. Add no more than 10 μL of the ligation mixture to the centrifuge tubes. To use the entire mixture, perform multiple transformations for each ligation mixture. Mix slowly by moving the pipet tip back and forth. Do not vortex or triturate. Incubate on ice for 30 min (*see* **Note 24**).
3. Heat-shock the transformation mixture by placing in a 42°C water bath for 1 min (timed). Do not use a heat block, since heat transfer is compromised. Return tubes to ice for another 2 min.
4. Add 2 mL of sterile NZCYM media (no antibiotic). Incubate with shaking for 1 h at 37°C to allow the cells to develop antibiotic resistance.
5. Concentrate cells by centrifuging tubes for 10 min at top speed in a table top centrifuge. Pipet and discard most of the supernatant except for 200 μL. Resuspend the cell pellet in the 200 μL media, and plate and spread on 100-mm ampicillin plates until moisture is absorbed. Invert plates and place in 37°C incubator overnight (*see* **Note 25** if using pBluescript).
6. Colonies should be apparent and greater in number than control plates (2–3×). If using the blue/white selection, recombinants will appear pure white. If not using

pBluescript or other vector that discriminates recombinants, the plasmids should be screened for insertion of DNA by restriction digestion before sequencing.

3.5.3. Sequencing

3.5.3.1. PLASMID PREPS FOR DOUBLE-STRANDED DNA SEQUENCING

1. Pick individual colonies with a sterile toothpick. Touch the toothpick onto an NZCYM agarose master plate (gridded and labeled for cataloging of numerous colonies), and drop the toothpick into a culture tube containing 2.5 mL sterile NZCYM. The NZCYM should contain 100 mg/mL ampicillin.
2. Shake tubes vigorously in a 37°C incubator overnight. Also, incubate the master plate in a 37°C incubator overnight.
3. The next morning, centrifuge culture tubes at 2000g for 10 min to pellet bacteria. Also, store the master plate at 4°C for later reference.
4. Aspirate medium, and resuspend bacterial cell pellets in 100 μL of solution 1. If you wish, transfer resuspended pellets to 1.5-mL Eppendorf tubes.
5. Incubate at room temperature for 5 min, and then add 200 μL solution 2. Mix by inversion—do not vortex.
6. Incubate on ice for 5 min, and then add 300 μL of ice-cold solution 3. Vortex gently for 10 s.
7. Incubate on ice for 5 min, and centrifuge for 5 min at maximum speed in a microfuge. Transfer supernatants to fresh Eppendorf tubes.
8. Phenol/chloroform-extract by adding 100 μL each of equilibrated phenol and chloroform. Vortex samples to mix completely, and centrifuge for 3 min at maximum speed in a microfuge. Transfer aqueous phase (upper phase) to fresh tubes.
9. Add 2 vol of 100% ETOH (equilibrated to room temperature) to recovered aqueous phase, vortex, incubate at room temperature for 10 min, and centrifuge at maximum speed for 10 min in a microfuge.
10. Decant supernatants, and wash pellets with 70% ETOH (200 μL). Drain all fluid from tubes, and air-dry DNA pellets for 15–30 min at room temperature. Resuspend pellets in 40 μL TE buffer.

DNA prepared using this alkaline lysis method is of sufficient quality for restriction analysis. However, for use in sequencing reactions, it is suggested that the DNA be further purified. Two common methods used for further purification are outlined below.

3.5.3.1.1. RNase A Method

1. Add 1 μL of RNase A (10 mg/mL stock) to each sample. Incubate at 37°C for 30 min.
2. Repeat **steps 8–10** of **Subheading 3.5.3.1.** to obtain DNA.

3.5.3.1.2. PEG Method

1. Add 50 μL of an ice-cold solution of 20% polyethylene glycol (PEG) + 2 M NaCl. Incubate on ice for 1 h or at 4°C overnight.

2. Centrifuge at maximum speed for 15 min in a microfuge. Decant.
3. Wash pellets with 70% ETOH (200 μL). Prior to washing, pellets may not be visible; the 70% ETOH wash may cause pellets to become visible.
4. Centrifuge at maximum speed in a microfuge for 10 min. Drain all fluid, let air-dry, and resuspend in 35 μL TE.

3.5.3.2. THE SEQUENCING REACTION

The sequencing reaction may be conducted using any one of a number of commercially available kits, but the protocol described below is optimized for the Sequenase version 2.0 DNA sequencing kit (Amersham). This kit, which employs the chain-termination method, includes a detailed instruction manual and a valuable troubleshooting guide. Use about 1–2 μg of the DNA for sequencing. The yield from a simple miniprep is sufficient for two sequencing reactions. Calculate the concentration of DNA before conducting the reaction (*see* **Note 22**).

1. Denature double-stranded template DNA. To 15-μL aliquots (1–2 μg) of stock plasmid DNA (prepared above), add 2 μL of 4 mM EDTA (pH 8.0) and 4 μL of fresh 1 N NaOH. Incubate at room temperature for 10 min. Place a 2-μL drop of 2 M ammonium acetate on the wall of each tube, and wash with 100 μL ice-cold 100% ETOH. Mix by inversion and incubate in a –20°C freezer for 10 min. Centrifuge tubes at maximum speed for 10 min in a microfuge, and dry pellet in a Speed-Vac for 10 min.
2. Anneal primer to single-stranded template DNA (if using pBluescript, use T7 or T3 primer). To the single-stranded DNA samples prepared in **step 1** above, add a cocktail containing 7 μL sterile H$_2$O, 2 μL of the 5X reaction buffer supplied with the kit, and 1 μL of a 1 pmol/μL stock of primer. Anneal primer to template by heating to 37°C for 10 min.
3. Elongation reaction: Label four tubes per clone and/or per primer with A, C, T, or G (termination tubes). Deposit 3.5 μL of stock dideoxy (dd) ATP, ddCTP, ddTTP, and ddGTP into appropriate tubes. Dideoxy nucleotide stock solutions are supplied with the kit. To annealed DNA, add a cocktail containing 1 μL of 0.1 M DTT, 2 μL of diluted labeling mix (supplied with kit), and 0.5 μL [^{35}S]deoxy ATP (1 μCi/μL, Amersham). Dilute polymerase 1:8 in polymerase dilution buffer (both supplied with kit), and add 2 μL of diluted enzyme to reaction mixture. Incubate at room temperature for 2 min, and immediately transfer 3.5 μL of reaction mixture to each termination tube. Incubate at room temperature for 10 min, add 4 μL stop solution, and store in a –20°C freezer until needed.

3.5.3.3. PREPARATION AND ELECTROPHORESIS OF SEQUENCING GEL

A number of sequencing apparatuses are commercially available. The method described below for pouring and conducting electrophoresis of a sequencing gel is optimized for the Bio-Rad Sequi-Gen sequencing apparatus, which includes a detailed set of instructions.

1. Assembly of gel: Thoroughly wash plates with glass cleaner and then with 100% ETOH. To ease removal of the outside plate electrophoresis, coat the gel-facing side of this plate with a solution of 2% silane (in chloroform) (**toxic**; **wear gloves**). Assemble the gel plate sandwich, and pour a plug consisting of 50 mL of a 6% sequencing gel stock solution (acrylamide/urea solution), 200 μL of a 20% ammonium persulfate stock solution, and 200 μL TEMED. Set the bottom edge of the assembled plates into the plug, and allow to polymerize for 45–60 min. Next, pour the gel solution, which consists of 100 mL of a 6% sequencing gel stock solution, 100 μL of a 20% ammonium persulfate stock solution (in H_2O), and 40 μL TEMED. Set comb to make wells, and allow gel to polymerize > 2 h (preferably overnight).

2. Loading and electrophoresis of gel: Assemble gel to running chamber and add 1X TBE buffer. Preheat gel to 50°C (temperature of the glass plate) by applying a current of 100 mA for about 30 min. Gel electrophoresis at high temperature (>60°C) may crack glass. Bio-Rad sells glass-attachable temperature gauges for sequencing. Heat sequencing reaction samples to 100°C for 3 min to ensure DNA is denatured, and then immediately place the samples on ice. Set comb. Wash (blow) out lanes just before loading since urea leaks out and disturbs uniform loading. Load 4–7 μL/well. Conduct electrophoresis at a constant current of 75 mA (or to maintain 50°C temperature) for desired length of time based on primer used.

3. Autoradiography: After electrophoresis is complete, disassemble gel plates by removing the outside plate. Gently rinse gel, which is still adhered to the inner gel plate, with 2 L of gel wash solution (10% MeOH, 10% HOAc) to remove urea. Rinse one more time with 1 L H_2O. Transfer (peel) gel to thick chromatography paper, and cover the exposed side of the blotted gel with plastic wrap. Trim edges to fit and dry in a gel dryer for 1 h. Remove plastic wrap, and apply to X-ray film ensuring that the dried gel is touching the film. Expose the film for 1–3 d.

4. Analyze sequence by computer programs and blast searches as outlined in **Subheading 3.6.** in Chapter 4 of this volume.

4. Notes

1. Which library to use is determined by the tissue distribution of the receptor clone in question. After probe isolation of a putative receptor subtype, conduct Northern analysis, RNase protection assay, or reverse transcriptase-polymerase chain reaction (RT-PCR) to determine tissue expressing the highest amount of target RNA. A cDNA library is then purchased or prepared from this corresponding tissue to increase the likelihood of isolating the entire coding region. cDNA libraries are a first choice, since they will express only the coding region, and the

inserts are generally small (1–2 kb) to allow faster subcloning and sequencing. Genomic libraries also contain introns possessed by some adrenergic receptor genes *(2)* and reported to be as large as 20 kb. This would increase the complexity of isolating a full-length clone. In addition, the typical size of an insert (15 kb) makes subcloning and analysis of the phage more difficult and time-consuming.

2. Southern blot analysis is recommended, since PCR can lead to false positives owing to contamination of template and primers with the probe or other receptor cDNA-containing plasmids. Minor contamination of PCR products or other plasmids used in the lab typically arise in the pipeters, buffers and other stock bottles, or equipment, and this is the major cause of false positives. If your laboratory does not possess other plasmids containing receptor cDNAs, the PCR method will be suitable.

3. Plating cells are specific for the type of library. λgt10 Libraries can only be screened via DNA probes and are plated on C600hfl cells (high frequency of lysogeny). This cell line allows only recombinant phage, which contains an insert to be expressed and form a plaque. λgt11 Libraries can be screened via DNA or by antibodies, and are plated on Y1090 cells. Other cDNA libraries (such as λZAP) are variations of these two types of libraries. When libraries are purchased, the appropriate cell line with instructions for stock plating are provided (i.e., C600hfl are plated on tetracycline plates). In general, if one does not have a high-titer, specific antibody for the receptor in question, it is advantageous to screen a λgt10 library, since the plaques are larger and it is easier to obtain more intense signals. The culturing of plating cells is always performed in antibiotic-free media (even though the stock plates have an antibiotic) with maltose and high magnesium (NZCYM contains high magnesium). Maltose is required to induce expression of cell-surface receptors for phage absorption. This interaction is magnesium-dependent. If either of these two reagents are limiting, there will be no cell lysis or plaque formation, a common mistake occurring in phage manipulation.

4. PEG (8000–10,000 mol wt) is used with the high salt to precipitate the large phage heads. One hour on ice is usually sufficient time, but longer incubations can sometimes increase yields. If no pellet is observed following centrifugation, the procedure can be stopped. A common cause of low phage yields is incorrect MOI, which is empirically determined for each type of library. *See* **Subheading 3.3.1.**

5. Premade equilibrated phenol can be purchased from Amersham. Alternately, solid phenol is bought, melted in a water bath at 60°C, and 1 *M* Tris base is added, mixed, and then settled. Repeat adding the Tris until a phase separation is achieved. The phenol will soak up the water and increase its volume until saturated at which the phase separation will appear. The lower phase is the equilibrated phenol. Test a small aliquot with pH paper. It should be ≥pH 8.0. If not, add more 1 *M* Tris base until this pH is achieved. Store in refrigerator with the Tris base still on top. Do not use old phenol, which has acquired a color. If pH is lower than 8.0, use of phenol may not give phase separation after mixing and centrifuging.

6. cDNA inserts will be subcloned into different sites in the phage vector DNA for specific libraries. See library information sheets to determine which restriction enzyme is required to release insert DNA. Many cDNA libraries use *Eco*RI. Set up the digestion reaction composed of DNA, buffer, and enzyme in a minimum volume (≥ 15 μL). To prevent nonspecific digestion, do not add restriction enzyme with volume that exceeds one-tenth of the entire reaction volume.

7. If the control tube yields a PCR product (any size), you have a contamination problem, and it is best to stop here and determine the source of the contaminant. Likely suspects are the pipets, water supply, buffers, hands, aerosols, or primers. The problem is particularly pertinent if you have pieces of cDNA from other receptors in the lab. It is best to start over and make fresh buffers, water solutions, change pipets to ones not used in PCR or in handling the cDNAs, or conduct the PCR setup in another lab.

8. You can experiment with different types of *Taq* polymerase, but since this is for identification purposes only, any type will suffice. Some of the other polymerases have a lower error rate in incorporation as well as higher stability, but have a slower polymerization rate. The first PCR condition consists of a longer initial denaturing step, because the DNA template is large. The longer extension time for one cycle allows a greater amount of first strand to be synthesized, increasing the likelihood of obtaining a product. Annealing temperature of 5–10°C below the estimated T_m of the primers is used.

9. If there is no product, the conditions and amounts of primers and template need to be adjusted empirically. You can increased the amount of DNA template and also increase the amount of primers. If this is not successful, try adding 20% glycerol or formamide to the PCR mixture. This will promote the dissociation of the double-stranded DNA, which may be impaired owing to high GC content in the gene of interest. Alternately, the DNA template can be restriction-digested with 6-base cutters to reduce the sizes of the DNA template stock. If a smear is obtained in the gel, the concentration of DNA, primers, or nucleotides may be too great. Reduce these amounts, and also try increasing the annealing temperature and reducing the time of elongation.

10. Centrifugation at high speeds ($>2000g$) might pellet the phage heads, resulting in lower yields.

11. The minigel (3 × 3 in.) is sufficient to separate insert (500–3000 bp) from phage arms (9–11 kb) and can be electrophoresed in approx 1 h. Larger gels can be used, but require a longer electrophoresis time. Electrophoresis should be stopped when the bottom of the gel contains fragments of >500 bp (xylene cyanol runs at 500 bp on a 4% polyacrylamide gel). The labeled probe fragments, which range from probe size and smaller (average size 200–500 bp) will be detected on autoradiography and provide a strong signal. It is best to electrophorese the probe off the gel, because it would ablate other signals. The use of a nondenaturing gel is critical. Denaturing gels would separate the annealed labeled probe from the cDNA insert, resulting in no detectable signal.

12. A kinase reaction exchanges the phosphate at the 5′ end of DNA, and uses the

enzyme T_4 kinase and γ^{32}P-ATP. This is a different reaction than the random primed labeling of the probe DNA which uses a DNA polymerase and α^{32}P-dCTP and incorporates a label several times in the sequence. The DNA ladder cannot be labeled by the random primed method because a statistical array of probe sizes are generated from the random priming. The end-labeling reaction with T_4 kinase preserves the correct molecular weight of the fragments.

13. Only cDNA inserts that hybridize with the probe will be apparent. If the probe hybridizes nonspecifically to the phage arms (9–11 kb) and also generates a smear in the insert size range, the temperature of the solution-phase hybridization should be increased to increase specificity. Also, the amount of digested λ DNA can be decreased if it appears to be overloaded and generate intense signal. Alternately, the salt concentration (XSSC) in the hybridization can be decreased to increase specificity. If this is not successful, and changing hybridization conditions does not alter specificity, the probe may not be of adequate length or specificity, or may be contaminated with other DNA fragments. However, if specific bands are achieved in the insert size range along with the background signal from the phage arms, one should not be concerned. It is common for probes to interact nonspecifically with large-sized DNA owing to the large hydrophobic contacts. It may also be owing to incomplete digestion of the phage DNA with the restriction enzyme.

14. Discrete, but dense plaques are critical for successful screening. A large number of plaques is required to achieve a high probability of isolating the correct clone. However, if the plaque number is too high on a single plate, the decreased plaque size limits signal intensity. If the plate appears to form no plaques, too much phage was added, lysing all of the bacteria. Alternately, the magnesium concentration may be too low, especially if NZCYM medium is not used. The top agarose may also be too hot, which will kill the plating cells, preventing plaque formation.

15. Boiling of the probe is required to redenature the double-stranded DNA, which reanneals on cooling. Hybridization probe can be reused for up to 1 wk. Each time, reheat the hybridization mixture in a microwave or heat plate to redenature the probe (at least to 90°C) and cool to hybridization temperature before adding filters.

16. One easy way to wash filters is to use metal bowls obtained from a kitchen store. The metal bowls are placed on top of a heating plate (use a stirring bar) and monitored closely until the temperature is maintained. Another option is to do the washes in a water bath in a plastic container, but care must be taken to monitor the temperature, since heat transfer is inefficient.

17. The filters should never be allowed to dry, since the probe will adhere irreversibly. If the filters are still moist, they can be rewashed or stripped for reuse. To remove excess liquid, merely blot the filters quickly with Whatman paper, and then quickly add to the old film backing and cover with plastic.

18. An under- or overlysed lawn will not provide sufficient phage DNA for future cloning purposes. Overlysis of the cells occurs before several cycles of

infection–lysis–reinfection (amplification) can happen, leading to a very low yield. An underlysed lawn does not produce enough phage. The best dilution of phage will yield a plate of subconfluent, but clearly visible plaques. It should be noted that it is easy to fall prey to generation of an overlysed lawn, which will look deceptively similar to a nonlysed lawn; in the former case, all of the bacterial cells have been depleted.

19. The DE 52 resin is polycationic and will bind to the negatively charged bacterial DNA released following cell lysis. The phage DNA is still contained in its protein head and, therefore, will not bind. Elimination or incompletion of this step will result in phage DNA that is mixed with degraded bacterial DNA, resulting in a smear when analyzed on an agarose gel.

20. Phosphatase treatment of the 5′ ends of the restriction-cut vector removes the phosphates that are needed for the vector to religate on itself. Therefore, removing the phosphates prevents self-ligation, which promotes recombination with inserts and gives little background of noninserted plasmids in the transformation and resulting colonies. It is not recommended to remove this step in the subcloning.

21. Take 1, 3, and 5 µL of the *Eco*RI/CIP-treated vector stock and transform into bacteria according to **Subheading 3.5.2.4.** Use the same aliquot in the ligation reaction that gives you some colonies, but under 10 colonies as background. If there are too many colonies, 30 or more with the lowest aliquot, your restriction digest and/or CIP treatment was inefficient, and it is best to restart the preparation and digest/dephosphorylate with longer incubation times. If no colonies were obtained, the DNA was probably lost in the manipulations. The cutting/dephosphorylation process is never complete to generate no colonies as background. Supercoiled DNA (uncut vector) transforms with 100× greater efficiency than linearized or circularized DNA. Once a suitable vector stock is prepared, isolate DNA fragment to achieve a sufficient stock for ligation at a 1:3 molar ratio.

22. Calculation of moles of either plasmid or insert is achieved by measuring the $OD260_{NM}$ of the sample and using the conversion 1 OD = 50 µg double-stranded DNA and 660 g/mol = 1 bp of DNA, or by approximating the concentration from the DNA markers (with established concentrations) electrophoresed in agarose gels.

23. The temperature of ligation is a trade-off between allowing a sufficiently low temperature to promote a 4-base overhang to anneal, but not too low to inhibit fully the reaction rate of the ligase whose optimum temperature is 37°C. The ligation can be performed at room temperature for 4 h, followed by the transformation. However, higher ligation efficiency can be achieved with the overnight incubation at 14°C.

24. The ligation mixture contains proteins, buffers, and other components that can inhibit transformation. Therefore, do not add the ligation mixture at more than 1/10 the volume of the competent cells, but separate the reaction into two aliquots if needed. The 30-min incubation is a minimum. Longer incubations (hours) can increase efficiency.

25. Another advantage of using pBluescript is blue/white selection. Recombinants disrupt the β-galactosidase gene in the vector, which negates production of the enzyme and its ability to cleave a substrate (turning blue). Therefore, if the inserts were incorporated, the colonies after transformation turn white. This selection reduces the amount of subsequent work needed to identify recombinants. To achieve this selection, add 50 μL of a 2% X-Gal (5-bromo-4-chloro-3-indolyl-β-D-galactoside; Amersham) solution in dimethylformamide and 50 μL of a 0.1 *M* isopropyl-β-D-thiogalactopyranoside (IPTG) (Amersham) solution in water and spread on the ampicillin plates before the cells are added. The next day, if blue (wild-type) colonies are not seen among the white colonies, incubate the plate for an additional day in the refrigerator to bring out the blue color.

References

1. Perez, D. M., Piascik, M. T., and Graham, R. M. (1991) Solution-phase library screening for the identification of rare clones: isolation of an α_{1D}-adrenergic receptor cDNA. *Mol. Pharmacol.* **40,** 876–883.
2. Ramarao, C. S., Kincade Denker, J. M., Perez, D. M., Gaivin, R. J., Riek, R. P., and Graham, R. M. (1992) Genomic organization and expression of the human α_{1B}-adrenergic receptor. *J. Biol. Chem.* **267,** 21,936–21,945.
3. Sambrook. J., Fritsch, E. F., and Maniatis, T. (1989) *Molecular Cloning: A Laboratory Manual.* Cold Spring Harbor Laboratory Press, Cold Spring Harbor, NY.

6

Genetic Polymorphisms of Adrenergic Receptors

Ian P. Hall, Amanda P. Wheatley, and Jane C. Dewar

1. Introduction

Single nucleotide polymorphisms exist at a high frequency in the human genome. Estimates of variability suggest that up to 1 in 1000 bp within coding regions of the genome are polymorphic, and the frequency of polymorphism within noncoding regions is even higher at about 1 in 500. Because of redundancy in amino acid coding, not all single-base substitutions result in an altered amino acid sequence within a gene. However, even degenerate polymorphisms can potentially alter transcriptional or translational activity. Polymorphisms within promoter or enhancer regions can interfere with transcription factor binding sites and, therefore, can also alter transcriptional activity.

Both the β_2- and β_3-adrenergic receptors (β_2-AR and β_3-AR) are known to be polymorphic, and disease-related associations have been described for polymorphisms in both receptors. Therefore, the arginine-glycine 16 (Arg-Gly 16) variant of the β_2-AR has been associated with nocturnal falls in FEV_1 (forced expiratory volume in 1 s) in asthmatic subjects and with bronchodilator subsensitivity (1–4). The glutamine-glutamate 27 (Gln-Glu 27) β_2-AR polymorphism was shown to be associated with reduced IgE levels in one study of asthmatic families (5), although in random populations this effect was not observed (6). Both these polymorphisms have functional effects. The glycine 16 variant produces increased receptor downregulation following agonist stimulation in both transformed and nontransformed cell systems, whereas the glutamate 27 variant reduces agonist-induced receptor downregulation (7,8). Two other rarer β_2-AR variants have been identified at codon 34 (valine-methionine) and at codon 164 (threonine-isoleucine) (1). Whereas the codon 34 polymorphism appears to be nonfunctional, the codon 164 polymorphism alters both agonist binding and receptor sequestration following agonist expo-

From: *Methods in Molecular Biology, vol. 126: Adrenergic Receptor Protocols*
Edited by: C. A. Machida © Humana Press Inc., Totowa, NJ

sure *(9)*. This polymorphism is rare, and to date, no homozygous individuals have been studied. Five other degenerate single-nucleotide polymorphisms have been identified in the β_2-AR coding region, although no functional properties have been ascribed to these polymorphisms *(1)*.

Single-nucleotide polymorphisms within the β_3-AR have also been identified. Disease associations with insulin resistance and obesity have been suggested for the Trp Arg 64 polymorphism *(10,11)*, although not all studies have seen association. To date, no β_1-AR polymorphisms have been described.

A number of methods have been identified that are able to genotype individuals for single-nucleotide polymorphisms, such as those described above. We have used allele-specific oligonucleotide (ASO) hybridization, which provides a high throughput assay and reliable genotype assignation. However, other methods, including restriction fragment-length polymorphism (RFLP) analysis and allele-specific polymerase chain reaction (PCR) amplification, can also reliably identify genotype. The reliability of each of these methods has not been formally compared for the determination of β_2-AR polymorphisms. However, in our hands, using direct sequencing as a gold standard, the error rate using ASO hybridization is under 2%.

Finally, as would be expected with polymorphisms, which are in such close approximation, there is linkage disequilibrium between the different β_2-AR polymorphisms. Thus, the Gly 16 variant is associated with the Glu 27 variant (**Fig. 1**). It is also worth noting that a previously ascribed RFLP, which produces a Ban 1 restriction site in the region of the β_2-AR locus, is owing to the degenerate (C-A) residue 523 polymorphism within the β_2-AR coding region *(6)*. The allelic distribution of the relevant β_2-AR polymorphisms is shown in **Table 1**.

2. Materials

2.1. Genomic DNA Extraction

Genomic DNA can be prepared from whole blood, cheek cells, or primary cell cultures using standard methods *(12)*. However, for large numbers of samples, kit-based DNA resin systems provide adequate amounts of good-quality DNA for genotype determination and provide a relatively rapid extraction method for multiple DNA. We use the Nucleon extraction system (Scot Lab UK) or for small-scale DNA extraction Quiagen columns (Quiagen UK).

2.2. Allele-Specific Oligonucleotide (ASO) Hybridization

The equipment needed for ASO hybridization is listed below.

1. Thermocycler.
2. Dot-blot apparatus.

Filter 1

	WT
Sample 1	● ●
Sample 2	● ●
Sample 3	

Filter 2

Mutant

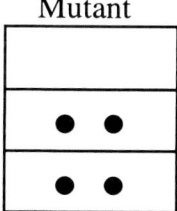

Fig. 1. Representative ASO for single nucleotide polymorphism in β_2-AR (e.g., Arg-Gly 16) (schematic). Each PCR aliquot from an individual DNA sample is applied to duplicate wells using a dot-blot apparatus (*see* **Subheading 3.**). If a hybridization signal is seen on both filters, the individual is a heterozygote: if a signal is seen on only one filter, the individual is a homozygote for the probe used on that filter (sample 1, homozygote wild type, sample 3 homozygote mutant). The expected frequency of homozygotes/heterozygotes can be estimated (assuming Hardy-Weinberg equilibrium) as a crosscheck for genotype data in random samples.

3. Hybridization oven and chambers.
4. X-ray cassettes and developing facilities (assuming [32]-P labeled probes are used).

2.3. Controls for Genotyping

It is advisable to use appropriate controls for all genotype determinations. We keep a stock of DNA from individuals of known genotype determined both by ASO hybridization and confirmed by direct sequencing as positive controls for homozygotes and heterozygotes at each locus.

2.4. Stock Solutions

The following stock solutions are required for the relevant stages.

Table 1
The Allelic Distribution of the Amino Acid 16, 27, 164, and 523 β_2-AR
polymorphisms in a Random Population of 228 Individuals[a]

Location	Wild-type homozygote	Heterozygote	Mutant homozygote
Codon 16	15%	49%	36%
Codon 27	24%	52%	24%
Codon 164	97%	3%	0%
Base 523 (RFLP)	71%	26%	3%

[a]Adapted from **ref. 6**.

2.4.1. Polymerase Chain Reaction (PCR)

1. dNTPs (200 mM).
2. 10X PCR buffer: 100 mM Tris-HCl, pH 9.0, 500 mM KCl, 1% Triton X-100 (Promega).
3. MgCl$_2$ (use at 1.5 mM final concentration in reaction).
4. Genomic DNA (50 ng/μL in H$_2$O) (*see* **Note 1**).

2.4.2. Labeling of Oligonucleotide Probes

1. Oligonucleotide probes (0.05 nmol in 6 μL H$_2$O) (*see* **Note 2**).
2. Polynucleotide kinase (PNK) (Stratagene).
3. 10X PNK buffer: 500 mM Tris-HCl, pH 7.5, 70 mM MgCl$_2$, 10 mM dithiothreitol (DTT) (Stratagene).
4. [^{32}P]dATP (9.25 MBq, 10.0 mCi/mL, Amersham).

2.4.3. Buffer for ASO Filter Loading

1. 0.4 M NaCl, 4X SSC (*see below* for SSC components), 2 μL bromophenol blue, 2 μL PCR product.
2. 2X SSC.

2.4.4. Hybridization

1. 5X SSPE (*see* **Subheading 2.4.5., item 2** for SSPE components): 1% sodium dodecyl sulfate (SDS).
2. 2X SSPE, 0.1 % SDS.
3. 5X SSPE, 0.1 % SDS.

2.4.5. Stock Buffers

1. 20X SSC: 175.3 g of NaCl, 88.2 g of sodium citrate. Make up to 1 L H$_2$O and pH to 7.0.
2. 20X SSPE: 175.3 g NaCl, 27.6 g NaH$_2$PO$_4$ H$_2$O, 7.4 g ethylenediamine tetra-acetic acid (EDTA). Make up to 1 L H$_2$O and pH 7.4.

Table 2
Primers Used for PCR Amplification

Codon 16 and 27
  Upstream CCC AGC CAG TGC GCT TAC CT
  Downstream CCG TCT GCA GAC GCT CGA AC
Codon 164 and nucleic acid residue 523
  Upstream CC AGC CAG TGC GCT TAC CT
  Downstream GAC ATG GAA GCG GCC CTC AG

3. Methods

3.1. PCR Amplification

1. Four β_2-AR polymorphisms at amino acid positions 16, 27, and 164 and nucleic acid residue 523 have been analyzed using the technique of ASO, and the PCR primers for each are shown in **Table 2**.
2. PCR amplification for codons 16 and 27:
   Using PCR, generate a 234-bp fragment from the 5' end of the β_2-AR gene spanning the two polymorphisms of interest. Set up a 50-μL PCR reaction using:
       50 ng of genomic DNA.
       200 mM of each deoxynucleotide.
       5 μL of 10X PCR buffer.
       1.5 mM MgCl$_2$.
       2 μM of each primer.
       X μL H$_2$O to make up the final volume.
       Use 1 U of *Taq* polymerase/reaction.
3. Set up the PCR reaction (36 cycles) using the following conditions:
   Melting temperature 94°C, 90 s.
   Annealing temperature 60°C, 90 s.
   Extension temperature 72°C, 90 s.
   We use 'hot-start' PCR to help reduce false priming. An initial period of 5 min at 94°C is used in the first cycle, followed by 1 min at 60°C, after which 1 U of *Taq* polymerase is added to each reaction. At the very end of the reaction, a 10-minute extension period at 72°C is employed.
4. PCR amplification for codon 164 and nucleic acid residue 523 polymorphisms:
   The PCR reaction is set up as described above (using primers shown in **Table 2**), but with the annealing temperature set at 64°C. Sufficient product for ASO can be obtained with lower cycle numbers (we have used 30 successfully).

3.2. Allele-Specific Oligonucleotide Hybridization

3.2.1. Introduction

ASO hybridization is based on the use of ASOs for the detection of single-point mutations. The oligonucleotides used are short segments of synthetic

DNA (approx. 20 nucleotides in length), which are designed to be complementary to a region of DNA containing a specific polymorphism. A single-base mismatch between an oligonucleotide probe and template DNA will substantially decrease its annealing affinity, thus allowing discrimination between different alleles. ASO is a competitive assay in which amplified DNA products are applied to duplicate filters (e.g., A and B) with a dot-blot apparatus. In brief, filter A is incubated initially with unlabeled ("cold") oligonucleotide probe homologous to the wild-type form of receptor, and filter B with "cold" (unlabeled) oligonucleotide probe homologous to the mutant form of receptor. Each filter is then hybridized with "hot" (^{32}P-labeled) probes in the opposite order, such that filter A is now exposed to the probe that is homologous to the mutant form of receptor, and filter B to the probe that is homologous to the wild-type form of receptor. The two filters are read in parallel: filter A thus displays signal for mutant homozygotes, and heterozygotes, but filter B displays signal for wild-type homozygotes and heterozygotes. The heterozygote signal is intermediate to the homozygote signal, and interpretation is aided by the application of control samples with known genotypes to the filters. This also verifies that filters have been appropriately labeled and probes added in the correct order.

3.2.2. Preparation of Filters for ASO

1. For the preparation of two filters for each DNA sample to be genotyped, add 2 μL of PCR product to 400 μL of 0.4 *M* NaOH, 400 μL 4X SSC, and 20 μL of bromophenol blue in a 1.5-mL microfuge tube.
2. Apply 200 μL of the sample to duplicate wells on two Hybond® N+ filters (*see* **Note 3**), using a dot-blot apparatus (e.g., Hybri-Dot model, Bethesda Research Labs).
3. For each filter, two pieces of blotting paper and one piece of Hybond® N+ membrane need to be cut to size and soaked in 20 mL of 2X SSC.
4. These materials are then secured on the dot-blot filter block with the Hybond® N+ membrane uppermost, suction applied via a manifold pump, and the sample solution applied to each well. Apply suction until all of the solution has disappeared from each well (*see* **Note 4**).
5. The filters are then labeled and allowed to air-dry. Care must be taken to orient the samples on the filter before hybridization (e.g., by cutting off one corner of the filter). If both the codon 16 and 27 polymorphisms are to be studied, four filters are required (two for each).

3.3. Labeling of Oligonucleotide Probes

1. The sequences for the ASO probes used are shown in **Table 3**. The position of each polymorphism is underlined. We have generally used HPLC-purified oligonucleotide probes, although this is probably not essential.

Table 3
β_2-AR ASO Hybridization Probes[a]

Amino acid	Oligonucleotide probe sequence
Gln 27	CAC GCA G<u>G</u>A AAG GGA CGG AG
Glu 27	CAC GCA G<u>C</u>A AAG GGA CGA G
Arg 16	GCA CCC AAT <u>A</u>GA AGC CAT G
Gly 16	GCA CCC AAT <u>G</u>GA AGC CAT G
Thr 164	CAG GCC TTA <u>C</u>CT CCT TCT T
Ile 164	CAG GCC TTA <u>T</u>CT CCT TCT T
523 Wild-type	CAC TGG TAC <u>C</u>GG GCC ACC C
523 Mutant	CAC TGG TAC <u>A</u>GG GCC ACC C

[a]The polymorphic nucleotide is underlined for each oligonucleotide.

2. Set up a labeling reaction with:
   0.05 nmol of oligonucleotide probe in 6 µL of water.
   1 µL 10X PNK buffer (Stratagene).
   5 U PNK (Stratagene).
   2 µL of [^{32}P] dATP (9.25 MBq, 10.0 mCi/mL, Amersham).
3. Incubate at 37°C for 1 h, and then heat for 5 min at 70°C to denature the PNK enzyme.

3.4. Hybridization

1. To determine genotype, one filter is hybridized as appropriate with "cold" (unlabeled) ASO probe homologous with the wild-type form of the receptor followed by " hot" (^{32}P-labeled) ASO probe homologous with the mutant form of the receptor. The duplicate filter is treated in an identical fashion, but using probes in the reverse order. We perform hybridization in a Biometra OV2 hybridization oven. Filters are rolled in gauze (*see* **Note 5**) and placed in Biometra hybridization chambers, such that the gauze is positioned to the edge of the chamber. Each chamber is clearly labeled.
2. Initiate hybridization with cold (unlabeled) probe (10-fold excess for codon 16, 30-fold excess for codon 27) (*see* **Note 6**) in 20 mL of 5X SSPE, and 1% SDS solution at 52°C for 60 min.
3. Conduct secondary hybridization with hot (labeled) probe in 20 mL of 5X SSPE and 1% SDS solution at 52°C for 60 min.
4. Wash filters with 20 mL of 2X SSPE and 0.1% SDS solution, at room temperature for 30 min, and then repeat. Conduct final wash with 20 mL of 5X SSPE and 0.1% SDS solution at 52°C for 15 min.

3.5. Reading Filters

Expose probed filters to autoradiographic film with intensifying screens overnight at −80°C. Filters are read in conjunction after correct control genotype is confirmed. It is probable that genotype determination could be performed with a similar degree of accuracy using nonradiolabeled probes (e.g. chemiluminescent probes), although we have not investigated this possibility (*see* **Notes 7–9**).

3.6. Conclusions

Using ASO hybridization, accurate genotype assignment for β_2-AR polymorphisms is relatively simple. The methods given above can be used for any single nucleotide polymorphism provided that an adequate PCR fragment can be generated spanning the region of interest and provided that the GC content of the polymorphic region is not very high (this leads to poor discrimination owing to less specific binding of the allele-specific probes). Although we have used the above method for a range of other single-nucleotide polymorphisms in other genes, it is sometimes necessary to alter the ratio of cold probe used or alter the wash conditions for any new polymorphism to be studied. The above method provides a reasonable initial protocol that may require further optimization.

4. Notes

1. In general, providing reasonable-quality genomic DNA is available, the ASO method is reliable and the reproducibility of genotype determination (which we have also assessed for other single-nucleotide polymorphisms in the interleukin 9 [IL-9] gene) is high (generally around 98%). Genomic DNA should be stored frozen in aliquots at –80°C (for long term): –20°C is adequate for shorter periods (<6 mo).
2. We generally use HPLC-purified probes, although this is not essential.
3. It is important to use the correct filters: Hybond® N+ provides the best results.
4. The position of each sample is visualized by the bromophenol blue in the loading buffer. However, this will disappear with the wash stages, and therefore, it is critical to label each filter for orientation.
5. Be careful to avoid direct contact between filters. Make sure the gauze separates each filter.
6. Cold probe concentrations: Owing to differing probe annealing stringencies, for the Arg-Gly 16 polymorphism, we use a 10-fold excess of cold probe in the initial part of the hybridization procedure, but use a 30-fold excess of cold probe for the Gln-Glu 27 polymorphism.
7. Although it is not essential to conduct agarose-gel electrophoresis for each PCR reaction (PCR amplification of the β_2-AR generally being reliable), DNA samples will occasionally not amplify owing to poor-quality template. If no signal is evident on the filter, conduct electrophoresis with 10 µL of the original PCR reaction and other PCR reactions where the samples provided signals on the filters. If no band is present on the agarose gel, the absence of a hybridization signal is probably owing to poor-quality DNA or to an error in the PCR stage (e.g., failure to add primer or *Taq* polymerase).

Solution: Repeat PCR using same DNA. Conduct agarose-gel electrophoresis; if still no band is evident, then repeat DNA extraction.

8. Occasionally, it can be difficult to be certain whether an individual is homozygous or heterozygous if there is some background (nonspecific) hybridization to the filter.

Solution: Repeating the ASO stage using the original PCR reaction usually clarifies the genotype. However, it is always desirable to sequence a random selection of samples to ensure genotype assignment is correct. This has the additional advantage of avoiding problems resulting from mislabeling of probes.

9. Other methods: Many single-nucleotide polymorphisms introduce new restriction sites or interrupt constitutive restriction enzyme recognition sites, and therefore, can be assessed by RFLP analysis. The base pair 523 (C-A) degenerate polymorphism in the β_2-AR is an example of this phenomenon. Therefore, RFLP analysis of PCR fragments can be used to assign genotype for single-nucleotide polymorphisms. Although this is a widely used method, we have encountered problems in reproducibility resulting from the difficulty in distinguishing between heterozygote individuals and incomplete digests of known homozygotes on agarose gels. Some groups have used allele-specific PCR to assign genotype. In experienced hands, this provides reliable genotype assignment, but probably has a lower specificity in inexperienced hands than ASO hybridization. Regardless of which method is utilized to determine genotype, it is always important to sequence a small number of samples randomly to ensure accuracy.

References

1. Reishaus, E., Innis, M., MacIntyre, N., and Liggett, S. B. (1993) Mutations in the gene encoding for the β_2-adrenergic receptor in normal and asthmatic subjects. *Am. J. Respir. Cell Mol. Biol.* **8,** 334–339.
2. Tan, K. S., Hall, I. P., Dewar, J. C., Dow, E., and Lipworth, B. J. (1997) β_2-adrenoceptor (β_2-AR) polymorphism determines susceptibility to bronchodilator desensitisation in asthmatics. *Lancet* **4, 350,** 995–999.
3. Turki, J., Pak, J., Green, S. A., Martin, R. J., and Liggett, S. B. (1995) Genetic polymorphisms of the β_2-adrenergic receptor in nocturnal and nonnocturnal asthma. Evidence that Gly16 correlates with the nocturnal phenotype. *J. Clin. Invest.* **95,** 1635–1641.
4. Martinez, F. D., Graves, P. E., Baldini, M., and Erickson, R. (1997) Association between genetic polymorphisms of the β_2-adrenoceptor and response to albuterol in children with and without a history of wheezing. *J. Clin. Invest.* **100,** 3184.
5. Dewar, J. C., Wheatley, A., Wilkinson, J., Thomas, N. S., Doull, I., Morton, N., et al. (1997) The glutamine 27 β_2-adrenoceptor polymorphism is associated with elevated IgE levels in asthmatic families. *J. Allergy Clin. Immunol.* **100,** 261–265.
6. Dewar, J. C., Wheatley, A. P., Venn, A., Morrison, J. F. J., Britton, J., and Hall, I. P. (1998) β_2 adrenoceptor polymorphisms are in linkage disequilibrium, but are not associated with asthma in an adult population. *Clin. Exp. Allergy* **28,** 442–448.

7. Green, S. A., Turki, J., Innis, M., and Liggett, S. B. (1994) Aminoterminal polymorphisms of the human β_2-adrenergic receptor impart distinct agonist promoted regulatory properties. *Biochemistry (USA)* **33,** 9414–9419.

8. Green, S. A., Turki, J., Bejarano, P., Hall, I. P., and Liggett, S. B. (1995) Influence of β_2-adrenergic receptor genotypes on signal transduction in human airway smooth muscle cells. *Am. J. Respir. Cell Mol. Biol.* **13,** 25–33.

9. Green, S. A., Cole, G., Jacinto, M., Innis, M., and Liggett, S. B. (1993) A polymorphism of the human β_2-receptor within the fourth transmembrane domain alters ligand binding and functional properties of the receptor. *J. Biol. Chem.* **268,** 23,116–23,121.

10. Mitchell, B. D., Blangero, J., Comuzzie, A. G., Almasy, L. A., Shuldiner, A. R., Silver, K., et al. (1998) A paired sibling analysis of the β_2-adrenergic receptor and obesity in Mexican Americans. *J. Clin. Invest.* **101,** 584–587.

11. Widen, E., Lehto, M., Kanninen, T., Walston, J., Shuldiner, A. R., and Groop, L. C. (1995) Association of a polymorphism in the β_3-adrenergic receptor gene with features of the insulin resistance syndrome in Finns. *N. Engl. J. Med.* **333,** 348–351.

12. Sambol, J., Fritsch, E. F., and Maniatis T. (1989) *Molecular Cloning: A Laboratory Manual*, 2nd. ed. Cold Spring Harbor Laboratory Press, Cold Spring Harbor, NY.

II

RNA ANALYSIS

7

Northern Blot Analyses Detecting Adrenergic Receptor mRNAs

Judith C. W. Mak

1. Introduction

Catecholamines exert their physiological effects via binding to cell-surface receptors known as adrenergic receptors (ARs). Based on the pharmacological and physiological effects of various agonists, the adrenergic receptors have been grouped into a single classification; several members have now been cloned and their corresponding deduced primary amino acid sequences established. The first adrenergic receptor to be cloned and sequenced was the hamster β_2-AR *(1)*. Of the nine adrenergic receptors that have been identified and cloned to date, only the β_1-AR and β_2-AR are found in human lung *(2)*. The adrenergic receptors are all members of the superfamily of seven-transmembrane domain, G-protein-coupled receptors; these receptors contain extracellular amino-termini, often glycosylated, and intracellular carboxyl-termini. Within each receptor are seven clusters of hydrophobic-rich amino acids, which are believed to represent transmembrane segments, each connected by extracellular and intracellular loops. Comparing different adrenergic receptors, the third intracellular loop and cytoplasmic tails can be highly variable in both length and amino acid composition *(3)*.

Northern blot analysis provides a determination of the steady-state levels of a specific receptor mRNA species within a complex RNA pool. The method described here is an adaptation of the method described by Chomczynski and Sacchi *(4)* using radiolabeled probes that contain a complementary sequence to that of the targeted mRNA. Total RNA or poly (A)$^+$ RNA is size-fractionated on a denaturing agarose gel and then transferred to a solid filter. The filter is then hybridized with the radiolabeled probe, and the specific RNA band is

From: *Methods in Molecular Biology, vol. 126: Adrenergic Receptor Protocols*
Edited by: C. A. Machida © Humana Press Inc., Totowa, NJ

visualized by autoradiography. The Northern blot analysis has several advantages, including:

1. Technical simplicity.
2. The ability to visualize the full-length mRNA species.
3. The flexibility to use either random primer-labeled DNA probes or riboprobes.

In Northern blot analysis, a constitutively expressed "common gene" controls must be employed to verify loading equivalent RNA amounts in all lanes of a single gel. Examples of "common gene" controls include glyceraldehyde-3-phosphate dehydreogenase (GAPDH), cyclophilin, and β-actin. However, caution must be taken, since all common gene probes may not be constitutively expressed in all tissue types or under all experimental conditions. This chapter describes the detailed protocols required to determine the expression of steady-state levels of β_1-AR and β_2-AR mRNAs using cDNA probes *(2,5,6)*.

2. Materials
2.1. RNA Isolation

1. Diethyl pyrocarbonate (DEPC): All solutions that will come into contact with RNA must be treated with 0.1% (v/v) DEPC. The DEPC-treated distilled water must be shaken vigorously, placed in a 37°C incubator overnight, and autoclaved.
2. Solution D: 4.0 M guanidinium thiocyanate, 25 mM sodium citrate, pH 7.0, 0.5% sarcosyl solution, and 0.1 M 2-mercaptoethanol, which is added immediately prior to use. To a bottle containing fresh 500 g guanidinium thiocyanate (Sigma, Poole, UK), add 586 mL of DEPC-treated water, 35.2 mL of 0.75 M sodium citrate, pH 7.0, and 52.8 mL of 10% sarcosyl solution. This solution is stable for several months at room temperature. Complete solution D is prepared by adding 0.36 mL of 2-mercaptoethanol to 50 mL prior to use.
3. 2 M sodium acetate, pH 4.0, autoclaved.
4. Water-saturated phenol—not neutralized, stored at 4°C.
5. Chloroform/isoamyl alcohol (49:1).
6. Isopropanol stored at 4°C.
7. 75 and 100% ethanol stored at 4°C.
8. Autoclaved RNase-free Oakridge tubes, pipet tips, and microcentrifuge tubes.
9. Oven-baked RNase-free Corex tubes.
10. Spectrophotometer.

2.2. Random Primed Labeling of DNA

1. Midi-prep plasmid DNA containing full length or fragment of target sequence.
2. Restriction enzymes that cut out fragment of interest with complete removal of vector DNA.
3. Multiprime DNA labeling system (Amersham International, Amersham, UK),

which includes dNTPs, reaction mixture containing hexanucleotide primers, and Klenow fragment. All of these reagents can be purchased separately.

4. $[\alpha\text{-}^{32}P]$dCTP (3000 Ci/mmol, Amersham, International, Amersham, UK).
5. 1 M Tris-HCl, pH 7.5, autoclaved.
6. 0.5 M ethylenediamine tetra-acetic acid (EDTA), pH 8.0, autoclaved.
7. 10% sodium dodecyl sulfate (SDS) in DEPC-treated water.
8. Sephadex G-50 in 10 mM Tris-HCl, pH 7.5, 1 mM EDTA, pH 8.0, and 0.1% SDS.
9. Tris-EDTA (TE) buffer: 10 mM Tris-HCl, pH 7.5, and 1 mM EDTA, pH 8.0.
10. Oven-baked RNase-free glass pipets.

2.3. RNA Formaldehyde Gel

1. Suitable RNase-free agarose (1.0%) (Promega, Southampton, UK).
2. 3-N-Morpholine-propanesulfonic acid (MOPS) 20X buffer: 400 mM MOPS, 100 mM sodium acetate, 20 mM EDTA. Autoclaved.
3. Gel-loading buffer: 75% formamide, 1.5X MOPS, 10% formaldehyde. Store in aliquots at −20°C.
4. 20X standard sodium citrate (SSC): 3 M NaCl, 0.3 M Na citrate, pH 7.0, autoclaved.
5. Ethidium bromide: generally available in a 10 mg/mL stock.
6. Orange G dye: 0.4% orange G, 50% glycerol, 1 mM EDTA, autoclaved.
7. Formaldehyde: 37% solution.

2.4. Capillary Transfer of RNA to Solid Support

1. Whatman 3MM filter paper.
2. Nylon membranes commonly used include Magna (Genetic Research Instruments Ltd., Telsted Dunmow, UK) and Hybond-N (Amersham International).
3. Plastic sandwich box, glass plates, sponge, weights.
4. UV Stratalinker-2400 (Stratagene, Cambridge, UK).

2.5. Hybridization

1. Hybridization buffer: 50% formamide, 4X SSC, 50 mM Tris-HCl, pH 7.5, 5X Denhardt's solution, 0.1% SDS, 5 mM EDTA, and 250 µg/mL denatured salmon sperm DNA. Stock solutions are listed below.
   a. 100% Deionized formamide.
   b. 20X SSC.
   c. 1 M Tris-HCl, pH 7.5.
   d. 50X Denhardt's solution: 1% bovine serum albumin, 1% polyvinyl-pyrrolidone (PVP-360), 1% Ficoll 400.
   e. 10% SDS.
   f. 0.5 M EDTA, pH 8.0.
   g. 10 mg/mL denatured salmon sperm DNA (Sigma).
2. Hybridization oven.

2.6. Posthybridization Washes

1. 20X SSC.
2. 10% SDS.
3. Cassettes, X-OMAT-S film (Kodak, Hemel, Hempstead, UK).

3. Methods
3.1. RNA Isolation

1. For preparation of RNA from tissue: The tissue should be removed quickly from the animal and frozen immediately in liquid nitrogen (*see* **Note 1**).
2. Pulverize larger frozen tissue with a prechilled mortar and pestle before homogenizing in denaturing solution D with a Polytron.
3. For preparation of RNA from cultured cells: The media are removed and washed with Hank's Balanced Saline Solution (HBSS). Ten microliters of denaturing solution D are added to the culture flask (T-75) containing ~10×10^6 cells.
4. Add (1/10 vol to solution D) 2 *M* sodium acetate, pH 4.0, to each sample and mix.
5. Add (equal vol to solution D) water-saturated phenol to each sample and mix.
6. Add (1/5 vol to solution D) chloroform/isoamyl alcohol mixture to each sample, mix vigorously for 15 s, and leave on ice for 15 min.
7. Spin for 15 min at 12,000g at 4°C.
8. Remove aqueous (i.e., RNA) phase and place into a fresh Corex tube. Add ice-cold isopropanol (equal volume to solution D), vortex, and either place on dry ice for 60 min or precipitate overnight at –20°C.
9. Spin for 20–30 min at 17,000g at 4°C.
10. Remove supernatant, and resuspend pellet in 75% ethanol.
11. Spin RNA for 15 min at 17,000g at 4°C.
12. Decant supernatant, freeze-dry the RNA pellet, and thoroughly resuspend in small volume of DEPC-treated water by repeat pipeting (*see* **Note 2**).
13. Carefully transfer RNA to microcentrifuge tube and keep on ice.
14. Quantitate RNA at OD_{260} (OD_{260} 1 = 40 µg/mL) (*see* **Notes 3** and **4**).
15. Store RNA at –70°C.
16. If required, isolate poly (A)$^+$ RNA from total cellular RNA using PolyATtract mRNA Isolation System (Promega) following the manufacturers' protocol.

3.2. Random Primed Labeling of DNA

1. DNA fragment to be random prime-labeled must be gel-purified. There are numerous kits to accomplish fragment purification. The JETsorb DNA Gel Extraction kit (AMS Biotechnology, Witney, UK) is recommended.
2. The DNA (50–100 ng) dissolved in distilled water must be denatured at 100°C for 3–5 min, followed by quick-cooling on ice.
3. Other components are added to the tube to give a final volume of 25 or 50 µL following the manufacturer's protocol.
4. Add 3–5 µL of the [α-^{32}P]dCTP and mix.

5. Add Klenow fragment (1 µL) to the tube.
6. Mix gently by repeated pipeting. Spin for a few seconds in a microcentrifuge.
7. Incubate reaction at 37°C for 45–60 min.
8. Stop reaction by addition of TE buffer.
9. Purify the probe from unincorporated radiolabel by Sephadex G-50 column chromatography, using TE buffer as the eluent.
10. Using an aliquot of the probe, determine the amount of incorporated radioactivity in a liquid scintillation counter.

3.3. RNA Formaldehyde Gel

1. Preparation of gel (*see* **Notes 5–7**): Microwave 1.0% agarose in 1X MOPS at medium setting until the agarose has been completely melted. When the flask is cool enough to grasp, add formaldehyde to a final concentration of 6.6% (27 mL of a 37% formaldehyde solution to a total volume of 150 mL). In a fume hood, pour mixture into casting tray with combs, and allow to solidify (>45 min).
2. Aliquot RNA (containing 20 µg of total RNA) into RNase-free microcentrifuge tubes, or resuspend lyophilized poly (A)$^+$ RNA pellets in 10 µL of DEPC-treated water.
3. Add 2 vol of gel-loading buffer and 1 µL of 400 µg/mL ethidium bromide, heat to 65°C for 5 min, and then chill on ice. Pulse spin tubes in microcentrifuge.
4. Place gel in electrophoresis tank filled with 1X MOPS, and load the RNA samples into the gel wells (*see* **Note 8**).
5. Conduct electrophoresis at 60–100 V (approx 100 mA) until the orange dye runs off the edge (*see* **Note 9**).
6. Place the gel onto Saran wrap (*see* **Note 10**), and photograph on UV transilluminator to visualize the ribosomal RNA (28S and 18S RNA, corresponding to approx 5 and 2 kb, respectively). This will also demonstrate the quality of the RNA (i.e., if degradation has occurred).

3.4. Capillary Transfer of RNA to Solid Support

1. Cut nylon membrane and 4 Whatman 3 MM filters to the same size as the gel.
2. Set up gel capillary transfer apparatus (i.e., the sandwich box) as follows (*see* **Note 11**): Fill reservoir with 20X SSC; place wick (i.e., the sponge) on a platform suspended above reservoir and submerge both ends of the wick in reservoir; prewet four Whatman filter papers in 20X SSC, and place two of them on top of the wick; place gel on top of the Whatman filter papers with the open side of the wells facing down; place nylon membrane on top of the gel; place another prewetted Whatman filter paper on top of the nylon membrane; add 8–10 absorbent filters (available from Sigma) and a stack of paper towels; place glass plate on top of transfer apparatus; on top of plate, place heavy objects, such as one 500-mL bottle containing 400 mL water.
3. Allow capillary transfer to proceed for approx 14–18 h.

4. Dismantle transfer apparatus, and view nylon membrane on transilluminator. With a pencil, mark the orientation of gel and location of ribosomal bands on the side of the nylon membrane.
5. Subject nylon membrane to UV crosslinking.
6. Store blot in Saran wrap in the refrigerator, or use immediately.

3.5. Hybridization

1. For hybridization in oven (*see* **Note 12**): Place blot in hybridization tube, and add 5–6 mL of hybridization buffer. Place in preheated rotating oven.
2. Prehybridization: 4–6 h at 42°C for cDNA probes.
3. Denaturation of probe: Heat cDNA probes at 95–100°C for 5 min. Probes should immediately be quick-cooled on ice before use.
4. Hybridization: Add probes to hybridization buffer without additional buffer (*see* **Notes 13** and **14**). Hybridization reaction should proceed overnight.

3.6. Posthybridization Washes

1. Following hybridization reaction, remove hybridization buffer, and discard into a radioactive disposal sink or container. Add 4X SSC, 0.1% SDS heated to hybridization temperature (*see* **Note 15**) to the hybridization tube, and return to the rotating oven for 30 min.
2. Additional washes (*see* **Note 16**): The remaining washes are in increasing stringent conditions (i.e., increasing temperature and reducing salt in the presence of 0.1% SDS) in the rotating oven. Blots hybridized with cDNA probes are washed successively in 2X SSC, 0.1% SDS at 50°C, 1X SSC, 0.1% SDS at 50°C, 0.5X SSC, 0.1% SDS at 55°C, and 0.1X SSC, 0.1% SDS at 55°C. Wash at each condition for 30 min, and check radioactivity remaining on blots after each wash, using a handheld Geiger counter.
3. Cover blot with Saran wrap, and expose to film. After exposure, strip blot in 50% formamide, 10 mM NaH$_2$PO$_4$, pH 6.8, at 65°C for 30–60 min. Rinse blot in 2X SSC, 0.1% SDS at room temperature. Store blot in Saran wrap for hybridization with additional probes.
4. The method of quantitation depends on the equipment available in the laboratory. Quantitate using the Protein and DNA Imageware systems by laser densitometry (The Discovery Series, Huntington Station, NY), and normalize the signal from the cDNA of interest with the control cDNA, such as glyceraldehyde-3-phosphate dehydrogenase (GAPDH).

4. Notes

1. In order not to exceed the capacity of the homogenization buffer, use a maximum of 2 g of lung tissue/15 mL of solution D. The tissue must be homogenized as quickly as possible to avoid degradation of RNA.
2. The resuspension volume of the final RNA pellet is dependent on the expected RNA yield. It is better to make concentrated samples that can be diluted follow-

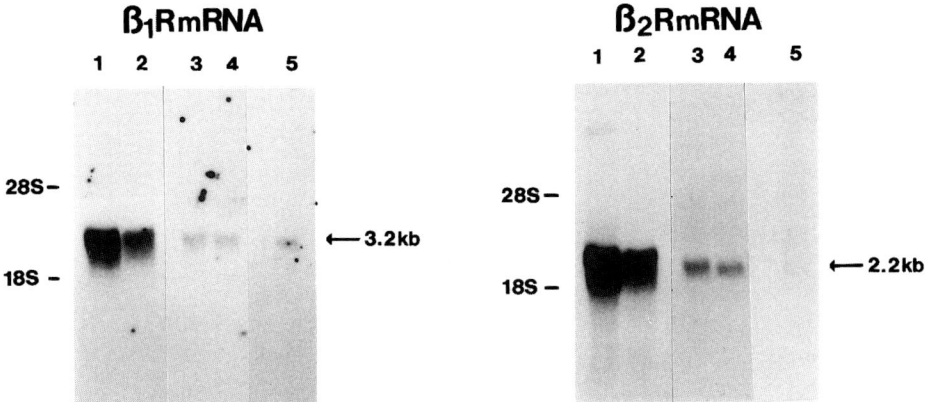

Fig. 1. Northern blot analysis. Human lung poly (A)$^+$ RNA (lanes 1 and 2), and total RNA (20 µg/lane) isolated from human lung (lanes 3 and 4), or human ventricle (lane 5) was submitted to electrophoresis, blot transfer, and hybridization to human β$_1$-AR or β$_2$-AR cDNA probes. The size of mRNA was estimated from 18S and 28S rRNA markers. Reprinted with permission from **ref. 2**.

    ing quantitation than to make samples that are too dilute to be useful. These latter samples would need to be lyophilized or reprecipitated with ethanol.

3. A large quantity of RNA should be diluted into several aliquots to avoid repeated freeze/thawing.
4. To check the quality of the RNA, an OD$_{260/280}$ ratio of >1.7 is acceptable. Values of <1.7 may indicate protein contamination.
5. Wash flasks, gel casters, combs, and gel apparatus in DEPC-treated water to ensure that they are RNase-free prior to use.
6. If feasible, a gel apparatus and tank should be dedicated solely for RNA gels.
7. RNA sample amounts: The amount of RNA necessary for message detection is dependent on the relative abundance of the target mRNA. Rare messages may require the isolation of poly (A)$^+$ RNA in order to detect a signal. This is often the case with G-protein-coupled receptors (*see* **Fig. 1**).
8. The outermost lanes on either side of the gel should not be used if at all possible. RNA in these lanes tends not to transfer as well.
9. Rapid electrophoresis may heat the agarose gel and distort the lanes of migrating RNA.
10. Prior to removing the gel from the tank, it is advisable to cut off the right-hand corner of the gel for orientation purposes.
11. It is important to remove all air bubbles between layers of blotting materials being added to the capillary transfer apparatus. Air bubbles will prevent uniform transfer of RNA to nylon membrane. A disposable 10-mL serological pipet works well to roll out the bubbles.

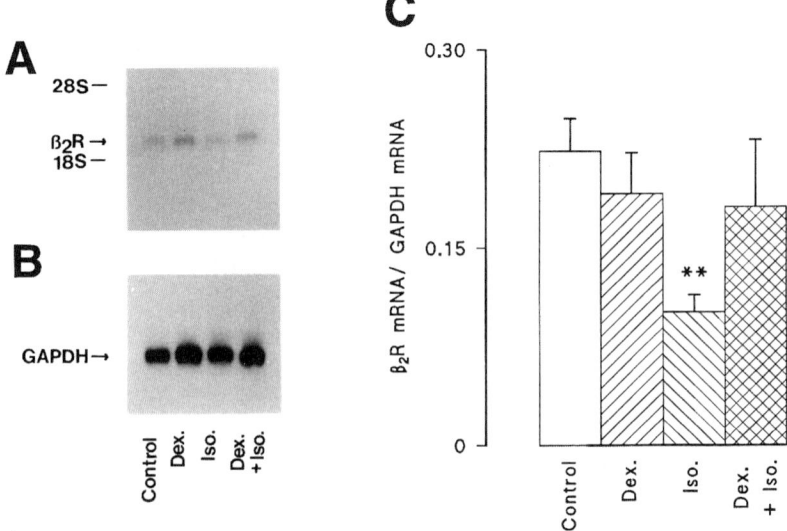

Fig. 2. Effects of treatment with dexamethasone (Dex) and/or isoproterenol (Iso) on β_2-AR mRNA expression in rat lung. **(A)** Representative autoradiogram from Northern blot of rat β_2-AR mRNA using rat β_2-AR cDNA probe. The size of mRNA was estimated from 18S and 28S rRNA markers. **(B)** Representative autoradiogram from Northern blot of rat GAPDH mRNA. The same membrane as used in (A). **(C)** Densitometric measurement of β_2-AR mRNA from control, Dex-treated, Iso-treated, and Dex plus Iso-treated rat lungs ($n = 7$ in each group). β_2-AR mRNA was normalized to that for GAPDH mRNA. Significance of difference from the control value. **$P < 0.01$. Reprinted with permission from **ref. 6**.

12. If using hybridization tubes, be sure that the blots are covered evenly with hybridization buffer. Confirm buffer coverage during the prehybridization step.
13. The amount of labeled probe added to the hybridization buffer depends on the relative abundance of the target RNA. A good starting point is $0.5–1 \times 10^6$ cpm/mL buffer for abundant messages and 2×10^6 cpm/mL for rare messages.
14. Internal standards: Quantitation by Northern blot analysis may require the use of an internal standard to normalize results caused by uneven RNA loading or transfer. If the size of the internal standard transcript is different from the target RNA, simultaneous hybridization with both probes can be conducted. It is important to remember that the level of RNA encoding the commonly used internal standards may far exceed the level of the target RNA, especially if studying G-protein-coupled receptors (*see* **Fig. 2**).
15. It is important to preheat the wash buffers to temperature prior to addition to the blot. This will ensure that the appropriate wash conditions have been reached.
16. The wash protocol outlined here is just a starting point. Additional or longer washes may be necessary.

References

1. Dixon, R. A. F., Kobilka, B. K., Strader, D. J., et al. (1986) Cloning of the gene and cDNA for mammalian β-adrenergic receptor and homology with rhodopsin. *Nature* **32,** 75–79.
2. Mak, J. C. W., Nishikawa, M., Haddad, E.-B., Kwon, O.-J., Hirst, S. J., Twort, C. H., et al. (1996) Localization and expression of beta-adrenoceptor subtype mRNAs in human lung. *Eur. J. Pharmacol.* **302,** 215–221.
3. Liggett, S. B. and Raymond, J. R. (1993) Pharmacology and molecular biology of adrenergic receptors, in *Catecholamines*, vol. 7 (Bouloux, P. M., ed.) Saunders, London, pp. 279–306.
4. Chomczynski, P. and Sacchi, N. (1987) Single-step method of RNA isolation by acid guanidinium thiocyanate-phenol-chloroform extraction. *Anal. Biochem.* **162,** 156–159.
5. Mak, J. C. W., Nishikawa, M., and Barnes, P. J. (1995) Glucocorticosteroids increase β2-adrenergic receptor transcription in human lung. *Am. J. Physiol.* **268,** L41–L46.
6. Mak, J. C. W., Nishikawa, M., Shirasaki, H., Miyayasu, K., and Barnes, P. J. (1995) Protective effects of a glucocorticoid on down-regulation of pulmonary β2-adrenergic receptor *in vivo. J. Clin. Invest.* **96,** 99–106.

8

Ribonuclease Protection Assay for the Detection of β_1-Adrenergic Receptor RNA

Yong-Feng Yang and Curtis A. Machida

1. Introduction

1.1. Molecular Characterization of the β-Adrenergic Receptors (β-ARs)

The β-ARs are members of a large family of neurotransmitter receptors, which interact with GTP binding proteins (G-proteins) to modulate second messenger systems *(1,2)*. The β-ARs mediate the physiological effects of the catecholamines epinephrine and norepinephrine; when ligand-activated, these receptors initiate the production of cyclic AMP. Three subtypes of β-ARs, β_1-AR, β_2-AR, and β_3-AR, have been identified on the basis of their pharmacological properties, physiological effects, tissue and cell-type specificity, and genetic structure *(1–2)*. All three β-AR subtypes have conserved structural features, including hydrophobicity profiles consistent with seven-transmembrane domains, sites of N-linked glycosylation near the amino-terminus, potential phosphorylation sites for protein kinases in presumed cytoplasmic domains, and highly conserved amino acid similarity within the transmembrane domains.

1.2. Modulation of β-AR mRNA Levels in Established Cell Lines

Several endocrine factors can serve as direct modulators of β-AR gene transcription in established cell lines *(1,3)*. For example, in the DDT1MF-2 muscle cell line, addition of glucocorticoids results in rapid elevation of β_2-AR mRNA, receptor product, and adrenergic responsiveness *(1)*. In addition, during adipose differentiation in the 3T3-F442A cell line, β_1-AR and β_2-AR mRNAs are differentially activated following dexamethasone treatment, undergoing tran-

From: *Methods in Molecular Biology, vol. 126: Adrenergic Receptor Protocols*
Edited by: C. A. Machida © Humana Press Inc., Totowa, NJ

scriptional repression and induction, respectively *(4)*. In other related studies, dexamethasone treatment of 3T3-L1 cells induces differential regulation of β_1-AR and β_2-AR mRNAs *(5)*. Interestingly, agonist-induced downregulation of β_2-AR mRNA can be reversed by dexamethasone treatment *(5,6)*.

Thyroid hormones increase the rate of β_1-AR transcription in cultured ventricular myocytes *(7)*. Addition of cyclic AMP analogs results in a transient induction (3–5×) in β_2-AR mRNA levels *(1)*.

1.3. Ribonuclease Protection Assay (RPA)

The detection and measurement of cellular RNA provide important information in understanding mechanisms of gene regulation and transcriptional control. The RPA is a simple, highly reproducible, and extremely sensitive procedure developed for the detection and quantitation of mRNA species in a complex mixture of total cellular RNA. The RPA is at least 10 times more sensitive than other currently available hybridization methods, including Northern and slot-blot hybridization *(8,9)*. In addition, the RPA is tolerant of partially degraded RNA, and allows the use of larger sample sizes to increase sensitivity.

In addition to detection and quantitation of RNA, the RPA can be utilized to identify RNA splice donor and acceptor sites, and transcriptional initiation and termination sites *(10–12)*. This assay can also be utilized to distinguish different mRNA forms encoded by distinct genes of a multigene family, which may not be easily discriminated by conventional blot hybridization approaches. Multiple probes can also be simultaneously used in a single assay, which is useful for providing normalization controls and for the coordinate analysis of mulitple RNA species within a single signaling pathway. When the probes are present in molar excess over the target RNA, the intensity of the protected fragment visualized following electrophoresis and autoradiography will be directly proportional to the amount of target RNA in the sample mixture.

The overall RPA procedure is described in **Fig. 1**. The RPA utilizes an antisense cRNA probe synthesized by in vitro transcription. This is easily performed by inserting the cDNA fragment (100- to 800-bp fragment; optimal length is 250–500 bp) into common transcription vectors controlled by bacteriophage promoters (such as the T3, T7, or SP6 promoters). Recombinants containing the target insert are linearized at the 5' end of the insert, generating templates with either a 5' overhang or blunt end. Linearized plasmids containing 3' overhangs cannot be used as template for in vitro transcription. Using suitable templates, RNA polymerase is then used to generate complementary RNA transcripts of high specific activity. Following liquid hybridization of probe to target RNAs in a complex mixture, hybrids are digested using RNase A/RNase T1 or RNase T1. RNase T1 alone is used if the recognized target

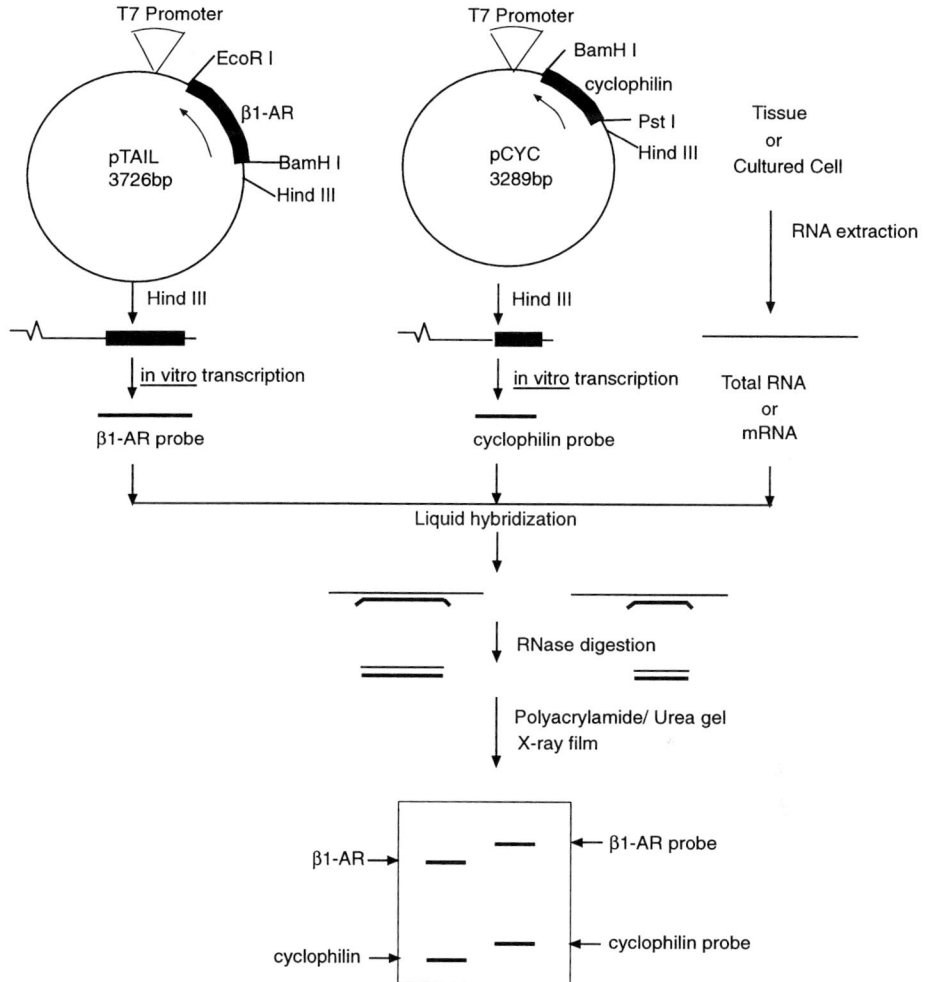

Fig. 1. RNase protection assay for detection of β₁-AR RNA.

sequence is AU-rich, such as in the polyadenylation region. The protected probe is then resolved in a denaturing polyacrylamide gel and identified following autoradiographic processing.

1.4. RNase Protection Assay in the Detection of β₁-AR mRNA

Examples of probes used for detection of the rat β₁-AR mRNA are described in **Fig. 2A**. These probes include pCoding, which extends from position −82 to +273 relative to the translational start site, and pRPT, a probe extending from position +2253 to +2797 in the rat β₁-AR sequence. pRPT is capable of identifying β₁-AR transcripts using either of the two polyadenylation sites located at

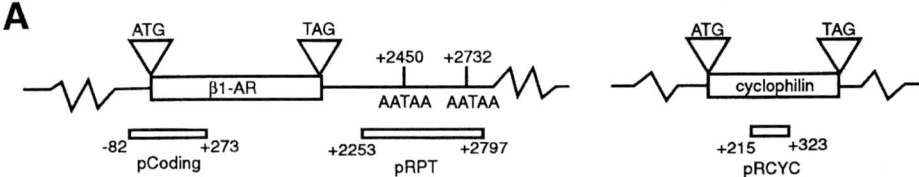

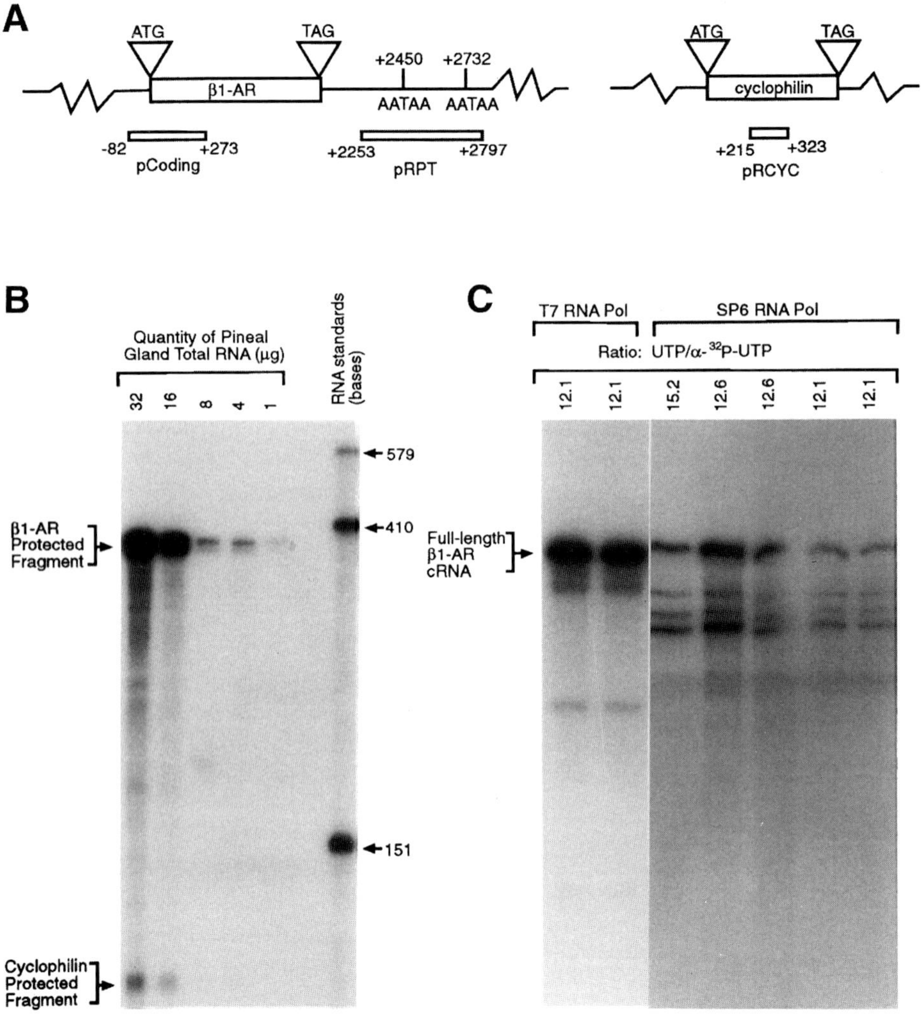

A

B

C

Fig. 2. (**A**) Probes utilized in β_1-AR RPA. Locations of β_1-AR probes (pCoding and pRPT) and cyclophilin probe (pRCYC) are diagrammed relative to positions in their appropriate genes. Templates are cDNA fragments inserted into pGEM 3Zf(+) vectors. (**B**) RPA detection of β_1-AR mRNAs in pineal gland. Probes pCoding (5×10^5 cpm) and pRCYC (1×10^3 cpm) were used. The amount of pineal gland RNA used in RPAs varies from 1 to 32 µg total RNA. RNA/probes mixtures were denatured at 95°C for 15 min, then transferred to a hot water bath (100°C), and placed within a preheated (46°C) hybridization incubator. The samples were then allowed to cool overnight (12 h) to 46°C. The hybrids were then digested with RNase A/T1 at room temperature for 30 min and the protected bands were separated on a 5% PAGE/7 M urea gel. Electrophoresis was conducted at 1700 V for 3 h, and the gel was exposed to X-ray film with double-intensifying screens for 24 h. Sizes of RNA and DNA mol-wt positions +2450 and +2732. The quantity of RNA used for the RPA detection of β_1-AR transcripts varies with the tissue source or cell line. The pineal gland is a rich source of β_1-AR mRNA; as little as 1 µg total RNA can be used from this tissue to detect β_1-AR transcripts readily (**Fig. 2B**). Other tissues (cortex or heart) or cell lines (C6 glioma cell line) that express β_1-ARs may require as much as 50–70 µg total RNA for detection of β_1-AR protected fragments.

2. Materials

2.1. Preparation of RNA Sample

2.1.1. Isolation of Total RNA from Tissue or Cultured Cells

1. QIA RNeasy Midi Kit, a product of QIAGEN Inc. (Valencia, CA).
2. 1X Phosphate-buffered saline (PBS) in H_2O/DEPC.
3. Ethanol.

2.1.2. Determination of RNA Quantitation and Sample Integrity

1. 1X TE: 10 mM Tris-HCl, pH 8.0, 1 mM EDTA in H_2O/DEPC.
2. Formamide (store at 4°C).
3. Formaldehyde (**toxic**; use in chemical hood).
4. 5X Formaldehyde RNA gel-running buffer: 0.1 M MOPS, pH 7.0, 40 mM sodium acetate, 5 mM EDTA, pH 8.0.
5. Diethylpyrocarbonate (DEPC).
6. 0.1 mg/mL Ethidium bromide (EtBr) in H_2O/DEPC (Store in light-tight bottle at room temperature).
7. Agarose.
8. Formaldehyde gel-loading buffer: 50% glycerol, 1 mM EDTA, 0.025% bromophenol blue.

Fig. 2. *(continued)* standards are illustrated. (**C**) β_1-AR cRNA probes synthesized by T7 or SP6 RNA polymerase. Probe synthesized was pRPT, which contains the AU-rich polyadenylation region of the β_1-AR mRNAs. In vitro transcription reactions driven by T7 RNA polymerase utilized 100 µCi (33 pmol) [α-^{32}P]UTP and 400 pmol nonradioactive UTP (UTP/[α-^{32}P]UTP ratio is 12:1) (duplicate samples in both lanes). For in vitro transcription reactions driven by SP6 RNA polymerase, varying combinations of radioactive and nonradioactive UTP were used. Ratios of 12:1, 12:6, and 15:2 correspond to (1) 100 µCi (33 pmol) [α-^{32}P]UTP and 400 pmol nonradioactive UTP, (2) 120 µCi (39.6 pmol) [α-^{32}P]UTP and 500 pmol nonradioactive UTP, and (3) 120 µCi (39.6 pmol) [α-^{32}P]UTP and 600 pmol nonradioactive UTP, respectively. Probes were electrophoresed on a 5% PAGE/7 M urea gel at 1700 V for 3 h. Gels were exposed for autoradiography for <1 min. Note higher degree of radioisoptope incorporation into full-length cRNAs when using T7 RNA polymerase in the in vitro transcription reaction.

2.2. Synthesis of High Specific Activity Radiolabeled Antisense cRNA Probes

2.2.1. Preparation of Template for In Vitro Transcription

1. Restriction endonuclease (RE) with 10X restriction enzyme buffer (**Table 1**). Enzymes are available from New England Biolabs (Beverly, MA), Boehringer Mannheim Corp. (Indianapolis, IN), or Promega (Madison, WI).
2. Phenol: Saturate with 0.1 M Tris-HCl, pH 8.0, until pH of phenolic phase is >7.8. Add hydroxyquinoline to a final concentration of 0.1%, and β-mercaptoethanol to a final concentration of 0.2%. Equilibrated phenol can be stored in a light-tight bottle at 4°C for 1 mo. Phenol is highly corrosive and will cause severe burns. All manipulations should be carried out in a chemical hood. Wear gloves, safety glasses, and protective clothing when handling phenol.
3. Chloroform: chloroform : isoamyl alcohol = 24 : 1.
4. 10 mg/mL EtBr in H_2O (store in light-tight bottle at room temperature).
5. 100% Ethanol (−20°C).
6. 5X Tris-boric/EDTA electrophoresis buffer (TBE): 0.45 M Tris, 0.45 M boric acid, 0.01 M EDTA, pH 8.3.
7. QIAquick gel extraction kit, QIAGEN Inc.
8. 1X TE: 10 mM Tris-HCl, pH 8.0, 1 mM EDTA.

2.2.2. In Vitro Transcription

1. RNA polymerase and 10X transcription buffer. Enzymes are available from Ambion Inc. (Austin, TX).
2. [α^{32}P]UTP (3000 Ci/mmol), ICN Biomedical (Irvine, CA) (store at −20°C).
3. 10 mM rA, G, C, UTP, Pharmacia Biotech Products (Piscataway, NJ) (store at −85°C).
4. 100 mM Dithiothriotol (DTT); dissolve in 0.01 M sodium acetate, pH 5.2 (store at −20°C).
5. RNasin, Ambion Inc. (Austin, TX) (store at −20°C).

2.2.3. Purification of cRNA Probe

2.2.3.1. REMOVAL OF DNA TEMPLATE FROM TRANSCRIPTION REACTION MIX

1. RQ RNase-free DNase I, available from Promega (Madison, WI) (store at −20°C).
2. Formamide loading buffer: 95% formamide, 0.5 mM EDTA, 0.02% sodium dodecyl sulfate (SDS), 0.025% xylene cyanol FF, 0.025% bromophenol blue.

2.2.3.2. FULL-LENGTH CRNA PROBE PURIFICATION

1. Acrylamide (neurotoxin).
2. *N,N*′-methylenebisacrylamide.
3. 40% Acrylamide (acrylamide: *N, N*′-methylenebisacrylamide =19 : 1).
4. 5X TBE prepared with H_2O/DEPC.

Table 1
Commonly Used REs Generating
3′ Protruding End

*Aat*II	*Cfo*I	*Pvu*I
*Apa*I	*Hae*II	*Sac*I
*Ban*II	*Hgi*A I	*Sac*II
*Bgl*I	*Hha*I	*Sfi*I
*Bsp*1286 I	*Kpn*I	*Sph*I
*BstX*I	*Pst*I	

5. RNase Away, available from Molecular Bio-Products, Inc. (San Diego, CA).
6. Elution buffer: 2 *M* ammonium acetate, 1% SDS, 25 µg/mL tRNA.

2.3. Liquid Hybridization and RNase Digestion

2.3.1. Liquid Hybridization

1. 5X Hybridization stock solution: 200 m*M* PIPES, pH 6.4, 2 *M* sodium chloride, 5 m*M* EDTA.
2. 1X Hybridization working solution: dilute 5X Hybridization stock solution in formamide (1 : 4 dilution).

2.3.2. RNase Digestion

1. RNA digestion buffer: 10 m*M* Tris-HCl, pH 7.5, 300 m*M* sodium chloride, 5 m*M* EDTA.
2. RNase A/T1, RNase T1, available from Ambion Inc.
3. 20% SDS.
4. 2 mg/mL proteinase K in proteinase K buffer (PKB): 30 m*M* Tris-HCl, pH 7.5, 100 m*M* sodium chloride, 5 m*M* EDTA.
5. Acid-phenol: Saturate with H_2O/DEPC.
6. Chloroform.
7. 3 *M* sodium acetate, pH 5.2.
8. 100% Ethanol.

2.4. Electrophoresis and Autoradiography

This is the same as in **Subheading 2.2.3.2.**

3. Methods

3.1. Preparation of RNA Sample

3.1.1. Isolation of Total RNA from Tissues or Cultured Cells

Total RNA is extracted from tissues or cultured cells using QIAgen RNeasy Midi Kit.

3.1.2. Determination of RNA Quantitation and Sample Integrity

3.1.2.1. QUANTITATION OF RNA SAMPLE

1. Dilute RNA sample in 1X TE (1:150).
2. Optical density (OD) is measured in spectrophotometer (use 1X TE as blank to adjust background). Readings should be measured at wavelengths of 260 and 280 nm. The ratio of the readings at 260 and 280 nm (A_{260}/A_{280}) for pure RNA preparations is 2.0 or above.
3. RNA concentration can be calculated using the formula:

$$\text{RNA } (\mu g/\mu L) = A_{260} \times 40 \times 150/1000 \tag{1}$$

3.1.2.2. DETERMINATION OF RNA SAMPLE INTEGRITY

1. Melt 0.4 g of agarose in 26 mL H_2O/DEPC. Microwave for 1.5 min.
2. Cool to ~60°C. Add 8 mL of 5X formaldehyde RNA gel-running buffer and 7.154 mL of formaldehyde.
3. Cast the gel, and allow solidification for 30 min in a chemical hood.
4. Prerun the gel for 5 min at 5 V/cm.
5. Mix x μL RNA sample (2–4 μg) with 2.0 μL 5X formaldehyde RNA gel-running buffer, 3.5 μL of formaldehyde, 10 μL of formamide, and adjust final volume to 20 μL with H_2O/DEPC. (RNA volume x varies with the concentration.)
6. Denature RNA at 65°C for 15 min. Chill in ice bath for 5 min.
7. Add 0.5 μL of 0.1 mg/mL EtBr.
8. Add 1 μL formaldehyde gel-loading buffer and load onto agarose gel (gel described in **step 4**).
9. Electrophoresis is conducted at 4 V/cm for 30 min. Examine gel using a UV transilluminator. RNA samples are aliquoted into mass amounts suitable (60 μg) for β_1-AR RPA and immediately vacuum-concentrated. Samples should be stored at −85°C.

3.2. Synthesis of High Specific Activity Radiolabeled Antisense cRNA Probes

3.2.1. Preparation of Template for In Vitro Transcription

3.2.1.1. TEMPLATE LINEARIZATION

1. Digest 20–30 μg plasmid DNA (CsCl preparation) overnight with 100 U of restriction endonuclease recognizing the 5′ end of the β_1-AR insert (*see* **Notes 1** and **2**).
2. Extract the restriction endonuclease digestion mixture with 1 vol of phenol: chloroform (1:1).
3. Transfer the aqueous phase to a fresh tube. Add NaCl to a final concentration of 0.2 *M*.
4. Precipitate plasmid DNA with 2.5 vol of 100% ethanol at −20°C for >1 h.

5. Centrifuge at 12,000g for 15 min. Dry and resuspend the plasmid in 40–60 µL of H_2O or 1X TE (*see* **Note 3**).

3.2.1.2. PURIFICATION OF THE LINEARIZED TEMPLATE

1. Electrophorese RE-digested plasmid in a 1% agarose gel. Electrophoresis is conducted at 5 V/cm for 30–60 min. Stain gel in EtBr for 20 min and destain in H_2O for 10 min.
2. Excise the linearized plasmid band following visualization with a UV transilluminator, and recover the plasmid from agarose using QIAquick gel extraction kit (*see* **Note 4**).
3. Quantitate the plasmid template by spectrophotometric measurement and aliquot into 0.5–1 µg/tube. Store at –20°C.

3.2.2. In Vitro Transcription (See **Note 5**)

Linearized plasmid (0.5–1 µg), 100 µCi [α-^{32}P]UTP (3000 Ci/mmol) (ICN), and 1 µL SP6 or T7 RNA polymerase (Ambion) are used for in vitro transcription reaction.

1. Mix the in vitro transcription reagents according to the order shown in **Table 2** (*see* **Note 6**).
2. Transfer the reaction mixture into a 1.5-mL tube containing 100 µCi vacuum-dried [α-^{32}P]UTP (3000 Ci/mmol). Mix gently and spin briefly in microcentrifuge. Incubate at 37°C (T7 RNA polymerase) or 41°C (SP6 RNA polymerase) for 5 min (*see* **Note 7**).
3. Add 1 µL of T7 RNA or SP6 RNA polymerase to the appropriate reaction, mix gently (do not vortex), and spin briefly in microcentrifuge. Incubate at 37 or 41°C for 30–90 min (depending on the length of the cRNA being synthesized).

3.2.3. Purification of cRNA Probe

3.2.3.1. REMOVAL OF DNA TEMPLATE FOLLOWING IN VITRO TRANSCRIPTION

1. Add 1 µL of RQ RNase-free DNase I to transcription reaction mixture. Vortex and spin briefly in microcentrifuge. Incubate at 37°C for 30 min (*see* **Note 8**).
2. Add 30 µL RNA loading buffer, denature at 95°C for 5 min, chill on ice bath for 5 min, and immediately load samples in a denaturing polyacrylamide gel.

3.2.3.2. FULL-LENGTH CRNA PROBE PURIFICATION (*SEE* **NOTES 9** AND **10**)

Urea polyacrylamide gels (7 *M*) are used to purify the full-length cRNA probe. The concentration of polyacrylamide used in gels is determined by the size of the probes to be resolved (*see* **Note 11**).

1. Electrophoresis is conducted at 250 V. Stop the electrophoresis when the marker dye (xylene cyanol and bromophenol blue) migrates to the expected positions (*see* **Notes 11** and **12**).

Table 2
In Vitro Transcription

	Stock	Volume, µL	Final
Template		X	0.5–1 µg
Buffer	10X	2	1X
DTT	100 mM	2	10 mM
A, G, CTP	2.5 mM/each	4	0.5 mM
UTP	0.1 mM	2	0.01 mM
RNasin	40 U/µL	1	40 U
H$_2$O		To 19	

2. Remove one glass plate from the electrophoretic assembly, and leave the gel on the other glass plate.
3. Label the plate with fluorescent dye. Cover the gel with polyvinyl film, and briefly expose the gel to X-ray film (10 s to 1 min).
4. Align the X-ray film to the gel using the fluorescent dye markers. Excise the probe from the polyacrylamide gel using the X-ray film as a guide. Chop the gel into small pieces and soak in 2–3X gel volumes of elution buffer (*see* **Note 13**).
5. Two methods are used to elute the cRNA probe from the polyacrylamide gel.
   a. Method 1: Incubate the tubes with cRNA probe gel in 37°C water bath for 1 h. Vortex every 10 min.
   b. Method 2: Repeatedly freeze (at –20°C) and thaw (at 50°C) the gel fragments containing the cRNA probe (three to four times) (*see* **Note 14**).
6. Centrifuge briefly and transfer the elution buffer into a fresh 1.5-mL microcentrifuge tube. Extract with phenol/chloroform once. Transfer the aqueous phase to a fresh tube, and add 2 vol of 100% ethanol to precipitate the probe.
7. Leave the tube on dry ice for 15 min, and centrifuge at 12,000g for 15 min.
8. Vacuum-centrifuge the probe for 5 min, and resuspend in 100 µL of 1X hybridization buffer.
9. Transfer 1 µL of probe into 3 mL of scintillation. Measure the amount of radioactivity.

3.3. Liquid RNA Hybridization and RNase Digestion

3.3.1. Liquid Hybridization

Probes used in RPA: β_1-AR cRNA probe (5×10^5 cpm) and cyclophilin cRNA probe (1×10^3 cpm) were used in each hybridization.

1. Resuspend RNA samples in (30 µL-X µL-Y µL) 1X hybridization solution (where X is the volume containing 5×10^5 cpm β_1-AR cRNA probe, and Y is the volume containing 1×10^3 cpm cyclophilin cRNA probe) (*see* **Note 15**).
2. Mix β_1-AR cRNA (X µL volume) and cyclophilin cRNA (Y µL volume) probes. Vortex vigorously, and centrifuge briefly (*see* **Note 16**).

3. Add $(X + Y)$ µL β_1-AR and cyclophilin probe mixture to each RNA sample. Vortex and centrifuge briefly. Cover with one drop of mineral oil (*see* **Note 17**).
4. Denature RNA sample and probe mixture at 95°C for 15 min. Transfer the tube to a container of boiling water. Place the container into a hybridization incubator that has been preheated to the hybridization temperature (46°C). Allow the water to equilibrate to the hybridization temperature, and incubate for 12 h (*see* **Note 18**).

3.3.2. RNase Digestion

1. RNase is diluted in RNase digestion buffer to provide designated enzyme activity. Aliquot 370 µL of diluted RNase into fresh tubes. Carefully transfer the hybridization mixture into the tube containing diluted RNase. Incubate at room temperature for 30 (RNase A/RNase T1) or 60 min (RNase T1) (*see* **Note 19**).
2. Add 20 µL of proteinase K (2 mg/mL) and 10 µL SDS (20%). Incubate at 37°C for 30 min.
3. Extract with 1 vol of phenol/chloroform (5/1 ratio). Transfer the aqueous phase to a fresh tube, add 1/10 vol of 3*M* sodium acetate, pH 5.2, and 2.5X vol of 100% ethanol.
4. Place the tube on dry ice for 15 min, and centrifuge at 12,000*g* for 15 min at 4°C.

3.4. Electrophoretic Separation and Autoradiography

1. Vacuum-centrifuge the undigested probe to dryness, and resuspend in 2 µL H$_2$O/DEPC. Repeatedly vortex, and centrifuge two to three times to dissolve the undigested probe completely.
2. Add 4 µL of formamide-loading buffer to each tube. Denature at 95°C for 3–5 min, and chill in ice bath for 5 min. The sample should be loaded onto prerun denaturing polyacrylamide gel immediately.
3. Electrophoresis is conducted at 1700 V for 2–4 h (time is dependent on the length of the protected probe) to separate each protected band.
4. After electrophoresis is completed, remove the gel from the glass plates. Expose the gel to X-ray film at −85°C (duration of exposure is dependent on the intensity of the protected band). Use double-intensifying screens if the signal of the protected band is weak. **Figure 2** shows the results of RPA using β_1-AR cRNA probe from +82 to +273 and cyclophilin cRNA probe from +210 to + 323.

4. Notes

1. It is important to digest the plasmid completely prior to purification of the linearized plasmid by agarose-gel electrophoresis. A small amount of undigested plasmid DNA will give rise to longer transcripts, which may incorporate larger proportions of the radiolabeled rNTP. This will significantly decrease the yield and specific activity of expected cRNA probe.
2. Extraneous transcripts will be produced when template DNA containing 3′ protruding ends are utilized for in vitro transcription reactions (*13*). The extraneous

transcripts may contain additional sequences complementary to the vector DNA. Therefore, restriction endonucleases generating 3′ protruding ends should not be used to linearize plasmid DNA template for in vitro transcription. These enzymes are listed in **Table 1**.

If there are no acceptable restriction endonuclease recognition sites, the 3′ protruding end must be converted to a blunt end prior to in vitro transcription. T4 or T7 DNA polymerase can be used to convert 3′ protruding ends into blunt ends by using the following method:
   a. Phenol-extract the DNA following RE digestion.
   b. Precipitate DNA by adding 1/25 vol of 5 M NaCl, 2.5X vol of 100% ethanol. Place sample at −20°C for 30 min.
   c. Centrifuge the sample at 12,000g for 10 min at 4°C.
   d. Carefully remove the supernatant, and wash the DNA pellet once with 70% ethanol.
   e. Resuspend the template DNA in 1X T4 DNA polymerase buffer (100 μM of each dNTP and 0.1 mg/mL bovine serum albumin [BSA]). Add 5 U of T4 DNA polymerase/μg of DNA and incubate at 37°C for 5 min.
   f. The enzyme can be heat-inactivated at 75°C for 10 min.
3. Plasmid DNA should be resupended at a concentration of 0.5–1 µg/µL. Resuspend the DNA in a small volume to minimize band diffusion during electrophoresis.
4. Plasmid should be washed twice with 70% ethanol. Allow complete evaporation of ethanol by leaving the column open for 10 min after washing. Elute plasmid DNA with H$_2$O/DEPC.
5. We recommend using T7 RNA polymerase instead of SP6 RNA polymerase. The T7 RNA promoter is more active and T7 RNA polymerase is more stable than the SP6 promoter and RNA polymerase (*see* **Fig. 2**).
6. The amount of transcription substrate rATP, rGTP, rCTP, and rUTP can be increased if the expected transcript size is large. However, the ratio between non-radioactive rUTP and [α-^{32}P]rUTP should not be too high if the anticipated levels of target RNA are limiting.
7. Radioisotope [α-^{32}P]rUTP is vacuum-centrifuged to dryness to minimize the overall reaction volume and to maximize the specific activity of the probe.
8. The probe cannot be effectively purified by polyacrylamide gel electrophoresis, if the DNA template is not completely digested following in vitro transcription. DNA template will migrate at the same electrophoretic position as the probe. DNA templates will hybridize with radiolabeled cRNA probes, creating false positives in the RPA.
9. In vitro transcription will produce some incompletely synthesized transcripts, which will competitively hybridize to the target RNA, decrease the sensitivity of the assay, and increase the background. Polyacrylamide gel purification is the best method for separation of the full-length probe from incompletely synthesized transcripts. We recommend gel purification of the probe following

Table 3
Effective Range of Separation of RNAs
in Polyacrylamide/Urea Gels

% Polyacrylamide/ urea gel	Size of nucleotide, nt
4	>230
6	55–200
8	35–110
10	20–55
12	10–45

Table 4
Migration Rates of Loading Dye in Denaturing Polyacrylamide Gels[a]

Polyacrylamide, %/urea gel	Xylene cyanol FF, nucleotides	Bromophenol blue, nucleotides
4	140	36
6	100	23
8	68	18
10	50	11
15	29	9
20	18	7

[a]If DNA ladders are used as mol-wt markers, note that single-stranded DNA migrates at a position approx 10% smaller than RNA of equivalent nucleotide length.

in vitro transcription, especially if multiple probes are used in a single RPA hybridization.

10. The glass plates, spacers, and comb used in preparing the polyacrylamide gel should be cleaned and treated with RNase Away. This will minimize potential RNase contamination. RNase Away will deplete the gel slick application; we recommend treatment of the glass plate with additional gel slick prior to pouring new polyacrylamide gels.

11. Selection of appropriate polyacrylamide gel concentrations are critical to achieving maximal separation of the full-length probe from incompletely synthesized transcripts. When two or more probes are purified on a single gel, select the gel concentration appropriate for the longer probe, and stop the electrophoresis before the shorter probe elutes from the gel. **Tables 3** and **4** may be used as reference guides to select appropriate gel concentration and electrophoresis times.

12. The TBE buffer should be prepared using DEPC-treated water. The wells need to be cleaned carefully and thoroughly. Prerun the gel at 250 V for 30 min to heat the acrylamide matrix (~50°C) before loading the probes.

13. Carrier tRNA is not necessary if the in vitro transcription reaction is efficient and labels the probe to a high specific activity. In some instances, tRNA may contribute to false postives, if it retains sequences complementary to the probe.

14. Do not freeze the gel at −85°C or on dry ice, which will damage the probe. **Step 5a** in **Subheading 3.2.3.2.** has a lower recovery efficiency from the gel (~50% if the probe is <400 nt, and ~40% if the probe is 400–550 nt) compared to **step 5b** in **Subheading 3.2.3.2.** This step has a higher recovery efficiency (10–20% higher than that of 5-A), but may damage the probe if it is longer than 450 nt.

15. RNA samples must be completely resuspended in the hybridization buffer before adding probes.

16. β_1-AR probe and cyclophilin probe must be mixed together before adding to the RNA sample. Adding these probes individually to the RNA sample will increase experimental variability.

17. The hybridization mixture must be covered with mineral oil to prevent evaporation of the solution. Changes in salt concentration in the hybridization buffer will affect the formation of probe/target hybrids.

18. Hybridization time is critical for successful RPAs. Short hybridization times will decrease the sensitivity of the assay, whereas overly long hybridization times will increase background by creating hybrids with partial cRNA probes damaged by radioactivity-induced photolysis. Thus, in light of this consideration, we also recommend using freshly eluted probes to minimize damage induced by photolysis.

19. RNase should be titrated with each new lot of enzyme. Remove as much mineral oil as possible prior to transfer of the RNA/probe hybrids to the RNase digestion tube. Mineral oil will be difficult to remove in subsequent steps and will interfere with the loading of samples for gel electrophoresis.

Acknowledgments

We acknowledge our laboratory colleagues Philbert Kirigiti, Li Biao, Tarsem Moudgil, Gail Marracci, and Howard Nichols for their support in the development of this contribution. We also thank Carol Houser for support in the preparation of the manuscript and for Philbert Kirigiti, Melissa Kirigiti, and Tarsem Moudgil for careful reading of the manuscript. C. A. M. is supported by NIH RR0163, HL42358, DK53462, and was a prior recipient of an American Heart Association Established Investigatorship. Y. F.-Y was a 1997 American Heart Association Oregon Affiliate Postdoctoral Fellow.

References

1. Collins, S., Lohse, M. J., O'Dowd, B., Caron, M. G., and Lefkowitz, R. J. (1991) Structure and regulation of G protein-coupled receptors: The β_2-adrenergic receptor as a model. *Vitam. Horm.* **46**, 1–37.

2. Strosberg, A. D. (1995) Structural and functional diversity of β-adrenergic receptors. *Ann. NY Acad. Sci.* **757**, 253–260.

3. Collins, S., Caron, M. G., and Lefkowitz, R. J. (1988) β_2-adrenergic receptors in hamster smooth muscle cells are transcriptionally regulated by glucocorticoids. *J. Biol. Chem.* **263,** 9067–9070.

4. Fève, B., Emorine, L. J., Briend-Sutren, M.-M., Lasnier, F., Strosberg, A. D., and Pairault, J. (1990) Differential regulation of β_1- and β_2-adrenergic receptor protein and mRNA levels by glucocorticoids during 3T3-F442A adipose differentiation. *J. Biol. Chem.* **265,** 16,343–16,349.

5. Guest, S. J., Hadcock, J. R., Watkins, D. C., and Malbon, C. C. (1990) β_1- and β_2-adrenergic receptor expression in differentiating 3T3-L1 cells. Independent regulation at the level of mRNA. *J. Biol. Chem.* **265,** 5370–5375.

6. Hadcock, J. R. and Malbon, C. C. (1988) Down-regulation of β-adrenergic receptors: Agonist-induced reduction in receptor mRNA levels. *Proc. Natl. Acad. Sci. USA* **85,** 5021–5025.

7. Hadcock, J. R. and Malbon, C. C. (1988) Regulation of β-adrenergic receptors by "permissive" hormones: Glucocorticoids increase steady-state levels of receptor mRNA. *Proc. Natl. Acad. Sci. USA* **85,** 8415–8419.

8. Brown, T. and Mackey, K. (1997) Analysis of RNA by Northern and slot blot hybridization, in *Current Protocols in Molecular Biology,* vol. 1 (Ausubel, F. M., Brent, R., Kingston, R. E., Moore, D. D., Seidman, J. G., Smith, J. A., et al., eds.), Greene and Wiley-Interscience, New York, pp. 4.9.1–4.9.16.

9. Zeller, R. and Rogers, M. (1989) *In situ* hybridization to cellular RNA, in *Current Protocols in Molecular Biology,* vol. 2 (Ausubel, F. M., Brent, R., Kingston, R. E., Moore, D. D., Seidman, J. G., Smith, J. A., et al., eds.), Greene and Wiley-Interscience, New York, pp. 14.3.1–14.3.14.

10. Calzone, F. J., Britten, R. S., and Davidson, E. H. (1987) Mapping of gene transcripts by nuclease protection assays and cDNA primer extension. *Methods Enzymol.* **152,** 611–632.

11. Kekule, A. S., Lauer, U., Meyer, M., Caselmann, W. H., Hofschneider, P. M., and Koshy, R. (1990) The Pre S2/s region of intgrated hepatitis B virus DNA encodes a transcriptional transactivator. *Nature* **343,** 457–461.

12. Melton, D. A., Krieg, P. A., Rebagliati, M. R., Maniatis, T., Zinn, K., and Green, M. R. (1984) Efficient *in vitro* synthesis of biologically active RNA and RNA hybridization probes from plasmids containing a bacteriophage SP6 promoter. *Nucleic Acids Res.* **12,** 7035–7056.

13. Schenborn, E. T. and Mierendorf, R. C., Jr. (1985) A novel transcription property of SP6 and T7 RNA polymerases: dependence on template structure. *Nucleic Acids Res.* **13,** 6223–6236.

9

Determination of Adrenergic Receptor mRNAs by Quantitative Reverse Transcriptase-Polymerase Chain Reactions

Stefan Engelhardt and Martin J. Lohse

1. Introduction

The quantification of adrenergic receptor (AR) mRNAs is an important tool in the study of the physiological and pathophysiological regulation of these receptors. Alterations of the levels of these mRNA represent one of the many mechanisms that regulate receptor signaling *(1,2)*. Such alterations can be triggered by stimulation of the receptors themselves, but also by a variety of other causes. In patients, reductions of receptor mRNA levels have been observed in response to treatment with receptor agonists; in pathophysiological states, the best-known example is the downregulation of cardiac β_1-ARs in heart failure. On the other hand, upregulation of receptor mRNAs has been observed in response to stimuli, such as corticosteroids and thyroid hormones.

Various techniques exist to detect and quantify receptor mRNAs. Since in most instances these mRNAs are of very low abundance, the methods used for their detection need to be extremely sensitive. This chapter describes the most sensitive of these methods, which is based on the polymerase chain reaction (PCR). The PCR, first described by Saiki et al. *(3)*, is a highly sensitive, specific, and rapid method for the enzymatic amplification of specific nucleic acids. Since its introduction in 1985, this technique has seen rapid development and has become indispensable in many fields of research. This technique soon proved to be valuable not only for the detection of DNA, but also for the detection and quantification of specific mRNAs by combining the amplification procedure with a preceding enzymatic step where RNA serves as a tem-

From: *Methods in Molecular Biology, vol. 126: Adrenergic Receptor Protocols*
Edited by: C. A. Machida © Humana Press Inc., Totowa, NJ

plate for reverse transcription into complementary DNA (cDNA) *(4)*. This technique has been variously termed, but the term reverse transcriptase-polymerase chain reaction (RT-PCR) is now widely used and will be used throughout this chapter.

Before setting up a quantitative PCR assay, one might also consider traditional techniques, such as Northern blotting, RNA dot/slot blotting, and ribonuclease protection assays. Although these methods are less sensitive, they are considerably less prone to assay-inherent variations, which can be difficult to control. We chose RT-PCR whenever the amount of RNA was too small to allow for replicate measurements with conventional methods. Owing to the low expression levels of the adrenergic receptor mRNAs in human biopsy tissue or certain cell lines, this is often the case.

1.1. Standardization

Since minute amounts of RNA are rapidly degraded by nucleases and because of the exponential nature of the amplification step, the RT-PCR is prone to considerable variation. Several different strategies to control for such variation have been developed. In principle, there are two different approaches. (1) Either an unrelated or nonregulated endogenous mRNA is quantified in the same RNA sample. Owing to their ubiquitous and (more or less) unregulated expression, various housekeeping genes are suitable for this application. These standards are also referred to as endogenous standards. They can be quantified both in parallel or in the same tube with the actual target sequence. (2) The alternative approach is to use various kinds of exogenous synthetic RNA or DNA templates as standards (exogenous standards). These are added either to the RNA preparation or after the reverse transcription step, respectively. This method is especially powerful when used as competitive PCR, as first described by Gilliland et al. *(5)*.

In the second approach, various known amounts of synthetic templates, which contain identical primer recognition sequences, but differ either in length or in sequence from the target, are added to the RNA or cDNA preparation. The main advantage of this method is to allow directly the absolute quantification of specific target sequences. However, this method does not control for variations in sample processing and RNA preparation when the RNA standard is added after the RNA preparation step. If the exogenous standard is added after cDNA synthesis, variations occurring during the reverse transcription step cannot be controlled. Thus, exogenous standards can only control the later steps of the entire quantification procedure. However, often the variations occurring during the early steps of the procedure—sample procurement and RNA preparation—are much larger than those of the later steps. This is especially important when sample tissue with potential RNA degradation (such as biopsy tissue) is investigated. We have recently studied this problem by determining the sources of variation in the quantification of human β_1-AR mRNA from human heart tissue *(6)*, and found that the major sources of variation were located in the treatment of the tissue and the RNA preparation.

Thus, when working with samples in which degradation is likely to occur before freezing or processing, it appears reasonable to standardize experiments with an endogenous RNA standard, which is present from the very beginning of the procedure. As mentioned above, various housekeeping genes have been advocated for this purpose. Glyceraldehyde-3-phosphate dehydrogenase (GAPDH) is probably the most frequently used endogenous standard. Endogenous standards can be determined either in the same tube as the target RNA or in parallel in a different tube. Both approaches yield reasonable results. Although determination in the same tube can in principle also control for tube-to-tube variations in the PCR, it is often difficult to achieve exponential amplification of two or more genes of interest in one tube.

2. Materials
2.1. RNA Preparation

1. TRIZOL (Life Technologies, Grand Island, NY) or equivalent.
2. Chloroform.
3. Isopropanol.
4. 75% Ethanol.
5. 55°C water bath or thermoblock.
6. Refrigerated microcentrifuge.
7. Homogenization device (e.g., Ultraturrax, IKA Labortechnik, Staufen, Germany or Polytron, Luzern, Switzerland).
8. UV spectrophotometer and 50-μL cuvets.

2.2. Reverse Transcription

1. RNA preparation from **Subheading 2.1.**
2. Deoxynucleotide triphosphates (dNTPs,10 mM each of dATP, dCTP, dGTP, and dTTP).
3. Oligo(dT)$_{15}$ primers (5 μM, e.g., Boehringer Mannheim, Mannheim, Germany).
4. 200 U/μL Reverse transcriptase (e.g., SuperScript, Life Technologies).
5. Incubation buffer for reverse transcriptase (5X, supplied by the manufacturer of the enzyme).
6. 0.1 M dithiothreitel (DTT) (also supplied).
7. 40 U/μL RNase inhibitor (e.g., RNaseOut from Life Technologies).
8. 42 and 70°C water bath or thermoblock.

2.3. Polymerase Chain Reaction (PCR)

1. Thermocycler with high-temperature uniformity (e.g., Perkin Elmer Applied Biosystems, Weiterstadt, Germany).
2. cDNA (10 ng/μL in H_2O).
3. 5 U/μL *Taq* polymerase (e.g., Boehringer Mannheim).
4. 10X Incubation buffer for *Taq* polymerase (supplied by manufacturer).
5. dNTPs as in **Subheading 2.2.**
6. 3000 Ci/mmol [α-^{32}P]dCTP (e.g., Amersham, Braunschweig, Germany—optional).
7. Polyethylene tubes (preferably thin-walled PCR tubes, e.g. Perkin Elmer Applied Biosystems, Weiterstadt, Germany).
8. Light-weight mineral oil (Sigma, Deisenhofen, Germany; only necessary if thermocycler has no heated lid).
9. PCR additives, such as glycerol and dimethyl sulfoxide (DMSO) (optional).
10. Forward and reverse primers (20 μ*M* concentration; *see* **Note 1**).

2.4. Quantification of PCR Products Using PAGE

1. Electrophoresis chamber (with plates, spacers, combs, and power supply, e.g., Hoefer Pharmacia, San Francisco, CA).
2. Acrylamide solution (38% acrylamide/2% *bis*-acrylamide, ready-to-use, e.g., Roth, Karlsruhe, Germany).
3. 10X TBE buffer: 890 m*M* Tris, 890 m*M* boric acid, 20 m*M* ethylenediamine tetra-acetic acid (EDTA).
4. *N,N,N',N'* tetramethylethylenediamine (TEMED) (Bio-Rad, Richmond, CA).
5. 10% Ammonium persulfate in H_2O (APS, Roth).
6. Gel-loading buffer: 20% (w/v) Ficoll 400 (Pharmacia, Uppsala, Sweden, 0.1 *M* Na_2EDTA, 1.0% [w/v] sodium dodecyl sulfate [SDS], 0.25% [w/v] bromophenol blue, 0.25% [w/v] xylene cyanol).
7. Vacuum gel dryer.
8. Phosphoimager exposure plates and phosphoimager (alternative: autoradiography films and quantitative densitometric analysis).

For **Subheadings 2.1.** and **2.2.**, it is absolutely essential that all reagents and instruments are free from RNases. Usually commercial sterile plasticware is free from RNases if stored in a clean place and used only for RNA work. The same is true for most commercial reagents. Distilled water should be RNase-free, but this needs to be confirmed. The presence of RNases should be suspected when no or diffuse bands are obtained after **Subheading 3.1.** (*see below*).

3. Methods

3.1. RNA Preparation

1. Add 1 mL of TRIZOL to 2-mL Eppendorf tubes, and incubate on ice.
2. Add fresh or frozen sample, and homogenize immediately at 13,000 rpm for 20 s. It is critical that frozen samples do not thaw in the TRIZOL solution before being

homogenized. Put the homogenized sample back on ice, and clean the homogenization device with a few strokes in TRIZOL (may be reused throughout the assay)
and then in H_2O. If cells are the starting material, directly apply 1 mL cooled
TRIZOL onto the phosphate-buffered saline- (PBS) washed cells, and maximize
yield by collecting all cells with a cell scraper. When isolating total RNA from
tissue-culture cells, no homogenization step is required.

3. Incubate all homogenized samples at room temperature for 5 min.
4. Add 200 µL chloroform, and shake thoroughly for 15 s.
5. Incubate at room temperature for 2 min.
6. Microcentrifuge at 12,000g, at 4°C for 15 min.
7. Remove 500 µL of the upper aqueous phase, and transfer to a fresh tube.
8. Add 1 vol of isopropanol, mix by inverting the tube several times, and incubate at
   room temperature for 10 min.
9. Microcentrifuge at 12,000g at 4°C for 15 min.
10. Carefully decant the supernatant, and wash the residual pellet with 75% ethanol.
    Do not shake, since this will disrupt the pellet.
11. Microcentrifuge at 7500g for 5 min at room temperature, carefully decant the
    supernatant, collect the residual fluid with use of an additional short spin, and
    pipet off the remaining fluid.
12. Mark the position of the pellet on the outside of the tube (to facilitate **step 13**),
    and dry the RNA pellet for 5–10 min on the benchtop. It is important not to
    overdry the RNA pellet.
13. Dissolve the pellet in 20 µL of H_2O by placing the tube in a thermoblock at 55°C
    for 10 min.
14. Dilute the RNA about 50-fold, and determine the absorbance at 260 and 280 nm.
    The concentration of RNA is calculated by using the following equation:

$$C \ (\mu g/mL) = A_{260} \times 40$$

The ratio A_{260}/A_{280} should be between 1.7 and 2.0.
15. Store RNA at –80°C.

3.2. Reverse Transcription

1. Thaw RNAs, oligo(dT) primers, 5X incubation buffer, and DTT, and place on ice
   together with the reaction tubes.
2. Pipet 2 µL (10 pmol) of the oligo(dT) solution in each tube and add $9-x$ µL of
   ddH_2O.
3. Add x µL of the total RNA containing 500 ng or 1 µg.
4. Heat to 70°C for 10 min; in the interim period, prepare the following master mix
   (multiply volumes by number of samples + 1):

4 µL	5X Incubation buffer
2 µL	DTT
1 µL	dNTPs
0.1 µL	RNase inhibitor
1 µL	reverse transcriptase

        0.9 µL H_2O
        9 µL   Total volume
5. Chill tubes on ice, and centrifuge the sample for 1 s to collect condensate.
6. Add 9 µL of master mix.
7. Incubate at 42°C for 60 min.
8. Stop the reaction by heating to 70°C for 10 min, and chill on ice.
9. Add H_2O to yield a final concentration of 10 ng/µL, and store at –20°C.

3.3. Polymerase Chain Reaction (PCR)

1. Thaw all reagents, and chill on ice, except the *Taq* polymerase.
2. Prepare the following master mix (multiply volumes by number of samples + 1):
        5 µl   10X incubation buffer
        1.25 µL   forward primer
        1.25 µL   reverse primer
        1 µl   dNTPs
        0.3 µL   [α-^{32}P]dCTP (optional)
        0.25 µL   *Taq* polymerase
        35.95 µL H_2O
        45 µL   Total
3. Pipet 45 µL master mix to each PCR tube.
4. Add 5 µL of cDNA solution (=50 ng reverse-transcribed RNA).
5. In cases where the thermal cycler has no heated lid, overlay with 50 µL light-weight mineral oil (two drops).
6. Place in PCR cycler preheated to 94°C.
7. Cycle for the appropriate number of cycles (*see* **Notes 2–5**), and store at 4°C.

3.4. Separation of PCR Products by PAGE and Quantification by Phosphoimager Analysis

1. Prepare the gel apparatus for casting the gel.
2. For a 5% gel, prepare a solution of 6.25 mL of acrylamide/*bis*-acrylamide, 5 mL 10X TBE, and 38.75 mL H_2O.
3. Add 50 µL TEMED and 500 µL 10% adenosine phosphosulfate (APS), mix immediately, pour the gel, and insert the comb.
4. Let the gel polymerize for at least 30 min.
5. During the interim period, add 5 µL of gel-loading buffer to your PCR products.
6. Assemble the gel apparatus, and fill with 1X TBE buffer.
7. Load 45 µL from the PCR tubes onto the gel.
8. Depending on the size of the PCR product, electrophorese the gel at 25 mA until the PCR product has migrated approx 50% of the gel length. This has to be determined empirically, but it can be expected that approx 1 h of electrophoresis time is needed for a 500-bp product. Sufficient resolution of the PCR product is often obtained even when the majority of nucleotides are still retained in the gel. Thus,

one can easily dispose of the unincorporated nucleotides by excising the lower portion of the gel.

9. Disassemble the gel apparatus, leave the gel on one glass plate, and excise the lower portion of the gel containing the unincorporated nucleotides.
10. Cover the gel with a layer of Whatman paper, and following attachment, carefully remove the paper with the attached gel from the glass plate.
11. Cover the gel with a sheet of household plastic wrap, and dry the gel using a vacuum gel dryer for approx 60 min at 80°C.
12. Expose the gel to a phosphoimager plate for a few hours, and scan the plate.
13. Calculate the intensity of the PCR bands with the phosphoimager software, and subtract appropriate background values.

3.5. Determination of the Exponential Phase of the PCR Assay and Validation of Its Accuracy

Before determination of the mRNA abundance of specific genes in sample tissue or cells, two essential steps of standardization should be performed. The first step is to determine the exponential phase of the specific PCR. This is essential, since every PCR reaches a plateau phase after an exponential phase of DNA replication. When reaching the plateau phase of the PCR assay, potential differences in mRNA expression will not be sufficiently detected with your PCR assay. We use the following approach: After the assay is optimized to yield a single band of the expected size on an agarose gel, a radioactive nucleotide tracer is included in the reaction mixture, and 9–15 identical PCR tubes are prepared from one cDNA as described in **Subheading 3.3.** All tubes are placed in the thermal cycler together, but are cycled for different numbers of cycles. The smallest number of cycles sufficient to produce a clearly visible band on a phosphoimager can be expected to be three to six cycles less than the cycle number necessary for a clearly visible band on an agarose gel. Beginning at that minimum cycle number, three tubes are removed from the cycler every three cycles. The tubes should be removed at the end of the respective 72°C extension step. Subsequently, the samples are subjected to polyacrylamide gel electrophoresis (PAGE), the gels are dried and the radioactivity incorporated into specific bands is quantified by phosphoimager analysis. **Figure 1** shows a typical experiment determining the exponential phase for the amplification of the β_2-AR mRNA from human alveolar macrophages. From this experiment, a cycle number is chosen for the subsequent assays that is well below the upper limit of the exponential phase and is sufficient to yield distinct bands when the gel is analyzed with a phosphoimager.

The second essential step in the validation of a PCR assay is to determine whether and how accurately the assay can measure differences in mRNA abun-

A

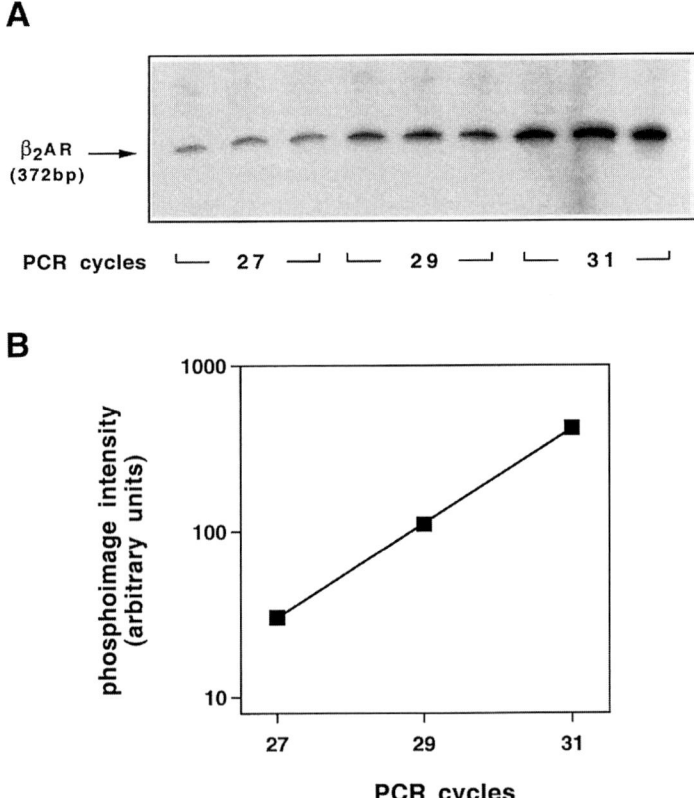

B

Fig. 1. Amplification of the β_2-AR mRNA from human alveolar macrophages. Determination of the exponential phase. Identical aliquots from one cDNA were amplified by PCR with the indicated number of cycles. The products were separated on polyacrylamide gels and quantified using phosphoimage analysis. (A) Phosphoimage with the radioactive PCR products as isolated bands. (B) Quantification in phosphoimage units. The half-logarithmic scale reveals that the reaction is exponential for at least 31 cycles. The symbols represent the mean of three determinations. The SEMs were smaller than the symbol size.

dance. For that purpose, varying amounts of cDNA are added to the reaction and amplified with the number of cycles that was determined in the first reaction. We routinely use 12.5, 25, 50, and 100 ng of reverse-transcribed RNA for this determination. A typical experiment is shown in **Fig. 2**.

After the determination of the optimal number of cycles and the appropriate amount of cDNA for the gene of interest and for an endogenous control gene, the mRNA abundance of the specific gene is quantified with the determined parameters.

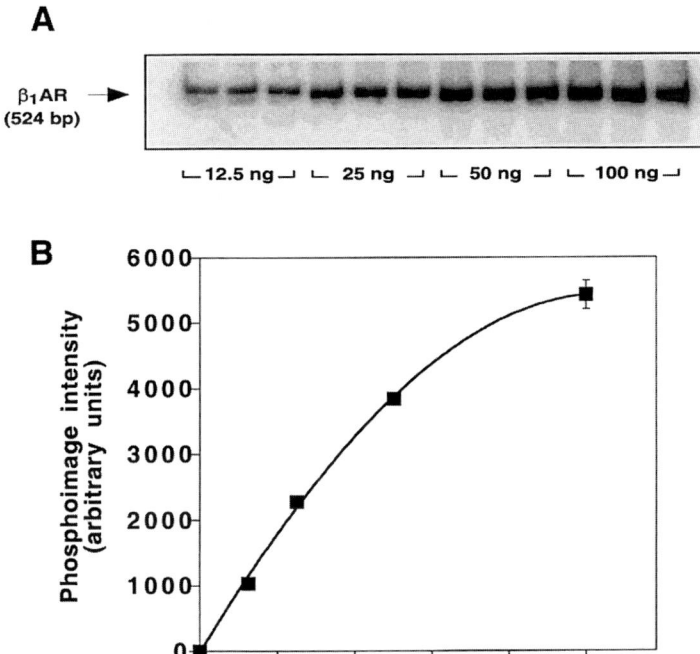

Fig. 2. Determination of the linear range of the PCR quantification. Different amounts of reverse-transcribed RNA were subjected to 29 rounds of amplification (**A**). Phosphoimage analysis revealed linearity up to 50 ng of reverse-transcribed RNA (**B**). Symbols represent means ± SEM (not shown when smaller than symbol size).

3.6. Normalization

When the PCR products are quantified as described above, this provides relative data in phosphoimage units. To allow a comparison between different samples, it is necessary to use a normalization standard as outlined at the beginning of the chapter. We prefer normalization to an endogenous standard mRNA, such as GAPDH. In the normalization, both the target mRNA (e.g., receptor mRNA) and the standard mRNA are quantified, and then the value for the target mRNA is divided by the value for the standard mRNA. Thus, the end result is dimensionless and is dependent on the specific assay conditions chosen.

Such results are sufficient to compare relative alterations, provided that all samples for such comparisons are treated identically and are preferably analyzed in the same experiment. Typical examples would be the agonist-induced reduction of a receptor mRNA in cell lines or the comparison of receptor mRNAs in samples from two groups of patients.

If absolute levels of an mRNA need to be determined, this requires the preparation and use of a synthetic target mRNA. This mRNA is synthesized with RNA polymerase using the target cDNA as the template. Defined amounts of this synthetic mRNA are then analyzed as described above. The simplest approach is to construct a standard curve with known amounts of synthetic mRNA and to examine the samples of interest in the same experiments. The procedures have been described by Engelhardt et al. *(6)*.

Alternatively, competitive PCR uses a slightly modified synthetic mRNA resulting in a PCR product that can be distinguished on gels from that of the target mRNA (either because it contains an additional restriction site allowing its digestion before PAGE or because it is different in length). For the quantitative analysis, reactions with different amounts of the synthetic mRNAs are established for each sample, and both bands are quantified for each reaction. This provides a standard curve for each sample, and the amount of target mRNA corresponds to the amount of standard mRNA at the point where target and standard mRNA result in equal amounts of PCR product, as described by Gilliland et al. *(5)*.

4. Notes

1. Primers for PCR: Whether certain primers for a given sequence are working or not is not very predictable, and some primers are better than others for no obvious reason. However, there are some rules that facilitate the selection of good primers. The end of this note contains a list of successfully used primers for adrenergic receptors.

   For the purpose of quantitative PCR, primers should have 100% sequence complementarity. Their length should be between 18 and 30 bp, and the expected product size should be <1000 bp. At higher product sizes, the synthesis efficiency may decrease *(7)*, and also extension times become significantly longer. The GC content of primers should be similar to that of the template, and any "unusual" sequences, such as repetitive single bases, should be avoided. Furthermore, attention should be given to the 3′ ends of primer pairs to avoid any 3′ end complementarity between two primers, since complementarity of the 3′ ends may lead to the formation of primer dimers that might compete with the specific PCR product. In addition to the above-mentioned rules, one might also consider the help of computer software, which is available from a variety of companies. If even after all considerations, primers exist that simply do not amplify, after some optimization effort (*see* **Note 2**), it is most economical to try another set of primers. **Table 1** provides an overview of selected primer sequences that have been used successfully for the amplification of adrenergic receptor mRNAs and for the housekeeping gene GAPDH.

Table 1
Primers Used in the Amplification of Adrenergic Receptor and GAPDH mRNAs

mRNA, species	Primer sequence	Position in coding sequence	Reference
α_{1B}-AR (rat)	5'- GCT CCT TCT ACA TCC CGC TCG -3'	629 Forward	9
	5'- AGG GGA GCC AAC ATA AGA TGA -3'	928 Reverse	
α_{1D}-AR (rat)	5'- CGT GTG CTC CTT CTA CCT ACC -3'	759 Forward	9
	5'- GCA CAG GAC GAA GAC ACC CAC -3'	1062 Reverse	
β_1-AR (human)	5'-CTC ACC AAC CTC TTC ATC ATG -3'	272 Forward	6
	5'-GAA ACG GCG CTC GCA GCT G -3'	795 Reverse	
β_2-AR (human)	5'-CCT CCT AAA TTG GAT AGG -3'	927 Forward	6
	5'-AGT CTG TTT AGT GTT CTG -3'	1298 Reverse	
β_1-AR (rat)	5'- CTC ACC AAC CTC TTC ATC ATG -3'	274 Forward	Unpublished
	5'- GAA GCG GCG CTC GCA GCT G -3'	795 Reverse	
β_2-AR (rat)	5'- CCT CCT TAA CTG GTT GGG -3'	927 Forward	Unpublished
	5'- AGT CTG GTT AGT GTC CTG -3'	1314 Reverse	
β_3-AR (mouse)	5'- ATG GCT CCG TGG CCT CAC -3'	1 Forward	10
	5'- CTG GCT CAT GAT GGG C -3'	519 Reverse	
β_3-AR (rat)	5'-TAG TCC TGG TGT GGA TCG TGT CCG C -3'	464 Forward	11
	5'- GCG ATG AAA ACT CCG CTG GGA ACT A -3'	980 Reverse	
GAPDH (human)	5'- GCT TTT AAC TCT GGT AAA GTG G -3'	64 Forward	12
	5'- TCA CGC CAC AGT TTC CCG GAG G -3'	593 Reverse	
GAPDH (rat)	5'- GCT GCC TTC TCT TGT GAC AAA -3'	55 Forward	Unpublished
	5'- CAC GCC ACA GCT TTC CAG A -3'	586 Reverse	
GAPDH (mouse)	5'- GCT GCC ATT TGC AGT GGC AAA -3'	55 Forward	Unpublished
	5'- ATC ACG CCA CAG CTT TCC AGA -3'	588 Reverse	

2. Cycling parameters: The recommended parameters for one PCR cycle are 1 min of denaturation at 94°C, 1 min of annealing of the primers to the template at the optimal annealing temperature, and 1 min of extension at 72°C. These relatively long cycling times can be significantly shortened when using thin-walled PCR tubes and fast PCR machines. For example, using PCR machines with heated lids from Perkin Elmer, we routinely use 15 s of denaturation and 30 s of annealing. Both processes should occur very rapidly once the target temperature is reached inside the tube. For PCR products composed of 100–250 bp in length, we use 30 s of extension at 72°C; for products composed of 250–1000 bp, we use 60 s of extension. In addition to the use of computer programs, there are several methods to calculate the optimal annealing temperature for a given set of primers. A traditional method, known as the "2 + 4 rule," is to estimate 2°C for every A or T and 4°C for every C or G within the primer sequence. This usually works well for

short oligonucleotides, but is less optimal for longer oligonucleotides. Also, the following simple rule (8) works well: Anneal at 55°C if the GC content of the primers is below 50%; if GC content is higher, anneal at 60°C. If any additional bands appear with the chosen annealing temperature, anneal at 2–5°C above that temperature and determine if this reduces or eliminates the nonspecific bands. In addition, we always preheat the PCR cycler to 94°C before placing the tubes into the thermoblock, thus reducing the initial ramp time where nonspecific annealing can occur. If this is not sufficient, the reaction components of the PCR should be optimized (*see* **Note 3**).

3. Optimization of reaction components: If a primer pair fails to produce a single band with the cycling parameters described in **Note 2**, we attempt to optimize the reaction components by using several MgCl$_2$ concentrations and, glycerol and DMSO as additives (modified from **ref. 8**). We routinely use the following approach:

   a. Prepare three master mixes according to **Table 2**.
   b. Prepare three MgCl$_2$ solutions of 10, 30, and 45 m*M*.
   c. Combine 5 µL of each of the three MgCl$_2$ solutions with 45 µL of each of the three master mixes, amplify with the appropriate number of cycles, and electrophorese the products on an agarose gel.

   The recommended preheating of the PCR cycler before placing the cooled tubes into the thermoblock cannot fully prevent nonspecific annealing and subsequent generation of nonspecific products. If this is a problem, try using *Taq* antibodies (Clontech), which keep the polymerases inactive until the antibodies are denatured with the first denaturation step, or AmpliTaqGold (Perkin Elmer), a *Taq* polymerase that becomes activated only after an initial incubation at 94°C. Both methods can increase the sensitivity and specificity of the PCR assay. We do not recommend the classic "hot-start" technique for a quantitative PCR assay, since this method requires a second opening of the PCR tubes after they have been positioned in the thermocycler; this represents an additional chance to introduce contaminations.

4. Cycling parameters for the amplification of β-AR mRNAs used by the authors: Amplification of β$_1$-AR mRNA from human and rat myocardium:
   Standard reaction components with 5% DMSO.
   30 Cycles with 30 s 94°C, 30 s 58°C, 1 min 72°C.
   Amplification of β$_2$-AR mRNA from human and rat myocardium:
   Standard reaction components.
   29 Cycles with 30 s 94°C, 30 s 50°C, 1 min 72°C.

5. Trouble-shooting: The two main problems with this procedure are the lack of a PCR product or the appearance of a PCR product where no amplification should occur. Both problems can be owing to a variety of reasons; however, a few causes are most likely and need to be controlled, preferably in each experiment. Apart from nonworking primers or incorrect amplification conditions, the two most

Table 2
Master Mixes Used in the Optimization of Reaction Components

Components	Volume per reaction, μL	Master mix 1, μL	Master mix 2, μL	Master mix 3, μL	Final concentration
PCR-buffer 10X (Mg²⁺-free)	5	20	20	20	1X
Forward primer	1.25	5	5	5	0.5 μ*M*
Reverse primer	1.25	5	5	5	0.5 μ*M*
cDNA	10	40	40	40	10 ng/μL
dNTP (10 m*M* each)	1	4	4	4	200 μ*M*
DMSO	2.5	—	10	—	5%
Glycerol	5	—	—	20	10%
H₂O 45 μL	18.75	105	95	85	—
Taq polymerase	0.25	1	1	1	2.5 U

important causes are degradation of RNA and contamination with the DNA template.

a. Degradation of RNA: RNAs are very easily degraded, and the RNases responsible for this process are very robust enzymes. They are likely to be carried on the investigator's hands, and also contaminate many laboratory reagents and instruments. The usual precautions are: work with gloves, keep special reagents for RNA work, and use sterilized disposable plasticware. Although this is counterintuitive, autoclaving does not always remove RNases, and it is not advised to autoclave commercial reagents or plasticware. The use of filter tips reduces the risk of contamination with RNases with DNA template.

b. Contaminations: The appearance of a PCR product where no amplification should occur is an often observed problem in RT-PCR, particularly when PCR of a given template is used frequently. The most obvious reason is contamination with DNA template. Because of the enormous levels of amplification, even minute amounts of contaminants can be sufficient to produce robust false-positive results. In the case of intronless genes—including most adrenergic receptor genes—genomic DNA is one source of contamination. The other source of contamination is with plasmids or PCR products. The precautions to be taken for the latter are essentially similar to those advised to prevent contamination with RNases. Spatial separation of the RNA preparation steps from work with the corresponding PCR products may be helpful. If the contaminant is genomic DNA, the RNA preparation step needs to be examined. When using our protocol for RNA preparation, genomic DNA most likely results from working with excessive amounts of tissue or cells. In order to verify lack of contaminants in RT-PCR, it is best to include a control reaction without reverse transcriptase in each experiment. If a PCR product is obtained in this tube, it indicates the presence of DNA contaminations. Experimental evaluation should be conducted only on those experiments where this control is negative.

References

1. Collins, S., Caron, M. G., and Lefkowitz, R. J. (1991) Regulation of adrenergic receptor responsiveness through modulation of receptor gene expression. *Annu. Rev. Physiol.* **53,** 497–508.
2. Lohse, M. J. (1993) Molecular mechanisms of membrane receptor desensitization. *Biochim. Biophys. Acta.* **1179,** 171–188.
3. Saiki, R. K., Scharf, F., Faloona, K. B., Mullis, K. B., Horn, G. T., Ehrlich, A. H., et al. (1985) Enzymatic amplfication of beta-globin genomic sequences and restriction site analysis for diagnosis of sickle cell anemia. *Science* **230,** 1350–1354.
4. Chelly, J., Kaplan, J. C., Gautron, S., and Kahn, A. (1988) Transcription of the dystrophin gene in human muscle and non-muscle tissues. *Nature* **333,** 858–860.
5. Gilliland, G., Perrin, S., Blanchard, K., and Bunn, H. F. (1990) Analysis of cytokine mRNA and DNA: detection and quantitation by competitive polymerase chain reaction. *Proc. Natl. Acad. Sci. USA* **87,** 2725–2729.
6. Engelhardt, S., Böhm, M., Erdmann, E., and Lohse, M. J. (1996) Analysis of β-adrenergic receptor mRNA levels in human ventricular biopsy specimens by quantitative polymerase chain reactions: progressive reduction of β$_1$-adrenergic receptor mRNA in heart failure. *J. Am. Coll. Cardiol.* **27,** 146–154.
7. Jeffreys, A. J., Wilson, V., Neumann, R., and Keyte, J. (1988) Amplification of human minisatellites by the polymerase chain reaction: Towards DNA fingerprinting of single cells. *Nucleic Acids Res.* **16,** 10,952–10,971.
8. Ausubel, F. M., Brent, R., Kingston, R. E., Moore, D. D., Seidman, J. G., Smith, J. A., et al. (1994–1998) *Current Protocols in Molecular Biology.* Wiley, New York.
9. Scofield, M. A., Liu, F., Abel, P. W., and Jeffries, W. B. (1995) Quantification of steady state expression of mRNA for alpha-1 adrenergic receptor subtypes using reverse transcription and a competitive polymerase chain reaction. *J. Pharmacol. Exp. Ther.* **275,** 1035–1042.
10. Feve, B., Elhadri, K., Quignard-Boulange, A., and Pairault, J. (1994) Transcriptional down-regulation by insulin of the β$_3$-adrenergic receptor expression in 3T3-F442A adipocytes: a mechanism for repressing the cAMP signaling pathway. *Proc. Natl. Acad. Sci. USA* **91,** 5677–5681.
11. Evans, B. A., Papaioannou, M., Bonazzi, V. R., and Summers, R. J. (1996) Expression of β$_3$-adrenoceptor mRNA in rat tissues. *Br. J. Pharmacol.* **117,** 210–216.
12. Ungerer, M., Böhm, M., Elce, J. S., Erdmann, E., and Lohse, M. J. (1993) Altered expression of β-adrenergic receptor kinase and β$_1$-adrenergic receptors in the failing human heart. *Circulation* **87,** 454–463.

10

Nuclear Run-On Assays for Measurement of Adrenergic Receptor Transcription Rate

Zhuo-Wei Hu and Brian B. Hoffman

1. Introduction

The functionally diverse group of G-protein-coupled receptors (GPCRs) is a superfamily of membrane receptors. They include receptors for many different signaling molecules, such as peptide and nonpeptide hormones, neurotransmitters, chemokines, prostanoids, and proteinases. The principal function of GPCRs is to transmit information about the extracellular environment to the interior of the cell by interacting with the heterotrimeric G-proteins and, thereby, participate in regulation of many cellular functions. In view of their major importance, it is not surprising that GPCR-mediated responses are subject to dynamic regulation by a number of mechanisms. These regulatory mechanisms have important roles in fine-tuning signals from multiple receptor signaling pathways. Multiple mechanisms contribute to the regulation of GPCRs and their transmembrane signaling. Posttranslational modifications of the receptors, such as phosphorylation, may modulate receptor function; in addition, changes in receptor gene expression can lead to alterations in sensitivity and responsiveness of cells to various signaling molecules *(1–3)*.

Adrenergic receptors are members of the superfamily of GPCRs. Adrenergic receptors and their downstream signaling pathways play important roles in regulation of many catecholamine-induced cellular responses, ranging from regulation of intracellular concentrations of cAMP to stimulation of gene transcription. Adrenergic receptors are prototypic models for the study of the relations between cell regulation and receptor regulation of GPCRs *(4–6)*. There are three major classes of adrenergic receptors (ARs), termed α_1-AR, α_2-AR,

From: *Methods in Molecular Biology, vol. 126: Adrenergic Receptor Protocols*
Edited by: C. A. Machida © Humana Press Inc., Totowa, NJ

and β-AR; each of these classes has three known subtypes of adrenergic receptors: α_{1A}-, α_{1B}-, and α_{1D}-AR; α_{2A}-, α_{2B}-, and α_{2C}-AR; and β_1-, β_2-, and β_3-AR.

The regulation of adrenergic receptor expression has been an area of intense investigative interest over the past several decades *(7–11)*. Considerable interest has focused on agonist regulation of receptor responsiveness, for example, desensitization of β-AR activation of downstream effects; posttranslation modifications in these receptors, especially phosphorylation, play major roles in these regulatory alterations. Perhaps less well known, but also important are examples where alterations in cellular responsiveness to catecholamines occur as a consequence of changes in adrenergic receptor expression; these may be owing to modification in the rates of receptor gene transcription. For example, a variety of second messengers and heterologous hormones are known to regulate expression of adrenergic receptors *(12–16)*.

Tissue-specific expression and the level of expression of receptor genes are generally determined by measuring the abundance of the corresponding mRNA. However, changes in abundance of mRNAs can result from alterations in the transcription rate or degradation rate of the mRNA. Whether a change in abundance of mRNA is owing to a change in the receptor gene transcription can be measured by the nuclear run-on assay through identification of the newly transcribed products of adrenergic receptor genes (*see* **Note 1**). Although rapid progress in molecular biological techniques has enabled the study of gene expression in a wide variety of tissues and cultured cells, a nuclear run-on assay is currently the most sensitive procedure for measuring specific gene transcription. There are two general methods for nuclear run-on assays, solution hybridization and filter hybridization, useful in determining transcription product mRNA of a specific adrenergic receptor gene. The two methods involve using transcriptionally active nuclei to generate new synthetic RNA in the presence of labeled nucleoside triphosphate; the newly synthesized RNA is then isolated. Both methods involve the use of a complementary nucleotide sequence to detect and quantify selectively the specific labeled mRNA of interest. For solution hybridization, the hybrids are isolated by filter binding and are quantitated by scintillation spectrophotometry. In the second method, hybridization is performed by incubating the labeled RNA probes with a filter which has been prebound with the unlabeled nucleotide strand. Quantitation is then performed either by film autoradiography or by a PhosphorImager System.

In this chapter, we will discuss the filter hybridization method only. We will present a standard protocol for a nuclear run-on assay using $\alpha[^{32}P]UTP$ to label nascent RNA transcripts. In recent years, replacing radioactive labeled probes by nonradioactive ones and detection by chemiluminescence provide an alternative method to avoid to the hazardous handling of RNA labeled to a high specific ^{32}P activity as in the standard procedure. We therefore will describe a method using nonradioactive dioxigenin- (DIG) labeled UTP instead of ^{32}P-labeled UTP to transcribe the RNA with subsequent detection with a chemiluminescent system. The standard protocol for nuclear run-on assays involves multiple steps and is relatively cumbersome and time-consuming. Consequently, we describe a simplified procedure, which will not only save at least 2–3 h of processing time, but also reduce exposure to ^{32}P and radioactive waste.

2. Materials (*see* Note 2)
2.1. Isolation of Nuclei from Tissues

1. Buffer A: 60 mM KCl, 15 mM NaCl, 0.15 mM spermine, 0.5 mM spermidine, 14 mM β-mercaptoethanol, 0.5 mM ethyleneglycol-bis-(β-aminoethylether) (EGTA), 2mM ethylenediamine tetra-acetic acid (EDTA), 10 mM N-2-hydroxyethylpiperazine-*N*-2-ethanesulfonic acid (HEPES) (pH 7.6), 1 mM dithiothreitol (DTT), 0.5 mM phenylmethylsulfonyl fluoride (PMSF), and 2 μg/mL each of aprotinin, leupeptin, and bestatin.
2. Homogenization buffer: 0.3 M sucrose in buffer A.
3. Cushion buffer: 2.2 M sucrose in buffer A.
4. Nuclei storage buffer: 50% glycerol, 50 mM HEPES pH 7.6, 2 mM MgCl$_2$, 0.1 mM EDTA, 1.0 mM DTT, 0.1 mM PMSF.

2.2. Isolation of Nuclei from Cultured Cells

1. Phosphate-buffered saline (PBS) wash buffer: 20 mM Tris-HCl, pH 7.5, 20% glycerol, 140 mM KCl, 10 mM MgCl$_2$.
2. Lysis buffer: 0.6 M Sucrose, 0.5% Nonidet P-40, and 0.5 mM DTT. Prepare fresh from stock solution.

2.3. Transcriptional Synthesis of RNA

1. 2X transcription reaction buffer: 10 mM Tris-HCl, pH 8.0, 5 mM MgCl$_2$, 0.3 M KCl, 0.2 mM EDTA, 1 mM DTT (add fresh).
2. Nucleotide mix: Prepare mixture with each 100 mM solution of ATP, CTP, and GTP (Pharmacia) at 1:1:1 ratio. Store frozen in aliquots at –20°C.
3. HBS buffer: 10 mM Tris-HCl, pH 7.4, 0.5 mM NaCl, 50 mM MgCl$_2$.
4. Ribonuclease inhibitor: human placental ribonuclease inhibitor (Promega, Inc.) 40 U/μL.
5. [α-^{32}P]UTP: Labeled UTP with SA of 800 Ci/mmol. Higher specific activities can be used if necessary.
6. DNase I: 10 mg/mL (RNase-free).
7. SET buffer: 5% sodium dodecyl sulfate (SDS), 50 mM EDTA, 100 mM Tris-HCl, pH 7.4. Prepare fresh.
8. Proteinase K: 10 mg/mL, store at –20°C.
9. RNA extraction buffer (NRO buffer II): 4 M guanidium thiocyanate, 25 mM sodium citrate, pH 7.0, 0.5 % Sarkosyl, and 0.1 M β-mercaptoethanol. Store at room temperature.

10. Sodium acetate: 2.0 M, pH 4.0.
11. RNA dilution buffer: diethylprocarbonate (DMPC)-treated H_2O, 20X standard saline citrate (SSC), and formaldehyde mixed in a volume ratio of $5:3:2$.
12. Phenol and chloroform: Salt-saturated phenol and a $49:1$ chloroform:isoamyl alcohol mix.
13. Yeast tRNA: 20 mg/mL solution.
14. Isopropanol and ethanol (70%).

2.4. Hybridization

1. 1 M HEPES (free acid).
2. Hybridization buffer: 20 mM PIPES, pH 7.0, 50% formamide, 2 mM EDTA, 0.8 M NaCl, 0.2% SDS, 0.02% Ficoll, 0.02% polyvinylpyrrolidone, 0.02% bovine serum albumin (BSA), and 500 g/mL denatured salmon sperm DNA. Make fresh before use.
3. 20X SSC: 3 M NaCl, 0.3 M tri-sodium citrate, 6X SSC, 2X SSC, and 0.5X SSC made by dilution of 20X SSC.
4. RNase A solution: 10 mg/mL.
5. Scintillation cocktail.
6. Glacial acetic acid.

2.5. Nonradioactive Detection

1. DIG-labeled UTP (Boehringer Mannheim).
2. NRO buffer: 4.0 M guanidinium isothiocyanate, 25 mM sodium citrate, pH 7.0, 0.5% Sarkosyl, 0.1 M 2-mercaptoethanol.
3. PCI: phenol-chloroform-isoamyl alcohol, $24.5:24.5:1$.
4. 0.5% SDS.
5. Alternate hybridization buffer: 0.5 M sodium phosphate, pH 7.2, 7%.
6. Wash buffers:
   a. 250 mM sodium phosphate, 1% SDS.
   b. 100 mM sodium phosphate, 1% SDS.

2.6. Equipment for Run-on Assays

1. Homogenizer: a glass or Teflon-glass Quick-fit (such as a 55-mL Potter) homogenizer is generally used for small amounts of tissue material. For larger amounts of tissue, use a conventional food mixer or, better, a Ystral homogenizer (Ystral Gmbh, Dottingen, Germany). Avoid homogenizers that generate ultrasound such as a Polytron.
2. Low-speed swinging-bucket centrifuge.
3. Glass and plasticware.
4. Scintillation counter.
5. Laser densitometer.
6. Phosphoimager.
7. Standard microscope with fluorescent excitation.

3. Methods

The standard procedure for run-on assays has been adapted from a protocol described previously by Greenberg and Bender *(17)*, and has been slightly modified and successfully used in our lab for measuring the transcription rate of several adrenergic receptor genes. The procedure can be divided into several steps. First, cells are harvested, and the transcriptionally active nuclei are isolated. The isolated nuclei can be used fresh or frozen in liquid nitrogen, where they can be stored for up to a year. Fresh or thawed nuclei are then used for transcription reactions wherein the newly transcribed RNA is labeled with ^{32}P-labled UTP. [^{32}P]UTP-labeled RNA is then purified and used to detect specific RNA transcripts by hybridization to cDNAs that have been prebound onto nitrocellulose filter or nylon membranes. Finally, the filter or membrane is exposed to X-film or used in a Phosphoimager System for quantitation of signals.

3.1. Isolation of Active Nuclei (see Note 3)

3.1.1. Preparation of Nuclei from Cultured Cells

1. Remove medium from tissue-culture flasks or dishes (1–5×10^7 cells/assay). Place cells on ice.
2. Wash flask or dish with 5 mL ice-cold PBS wash buffer. Gently collect cells, and centrifuge cells for 5 min at $500g$ (1500 rpm in JS-4.2 rotor) at 4°C. Remove supernatant completely.
3. Loosen the cell pellet by vortexing at half-maximal speed for 5 s prior to the addition of 4 mL NP-40 lysis buffer. Continue vortexing as the buffer is added. Once the lysis buffer is completely added, vortex the cells at half-maximal speed for an additional 10 s.
4. Allow cells to sit on ice for 10 min. Centrifuge for 5 min at $1500g$ at 4°C.
5. Discard the supernatant and resuspend the pellet in 4 mL of NP-40 lysis buffer as in **step 3**. Centrifuge as in **step 4**.
6. Discard the supernatant, and resuspend the pellet in 100 μL of storage buffer by gently vortexing. The resuspended nuclei can be either processed directly in the next step or frozen in liquid nitrogen and stored at –70°C. The nuclei can be stored for up to 1 yr.

3.1.2. Isolation of Active Nuclei from Intact Tissues

The method for preparation of transcriptionally active nuclei from rat liver tissue outlined below is based on previous methods described by Hattori et al. *(18)* (*see* **Note 3**).

1. Hepatectomies are performed, and the livers (from two rats, body wt about 150–250 g) are cut with scissors into approx 3-mm cubes in 10 mL of homogenization buffer.

2. Homogenize tissue gently using a hand or powered homogenizer. Half of the minced tissue is added to 25 mL of homogenization buffer in a 55-mL Potter homogenizer and is homogenized with two or three strokes by a motor-driven Teflon pestle.

3. The homogenate is transferred to a precooled 250-mL cylinder by filtering through a cheesecloth to remove debris.

4. The remaining half of the tissue is then homogenized following the same procedure.

5. The homogenized tissue (50 mL in solution) is then mixed with 100 mL of cushion buffer (identical to homogenization buffer, except that the concentration of sucrose is 2.2 M), resulting in a final sucrose concentration of 1.57 M.

6. The homogenate is then laid over 10 mL of cushion buffer in 38-mL polyallomer ultracentrifuge tubes. The tubes are spun at 18,000g for 50 min at 1°C in an SW28 ultracentrifuge rotor.

7. The supernatant, including an upper layer of lipids and intact cells, is removed, and the tube inverted for 10 min in ice to drain the remaining buffer from the nuclear pellet.

8. At this point, the resulting hepatic nuclei may be flash-frozen in liquid nitrogen after resuspending the nuclear pellets in 20 mL of nuclear storage buffer.

3.2. The Standard Protocol for Run-On Assays

3.2.1. Transcription Reaction Using [α-^{32}P]UTP to Label Nascent RNA Transcripts

1. Prepare nuclei fresh or thaw a frozen nuclear suspension at room temperature or on ice. To a total volume of 200 µL, add 100 µL of 2X transcription reaction buffer, 2 µL each of 0.1 M ATP, CTP, and GTP, 1 µL 1 M DTT, 100 µL nuclear suspension (1 × 10^7 nuclei), 1 µL (40 U) RNase inhibitor, and 10 µL (100 µCi) [α-^{32}P]UTP.

2. Allow the transcription reaction to proceed at 30°C for 30 min and then stop by addition of 0.3 mL HSB buffer containing 200 U DNase I (RNase-free). Mix with a Pasteur pipet 10 times, and incubate for an additional 10 min at 30°C.

3. (Optional): To determine exclusively the RNA polymerase II-dependent transcription, transcription is also performed in the presence of α-amanitin (2 µg/mL), and the counts obtained are subtracted from the total counts.

4. Add 20 µL SET buffer and 2 µL proteinase K (20 mg/mL) for 30 min at 37°C.

5. Add 400 µL of extraction buffer and 80 µL of sodium acetate (2.0 M, pH 4.0), and vortex for 10 s.

6. Add 700 µL of water saturated phenol and 150 µL of chloroform:isoamyl alcohol, vortex for 10 s, and allow to sit on ice for 15 min.

7. Centrifuge at 4°C and 12,000g for 5 min. Transfer the aqueous layer to a clean tube (e.g., Falcon #2059 tube), and add 1–2 µL of tRNA (20 mg/mL).

8. Add an equal volume of cold isopropanol to the tube, mix the tube, and incubate at −20°C for 20 min. Centrifuge at 12,000g at 4°C for 15 min.

9. Wash the resulting pellet with 70% ethanol.
10. Lyophilize pellet and redissolve in 100 μL of hybridization buffer. Count 1-μL aliquot (usually typical count is $1–2 \times 10^8$ cpm/reaction).

3.2.2. Filter Hybridization Using [α-^{32}P]UTP-Labeled Transcripts

1. Linearize the single- or double-stranded recombinant plasmid containing the cDNA fragment of interest with an appropriate restriction enzyme (cut at least one site; do not digest within cDNA sequence) (*see* **Note 4**).
2. Plasmid containing cDNA fragment of a gene whose transcription does not change in response to the stimulus is subjected to the same treatment and used as an internal control.
3. Extract plasmid cDNA with phenol, phenol/chloroform, and precipitate with ethanol.
4. Dissolve and denature 50–100 μg (5–10 μg/dot) of the resulting cDNA sample with 0.5 mL of 0.2 M NaOH /2 mM EDTA, and incubate for 15 min at 37°C.
5. Add 5 mL of 6X SSC to neutralize the cDNA sample.
6. Spot 5 μg of the denatured plasmid DNA onto nitrocellulose filter or nylon filter using slot apparatus or dot apparatus (using Minifold II from Schleicher & Schuell, Keene, NH, if possible, which produces a much better result than dot-blot apparatus).
7. Rinse the slots with 500 μL of 6X SSC, air-dry the filter, and bake for 2–3 h at 80°C in a vacuum.
8. Prehybridize the filter for 2 h in 400 μL of hybridization buffer.
9. Remove buffer from the tube, and replace with 250–500 μL hybridization buffer containing the labeled RNA probe (10^7 cpm/mL; if signal is faint, use 5×10^7 cpm/mL).
10. Cover the hybridization mixture plus filter with 100 μL of mineral oil and hybridize for 36–48 h with shaking at 42°C.
11. After hybridization, wash the filter twice in 50–100 mL of 2X SSC, and 0.1% SDS at 65°C for 15 min, followed by wash with 0.5X SSC and in 0.1% SDS at 65°C for 15 min.
12. Rinse the filter with 2X SSC (no SDS) several times, and transfer the filter to a glass vial containing 2X SSC, 2.5 μg/mL RNase A plus 5 U/mL RNase T1. Incubate the filter for 30 min at 37°C. This step removes nonspecific binding and unhybridized RNA.
13. Wash the filter with 1 mL of 2X SSC for 15 min at 37°C.
14. Dry the filter on Whatmann 3MM paper. Expose the filter to Kodak XAR-5 film for 1–2 d at −70°C.
15. To elute the hybridized RNA, incubate the filters with 200 μL of 0.3 M NaOH for 15 min at 65°C followed by the addition of 50 μL glacial acetic acid and 4 mL of scintillation cocktail. Count ^{32}P radioactivity in a liquid scintillation counter. This value provides the relative rates of transcription of a specific mRNA.

3.3. The Protocol for Run-On Assay Using Nonradioactive (Chemiluminescent) Detection

The major disadvantage of the standard procedure for nuclear run-on assays is the exposure to handling RNA labeled to a high specific activity during the procedure. A protocol using DIG-labeled UTP instead of ^{32}P-labeled UTP to transcribe the RNA with chemiluminescent detection after hybridization has been developed and described previously (19). Those authors found that the nonradioactive detection protocol yields reproducible results with low background signals and is at least as sensitive as the standard run-on procedure (see **Notes 4** and **5**).

3.3.1. Transcription Reaction Using Nonradioactive (Chemiluminescent) Detection

1. Frozen nuclei (100 μL; DNA concentration = 1–2 mg/mL; approx 2–5 × 10^7 nuclei/reaction) are thawed and mixed with 1 vol 2X reaction buffer, 4 mM ATP, GTP, CTP, and 3 μL DIG-UTP (Boehringer Mannheim), 10 nmol/mL.
2. The mixture is incubated for 20 min at 30°C.
3. Ten milliliters of DNase I, RNase-free (10 mg/mL), and 10 μL CaCl$_2$ are added, and then incubated for an additional 5 min at 26°C.
4. After addition of 2 μL proteinase K (10 mmol/mL), 25 μL of SET buffer, and 5 μL tRNA, the solution is incubated for 30 min at 37°C.
5. 550 μL NRO buffer II and 90 μL 2 M sodium acetate, pH 4.0, are added and mixed well.
6. The reaction mixture is extracted with phenol-choroform-isoamyl alcohol, (24.5:24.5:1) (PCI). The aqueous phase is retained for the next step.
7. The aqueous phase is precipitated with 1 vol isopropanol or 2.5 vol ethanol.
8. The pellet is resuspended in 300 μL NRO II precipitated in 0.2 M NaHc, pH 4.0, and washed in 70% ethanol.
9. The pellet is then resuspended in 20 μL 0.5% SDS.
10. One microliter of probe and of labeled control RNA (from Boehringer Mannheim) are serial-diluted in 1:10 steps down to a dilution of 1:10,000 with RNA dilution buffer.
11. One microliter of the various dilutions are spotted onto a positively charged nylon membrane and processed by the direct detection assay according to the manufacturer's instructions (DIG System User's Guide for Filter Hybridization, Boehringer Mannheim).
12. Spot intensities of the control and experimental dilutions are compared to estimate the concentration of the experimental probe. If a signal is visible after chemiluminescent detection in the 1:10,000 dilution, the probe can be used for hybridization. Otherwise, start again from **step 1** of this section.
13. Five microliters of linerarized plasmid DNA are denatured by heat for 10 min at 90°C in 200 μL 6X SSC and transferred to nylon membrane, positively charged according to procedure described below.

14. Blots are hybridized in 10 mL of alternate hybridization solution containing 0.5 *M* sodium phosphate buffer, pH 7.2, 7% SDS for 12 h at 65°C. Blots are washed once in 250 m*M* sodium phosphate buffer, 1% SDS, and two times in 100 m*M* sodium phosphate buffer, 1% SDS for 15 min at 65°C each.

15. The DIG-chemiluminescent detection procedure is performed following the instructions in the DIG Nonradioactive Nucleic Acid Labeling and Detection System (Boehringer Mannheim).

16. Optimal exposure times of X-ray films are generally between 1 and 6 h.

3.4. Simplified Protocol for the Nuclear Run-On Transcription Assay (See Note 6)

1. After run-on assay is conducted as described in **Subheading 3.2.1., step 1**, the reaction is stopped by addition of 0.3 mL HBS buffer containing 200 U DNase I (RNase-free). Mix with Pasteur pipet 10 times before incubation, and incubate for an additional 10 min at 30°C.

2. Add 2.5 µL of yeast tRNA (20 mg/mL).

3. Nuclear RNA is isolated by adding µL of RNA extraction solution (NRO buffer II), 700 µL of water-saturated phenol, and 150 µL of chloroform isoamyl alcohol (49:1), vortex for 10 s, and allow to sit on ice for 15 min.

4. Centrifuge at 4°C and 12,000*g* for 10 min. Transfer the aqueous layer to a clean tube, and add an equal volume of cold isopropanol to the tube, mix the tube, and incubate at –20°C for 1 h. Centrifuge at 12,000*g* at 4°C for 15 min.

5. Wash the resulting pellet with 70% ethanol.

6. Lyophilize pellet and redissolve in 100 µL of hybridization buffer. Count 1-µL aliquot (usually typical count is $1–3 \times 10^6$ cpm/reaction).

7. The remaining steps for filter hybridization using the ^{32}P-labeled RNA probe are performed as described in **Subheading 3.2.2.**

4. Notes

1. Two terms "run-off" and "run-on" have often been used to describe this method, namely, nuclear run-on and nuclear run-off assays. In most publications, authors have used either of the two terms with similar meanings.

2. Care must be taken to avoid disruption of lysosomes and the release of ribonucleases. It should be remembered that the procedure involves the isolation of RNA, and therefore, all appropriate precautions must be taken to avoid RNase contamination of the sample and of the buffers, e.g., use of sterile plasticware or baked glassware, continual wearing of gloves, preparation of buffers in RNase-free containers, and using RNase-free salts and RNase-free water. If buffer solutions are treated with diethyl pyrocarbonate (DEPC) to inactivate RNase, all traces of the DEPC must be removed from the solutions to prevent inactivation of RNA polymerase, e.g., autoclave buffers for two cycles.

3. It is important that nuclei be prepared with minimal damage to maintain them in the transcriptionally active state. Procedures for isolation of nuclei will vary slightly from tissue to tissue or from cell to cell, and must be derived empirically.

4. One of two major disadvantages of the standard protocol described above is the handling of RNA labeled to a high specific activity throughout the whole procedure. To avoid this situation, the use of a nonradioactive method to label RNA is attractive. Other advantages of using the nonradioactive protocol include the long-term storage of the labeled RNA probe (up to 1 yr) and the reuse of hybridization solutions. The results illustrated in **ref. *19*** described by Merscher et al. demonstrate that a nonradioactive run-on assay using chemiluminescent detection offers a safe and highly sensitive technique for measurement of the transcription rate of gene expression.

5. In a previous description of probe preparation, we employed a protocol for preparation and use of a double-stranded DNA probe *(17)*. It has been reported that the use of double-stranded probes may give misleading results owing to the presence of antisense transcripts of uncertain importance *(20,21)*. Consequently, it is important to distinguish between sense and antisense transcription in a nuclear run-on assay. For this reason, the use of single-stranded probes should be the procedure of choice in nuclear run-on assays *(22)*.

6. Several simple procedures of the nuclear run-on assay have been described by different investigators *(23,24)*. The major modification of the simple procedure in comparison to the standard procedure is the elimination of proteinase K digestion and second-round guanidine thiocyanate solution treatment. This modification greatly facilitates the performance of the transcription assay by reducing the labor-intensive process of nuclear RNA isolation. The described protocol is mainly based on the procedure described in **ref. *23***, and has been modified and conducted in our laboratory.

Acknowledgments

This work was supported in part by an NIH grant (HL41315) and by the Research Service of the Veterans Administration.

References

1. Bohm, S. K., Grady, E. F., and Bunnett, N. W. (1997) Regulatory mechanisms that modulate signalling by G-protein-coupled receptors. *Biochem. J.* **322,** 1–18.
2. Ferguson, S. S. and Caron, M. G. (1998) G protein-coupled receptor adaptation mechanisms. *Semin. Cell. Dev. Biol.* **9,** 119–127.
3. Collins, S., Caron, M. G., and Lefkowitz, R. J. (1992) From ligand binding to gene expression: new insights into the regulation of G-protein-coupled receptors. *Trends Biochem. Sci.* **17,** 37–39.
4. Scanga, D. R. and Schwinn, D. A. (1998) Transcriptional regulation of alpha-1 adrenergic receptors. *Front Biosci.* **3,** d348–353.
5. Heck, D. A. and Bylund, D. B. (1998) Differential down-regulation of alpha-2 adrenergic receptor subtypes. *Life Sci.* **62,** 1467–1472.

6. Lohse, M. J., Engelhardt, S., Danner, S., and Bohm, M. (1996) Mechanisms of beta-adrenergic receptor desensitization: from molecular biology to heart failure. *Basic Res. Cardiol.* **91 (Suppl. 2),** 29–34.

7. Collins, S., Altschmied, J., Herbsman, O., Caron, M. G., Mellon, P. L., and Lefkowitz, R. J. (1990) A cAMP response element in the beta 2-adrenergic receptor gene confers transcriptional autoregulation by cAMP. *J. Biol. Chem.* **265,** 19,330–19,335.

8. Sakaue, M. and Hoffman, B. B. (1991) cAMP regulates transcription of the alpha 2A adrenergic receptor gene inHT-29 cells. *J. Biol. Chem.* **266,** 5743–5749.

9. McGraw, D. W., Chai, S. E., Hiller, F. C., and Cornett, L. E. (1995) Regulation of the beta 2-adrenergic receptor and its mRNA in the rat lung by dexamethasone. *Exp. Lung Res.* **21,** 535–546.

10. Thomas, R. F., Holt, B. D., Schwinn, D. A., and Liggett, S. B. (1992) Long-term agonist exposure induces upregulation of beta 3-adrenergic receptor expression via multiple cAMP response elements. *Proc. Natl. Acad. Sci. USA* **89,** 4490–4494.

11. Sakaue, M. and Hoffman, B. B. (1991) Glucocorticoids induce transcription and expression of the alpha 1B adrenergic receptor gene in DTT1 MF-2 smooth muscle cells. *J. Clin. Invest.* **88,** 385–389.

12. Kiely, J., Hadcock, J. R., Bahouth, S. W., and Malbon, C. C. (1994) Glucocorticoids down-regulate beta 1-adrenergic-receptor expression by suppressing transcription of the receptor gene. *Biochem. J.* **302,** 397–403.

13. Hu, Z. W., Shi, X. Y., and Hoffman, B. B. (1996) Insulin and insulin-like growth factor I differentially induce alpha1-adrenergic receptor subtype expression in rat vascular smooth muscle cells. *J. Clin. Invest.* **98,** 1826–1834.

14. Gong, G., Johnson, M. L., and Pettinger, W. A. (1995) Testosterone regulation of renal alpha 2B-adrenergic receptor mRNA levels. *Hypertension* **25,** 350–355.

15. Devedjian, J. C., Fargues, M., Denis-Pouxviel, C., Daviaud, D., Prats, H., and Paris, H. (1991) Regulation of the alpha 2A-adrenergic receptor in the HT29 cell line. Effects of insulin and growth factors. *J. Biol. Chem.* **266,** 14,359–14,366.

16. Hu, Z.-W., Shi, X.-Y., Okazaki, M., and Hoffman, B. B. (1995) Angiotensin II induces transcription and expression of alpha$_1$ adrenergic receptors in cultured rat vascular smooth muscle cells. *Am. J. Physiol.* **268,** H1006–1014.

17. Greenberg, M. E. and Bender, T. P. (1997) Identification of newly transcribed RNA, in *Current Protocols in Molecular Biology*, (Ausubel, F. M., et al., eds.), John Wiley, New York, pp. 4.10.1–4.10.11.

18. Hattori, M., Tugores, A., Veloz, L., Karin, M., and Brenner, D. A. (1990) A simplified method for the preparation of transcriptionally active liver nuclear extracts. *DNA Cell. Biol.* **9,** 777–781.

19. Merscher, S., Hanselmann, R., Welter, C., and Dooley, S. (1994) Nuclear runoff transcription analysis using chemiluminescent detection. *Biotechniques* **16,** 1024–1026.

20. Haley, J. D. and Waterfield, M. D. (1991) Contributory effects of de novo transcription and premature transcript termination in the regulation of human epidermal growth factor receptor proto-oncogene RNA synthesis. *J. Biol. Chem.* **266,** 1746–1753.
21. Kindy, M. S., McCormack, J. E., Buckler, A. J., Levine, R. A., and Sonenshein, G. E. (1987) Independent regulation of transcription of the two strands of the c-myc gene. *Mol. Cell. Biol.* **7,** 2857–2862.
22. Sell, C., Chen, H. M., and Baserga, R. (1992) A simple method to generate single-stranded probes for run-on transcription assays. *Biotechniques* **12,** 692–694.
23. Fei, H. and Drake, T. A. (1993) A rapid nuclear runoff transcription assay. *Biotechniques* **15,** 838.
24. Celano, P., Berchtold, C., and Casero, R. A. Jr. A simplification of the nuclear run-off transcription assay. *Biotechniques* **7,** 942–944.

11

Primer Extension Methods for Determination of β_1-Adrenergic Receptor mRNA Start Sites

Yi-Tang Tseng and James F. Padbury

1. Introduction

Primer extension is often used to map the 5′ end of RNA *(1)*. A single-stranded, end-labeled DNA primer is hybridized to RNA first. Using an RNA-dependent DNA polymerase (reverse transcriptase [RT]) and nonradioactive deoxynucleotides, the primer is extended to yield cDNA. The cDNA is then analyzed on a sequencing gel to determine nucleotide length. The length of the cDNA reflects the distance between the primer and the 5′ end of RNA, and hence, maps the transcription start sites.

The transcription start sites of the β_1-adrenergic receptor (β_1-AR) gene in mouse and sheep have been reported *(2,3)*. Probably owing to the extreme G-C-rich nature of the 5′ flanking region of the β_1-AR gene, the mapping of exact transcription start sites of this gene has eluded many investigators. With improper primer design, primer extension experiments often fail to yield discrete results, and sometimes lead to conflicting or false results. In our laboratory, we used RNase protection assays to assess the approximate location of transcription start sites *(3)*. This allowed flexibility in choosing a location for potential primers and for optimization of the length of the cDNA product. Data obtained from RNase protection assays were used to help design the primers in primer extension experiments (**Fig. 1**). The detailed protocol for RNase protection assay has been described elsewhere in this volume. This chapter will focus only on the protocol for primer extension.

From: *Methods in Molecular Biology, vol. 126: Adrenergic Receptor Protocols*
Edited by: C. A. Machida © Humana Press Inc., Totowa, NJ

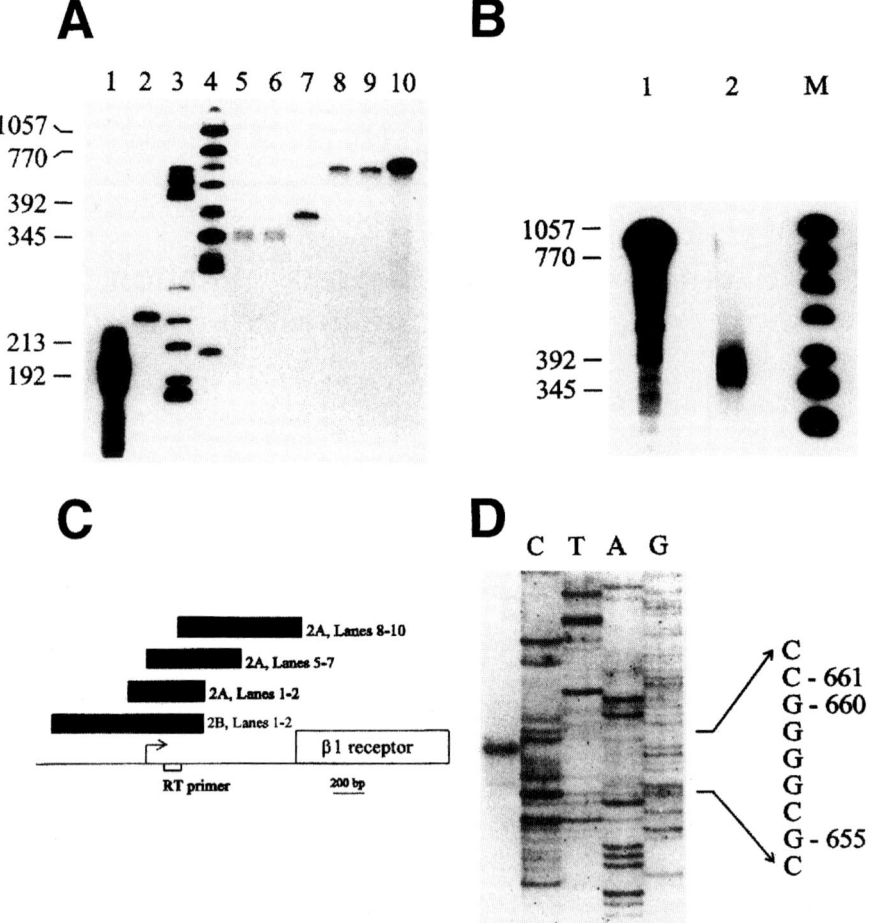

Fig. 1. RNase protection assay was first used to estimate the approximate location of the transcription start sites of β_1-AR gene. **(A)** Three ovine β_1-AR nested deletion constructs, spanning the 5′ sequence from –1157 to +175 relative to the initiator methionine (lanes 2, 7, and 10; B, lane 1), were linearized and used as the templates for in vitro transcription using T7 RNA polymerase. The larger probes (lanes 7 and 10) did not overlap the transcription start sites and were fully protected except for deletion of vector sequences generated during in vitro transcription (lanes 5–6 and 8–9). The smallest probe (lane 2) overlapped the transcription start sites and was not fully protected (lane 1). Lane 3, Boehringer Mannheim DNA mol-wt marker V. Lane 4, *Hinc*II digest of φx174 DNA mol-wt marker. **(B)** Another probe overlapping the transcription start sites was used to confirm the results from (A) lanes 1–2. Both predicted the same transcription start sites more than 550 bp upstream from the initiator methionine. **(C)** A schematic of the RNase protection assay results. **(D)** To map the precise location of transcription initiation, a primer extension experiment was performed using a 30-mer primer approx 100 bp upstream from the estimated transcription start sites. The predominant start site was at –660 bp relative to the translation start site with less prevalent start sites at –661 and –665.

2. Materials

1. Primer, 100 µg/mL, (*see* **Note 1**).
2. [γ-^{32}P]ATP, 10 mCi/mL, >5000 Ci/mmol (Amersham, Piscataway, NJ).
3. T4 Polynucleotide kinase, 10 U/µL (Promega, Madison, WI).
4. Phenol/chloroform, TE-saturated phenol, pH 8.0:chloroform:isoamyl alcohol = 50:49:1.
5. 4 *M* Ammonium acetate.
6. 3 *M* Sodium acetate, pH 5.5.
7. 0.5 *M* ethylenediamine tetra-acetic acid (EDTA), pH 8.0.
8. Ethanol, 100 and 70%.
9. Total RNA, prepared from sheep heart.
10. RNasin ribonuclease inhibitor, 20 U/µL (Promega).
11. dNTP, 10 m*M* each.
12. 10X Kinase buffer: 700 m*M* Tris-HCl, pH 7.5, 100 m*M* MgCl$_2$, 50 m*M* dithiothreitol, 1 m*M* spermidine HCl, 1 m*M* EDTA, pH 8.0.
13. Hybridization buffer: 0.4 *M* NaCl, 20 m*M* Tris-HCl, pH 7.6, 1 m*M* EDTA, pH 8.0, 0.1% sodium dodecyl sulfate (SDS).
14. Moloney Murine Leukemia Virus (M-MLV) RT, 200 U/µL (*see* **Note 2**).
15. 5X RT buffer: 0.25 *M* Tris-HCl, pH 7.6, 0.375 *M* KCl, 15 m*M* MgCl$_2$, 50 m*M* dithiothreitol.
16. Formamide gel-loading buffer: 80% deionized formamide, 10 m*M* EDTA, pH 8.0, 1 mg/mL xylene cyanole FF, 1 mg/mL bromophenol blue.
17. DNase-free bovine pancreatic RNase, 0.5 mg/mL (Boehringer Mannheim, Indianapolis, IN).
18. Tris-EDTA (TE) buffer.
19. Autoclaved deionized water.
20. A benchtop microcentrifuge capable of up to 12,000*g*, e.g., Eppendorf 5415C.
21. Polyacrylamide gel electrophoresis apparatus and reagents.

3. Methods

3.1. Primer Labeling

1. In an Eppendorf tube, mix the following:

20 µL	[γ-^{32}P]ATP
1 µL	primer
3 µL	10X kinase buffer
1 µL	T4 polynucleotide kinase
5 µL	water

    Mix gently and incubate for 30 min at 37°C.
2. Heat for 5 min at 65°C to inactivate the kinase.
3. Phenol/chloroform extraction: Add 30 µL phenol/chloroform. Vortex for 1.5 min and centrifuge at 12,000*g* in a benchtop microcentrifuge for 3 min. Transfer 28 µL of the upper layer to a fresh tube. Add 28 µL 4 *M* ammonium acetate and 56 µL of 100% ethanol. Mix and precipitate for 30 min at −70°C.
4. Centrifuge at 12,000*g* in a microcentrifuge for 15 min at 4°C. Decant the super-

natant. To minimize the unbound [γ-^{32}P]ATP, an ammonium acetate/ethanol precipitation step is repeated. The pellet is washed once with 70% ethanol.

5. Dissolve the final pellet in 50 μL TE buffer. Count 1 μL by liquid scintillation counting. Dilute with TE buffer to a final concentration of 10,000 cpm/μL.

3.2. Hybridization

1. The labeled primer is first coprecipitated with the ovine heart total RNA: Mix 10 μL of the labeled primer (total = 100,000 cpm) with 50 μL RNA and 3 μL 3 *M* sodium acetate, and then add water to a final volume of 30 μL. Add 60 μL 100% ethanol, and store for 30 min at −70°C.
2. Centrifuge at 12,000*g* in a microcentrifuge for 15 min at 4°C. Wash once with 70% ethanol.
3. Decant the supernatant, and dry the pellet. Resuspend in 30 μL hybridization buffer (*see* **Note 3**).
4. Heat for 10 min at 85°C, and then immediately quench on ice to denature RNA.
5. Incubate for 4 h at 37°C (*see* **Note 4**).
6. The RNA/DNA hybrid is then precipitated: Add 3 μL 3 *M* sodium acetate and 200 μL 100% ethanol. Mix and store for 30 min at −70°C.
7. Centrifuge at 12,000*g* in a microcentrifuge for 15 min at 4°C. Wash once with 70% ethanol. Dry the pellet, and prepare for primer extension step.

3.3. Primer Extension

1. Place the tube with the pellet on ice and add the following:

2 μL	dATP, 10 m*M*
2 μL	dTTP, 10 m*M*
2 μL	dGTP, 10 m*M*
2 μL	dCTP, 10 m*M*
4 μL	5X RT buffer
1 μL	RNasin ribonuclease inhibitor
1 μL	M-MLV RT
6 μL	Water.

2. Mix gently and incubate for 90 min at 42°C.
3. Add 1 μL 0.5 *M* EDTA and 1 μL DNase-free bovine pancreatic RNase.
4. Incubate for 30 min at 37°C.
5. Perform one phenol/chloroform extraction (*see* **Subheading 3.1.**, **step 3**).
6. Dissolve the pellet in 4 μL TE, and add 4 μL formamide-loading buffer.
7. Boil for 3 min, and quench on ice.
8. Load 3.5 μL onto a sequencing gel along with a sequencing ladder created by using the same primer.

4. Notes

1. The primer is a synthetic 30-mer oligonucleotide. Primer location and sequence are dictated by the results of the RNase protection assay. A rule of thumb is to avoid the region of RNA that may form difficult secondary structures. To avoid

premature termination of primer extension caused by high G-C content in the β_1-AR promoter, it is important to use a primer located within 100 nucleotides of the start sites.

2. For best results, M-MLV RT is the first choice of enzyme compared to AMV RT because of the former enzyme's weaker ribonuclease H activity. There are commercially available M-MLV RTs, which are free of intrinsic ribonuclease H activity.

3. Do not overdry the RNA pellet. Overdried RNA can be very difficult to resuspend. After decanting, place the tube upside down on several layers of absorbing paper to drain most of the excess ethanol. Any remaining liquid adhered to the tube wall can be centrifuged to the tube bottom using a brief spin. Use a fine pipet tip to drawn the liquid out. Resuspend the pellet right before the pellet turns invisible.

4. Alternatively, hybridization can be performed overnight at 30°C. However, the optimal condition of hybridization will vary depending on the primer length and G-C content. A preliminary experiment can be used to determine optimal conditions.

References

1. Sambrook, J., Fritsch, E. F., and Maniatis, T. (1989) Analysis of RNA by primer extension, in *Molecular Cloning, A Laboratory Manual*, Cold Spring Harbor Laboratory, Cold Spring Harbor, NY, pp. 7.79–7.83.
2. Cohen, J. A., Baggott, L. A., Romano, C., Arai, M., Southerling, T. E., Young, L. H., et al. (1993) Characterization of a mouse β_1-adrenergic receptor genomic clone. *DNA Cell Biol.* **12**, 537–547.
3. Padbury, J. F., Tseng, Y.-T., and Waschek, J. A. (1995) Transcription initiation is localized to a TATAless region in the ovine β_1 adrenergic receptor gene. *Biochem. Biophys. Res. Comm.* **211**, 254–261.

III

EXPRESSION ANALYSIS

12

Use of Eukaryotic Vectors for the Expression of Adrenergic Receptors

Stéphane Schaak, Jean-Christophe Devedjian, and Hervé Paris

1. Introduction

As in many other fields of research in biology, the technology of expressing cloned genes in eukaryotic cells has become an increasingly important method for the study of adrenergic receptors (ARs). As developed in Chapter 14, substantial expression of adrenergic receptors can be achieved using different host organisms, including bacteria and yeast. Expression in higher eukaryotic cells is, however, of particular interest because these cells: (1) possess the machinery for proper processing, delivery, and insertion of the recombinant receptors into the membrane; and (2) generally express the GTP binding proteins and the other effector proteins responsible for propagation and amplification of the signal triggered by receptor agonists.

Mammalian cells transfected with expression vectors containing the cDNA encoding wild-type or mutated forms of the different adrenergic receptor subtypes are then used not only to produce large quantities of well-defined receptor material suitable for the screening of new subtype-selective adrenergic drugs, but also to examine:

The interaction of the adrenergic receptor subtypes with the various forms of G-proteins (1) and their coupling to different signal transduction pathways (2,3),

The properties of constitutively active mutants of receptors (4,5),

The subcellular localization and the metabolism of the receptor protein within differentiated or undifferentiated cells (6,7),

The role of intracellular domains and/or of specific amino acid residues in receptor phosphorylation, desensitization, and internalization (8,9),

The importance of posttranslational modifications, such as glycosylation and acylation, in receptor membrane delivery (10).

From: *Methods in Molecular Biology, vol. 126: Adrenergic Receptor Protocols*
Edited by: C. A. Machida © Humana Press Inc., Totowa, NJ

The purpose of this chapter is to describe the tools and techniques that have been used to express adrenergic receptors in mammalian cells. First, the characteristics of the different expression vectors and the principles of the various transfection techniques that are necessary to achieve this goal are reviewed (*see* **Note 1**). Next, the construction of bicistronic vectors for expression of human α_2-AR subtypes is detailed, and the DEAE–dextran and calcium–phosphate transfection methods are described. Finally, some of the functional characteristics of α_2-ARs expressed in COS-7 and 3T3F442A cells will be briefly reported.

1.1. The Eukaryotic Vectors

A large variety of vectors for expression in eukaryotic cells are described in the literature. The different types of plasmids that have been used to express adrenergic receptor in eukaryotic cells can be classified into three groups, specifically: monogenic, bigenic, and bicistronic vectors. The general characteristics of these plasmids are depicted in the **Fig. 1**.

1.1.1. Monogenic Vectors

Monogenic vectors are the simplest form of expression plasmids. Briefly, they consist of a single transcription unit containing a strong promoter (SV40, CMV or RSV) for high-level constitutive expression in a large variety of mammalian cell lines, a series of unique restriction sites for insertion of the gene of interest, and a polyadenylation signal and site for transcriptional termination. They also generally contain a sequence responsible for mRNA stabilization (β-globin intron, for instance). Examples of monogenic vectors used for expression of adrenergic receptors include pSVL *(11)*, pRK5 *(12)*, pBC12BI *(7)*, pCMV4 *(10)*, or pSG5 *(13)*. By themselves, such vectors allow transient expression only. Generation of stable transfectants requires cotransfection with a second vector carrying an antibiotic resistance gene.

1.1.2. Bigenic Vectors

In contrast to monogenic vectors, bigenic vectors are designed to allow both transient and stable transfection. This feature is owing to the fact that these plasmids contain two transcription units, which encode the protein of interest and a selectable marker. Typical examples of bigenic vectors used for adrenergic receptor expression are pcDNA3 and pREP4 *(7)*. In pcDNA3, the expression of the gene of interest and that for neomycin resistance is under the control of CMV and SV40 promoters, respectively. There are now several variants of pcDNA3 (pcDNA3.1/Zeo, pcDNA3.1/Hygro, pcDNA6, pcDNA3.1/Myc-His) that differ from the original vector by incorporating different antibiotic resistance genes (zeocyn, hygromycin, blasticidin) or by the insertion of C-terminal

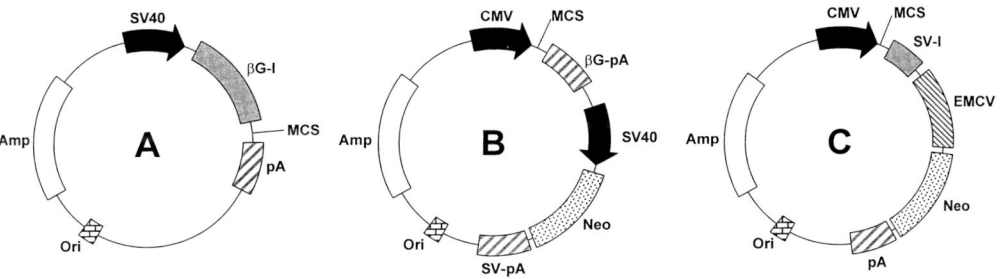

Fig. 1. Schematic organization of the three classes of eukaryotic expression vectors. All expression vectors contain an origin of replication (Ori) and the β-lactamase gene conferring resistance to ampicillin (Amp), for propagation in *E. coli*. They also contain a strong promoter (SV40 or CMV), a multiple cloning site (MCS) to insert the gene of interest and polyadenylation site and signal (pA) for eukaryotic expression. **(A)** Monogenic vector pSG5: The single transcription unit is driven by the SV40 promoter and contains a sequence for the stabilization of the mRNA (intron of the β-globin gene, βG-I). **(B)** Bigenic vector pcDNA3: The gene of interest is inserted into the first transcription unit, driven by the CMV promoter. The second transcription unit is driven by the SV40 promoter and contains the aminoglycoside phosphotransferase gene conferring the resistance to G418 (Neo). There is no stabilization sequence to minimize the size of the vector. Moreover, the polyadenylation sites of the two genes are from different origins (SV40 and β-globin, respectively, SV-pA and βG-pA) to avoid potential homologous recombination. **(C)** Bicistronic vector pIRES1neo: The gene of interest and the Neo gene belong to the same transcription unit driven by the CMV promoter. They are separated by an Internal Ribosome Entry Site (EMCV). The stability of the mRNA is enhanced by the presence of the SV40 intron (SV-I).

sequences suitable for detection with antibodies (c-*myc* or V5 epitope) and for affinity purification (His-tag). Another variant of pcDNA3 is pMAMNeo. This vector, which contains the RSV LTR-enhancer linked to glucocorticoid-inducible mouse mammary tumor virus long terminal repeat (MMTV LTR), was used for dexamethasone-controlled expression of α_2-ARs *(14)*. Other examples of bigenic vectors that were successfully used to produce recombinants expressing high levels of adrenergic receptors include: pZIP-NeoSV *(4)* and pJM16 *(5)*, in which expression is under the control of the MLV LTR and of the human β-actin promoter, respectively, or pBMT3X *(15)*, a vector conferring cadmium-resistance.

1.1.3. Bicistronic Vectors

The key feature of bicistronic vectors lies in the fact that the gene to be expressed and the antibiotic resistance gene are both contained within the same transcription unit. Expression of the second gene is possible because of the insertion of an internal ribosomal entry site (IRES) following the stop codon of the first open reading frame (ORF). Different types of IRES have been characterized and may be used, but the IRES derived from encephalomyocarditis virus (EMCV) appears to be the most efficiently utilized IRES in a large variety of mammalian cells *(16)*. Such expression vectors are of particular interest for the development of permanently transfected cells, because they yield small proportions of false positives, thus minimizing the number of colonies to be screened to find functional clones. Indeed, every clone that is resistant to the antibiotic obligatorily contains the mRNA for the gene of interest, and is thus very likely to express the protein. The use of bicistronic vectors also leads to a better correlation between the concentration of G418 used and the level of expression achieved than with other vectors. A series of bicistronic vectors (pIRES1) are now commercially available from Clontech (Palo Alto, CA). These vectors allow either antibiotic-based (neomycin or hygromycin) or FACS-based (green fluorescent protein; GFP) selection of transfectant populations.

1.2. The Mammalian Cell Lines

A wide panel of mammalian cell types were used for expression of recombinant adrenergic receptors. This list includes: SV40-transformed African green monkey kidney (COS), chinese hamster ovary (CHO), chinese hamster fibroblast (CHW), chinese hamster lung fibroblast (CCL39), human B-lymphoblastoid cells (IBW4), rat fibroblasts (Rat-1), mouse fibroblasts (NIH-3T3), neuroblastoma × glioma hybrid cell (NG 108-15), Madin-Darby canine kidney (MDCK), pig kidney epithelial cell (LLC-PK1), mouse pituitary cell (AtT20), human embryonic kidney fibroblasts (HEK 293), human epidermoid carcinoma (A431), rat pheochromocytoma (PC12), and mouse mammary tumor (S115 and C127).

1.3. The Transfection Methods

Methods developed to introduce foreign DNA into mammalian cells can be schematically classified into three groups, specifically charge neutralization, lipidic interaction, and physical techniques. A fourth method termed transvection is based on the use of retroviruses. This very efficient method will not be discussed here, because it has not yet been utilized for adrenergic receptor expression and because it requires special vectors and equipment to prepare viruses.

1.3.1. Charge Neutralization Techniques

The general principle of these transfection methods is to neutralize the negative charges of the DNA either with polycationic compounds or by inorganic precipitation. The most largely used polycations are the DEAE–dextran (a carbohydrate interacting with the glycosylated components of the membrane) and the polylysine–transferrin conjugate (a complex binding to the transferrin receptors expressed in dividing cells). In the DEAE–dextran procedure, the entry of the adsorbed complex into cells is provoked by a brief osmotic shock (generally using dimethyl sulfoxide [DMSO]), whereas the polylysine– transferrin technique exploits the property of transferrin receptors to be spontaneously internalized. Another method belonging to this group is based on the use of polyethylenimine (PEI). PEI is highly protonable and is therefore able both to interact with negatively charged DNA and inhibit degradation in lysosomes, where it acts as a "proton sponge." Lysosomic degradation is indeed a limitation to the expression of exogenic DNA. Different systems were therefore developed to inhibit this event. They include treatment with chloroquine or use of viral particles for lysosome neutralization or disruption.

On the other hand, the calcium–phosphate transfection method is based on coprecipitation of the DNA with calcium and phosphate. The resulting precipitate is adsorbed onto the cell surface and enters the cell either by spontaneous endocytosis or osmotic shock (generally using glycerol). This procedure ensures the entry of a very high number of plasmid copies, and therefore, allows high-level expression and augments the probability of genomic insertion when performing stable transfection.

1.3.2. Lipidic Interaction Techniques

In the transfection process utilizing lipidic compounds (lipofection), the DNA is packaged into lipidic particles or liposomes, which will fuse with the plasma membrane and deliver the DNA into the cytoplasm. This method is generally regarded as the most "physiological" transfection procedure. It is very simple to implement and works on a larger panel of cell lines than charge neutralization techniques. However, for investigators lacking expertise in lipid synthesis, lipofection reagents must be purchased from commercial companies, thus making the technique somewhat expensive. Numerous lipofection kits (lipofectin, lipofectamin, transfectam, TFX, fugene-6) containing either positively charged lipids or neutral lipids, or a combination of both are now available. If working with refractory cells, the development of the good cocktail giving satisfactory results may be difficult to determine. Interestingly, lipofection often permits reaching high levels of transfection efficiency and, thus, can significantly minimize the quantities of cells and DNA used.

1.3.3. Physical Techniques

Electroporation is the principal technique belonging to this group. This method creates holes in the plasma membrane by applying a brief high-voltage electric field to the cells. Electroporation offers at least two advantages: it is virtually applicable to all cells, and it can be used to introduce other macromolecules (RNA, proteins, antibodies). Unfortunately, this technique is rather expensive, because it requires costly disposable material and necessitates very large quantities of DNA and cells. Moreover, electroporation is frequently of low efficiency (except for easily transfectable cells) and generates a high cellular mortality. Consequently, this method is suitable for some transfections (generation of stable transformants for instance), but is less adapted to the study of large series of constructs (as for promoter analysis or methodical mutagenesis). Generally, only a few molecules of DNA penetrate the cell; thus, the expression levels are below those reached with the calcium phosphate technique, especially in stable transfection experiments. Therefore, electroporation should be considered only after the other above-cited techniques have failed. Another physical technique is the "DNA gun," which shoots colloidal gold particles containing adsorbed foreign DNA into cells.

2. Materials

2.1. Vector Construction

1. Chemically competent *Escherichia coli* (strain DH5α).
2. Growth medium (LB and LB-agar plates) and antibiotics (ampicillin, methicillin).
3. Restriction enzymes and T4-DNA ligase.
4. DNA purification kits were from Qiagen (Chatsworth, CA).
5. Other required laboratory equipment includes gel boxes and power supply for agarose electrophoresis, and UV illuminator for BET-stained gel visualization.
6. Plasmids containing the entire ORFs of human α_2-AR genes were kindly provided by R. J. Lefkowitz, Duke University, Durham, NC (pBCα2C2, pSPα2C4) or purchased from ATCC (HPalpha2GEN).
7. The plasmid pEN was constructed on the pBluescript II KS+ backbone (pKS, Stratagene, La Jolla, CA) and was a gift from H. Prats, Institut Louis Bugnard, Toulouse, France *(17)*. As indicated in the map presented in the **Fig. 2**, it contains an expression cassette comprising:
   The human cytomegalovirus major immediate early promoter/enhancer (CMV).
   A multiple cloning site.
   The EMCV internal ribosome entry site.
   The neomycin phosphotransferase gene.
   The rabbit β-globin genomic sequence (nt + 905 to +2080 referring to β-globin map) containing an intron for mRNA stabilization and a signal for polyadenylation (IVS2β).
   The SV40 origin, which allows episomal replication in cells transformed by the large T-antigen of SV40.

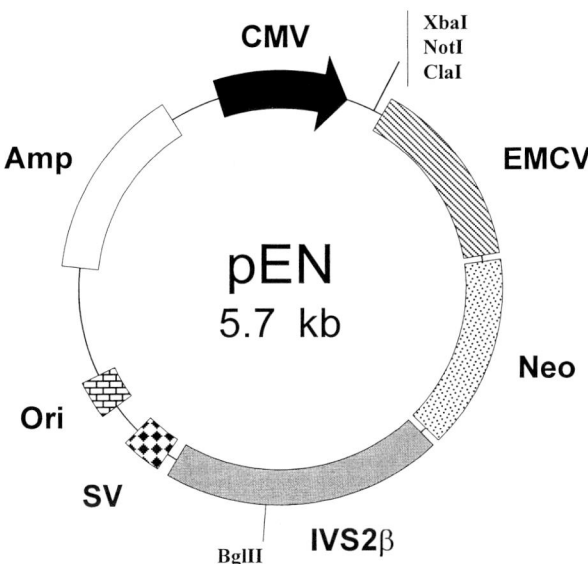

Fig. 2. Map of the bicistronic expression vector pEN. The pEN vector is 5.7 kb in size. It contains a bacterial origin of replication (Ori) and the β-lactamase gene (Amp). For eukaryotic expression and selection, pEN also contains a cassette comprising the human cytomegalovirus major immediate early promoter/enhancer (CMV), a multiple cloning site, the internal ribosomal entry site derived from encephalomyocarditis virus (EMCV), the neomycin gene (Neo), and a genomic fragment of the rabbit β-globin (IVS2β). In addition, pEN contains the SV40 origin of replication (SV).

2.2. Cell Culture and Transfection

2.2.1. COS-7 and 3T3-F442A Cell Lines

COS-7 cells are routinely subcultured in Dulbecco's Modified Eagle Medium (DMEM) containing 4.5 mg/mL glucose, 100 U/mL streptomycin, 100 μg/mL penicillin, and supplemented with 5% heat-inactivated fetal calf serum. 3T3-F442A cells are grown in the same culture medium but supplemented with 10% donor calf serum.

2.2.2. DEAE–Dextran Transfection Method

The DEAE–dextran transfection method was initially described by McCutchan and Pagano *(18)* and is specifically tailored for COS cell transfection. There are several modified versions of the original technique, and kits are commercially available. This transfection procedure is, however, simple and works very well with in-house-prepared reagents.

1. DEAE–dextran, chloroquine, DMSO, and all other chemicals were from Sigma (St. Louis, MO).
2. Phosphate-buffered saline (PBS): for 1 L, 8 g NaCl, 0.2 g KCl, 1.75 g Na$_2$HPO$_4$ · 12 H$_2$O, and 0.24 g KH$_2$PO$_4$, adjust the pH to 6.95 with HCl. Sterilize by autoclaving, and store at room temperature.
3. Plasmid DNA was purified on Qiagen columns.
4. Chloroquine stock (prepare fresh): 20 mg/mL in PBS, sterilize by filtration.
5. DEAE dextran stock (prepare fresh): 10 mg/mL in PBS (DEAE-dextran is difficult to dissolve; mix several time; do not heat).
6. Culture media (prepare fresh):
   a. Serum-free DMEM.
   b. Regular culture medium: DMEM + 5% heat-inactivated fetal calf serum (FCS).
   c. DMEM + chloroquine 80 μ*M*: 2 μL of chloroquine stock per mL of serum-free DMEM
   d. Shock medium: DMSO 10% (v/v) in serum-free DMEM

2.2.3. Calcium Phosphate Transfection Method

The calcium phosphate transfection method was first reported by Graham and Van der Eb *(19)*. It is routinely used for both transient and stable transfection of a large variety of mammalian cell lines. As for the DEAE–dextran method, transfection kits can be purchased from several companies. However, the technique does not present any particular difficulty, and the protocol detailed below works well with in-house prepared solutions.

1. CaCl$_2$ and *N,N-bis*(2-Hydroxyethyl)-2-aminoethanesulfonic acid; 2-(*bis*[2-Hydroxyethyl]amino)-ethanesulfonic acid (BES) are from Sigma cell culture.
2. CaCl$_2$ solution: prepare a 2 *M* CaCl$_2$ solution in ultrapure H$_2$O; sterilize by filtration and store at −20°C into 1-mL aliquots.
3. 2X concentrated BBS: for 1 L, 10.66 g BES, 16.36 g NaCl, and 0.402 g Na$_2$HPO$_4$ · 7 H$_2$O. The critical parameter of this technique is the pH of BBS. Optimum pH is between 6.95 and 7.05. It is thus preferable to prepare five 200-mL aliquots of the same solution at different pHs (6.90, 6.95, 7.00, 7.05, and 7.10). Sterilize the solutions by autoclaving and store them at 4°C.

   The quality of the solutions can be tested as follows: dilute 100 μL of 2 *M* CaCl$_2$ solution within 700 μL of H$_2$O. Then add dropwise 500 μL of this 250 m*M* CaCl$_2$ into 500 μL of 2X concentrated BBS, and mix well by vortexing. Leave for 30 min at room temperature to allow precipitate formation. The fineness of the precipitate can be assessed under the microscope. Test satisfactory batches by transfecting cells (*see* **Subheading 3.2.2.**) with an expression vector containing a strong promoter and an easily detectable reporter gene (β-galactosidase,

luciferase, or GFP. Analyze 48 h later with appropriate assay (X-Gal staining, enzymatic activity or fluorescent microscopy).
4. Glycerol shock solution: glycerol 15% (v/v) in PBS.

3. Methods

3.1. Construction of Bicistronic Vectors for Expression of α_2-Adrenergic Receptor Subtypes

The design of bicistronic plasmids for expression of human α_2-AR subtypes (α_{2B}-AR, α_{2C}-AR and α_{2A}-AR) is described in the following sections. By following some basic recombinant construction rules (*see* **Note 2**), similar constructs containing other adrenergic receptor subtypes can be obtained using adapted cloning strategies and commercially available bicistronic vectors (*see* **Subheading 1.1.3.**).

The construction of the vectors requires molecular biology techniques, including transformation of *E. coli*, plasmid purification, restriction enzyme digestion, agarose gel electrophoresis, DNA fragment isolation and ligation. These standard methods are detailed in other manuals *(20,21)*.

3.1.1. Construction of pα2C2ENeo

The PBCα2C2 is a monogenic expression vector that was constructed in R. J. Lefkowitz's laboratory *(22)*.

1. pBCα2C2 was digested with *Nae*I and *Hin*dIII and the fragment containing the α_{2B}-AR gene was subcloned into pKS predigested with *Sma*I and *Hin*dIII. This subcloning step results in the construction of the pKSα 2C2 recombinant.
2. The insert was then excised from pKSα2C2 by digestion with *Xba*I and *Cla*I, and subsequently ligated in the corresponding sites of pEN to provide the final expression vector, pα2C2ENeo. This bicistronic vector thus contains 40 bp of the 5'NC region of the human β_2-AR, the α_{2B}-AR ORF (1353 bp), and 184 bp of 3'NC of α_{2B}-AR gene.

3.1.2. Construction of pα2C4ENeo

The original plasmid pSPα2C4 *(23)* contained the coding region of α_{2C}-AR in addition to 40 bp of the 5' and the entire 3' untranslated region of the human β_2-adrenergic receptor.

1. The pSPα2C4 construct was digested either by *Mae*III or by *Nco*I and *Sma*I.
2. The *Mae*III-*Mae*III fragment corresponding to nt +457/+1387 of the α_{2C}-AR gene was purified, blunt-ended by the Klenow fragment of the DNA polymerase, and inserted into dephosphorylated *Eco*RV-cut pKS, providing pKSα2C4(+457/ +1387). The correct orientation of this construct contains the 3' *Mae*III site of the insert in close proximity to the *Hin*dIII site of the vector.

3. In parallel, the *NcoI-SmaI* fragment corresponding to nt –2/+1013 of the α_{2C}-AR gene was inserted at the same sites of a pKS vector that was previously engineered to create an *NcoI* site next to *BamHI*, giving pKSα2C4(–2/+1013).

4. Next, the *BamHI-SmaI* fragment of pKSα2C4(–2/+1013) was ligated into pKSα2C4(+457/+1387) digested by the same enzymes, providing the pKSα2C4 (-2/+1387) vector.

5. Finally, the *XbaI-ClaI* fragment of pKSα2C4(–2/+1387) was inserted in the corresponding sites of pEN to provide the pα2C4ENeo vector. This expression vector thus contains the entire ORF (1386 bp) and 39 bp of 3′NC of the α_{2C}-AR gene.

3.1.3. Construction of pα2C10ENeo

The HPalpha2GEN construct contains the 5.5-kb *BamHI-BamHI* fragment of the human α_{2C}-AR gene cloned into pUC18 *(24)*. This recombinant contains 2078 bp of 5′NC, the ORF, and about 1.9 kb of 3′NC region of the α_{2C}-AR gene.

1. The *KpnI-HindIII* restriction fragment of this vector (corresponding to nt –1208/ +1526 of the α2C10 gene) was first subcloned at the *KpnI* and *HindIII* sites of pKS to provide the pKSα2C10 construct.

2. This construct was then digested by *NheI* at position –201 of the α2C10 sequence and *NotI* in the pKS polylinker. The purified insert was ligated with the *NotI-BglII* and the *BglII-XbaI* fragments of pEN. The resulting pα2C10ENeo vector thus contains 201 bp of 5′NC, the entire ORF (1353 bp), and 175 bp of 3′NC region of the α_{2C}-AR gene.

3.2. Transfection of COS-7 and 3T3-F442A Cells

3.2.1. Transient Expression of α₂-Adrenergic Receptors in COS–7 Cells

1. Plate the cells 24 h before the transfection experiment. Seeding density should be adjusted to ensure that COS-7 cells will be 40–60% confluent on the day of transfection. As a general guideline, plate approx 1×10^6 cells/100-mm culture dish.

2. Prepare DNA/DEAE-dextran premix: for one 100-mm diameter culture dish, pipet 3.6 mL of PBS into a sterile polypropylene tube, add 4 µg of plasmid DNA, mix, add 400 µL of DEAE–dextran stock solution, and mix again (it is important to dilute the concentrated DNA in PBS before adding the DEAE–dextran to avoid DNA flocculation).

3. Aspirate the medium, and rinse the cell layer twice with 10 mL of PBS.

4. Add the DNA/DEAE–dextran premix (4 mL/dish), and disperse it evenly over the cells by gently rocking the plates.

5. Incubate at 37°C for about 20 min. At the end of this treatment, a large number of cells will become rounded. Morphological changes are evident and can be checked under microscope.

6. Gently add 12 mL of DMEM+ 80 µ*M* chloroquine (be careful not to detach the

cells), and incubate the plates 4 h at 37°C. During this period, the cells will spread again and will accumulate chloroquine.

7. Remove the medium, add 5 mL of DMSO shock solution and incubate for 2.5 min at room temperature.
8. Aspirate the shock solution, wash the cell layer twice with 10 mL of PBS, once with 10 mL of serum-free DMEM, and refeed the cells with 10 mL of regular culture medium.
9. Return the plates to the CO_2 incubator, and harvest the cells 48 h after the transfection experiment.

3.2.2. Protocol for Permanent Expression of α_{2A}-Adrenergic Receptor in 3T3-F442A Preadipocytes

1. At 1 d prior to transfection, seed a 60-mm culture dish with 3.6×10^4 3T3-F442A cells in 5 mL of culture medium.
2. $CaCl_2$/DNA premix: in a sterile tube, successively add 250 μL H_2O, 37.5 μL 2 M $CaCl_2$ (mix), and 10 μg pα2C10ENeo in 12.5 μL (mix thoroughly).
3. Prepare a second sterile tube containing 300 μL of 2X concentrated BBS.
4. While vortexing at medium speed, add the $CaCl_2$/DNA mix drop by drop into the BBS buffer, and then mix well.
5. Incubate for 30 min at room temperature, vortex again, and dispense the precipitate drop by drop onto the culture dish.
6. Rock the plate to avoid localized acidification.
7. Return the cells to the CO_2 incubator.
8. Four hours later, remove the medium, and replace with 1 mL of glycerol shock solution (*see* **Note 3**).
9. Incubate for 3 min at room temperature, aspirate the glycerol, and rinse cells twice with 4 mL of PBS.
10. Add 5 mL of culture medium, and return plates to incubator.
11. At 48 h posttransfection, harvest the cells by treatment with trypsin–EDTA and subculture at the density of 1.2×10^5 cells/100-mm dish in culture medium containing 0.6 mg/mL G418 (Geneticin, Gibco-BRL, Rockville, MD) (*see* **Note 4**).
12. Selection medium is replaced every 2 d. Antibiotic-resistant colonies are visible approx 2–3 wk posttransfection and can be individually isolated by trypsin–EDTA treatment using cloning cylinders.

3.3. Pharmacological Characteristics and Functional Properties of α_2-Adrenergic Receptors Expressed in COS-7 and in 3T3-F442A Cells

3.3.1. Binding of [^{3}H]MK912 to Receptors Expressed in COS-7 Cells

The pharmacological characteristics of recombinant α_2-ARs expressed in COS-7 cells have been previously investigated using [^{3}H]yohimbine *(22)*, [^{3}H]rauwolscine *(23)*, or [^{3}H]RX821002 *(13)*. **Figure 3** presents the results from the study of [^{3}H]MK912 binding onto membranes prepared from cells

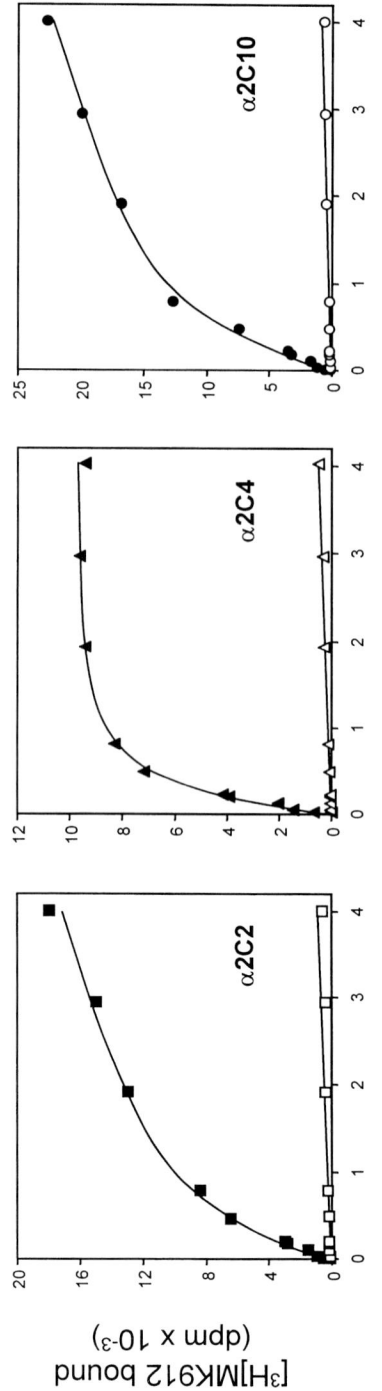

Fig. 3. Binding of [³H]MK912 to human α_2-ARs transiently expressed in COS-7 cells. COS-7 cells were transfected with the bicistronic vectors pα2C2Eneo (squares), pα2C4Eneo (triangles) and pα2C10Eneo (circles). Membranes were prepared and incubated (10 µg/point) for 45 min at 25°C in Tris/Mg buffer (Tris-HCl 50 mM, MgCl$_2$ 0.5 mM, pH 7.5) containing various concentrations of [³H]MK912. Bound radioligand was separated from free by filtration on glass fiber filters. Total binding is designated by the black symbols, whereas open symbols represent nonspecific binding measured in the presence of 10^{-4} M phentolamine.

transfected with pα2C2ENeo, pα2C4ENeo, and pα2C10ENeo plasmids. The nonspecific binding defined with 100 μM phentolamine was a linear function of [^{3}H]MK912 concentration; it was low and caused entirely by binding of the radioligand onto the filters. Saturation isotherms were monophasic and consistent with the existence of a single population of sites. Computer-assisted analysis of the data from five independent transfection experiments indicated that equilibrium dissociation constants for [^{3}H]MK912 (K_d) are 1.5 ± 0.4, 0.13 ± 0.02, and 1.1 ± 0.2 nM for α2C2, α2C4, and α2C10, respectively. The levels of receptor expression are extremely high 9.8 ± 2.5, 4.7 ± 0.8, and 13.6 ± 1.4 pmol of binding sites/mg of membrane proteins, so that a single transfected dish provided enough material to perform large series of binding points. Antagonists inhibited [^{3}H]MK912 binding with rank orders agreeing with what was expected for each subtype. However, in contrast to what was observed in native receptor preparations, agonists yielded inhibition curves with Hill coefficient values near unity (0.85–0.95) and did not allow discrimination of receptor subpopulations exhibiting high and low affinity for agonists. The reasons for this failure are not fully elucidated, but very similar observations were obtained with other radioligands *(23)* and with other transfected cells expressing high receptor levels *(15)*.

3.3.2. Functions of α_{2A}-Adrenergic Receptors in 3T3-F442A Cells

Transfection of 3T3-F442A preadipocytes with the pα2C10Eneo construct yielded a series of clones expressing the human α_{2A}-AR subtype at densities varying from 50 to 5000 fmol of receptor/mg of membrane protein. As assessed by competition binding studies with UK14304 and by measurement of intracellular cAMP level, the receptors are functionally coupled to G-proteins and their stimulation significantly inhibited forskolin-induced cAMP accumulation *(25)*. Expression of the receptor remains constant over several cell passages and is independent of the stage of adipocyte differentiation *(26)*. These models demonstrated that receptor activation by agonists provoked a significant increase in the rate of cell proliferation and restored cell spreading in serum-free culture conditions *(25,27)*. As shown in **Fig. 4**, the mitogenic and morphogenic actions of α_{2A}-ARs are associated with a rapid tyrosyl phosphorylation of the mitogen-activated protein kinases (ERK1/ and ERK2/MAPK, p42 and p44) and of the focal adhesion kinase (FAK, pp125). Study of the molecular mechanisms underlying these effects demonstrated that phosphorylation of FAK, but not of MAPKs was abolished by pretreatment of the cells with the C3 exoenzyme, a toxin that impairs p21^{rhoA} activity. Conversely, phosphorylation of MAPKs, but not of FAK was strongly inhibited by overexpression of the COOH-terminal domain of βARK1 *(28)*, a result that agrees with previous observations demonstrating that α_2-AR dependent

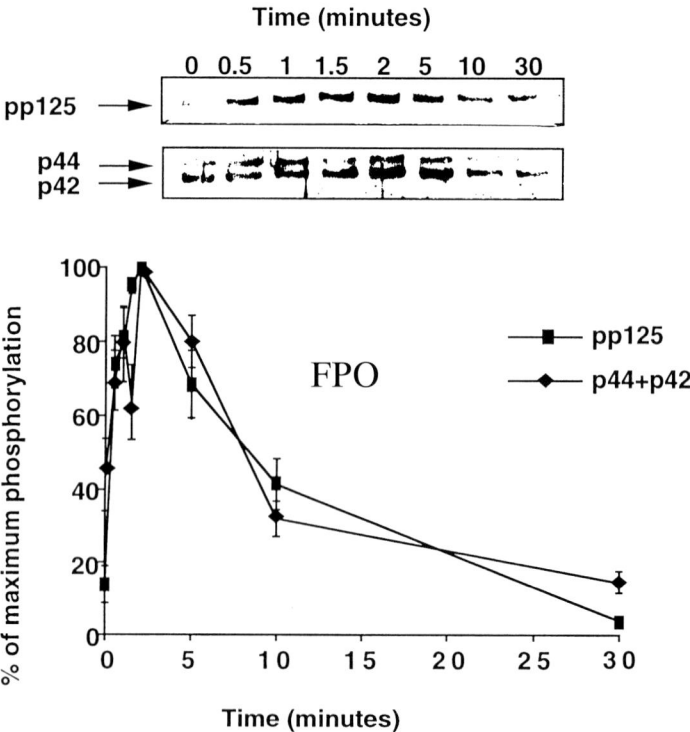

Fig. 4. Cellular responses following α_2-AR stimulation in a 3T3-F442A clone permanently transfected with the pα2C10Eneo vector (according to **ref. 27**), 3T3-F442A
preadipocytes expressing 2.5 pmol of human α_{2A}-AR/mg of membrane proteins
were placed in serum-free medium to induce cell retraction. Twenty minutes later,
cells were stimulated with 10^{-6} M UK14304, and the tyrosyl phosphorylation state
of pp125FAK and p42 and p44 MAPKs was monitored over a 30-min period.
Tyrosine-phosphorylated proteins were immunoprecipitated with an antiphospho-
tyrosine antibody (PY20) and protein G-Sepharose-coupled beads, separated on an
8% polyacrylamide-SDS gel and electrotransferred onto nitrocellulose membrane.
From the same cellular extract, both antifocal adhesion kinase antibody (FAK C-20)
and a mixture of anti-MAPKs ERK1 (C-16) and ERK2 (C-14) were used for
immunodetection by ECL.

activation of the p21^{ras}/MAPK pathway was mediated by the $\beta\gamma$-subunits of
G-proteins *(29)*. Future studies will be necessary to elucidate the precise
mechanisms of p21^{rhoA} activation, but the results obtained in these transfected
cells demonstrated that the combined activation of these two small GTPases
are responsible for the mediation of α_2-AR effect on preadipocyte proliferation
and/or differentiation.

4. Notes

1. The advantages, utilities, and disadvantages of the different vector systems and transfection methods will be discussed. The choice of the vector and of its promoter depends on the goal to be achieved and on the cell type to be transfected. Activity of the promoters generally used (RSV, SV40, and CMV) and considered "strong" is highly variable depending on the host cell. Similarly, there is no universal transfection procedure that can be applied indistinguishably to all cell lines. Each method has its advantages and disadvantages, and should be selected based on the experimental or cellular model. For example, the DEAE–dextran-mediated transfection method is suitable for transient, but not for stable transfection because of the toxicity of DEAE–dextran. In addition, the calcium phosphate transfection method and electroporation can be used on cells growing in suspension.

2. When inserting the cDNA of interest into the expression vector, care must be taken with several points that can result in attenuated or null expression. In most cases, the ORF of the gene of interest is inserted with fragments of its 5′ and 3′ untranslated regions. When defining the cloning strategy, ensure that the 3′NC region is devoid of canonical signals, such as polyadenylation (AAUAAA) or mRNA destabilization (AUUUA, AU-rich elements). Also check that the 5′ noncoding (NC) region of the gene does not contain any start site ATG or micro-ORF that will be inhibitory for translation. In the same way, certain 5′NC-regions having high GC content will form excessive secondary structures that may impede mRNA translation. In this situation, the best solution is to take the smallest fragment of 5′NC as possible, if necessary by mutating the wild-type cDNA in order to introduce a convenient restriction site. When constructing an expression vector, do not artificially create an upstream ATG by cloning artifact. For example, the introduction of an *Sma*I-X insert into an *Eco*RV-X-digested vector will create an initiation codon, as follows. The *Eco*RV (GAT/ATC) digested vector ends with "GAT" and the *Sma*I (CCC/GGG) digested insert begins with "GGG"; after ligation, a sequence "**GAT**GGG" is obtained introducing an ATG, which may be in a favorable initiation context. When using blunt sites, an ATG will be also created if ligating *Sma*I, *Nae*I, *Fsp*I, *Eco*47III, *Pml*I, or *Sna*BI at the 5′ end of the insert with *Eco*RV or *Ssp*I sites of the vector. Also, if using methylation-sensitive sites, such as *Cla*I, it is necessary to amplify the plasmid into a dam⁻ bacterial strain (JM110 or SCS110) to obtain unmethylated DNA.

3. Variants of this method use DMSO or butyrate to shock the cells. However, the shocking step may be omitted depending on the cell type to be transfected.

4. Cell lines exhibit different sensitivity to G418. Primary experiments should define the minimum level of antibiotic that kills untranfected cells.

References

1. Duzic, E., Coupry, I., Downing, S., and Lanier, S. M. (1992) Factors determining the specificity of signal transduction by guanine nucleotide-binding protein-coupled receptors. I. Coupling of α_2-adrenergic receptor subtypes to distinct G-proteins. *J. Biol. Chem.* **267,** 9844–9851.

2. Cotecchia, S., Kobilka, B. K., Daniel, K. W., Nolan, R. D., Lapetina, E. Y., Caron, M. G., et al. (1990) Multiple second messenger pathways of alpha-adrenergic receptor subtypes expressed in eukaryotic cells. *J. Biol. Chem.* **265,** 63–69.
3. Alblas, J., Corven, E. J. V., Hordijk, P. L., Milligan, G., and Moolenaar, W. H. (1993) Gi-mediated activation of the p21ras-mitogen-activated protein kinase pathway by alpha2-adrenergic receptors expressed in fibroblasts. *J. Biol. Chem.* **268,** 22,235–22,238.
4. Allen, L. F., Lefkowitz, R. J., Caron, M. G., and Cotecchia, S. (1991) G-protein-coupled receptor genes as protooncogenes: Constitutively activating mutation of the α_{1B}-adrenergic receptor enhances mitogenesis and tumorigenicity. *Proc. Natl. Acad. Sci. USA* **88,** 11,354–11,358.
5. McEwan, D. J. and Milligan, G. (1996) Inverse agonist-induced up-regulation of the human β_2-adrenoceptor in transfected neuroblastoma x glioma hybrid cells. *Biochem. J.* **50,** 1479–1486.
6. Wozniak, M. and Limbird, L. E. (1996) The three α_2-adrenergic receptor subtypes achieve basolateral localization in Mardin-Darby Canine Kidney II cells via different targeting mechanisms. *J. Biol. Chem.* **271,** 5017–5024.
7. Daunt, D. A., Hurt, C., Hein, L., Kallio, J., Feng, F., and Kobilka, B. K. (1997) Subtype intracellular trafficking of α_2-adrenergic receptors. *J. Biol. Chem.* **51,** 711–720.
8. Bouvier, M., Colins, S., O'Dowd, B. F., Campbell, P. T., de Kobilka, B. K., Mac Gregor, C., et al. (1989) Two distinct pathways for cAMP-mediated down-regulation of the beta2-adrenergic receptor. Phosphorylation of the receptor and regulation of its mRNA level. *J. Biol. Chem.* **264,** 16,786–16,792.
9. Gabilondo, A. M., Hegler, J., Krasel, C., Boivin-Jahns, V., Hein, L., and Lohse, M. J. (1997) A dileucine motif in the C terminus of the beta2-adrenergic receptor is involved in receptor internalization. *Proc. Natl. Acad. Sci. USA* **94,** 12,285–12,290.
10. Keefer, J. R., Kennedy, M. E., and Limbird, L. E. (1994) Unique structural features important for stabilization versus polarization of the alpha2A-adrenergic receptor on the basolateral membrane of Madin-Darby canine kidney cells. *J. Biol. Chem.* **269,** 16,425–16,432.
11. Ogino, Y., Fraser, C. M., and Costa, T. (1992) Selective interaction of beta2- and alpha 2-adrenergic receptors with stimulatory and inhibitory guanine nucleotide-binding proteins. *Mol. Pharmacol.* **42,** 6–9.
12. Ren, Q., Kuroze, H., Lefkowitz, R. J., and Cotecchia, S. (1993) Constitutively active mutants of the alpha2-adrenergic receptor. *J. Biol. Chem.* **268,** 16,483–16,487.
13. Devedjian, J. C., Esclapez, F., Denis-Pouxviel, C., and Paris, H. (1994) Further characterization of human alpha2-adrenoceptor subtypes: [^{3}H]RX821002 binding and definition of additional selective drugs. *Eur. J. Pharmacol.* **252,** 43–49.
14. Marjamäki, A., Ala-Uotila, S., Luomala, K., Parälä, M., Jansson, C., Jalkanen, M., et al. (1992) Stable expression of recombinant human α_2-adrenoceptor subtypes in two mammalian cell lines: characterization with [^{3}H]rauwolscine,

inhibition of adenylate cyclase and RNase protection assay. *Biochim. Biophys. Acta* **1134,** 169–177.

15. Bresnahan, M. R., Flordellis, C. S., Vassilatis, D. K., Makrides, S. C., Zannis, V. I., and Gavras, H. (1990) High level of expression of functional human platelet α_2-adrenergic receptors in a stable mouse C127 cell line. *Biochim. Biophys. Acta* **1052,** 439–445.

16. Borman, A. M., Le Mercier, P., Girard, M., and Kean, K. M. (1997) Comparison of picornaviral IRES-driven internal initiation of translation in cultured cells of different origins. *Nucleic Acids Res.* **25,** 925–932.

17. Maret, A., Galy, B., Arnaud, E., Bayard, F., and Prats, H. (1995) Inhibition of fibroblast growth factor 2 expression by antisense RNA induced a loss of the transformed phenotype in a human hepatoma cell line. *Cancer Res.* **55,** 5075–5079.

18. McCutchan, J. H. and Pagano, J. S. (1968) Enhancement of the infectivity of simian virus 40 deoxyribonucleic acid with diethylaminoethyl-dextran. *J. Natl. Cancer Inst.* **41,** 351–357.

19. Graham, F. L. and van der Eb, A. J. (1973) A new technique for the assay of infectivity of human adenovirus 5 DNA. *Virology* **52,** 456–467.

20. Sambrook, J., Fritsch, E. F., and Maniatis, T. (1989) *Molecular Cloning: A Laboratory Manual* (Ford, N., ed.), Cold Spring Harbor Laboratory, Cold Spring Harbor, NY.

21. Ausubel, F. M., Brent, R., Kingston, R. E., Moore, D. D., Seidman, J. G., Smith, J. A., et al. (eds.), *Currents Protocols in Molecular Biology*. Greene Publishing and Wiley-Interscience, New York.

22. Lomasney, J. W., Lorenz, W., Allen, L. F., King, K., Regan, J. W., Yang-Feng, T. L., et al. (1990) Expansion of the alpha2-adrenergic receptor family: cloning and characterization of a human alpha2-adrenergic receptor subtype, the gene for which is located on chromosome 2. *Proc. Natl. Acad. Sci. USA* **87,** 5094–5098.

23. Regan, J. W., Kobilka, T. S., Yang-Feng, T. L., Caron, M. G., Lefkowitz, R. J., and Kobilka, B. K. (1988) Cloning and expression of a human kidney cDNA for an alpha2-adrenergic receptor subtype. *Proc. Natl. Acad. Sci. USA* **85,** 6301–6305.

24. Kobilka, B. K., Matsui, H., Kobilka, T. S., Yang-Feng, T. L., Francke, U., Caron, M. G., et al. (1987) Cloning, sequencing, and expression of the gene coding for the human platelet alpha2-adrenergic receptor. *Science* **238,** 650–656.

25. Bouloumie, A., Planat, V., Devedjian, J. C., Valet, P., Saulnier-Blache, J. S., Record, M., et al. (1994) Alpha2-adrenergic stimulation promotes preadipocyte proliferation. Involvement of mitogen-activated protein kinases. *J. Biol. Chem.* **269,** 30,254–30,259.

26. Betuing, S., Valet, P., Lapalu, S., Peyroulan, D., Hickson, G., Daviaud, D., et al. (1997) Functional consequences of constitutively active alpha2A-adrenergic receptor expression in 3T3F442A preadipocytes and adipocytes. *Biochem. Biophys. Res. Commun.* **235,** 765–773.

27. Betuing, S., Daviaud, D., Valet, P., Bouloumie, A., Lafontan, M., and Saulnier-Blache, J. S. (1996) alpha2-Adrenoceptor stimulation promotes actin polymeriza-

tion and focal adhesion in 3T3F442A and BFC-1β preadipocytes. *Endocrinology* **137,** 5220–5229.

28. Betuing, S., Daviaud, D., Pages, C., Bonnard, E., Valet, P., Lafontan, M., et al. (1998) Gβγ-independent coupling of α_2-adrenergic receptor to p21rhoA in preadipocytes. *J. Biol. Chem.* **273,** 15,804–15,810.

29. Della Rocca, G. J., Van Biesen, T., Daaka, Y., Luttrell, D. K., Luttrell, L. M., and Lefkowitz, R. J. (1997) Ras-dependent mitogen-activated protein kinase activation by G-protein-coupled receptors. Convergence of Gi- and Gq-mediated pathways on calcium/calmodulin, Pyk2 ans Src kinase. *J. Biol. Chem.* **272,** 19,125–19,132.

13

Expression of β-Adrenergic Receptors in Recombinant Baculovirus-Infected Insect Cells

A. Donny Strosberg and Jean-Luc Guillaume

1. Introduction

Ever since the initial discovery of the structurally homologous G-protein-coupled receptors, the search has been ongoing for identifying suitable expression systems. Such systems should be easy to use, inexpensive, and appropriate for mass production. In addition, these systems should permit analysis of function, including ligand binding and G-protein-regulated effector activation. Although expression in *Escherichia coli* (*see* Chapter 14) provides several of these features, only eukaryotic cells contain the appropriate complement of endogenous G-proteins and effectors to study second messenger activation. Mammalian cell culture remains expensive, is not amenable to large-scale amplification, is prone to infection by contaminant microorganisms, and can undergo loss of receptor-bearing plasmids.

A suitable alternative is the use of the baculovirus expression system introduced into appropriate insect cells. Even though in some applications up to 90% of the receptor protein generated may be nonfunctional, potentially owing to differences in glycosylation and inappropriate protein folding, the baculovirus system has been successfully utilized in the expression of receptors. In most cases, functional receptors can be purified by ligand-containing gel chromatography and used for the determination of pharmacological and structural characteristics. Loisel et al. (*1*) have recently described a method to separate functional receptors present in the viral particles from nonfunctional receptors contained primarily in the cell membrane.

The first G-protein-coupled receptor expressed in recombinant baculovirus-infected Sf9 insect cells were the β_1- and β_2-adrenergic receptors (β_1- and β_2-ARs) (*2*). The expressed protein retains all the expected properties of the

From: *Methods in Molecular Biology, vol. 126: Adrenergic Receptor Protocols*
Edited by: C. A. Machida © Humana Press Inc., Totowa, NJ

native receptor, including ligand binding, adenylyl cyclase activation, and phosphorylation by the GRK2 kinase *(1,3–5)* (*see* **Note 1**). Solubilized receptor can be affinity-purified and reconstituted with partner proteins to study interactions in a simplified model *(2)*.

Subsequently, several other G-protein-coupled receptors have been expressed in the baculovirus/insect cell system and characterized at the protein level. These include receptors for serotonin *(6,7)*, dopamine *(8,9)*, histamine *(10)*, acetylcholine *(11,12)*, opioids *(13)*, neuropeptide Y *(14)*, substance P *(15)*, choriogonadotropin *(16)*, and odorant molecules *(17)*.

Some investigators have chosen to bypass the use of the recombinant virus and inserted the β_2-AR gene directly in the insect cell genome; this manipulation avoids cell death resulting from infection by the baculovirus *(18)*.

To facilitate screening of recombinant virus, Ravet et al. *(4)* developed an *in situ* method by which binding of radiolabeled ligand to recombinant receptor expressed on the infected cells was identified by autoradiography. This constitutes a significant improvement in the identification step compared to the previous tedious selection of clones by microscope examination. Recent advances in the field have led to the development of engineered baculovirus facilitating the integration of the target receptor gene into the virus DNA and allowing easier clone selection of recombinant virus.

2. Materials

2.1. Cell Growth

1. Cell-culture material: culture hood, culture incubator regulated at 27°C, Petri dishes, spinners (*see* **Notes 2** and **3**).
2. Spodoptera frugiperda cells: both Sf9 and its parent strain Sf21 are convenient.
3. Insect cell-culture medium: use a rich medium, such as "supplemented Grace's medium" from Gibco-BRL.
4. Fetal calf serum (FCS) tested for insect cells L (Gibco-BRL), penicillin, streptomycin, and pluronic acid (Pluronic F68, Sigma).
5. For cultivation of attached cells, our medium is made of 5% serum in supplemented Grace's medium containing 50 IU/mL penicillin and 50 µg/mL streptomycin; for cells grown in suspension, we add 0.1% pluronic acid to the medium.

2.2. Genetic Constructs

1. Several companies offer commercially available baculovirus-driven expression systems composed of engineered virus and plasmids, allowing by cotransfection and recombination, the insertion of the target receptor gene. We routinely use the Clontech products, but any other system should be convenient.
2. BacPAK6 virus DNA linearized/digested by Bsu36 I. This parent virus is unable

to form viable progeny because of the disruption of an essential gene, which has to be reinserted by recombination with the plasmid DNA carrying the target receptor gene. The gene coding for the human β_2-AR is inserted in the BacPAK8 plasmid.
3. Lipofectin, 3% agar, neutral red solution.

All solutions and their dilutions should be made sterile by filtration through a 0.22-μm membrane.

2.3. Membrane Cell Preparation

1. Proteases inhibitors: leupeptin, benzamidine, pepstatin, 4-2-(aminoethyl) benzinesulfonyl fluoride (AEBSF).
2. Membrane resuspension buffer: 25 mM Tris-HCl, pH 7.4, 2 mM ethylenediamine tetra-acetic acid (EDTA), 5 μg/mL leupeptin, 5 μg/mL pepstatin, 2 mM AEBSF, 5 μg/mL benzamidine.

3. Methods
3.1. Recombination

1. Seed 2 mL of a Sf9 cell suspension in culture medium without serum (10^6 cells/ mL) in a 20-cm^2 Petri dish. Allow adherence of cells at room temperature (about 60 min).
2. In a 15-mL Falcon tube, add 40 μL of water, 5 μL (500 ng) of the BacPAK8 plasmid containing the receptor gene, 5 μL of BacPAK6 linearized virus solution, and 50 μL of lipofectin (100 μg/mL in water). Mix gently and allow to stand at room temperature for 15 min.
3. Remove the medium from the cells. Add 1.5 mL of culture medium without serum to the DNA/lipofectin preparation, and gently overlay the cells with the mixture. Incubate for 5 h at 27°C.
4. Remove the medium from the cells, and replace with 1.5 mL culture medium containing 5% calf serum (all subsequent culture steps are performed in supplemented Grace's medium containing 5% calf serum). Grow the cells for 3 d at 27°C.
5. Retain the supernatant, which will constitute your crude recombinant virus stock.

3.2. Clone Selection

1. Prepare three dilutions (1/10, 1/100, and 1/1000) of the crude recombinant virus stock in 3 mL of culture medium. Retain the remaining virus stock at –80°C. Seed 5 mL of a Sf9 cell suspension (10^6 cells/mL) in three 50-cm^2 Petri dishes (5×10^6 cells/dish). Allow the cells to adhere (60 min). Remove the medium from a dish, and overlay the cells with a virus dilution. Carefully rotate the dish to ensure a homogenous distribution of the virus. Repeat this step on the remaining dishes with the two other virus dilutions. Allow the infection to occur for 60 min at room temperature.

2. Prepare a solution of agar in culture medium: incubate 8 mL of a 3% agar solution (preheated to 80°C) with 16 mL of the culture medium. Equilibrate to 40°C. Completely remove the medium from the dishes, and overlay the cells with 7 mL of the agar-containing medium. Allow to solidify, and overlay with 5 mL of culture medium. Incubate the infected cells for 5 d at 27°C (*see* **Note 4**).

3. Dilute 1 mL of neutral red solution (0.33% in water) in 10 ml phosphate-buffered saline (PBS). Remove the medium from the cells, and replace with 2.5 mL of the neutral red staining dilution. Incubate for 3 h at 27°C to allow the stain to penetrate the agar. Then remove the supernatant and incubate the inverted (upside down) Petri dishes for an additional 5 h at 27°C (*see* **Note 5**).

4. Examine the dishes under oblique light to identify lysis plaques appearing as small (about 2 mm diameter) unstained circles on a pinkish background.

5. Seed a 24-well culture plate with a 10^6 cells/mL suspension (1 mL/well), and allow the cells to attach for 60 min. With a 1-mL pipet tip, remove an agar plug corresponding to a well-isolated plaque, and transfer it to a well, gently pipeting the medium several times to expel the agar. Repeat this step to transfer plaques to other wells. Incubate for 3 d at 27°C.

6. Retain the supernatant of individual wells (which contains the virus) and screen the corresponding cells for receptor expression. This can be conducted using either ligand binding or Western blotting approaches when antibodies against the receptor are available (*see* **Notes 6–9**).

7. Filter the supernatants from positive clones using a 0.45-μm membrane, and retain at −80°C for long-term storage.

3.3. Amplification and Receptor Production

1. Using Sf9 cells grown to 80% confluence in a 170-cm² culture flask, add 0.5 mL suspension of a selected virus clone into 50 mL of culture medium. Allow the infection to develop for 6 d at 27°C; then collect the culture medium. Pellet the cell debris by centrifugation ($3000g$/10 min). The resulting supernatant constitutes the preamplification supernatant used in subsequent steps.

2. In a culture spinner containing 250 mL culture medium (supplemented with 0.1% pluronic acid), infect an insect cell suspension (1.5–2.5×10^6 cells/mL) with 25 mL of the preamplification supernatant (retain the remaining 25 mL at −20°C). Allow the infection to proceed for 5 d. Collect the culture medium and subject to centrifugation ($3000g$; 5 min at room temperature). Retain the resulting supernatant at 4°C in the dark for routine receptor production; this constitutes the working stock of amplified virus (*see* **Notes 10–12**).

3. Determine the time-course of receptor production; seed cells in five 65-cm² tissue-culture flasks (10 mL culture medium); when cell density reaches 80% confluence, infect each flask with 1 mL of the amplified virus stock; at 24-h intervals, remove the cells from one flask, wash the pellet with PBS, and either count the cells and perform a direct ligand binding assay or prepare plasma membranes from the cells.

Membrane preparation is easily performed by resuspending the cell pellet in 5 mL ice-cold membrane resuspension buffer; disrupt the cells on ice using Ultra-Turrax (3 passages of 10 s at maximum speed), centrifuge ($3000g$/10 min at 4°C) to remove cell debris, and then centrifuge ($50,000g$/30 min at 4°C) to pellet membranes; resuspend the pellet in 0.5 mL membrane resuspension buffer, and store at –20°C until use. Membrane preparations from the different times of infection can be examined either by ligand binding or by Western blotting experiments.

4. For protein production, infect either (for small-scale production) attached cells grown in 170-cm^2 culture flasks (25 mL culture medium) with 2.5 mL of the amplified virus, or (for larger-scale production) 200 mL batches of Sf9 cells grown in suspension in spinner (10^6 cells/mL) in culture medium containing 0.1% pluronic acid with 10 mL of the amplified virus (*see* **Notes 13** and **14**).

4. Notes

1. There are many examples of successful second messenger production induced by recombinant receptor activation in insect cells; however, in some cases, the occurrence of poor activation has been reported owing to the absence of endogenous elements of the signaling cascade *(19)*. Multiple-target-gene expression plasmids have been commercially developed, allowing the insertion of several genes and their subsequent coexpression during virus infection. It is thus possible to express simultaneously the receptor and a potentially missing component of the signaling pathway (for instance, particular G-protein subunit).

2. Insect cells adhere readily to plastic supports, but not tightly, so it is not necessary to use EDTA or trypsin when passaging; cells are easily resuspended by hitting the side of the culture flask or by repeated pipeting and expulsion of the culture medium onto the cells.

3. Optimum cell culture temperature is 27°C. However, cell growth occurs at temperatures between 23°C and 28°C. It is wise to lower the temperature when the number of binding sites is low compared to the amount of expressed receptor; reduction in temperature may result in improved folding of the protein.

4. The agar overlay is a critical step of the cloning procedure: the temperature of the agar solution must be below 40°C to avoid virus damage, but sufficiently high to allow pipeting without solidification of the medium; the overlay should consist of homogenous thickness to prevent regional overstaining of the gel during the neutral red incubation.

5. The neutral red staining is sometimes too light to observe the plaques after 5 h; in this case, increase the incubation time (by 16 h) to enhance the contrast between lysis plaques and uninfected cells.

6. An antireceptor antibody to determine receptor protein expression during infection or purification is an invaluable tool *(20,21)*. If such a reagent is not available, it is desirable to insert an epitope tag at the extreme end of the receptor sequence during plasmid construction. In our hands, the c-myc epitope, recog-

nized by the monoclonal antibody (MAb) 9E10 *(22)*, provides the best results with proteins produced in the baculovirus/insect cell system.

7. To test the binding activity of the produced receptor during the "clone selection" step, cells from individual wells are resuspended in 1 mL PBS. For β-adrenergic receptors (β-AR), the ^{125}I-iodocyanopindolol binding activity is assayed on 250-μL fractions as described in Chapter 14 using PBS instead of Tris buffer.

8. For Western blotting experiments, approx 20 μg of membrane proteins are loaded per well (for a minigel). After transfer, the membrane is incubated for several hours in a blocking buffer containing 5% skimmed milk powder and 0.2% Tween-20. There are some crucial points in sample preparation for correct gel separation of the membrane proteins: the amount of SDS in the loaded sample should be 5% (use a 2X modified Laemmli sample buffer containing 10% sodium dodecyl sulfate [SDS]), and the temperature of incubation prior to loading should not be over 37°C (use 37°C for at least 2 h or room temperature for 6–16 h; boiling should be absolutely avoided, because it induces the aggregation of hydrophobic proteins, which are unable to penetrate the gel during migration).

9. If no activity is found in infected cells and no antibody to detect the receptor by Western blotting is available, it is a good idea to try to detect receptor mRNA in the cells and/or detect the DNA of the recombinant virus in the culture supernatant by molecular biology techniques.

10. The recombinant baculovirus stock is very stable at 4°C (we have infected cells with stocks retained at 4°C for several years without noticing any appreciable decrease in the virus infectivity or receptor production). The only point to observe is to keep the stocks in the dark *(23)*. However, for convenience, we prefer to keep intermediate amplification stocks below 0°C.

11. Overpassaging the virus to produce infection stocks should be avoided because of risks of point mutations in the target gene. When the working amplification stock is depleted, a new stock is prepared using the remaining 25 mL of the preamplified solution (**step 2** in **Subheading 3.3.**). When this second stock becomes depleted, a new preamplified solution of the selected clone will need to be developed (**step 1** in **Subheading 3.3.**). When the first selected clone becomes depleted, develop a working solution starting from one of the positive clones retained at –80ºC (**step 7** in **Subheading 3.2.**).

12. Using optimized amplification and infection conditions described above, titration of virus stock is unnecessary.

13. The membrane protein yield is about 0.5–1 mg/65-cm^2 flask. The receptor production generally peaks at 48–72 h postinfection, but with certain receptor constructs, we have experienced an increase in receptor concentration until 120 h postinfection.

14. When adherent cells are infected, the amount of functional human β$_2$-AR routinely reaches 10–15 pmol/mg of membrane proteins. In suspension culture, these quantities are generally increased to 25–35 pmol/mg. However, the quantities of active sites can vary greatly, depending on the expressed receptor: similar amounts are obtained with the human β$_1$-AR; in contrast, no detectable ligand

binding activity is observed when the β_3-AR is expressed, although the protein is correctly translated and present in large amounts in the cell membrane (J.L.G., unpublished data).

Acknowledgments

This work was supported by the Centre National de la Recherche Scientifique, the Institut National de la Santé et de la Recherche Médicale, the University Paris-VII, the French Ministry for National Education and Research, the Fondation pour la Recherche Médicale, and the Ligue Nationale contre le Cancer.

References

1. Loisel, T. P., Ansanay, H., St-Onge, S., Gay, B., Boulanger, P., Strosberg, A. D., et al. (1997) Recovery of homogeneous and functional beta 2-adrenergic receptors from extracellular baculovirus particles. *Nat. Biotechnol.* **15,** 1300–1304.
2. George, S. T., Arbabian, M. A., Ruoho, A. E., Kiely, J., and Malbon, C. C. (1989) High-efficiency expression of mammalian beta-adrenergic receptors in baculovirus-infected insect cells. *Biochem. Biophys. Res. Commun.* **163,** 1265–1269.
3. Reilander, H., Boege, F., Vasudevan, S., Maul, G., Hekman, M., Dees, C., et al. (1991) Purification and functional characterization of the human beta 2-adrenergic receptor produced in baculovirus-infected insect cells. *FEBS Lett.* **282,** 441–444.
4. Ravet, V., Blin, N., Guillaume, J. L., Petitjean, F., Cabanie, L., and Strosberg, A. D. (1993) High level functional expression of human beta 1-adrenergic receptor in baculovirus-infected cells screened by a rapid in situ procedure. *J. Recept. Res.* **13,** 541–558.
5. Mouillac, B., Caron, M., Bonin, H., Dennis, M., and Bouvier, M. (1992) Agonist-modulated palmitoylation of beta 2-adrenergic receptor in Sf9 cells. *J. Biol. Chem.* **26,** 21,733–21,737.
6. Parker, E. M., Kameyama, K., Higashijima, T., and Ross, E. M. (1991) Reconstitutively active G protein-coupled receptors purified from baculovirus-infected insect cells. *J. Biol. Chem.* **266,** 519–527.
7. Ng, G. Y., George, S. R., Zastawny, R. L., Caron, M., Bouvier, M., Dennis, M., et al. (1993) Human serotonin1B receptor expression in Sf9 cells: phosphorylation, palmitoylation, and adenylyl cyclase inhibition. *Biochemistry* **32,** 11,727–11,7233.
8. Mills, A., Allet, B., Bernard, A., Chabert, C., Brandt, E., Cavegn, C., et al. (1993) Expression and characterization of human D4 dopamine receptors in baculovirus-infected insect cells. *FEBS Lett.* **320,** 130–134.
9. Boundy, V. A., Lu, L., and Molinoff, P. B. (1996) Differential coupling of rat D2 dopamine receptor isoforms expressed in Spodoptera frugiperda insect cells. *J. Pharmacol. Exp. Ther.* **276,** 784–794.

10. Kuhn, B., Schmid, A., Harteneck, C., Gudermann, T., and Schultz, G. (1996) G proteins of the Gq family couple the H2 histamine receptor to phospholipase C. *Mol. Endocrinol.* **10,** 1697–1707.

11. Kukkonen, J. P., Nasman, J., Ojala, P., Oker-Blom, C., and Akerman, K. E. (1996) Functional properties of muscarinic receptor subtypes Hm1, Hm3 and Hm5 expressed in Sf9 cells using the baculovirus expression system. *J. Pharmacol. Exp. Ther.* **279,** 593–601.

12. Vasudevan, S., Premkumar, L., Stowe, S., Gage, P. W., Reilander, H., and Chung, S. H. (1992) Muscarinic acetylcholine receptor produced in recombinant baculovirus infected Sf9 insect cells couples with endogenous G-proteins to activate ion channels. *FEBS Lett.* **311,** 7–11.

13. Obermeier, H., Wehmeyer, A., and Schulz, R. (1996) Expression of mu-, delta- and kappa-opioid receptors in baculovirus-infected insect cells. *Eur. J. Pharmacol.* **318,** 161–166.

14. Munoz, M., Sautel, M., Martinez, R., Sheikh, S. P., and Walker, P. (1995) Characterization of the human Y1 neuropeptide Y receptor expressed in insect cells. *Mol. Cell. Endocrinol.* **107,** 77–86.

15. Nishimura, K., Frederick, J., and Kwatra, M. M. (1998) Human substance P receptor expressed in Sf9 cells couples with multiple endogenous G proteins. *J. Recept. Signal Transduct. Res.* **18,** 51–65.

16. Narayan, P., Gray, J., and Puett, D. (1996) Expression of functional lutropin/choriogonadotropin receptor in the baculovirus system. *Mol. Cell. Endocrinol.* **117,** 95–100.

17. Raming, K., Krieger, J., Strotmann, J., Boekhoff, I., Kubick, S., Baumstark, C., et al. (1993) Cloning and expression of odorant receptors. *Nature* **361,** 353–356.

18. Kleymann, G., Boege, F., Hahn, M., Hampe, W., Vasudevan, S., and Reilander, H. (1993) Human beta 2-adrenergic receptor produced in stably transformed insect cells is functionally coupled via endogenous GTP-binding protein to adenylyl cyclase. *Eur. J. Biochem.* **213,** 797–804.

19. Butkerait, P., Zheng, Y., Hallak, H., Graham, T. E., Miller, H. A., Burris, K. D., et al. (1995) Expression of the human 5-hydroxytryptamine1A receptor in Sf9 cells. Reconstitution of a coupled phenotype by co-expression of mammalian G protein subunits. *J. Biol. Chem.* **270,** 18,691–18,699.

20. Guillaume, J. L., Petitjean, F., Haasemann, M., Bianchi, C., Eshdat, Y., and Strosberg, A. D. (1994) Antibodies for the immunochemistry of the human β3-adrenergic receptor. *Eur. J. Biochem.* **224,** 761–770.

21. Luxembourg, A. (1995) Anti-peptide monoclonal antibodies to the beta-adrenergic receptor: use in purification of beta receptor. *Hybridoma* **14,** 261–264.

22. Evan, G. I., Lewis, G. K., Ramsay, G., and Bishop, J. M. (1985) Isolation of monoclonal antibodies specific for human c-myc proto-oncogene product. *Mol. Cell. Biol.* **5,** 3610–3616.

23. Jarvis, D. L. and Garcia, A. (1994) Long-term stability of baculovirus stored under various conditions. *BioTechniques* **16,** 508–513.

14

Expression of β-Adrenergic Receptors in *E. coli*

Ralf Jockers and A. Donny Strosberg

1. Introduction

G protein-coupled receptors (GPCRs) are among the most important targets for pharmacological compounds used in humans. During the last 10 years, more than 1000 genes coding for GPCRs, representing approx 0.1% of the human genome, have been identified, of which at least 200 have been demonstrated to correspond to active receptors. Cloning of receptor cDNAs allowed expression in heterologous cell systems. Among the available expression systems, considerable interest has focused on microorganisms, such as *Escherichia coli* (*E. coli*), that provide a reproducible and inexpensive source of receptor, combined with ease of use, low background binding, and the absence of endogenous GPCR. All these characteristics are prerequisites for the successful application of GPCRs in high-throughput screening assays for the discovery of new drugs *(1)*.

GPCRs are integral membrane proteins with seven transmembrane-spanning regions. Thus, technical adaptations are necessary for the expression of suitable amounts of properly folded GPCRs in *E. coli*. The use of a fusion protein, including the receptor and a bacterial protein, has proven successful, as demonstrated for the first time for the β_2-adrenergic receptor (β_2-AR) *(2)*. Initially, this receptor was fused to the cytoplasmic β-galactosidase protein *(2)*. Fusion to the outer membrane protein LamB and later to the inner membrane protein MalE further increased the number of functionally expressed receptors *(3,4)*.

Subsequently other GPCRs have been successfully expressed in *E. coli*, including the human β_1-AR *(3)*, the human serotonin $5HT_{1A}$ receptor *(5)*, the adenosine A_1 receptor *(6)*, the rat neurotensin receptor *(7,8)*, the human endothelin ET_B receptor *(9)*, the human dopamine receptor *(10)*, the rat neuro-

From: *Methods in Molecular Biology, vol. 126: Adrenergic Receptor Protocols*
Edited by: C. A. Machida © Humana Press Inc., Totowa, NJ

kinin A receptor *(11)*, and the human neuropeptide Y1 receptor *(12,13)*. Not all receptors expressed in *E. coli* retain their functional properties: we have never been able to demonstrate ligand binding activity for the β_3-AR expressed in *E. coli* using various fusion partners (Guillaume and Strosberg, unpublished results).

Today's applications of GPCRs expressed in *E. coli* go far beyond the initial purpose for multidrug screening. GPCRs have been used to analyze receptor topology, the ligand binding pocket, and G protein signaling, as well as providing an abundant source of receptor for structural purposes *(14–18)*. The present protocol describes the expression of β-ARs as an MalE fusion protein in *E. coli*.

2. Materials

1. *E. coli* bacterial strain KS303, which limits degradation of abnormal periplasmatic proteins *(19)*.
2. Expression plasmid pMES *(20)* containing the MalE promoter followed by the first 56 codons of the MalE coding sequence, including its signal sequence, in phase with the coding region of the β-AR.
3. Ampicillin stock solution (50 mg/mL) in water stored at –20°C.
4. Luria-Bertani Medium (LB medium): 10 g Bacto-tryptone, 5 g Bacto-yeast extract, 10 g NaCl prepared to 1 L with deionized water. Store LB medium at room temperature after sterilization by autoclaving (30 min, 120°C).
5. LB/amp medium: LB medium supplemented with ampicillin (50 µg/mL).
6. Medium for agarose Petri dishes: 15 g agarose/L LB medium supplemented with 50 µg/mL ampicillin and 1% glucose.
7. 50% Stock solution of D-(+)-glucose in deionized water. Store at room temperature after sterilization by autoclaving (20 min, 110°C).
8. 20% Stock solution of maltose in deionized water. Store at room temperature after sterilization by filtration through 0.22-µm filter (*see* **Note 1**).
9. STE buffer: 100 m*M* NaCl, 10 m*M* Tris-HCl, pH 8.0, 1 m*M* ethylenediamine tetra-acetic acid (EDTA).
10. Standard sonicator with small tip (we use Vibra cell™ from Sonics materials INC, Danbury, CT).
11. Membrane buffer: 1 m*M* EDTA, 1 m*M* ethyleneglycol-bis-(β-aminoethylether)-*N,N,N,'N,'*-tetra-acetic acid (EGTA), 20 m*M* Tris-HCl, pH 8.0, 0.5 m*M* 4-(2-aminoethyl)-benzenesulfonyl fluoride hydrochloride (AEBSF), 1 mg/100 mL benzamidine, 0.5 mg/100 mL aprotinin, 0.5 mg/100 mL leupeptin.
12. TEM buffer: 50 m*M* Tris-HCl, pH 8.0, 1 m*M* EDTA, 5 m*M* MgCl$_2$.
13. Alprenolol stock solution (10 m*M*) in deionized water, stored at –20°C. Multiple freeze–thaw cycles of the stock solution are possible.
14. [125I]cyanopindolol ([125I]CYP) (commercially available) stored at –20°C.
15. 30% Stock of polyethyleneimine (30% w/v) stored at 4°C (Amersham Life Science Products).

16. Whatman GF/C glass fiber filters. Adapt filter size to the corresponding filtration unit.
17. Filtration unit for radioligand binding assay. We use a filtration unit from Brandel for 48 samples, but any other filtration unit adapted for radioactive compounds should work as well.
18. Two incubators with rotary shaker at 30°C and 37°C.

3. Methods

1. Transform competent *E. coli* KS303 bacteria following standard procedures *(21)* with pMES plasmid containing the coding region of the β-AR of interest.
2. Spread transformed bacteria with a sterile bent glass rod gently over the surface of an agar plate containing LB/amp medium supplemented with 1% glucose, and incubate overnight at 37°C.
3. Take one to two colonies from the overnight agar plate, inoculate 20 mL of LB/amp liquid medium supplemented with 1% glucose, and incubate with vigorous shaking (180 cycles/min in a rotary shaker) for 5–6 h at 37°C.
4. Store preculture overnight at 4°C. Longer storage of the preculture may reduce receptor expression levels (*see* **Note 2**).
5. Dilute the preculture in 1 L of LB/amp medium supplemented with 1% glucose, and incubate at 37°C with vigorous shaking until the optical density at 600 nm reaches 0.8–1.0. Higher optical densities decrease the amount of expressed receptors.
6. Centrifuge bacterial suspension at 3800*g* (5000 rpm with JA-14 rotor from Beckmann) for 7 min at 4°C.
7. Wash the bacterial pellet free of glucose in 250 mL of LB medium, and centrifuge as described.
8. Resuspend bacterial pellet in 1 L of LB/amp medium supplemented with 1% maltose to induce receptor expression. Incubate 5 h at 30°C with vigorous shaking (*see* **Notes 3** and **4**).
9. Centrifuge bacterial suspension as described above, and wash the bacterial pellet once in STE buffer (*see* **Note 5**).
10. Resuspend the bacterial pellet in Laemmli loading buffer if receptor expression is to be verified by sodium dodecyl sulfate-polyacrylamide gel electrophoresis (SDS-PAGE) and Western blot analyses (*see* **Note 6**). Homogenize samples by sonication before loading onto SDS-PAGE. SDS-PAGE is performed according to standard procedures *(21)*.
11. If receptor expression is to be verified by radioligand binding analyses, resuspend the bacterial pellet in 40 mL membrane buffer, and lyse enzymatically with 5 mg lysozyme and 1 mg DNase I for 30–60 min at room temperature with stirring.
12. For radioligand binding assays, sonicate the bacterial suspension for approx 3 min on ice at settings: 50%, level 6 with a small tip.
13. Centrifuge at 1500*g* for 15 min at 4°C.

14. Centrifuge the supernatant at 48,000g (20,000 rpm with JA-20 rotor from Beckmann) for 30 min at 4°C.

15. Resuspend the pellet in approx 20 mL of membrane buffer.

16. Centrifuge again at 48,000g for 30 min at 4°C, and homogenize the pellet in approx 5 mL of TEM buffer. Homogenization using a ceramic pestle may be necessary.

17. Determine the protein content, freeze aliquots in liquid nitrogen, and store at −80°C.

18. Radioligand binding assay: Thaw frozen membranes rapidly. Use between 5 and 100 µg of protein in a final volume of 0.5 mL TEM buffer containing 50 mg/100 mL bovine serum albumin (BSA), and [^{125}I]CYP as radioligand (200 pM is a saturating concentration for β_1- and β_2-ARs). Specific binding is defined as binding displaced by 10 µM alprenolol. Perform each data point at least in duplicate (two in the presence and two in the absence of alprenolol). Incubate for 45–90 min at 25°C, and terminate by rapid filtration through Whatman GF/C glass fiber filters previously soaked in TEM containing 50 mg/100 mL BSA and 0.3% polyethyleneimine (*see* **Note 8**).

4. Notes

1. Do not sterilize maltose stock solution by autoclaving, since this action will destroy the sugar.

2. As an alternative to **steps 4** and **5** in **Subheading 3.**, 2 mL of preculture may be diluted in 1 L LB/amp and incubated overnight at 23°C. Following the overnight incubation, transfer the culture to 37°C, and incubate until optical density at 600 nm reaches 0.8–1.0.

3. The described method generally yields 6–12 pmol of receptor/mg of membrane protein. However, conditions for induction of receptor expression as described in **step 8** of **Subheading 3.** must be optimized for each receptor gene construct. An induction time between 0 and 7 h is recommended. The induction temperature may also be varied (for example 20, 25, 30°C). Simple glucose withdrawal (without addition of maltose) may be sufficient to induce receptor expression in some cases.

4. Receptor expression is often accompanied by growth arrest of bacteria. Therefore, a preliminary and simple test to verify receptor expression is to compare optical densities (at 600 nm) of two bacterial cultures, one from bacteria containing the pMES plasmid alone and one from bacteria containing the plasmid with the coding region of the receptor.

5. The bacterial pellet obtained in **step 9** of **Subheading 3.** may be resuspended in 150 mL STE buffer supplemented with 15% glycerol, frozen in liquid nitrogen, and stored at −80°C prior to verification of receptor expression by radioligand binding (**steps 1–18**).

6. Verification of receptor expression by Western blotting depends on the availability of suitable antibodies. Since in the case of β-ARs a radioligand is available, this functional assay is the method of choice.

7. We use the commercially available iodinated radioligand [^{125}I]CYP in radioligand binding assays, since this radioligand has a high affinity to β-ARs and labeling with 125-iodine provides a high specific activity. Other commercially available radioligands, such as [^{3}H]alprenolol or [^{3}H]CGP12177A, may also be used; however, the assay using tritiated radioligands may be less sensitive.

8. The addition of 0.3% polyethyleneimine may improve the ratio of nonspecific to specific radioligand binding by decreasing the nonspecific binding. This compound is not an essential component of the binding assay.

References

1. Luyten, W. H. and Leysen, J. E. (1993) Receptor cloning and heterologous expression—towards a new tool for drug discovery. *Trends Biotechnol.* **11,** 247–254.

2. Marullo, S., Delavier-Klutcho, C., Eshdat, Y., Strosberg A. D., and Emorine, L. J. (1988) Human β$_2$-adrenergic receptors expressed in *Escherichia coli* membranes retain their pharmacological properties. *Proc. Natl. Acad. Sci. USA* **85,** 7551–7555.

3. Marullo, S., Delavier-Klutchko, C., Guillet, J.-G., Charbit, A., Strosberg, A. D., and Emorine, L. J. (1989) Expression of human beta1 and beta2 adrenergic receptors in *E. coli* as a new tool for ligand screening. *Bio/Technology* **7,** 923–927.

4. Chapot, M.P., Eshdat, Y., Marullo, S., Guillet, J.G., Charbit, A., Strosberg, A.D., et al. (1990) Localization and characterization of three different beta-adrenergic receptors expressed in *Escherichia coli. Eur. J. Biochem.* **187,** 137–144.

5. Bertin, B., Freissmuth, M., Breyer, R. M., Schütz, W., Strosberg, A. D., and Marullo, S. (1992) Functional expression of the human serotonin 5HT$_{1A}$ receptor in *Escherichia coli. J. Biol. Chem.* **267,** 8200–8206.

6. Jockers, R., Linder, M. E., Hohenegger, M., Nanoff, C., Bertin, B., Strosberg, A. D., et al. (1994) Species difference in the G protein selectivity of the human and bovine A(1)-adenosine receptor. *J. Biol. Chem.* **269,** 32,077–32,084.

7. Grisshammer, R., Duckworth, R., and Henderson, R. (1993) Expression of a rat neurotensin receptor in *Escherichia coli. Biochem. J.* **295,** 571–576.

8. Tucker, J. and Grisshammer, R. (1996) Purification of a rat neurotensin receptor expressed in *Escherichia coli. Biochem. J.* **317,** 891–899.

9. Haendler, B., Hechler, U., Becker, A., and Schleuning, W. D. (1993) Expression of human endothelin receptor ET$_B$ by *Escherichia coli* transformants. *Biochem. Biophys. Res. Commun.* **191,** 633–638.

10. Vanhauwe, J., Luyten, W. H. M. L., Josson, K., Fraeyman, N., and Leysen, J. E. (1996) Expression of the human dopamine D$_3$ receptor in *Escherichia coli*: investigation of the receptor properties in the absence and presence of G proteins, in *The Eighth International Catecholamine Symposium.* Asilomar Conference Center, Pacific Grove, CA.

11. Grisshammer, R., Little, J., and Aharony, D. (1994) Expression of rat NK-2 (neurokinin A) receptor in *E. coli. Receptors Channels* **2,** 295–302.

12. Herzog, H., Münch, G., and Shine, J. (1994) Human neuropeptide Y1 receptor

expressed in *Escherichia coli* retains its pharmacological properties. *DNA Cell. Biol.* **13,** 1221–1225.

13. Münch, G., Walker, P., Shine, J., and Herzog, H. (1995) Ligand binding analysis of human neuropeptide Y1 receptor mutants expressed in *E coli. Receptors Channels* **3,** 291–297.

14. Strosberg, A. D. and Leysen, J. E. (1991) Receptor-based assays. *Curr. Opinion Biotechnol.* **2,** 30–36.

15. Marullo, S., Emorine, L., Strosberg, A., and Delavier-Klutchko, C. (1990) Selective binding of ligands to β_1-, β_2- or chimeric β_1-/β_2-adrenergic receptors involves multiple subsites. *EMBO J.* **9,** 1471–1476.

16. Breyer, R. M., Strosberg, A. D., and Guillet, J. G. (1990) Mutational analysis of ligand binding activity of beta-2-adrenergic receptor expressed in *Escherichia coli. EMBO J.* **9,** 2679–2684.

17. Freissmuth, M., Selzer, E., Marullo, S., Schütz, W., and Strosberg, A. D. (1991) Expression of two human β-adrenergic receptors in *Escherichia coli*: Functional interaction with two forms of the stimulatory G protein. *Proc. Natl. Acad. Sci. USA* **88,** 8548–8552.

18. Grisshammer, R. and Tate, C. G. (1995) Overexpression of integral membrane proteins for structural studies. *Q. Rev. Biophys.* **28,** 315–422.

19. Strauch, K. L. and Beckwith, J. (1988) An *Escherichia coli* mutation preventing degradation of abnormal periplasmic proteins. *Proc. Natl. Acad. Sci. USA* **65,** 1576–1580.

20. Szmelcman, S., Clément, J. M., Jehanno, M., Schwatz, O., Montagnier, L., and Hofnung, M. (1990) Export and one-step purification from *Escherichia coli* of a MalE-CD4 hybrid protein that neutralizes HIV in vitro. *J. Acquir. Immune Defic. Syndr.* **3,** 859–872.

21. Sambrook, J., Fritsch, E. F., and Maniatis, T. (1989). *Molecular Cloning— a Laboratory Manual.* Cold Spring Harbor Laboratory Press, Cold Spring Harbor, NY.

15

Use and Pharmacological Analysis of Established and Transfected Cell Lines Expressing Adrenergic Receptors

Hongying Zhong and Kenneth P. Minneman

1. Introduction
1.1. Why Use Cell Lines Expressing Adrenergic Receptors?

A large number of receptors for neurotransmitters and hormones have been identified by pharmacological studies and by molecular cloning. These receptors are divided into homologous families, which consist of a number of even more homologous subtypes. At least nine distinct adrenergic receptor (AR) subtypes have so far been identified, each of which is the product of a separate gene *(1)*, and some of these are known to have additional splice variants *(2,3)*. Studies of these receptor subtypes in tissues where they are endogenously expressed are often complicated by the presence of many different cell types and coexisting receptor subtypes. The use of cell lines that endogenously express known complements of AR subtypes, or are stably transfected with receptor cDNAs, provides valuable tools for studying the pharmacological properties and functional characteristics of individual receptor subtypes *(4)*. Cell lines expressing a single AR subtype can be identified and used to study their signaling properties in the absence of other subtypes. Receptor cDNAs can be heterologously expressed in cells that natively express no AR subtypes, minimizing potential complications and interference from closely related subtypes. Different receptor subtypes can be expressed in the same cell line, allowing direct comparison of their pharmacology and signaling properties in the same cellular phenotype. Using established cell lines expressing receptor cDNAs makes it possible to obtain high levels of receptor expression, facilitating receptor purification. Once purified receptors are available, specific antibodies

From: *Methods in Molecular Biology, vol. 126: Adrenergic Receptor Protocols*
Edited by: C. A. Machida © Humana Press Inc., Totowa, NJ

can be generated and characterized, and used to study the distribution of receptor proteins by immunohistochemical methods *(5)*. Expression of receptor cDNAs containing point mutations also allows identification of key amino acid residues involved in agonist and antagonist binding or signaling specificity, whereas expression of chimeric receptors can identify larger domains critical for ligand recognition and/or signal initiation. Finally, two or more ARs can be coexpressed in the same cells, either natively or by cotransfection and selection, to examine potential interactions between the receptors and the signals that they generate *(6)*.

An important advance is the recent availability of expression vector systems that allow direct control of receptor expression. These include systems where cDNA expression can be either induced or repressed *(7,8)*. Once stably transfected into cell lines, inducible or repressible vector systems carrying receptor cDNAs can be induced to titrate receptor density over a defined range. This allows direct control of receptor density within a single cell line, by graded induction or repression of receptor expression in response to increasing time or concentration of exogenous compounds. This permits direct quantitation of receptor density–response relationships, comparison of coupling efficiencies of closely related subtypes, and direct tests of occupancy response theory. Although there are disadvantages to these heterologous expression systems, the use of stable cell lines expressing various receptor subtypes has provided an extremely valuable system for studying many important aspects of receptor structure and function.

1.2. Which Cell Lines Endogenously Express Adrenergic Receptors?

Many cell lines endogenously express one or more AR subtypes. Some that have been carefully characterized and/or used extensively are listed in **Table 1**. It is clear that most cell lines express β_2-ARs, whereas fewer cell lines endogenously express the other eight AR subtypes. Some cell lines natively express two or more subtypes, particularly SK-N-MC neuroepithelioma cells *(9,10)*, which are known to coexpress seven of the nine known AR subtypes (**Table 1**). Interestingly, most cell lines that endogenously express α_2-ARs do not express other AR subtypes, but cell lines expressing α_1-ARs commonly coexpress β_2-ARs (**Table 1**). This is not universally true, however, and it is not clear if any conclusions about regulation of subtype expression can be drawn from the data compiled.

1.3. Which Cell Lines Should Be Used for Heterologous Expression of Adrenergic Receptors?

It is difficult to choose among the many different cell lines available for transient or stable expression of AR cDNAs. Commonly used cell lines include

Table 1
Selected Continuous Cell Lines Known to Express
Significant Levels of Particular AR Subtypes Endogenously

Species cell line	Source	α_1-ARs	α_2-ARs	β-ARs	Refs.
Human					
SK-N-MC	Neuroepithelioma	α_{1A}, α_{1B}, α_{1D}	α_{2A}, α_{2C}	β_1, β_3	*9,10,14*
HT29	Colon carcinoma	N.R.[a]	α_{2A}	N.R.	*15*
HEPG2	Liver	N.R.	α_{2C}	N.R.	*14*
HEL	Erythroleukemia	N.R.	α_{2A}	N.R.	*16*
HEK 293	Embryonic kidney	N.R.	N.R.	β_2	*17*
HL60	Promyocytic leukemia	N.R.	N.R.	β_2	*18*
HeLa	Cervical carcinoma	N.R.	N.R.	β_2	*19*
1321N1	Astrocytoma	N.R.	N.R.	β_2	*20*
Rat					
FRTL-5	Thyroid	α_{1B}	N.R.	N.R.	*21*
RMTC 6-23	Thyroid	α_{1B}, α_{1D}	N.R.	N.R.	*22*
C6	Glioma	N.R.	N.R.	β_1, β_2	*23*
KNRK	Kidney	N.R.	N.R.	β_2	*24*
FRSK	Fetal keratinocytes	N.R.	N.R.	β_2	*25*
L6	Myoblast	N.R.	N.R.	β_2	*26*
Mouse					
NB41A3	Neuroblastoma	α_{1B}	N.R.	N.R.	*10*
BC3H1	Brain	α_{1B}	N.R.	β_2	*27*
NG108	Neuro/Glioma	N.R.	α_{2B}	N.R.	*15*
NCB20	Neuroblastoma	N.R.	α_{2B}	N.R.	*28*
3T3-L1	Fibroblasts	N.R.	N.R.	β_1, β_2	*29*
3T3-F442A	Adipocyte	N.R.	N.R.	β_1, β_2, β_3	*30*
S49	Lymphoma	N.R.	N.R.	β_2	*31*
Hamster					
DDT$_1$-MF2	Vas deferens	α_{1B}	N.R.	β_2	*32*
Dog					
MDCK	Kidney	α_{1B}	N.R.	β_2	*33*
Monkey					
COS	Kidney	N.R.	N.R.	β_2	*34*
Opossum					
OK	Kidney	N.R.	α_{2C}	N.R.	*35*

[a]N.R. Not reported at present. In some cases, the existence of these subtypes has not been studied, whereas in other cases, no significant expression has been found. Note that subclones of the same parent cell line may vary in types of receptors expressed endogenously.

Chinese hamster ovary (CHO), COS 7 (from monkey kidney), human embryonic kidney (HEK) 293, and NIH-3T3, among many others. Among the cell lines that we have screened in our laboratory, only a few contain no detectable endogenous ARs. These include rat PC12 pheochromocytoma and rat GH_3 pituitary cells. Almost all other cell lines we have screened endogenously express at least one AR subtype, most commonly the β_2-AR subtype. As discussed above, some cell lines express multiple endogenous ARs. COS and CHO cells are most commonly used for transient, high level of receptor expression, but a variety of other cell lines are used for stable expression of lower receptor densities.

Factors to be considered in choosing a cell line include:

1. Which endogeneous receptors are expressed? Ideally, a cell line should not endogenously express the receptor you are interested in, and not express other subtypes in the same family that might interfere with the characterization of the expressed receptors.
2. Transfection efficiency: Although the efficiency of transfection can be improved by using different methods, some cell lines are much easier to transfect than others. There are substantial differences between cell lines in the efficiency with which plasmid DNA is integrated into host cell chromosomes.
3. Appropriate signaling machinery: The cell line should express the endogenous G-proteins and other effector molecules necessary to reconstitute the normal signal transduction cascades activated by the expressed receptors.
4. Rate of growth: For most efficient expression and establishment of individual subclones, cells to be transfected should grow quickly and not require much care. If they grow too slowly or die too easily, the transfection procedure can be toxic with few cells surviving.

Although these are common factors that should be taken into account when choosing a cell line, the final decision usually depends on combinations of many of these factors. The purpose for which a receptor is being expressed will play a key role in determining which cell lines to use.

1.4. Which Expression Vector—Constitutive or Inducible?

AR cDNAs have been expressed using many commonly available vectors. Choice of a vector is influenced by the promoter in the vector, which must be active in the cell line to be transfected. Different vectors with different promoters may achieve their maximal expression in different cell lines. Vectors with a CMV promoter, SV-40 promoter, or RSV LTR promoter are the most commonly used, and can achieve high-level constitutive expression in a variety of cell lines. Another important factor to be considered in choosing a vector is the selectable marker(s) encoded by the vector, which will provide antibiotic resistance for obtaining stably transfected cells. One must also decide whether to use constitutively active, inducible, or repressible expression vectors. Although constitutively active vectors are suitable for many purposes, such as characterization of receptor pharmacology, ligand binding, and large-scale receptor purification, inducible or repressible vectors allow one to control expression levels. Inducible vector systems now widely available include those responsive to metals, glucocorticoids, tetracycline, or isopropylthiogalactose (IPTG) (a galactose analog). Most repressible vectors now in use are responsive to tetracycline *(8)*. Since glucocorticoids and metal ions often affect expression of many endogenous genes, we have found the LacSwitch-inducible expression vector system *(7)* to be very useful for inducible expression of adrenergic receptors in various cell lines. This system has been used in our lab for β_1- and β_2-ARs in C_6 cells *(6)*, α_{1B}-ARs in DDT_1-MF_2 cells *(11)*, α_1-AR subtypes in SK-N-MC cells *(12)*, and multiple AR subtypes in PC12 cells. In these studies, stable subclones with low constitutive and highly inducible receptor expression have been obtained and used for studies of pharmacological and signaling properties of these receptors.

1.5. Which Transfection Method?

Calcium phosphate-, diethylaminoethyl (DEAE) dextran-, electroporation-, liposome-, and virus-mediated transfection are all commonly used *(13)*. However, different cell lines may show different tolerances to different methods, and different methods may give different transfection efficiencies. Therefore, the best method must usually be determined experimentally. Calcium phosphate-mediated transfection is widely used in COS and HEK 293 cells, where it results in good transfection efficiency. In many other cell lines, such as PC12 cells, various liposome methods (such as lipofectamine) or electroporation produce higher efficiencies of transfection than does calcium phosphate. For cells highly resistant to plasmid transfection, such as primary cultures, retroviral methods may be the best choice. Generally, retroviral methods give the highest transfection efficiency, since the virus attaches to cell-surface receptors and transfers its DNA directly into the recipient cells.

In this chapter, we present methods for characterizing which receptor subtypes are expressed in cell lines, either endogenously or after transfection. We also present methods for obtaining stable subclones of PC12 cells inducibly expressing α_1-AR cDNAs. We have found that using the LacSwitch-inducible vector system with lipofectamine-mediated transfection produces excellent results in this cell line.

2. Materials

2.1. Equipment

Standard cell-culture equipment is used, including a laminar flow hood, 37°C, 5–7% CO_2 incubator, and an inverted microscope.

2.2. Plasticware

Standard tissue-culture materials are used, including disposable pipets and tips; 100-mm culture dishes; 6- and 24-well cell-culture plates; sterile test tubes, polypropylene tubes; cloning cylinders; and autoclaved sterile vacuum grease.

2.3. Reagents

1. Rat tail collagen solution: Rat tail collagen was purchased from Biomedical Technologies Inc. (Stoughton, MA) (cat. no. 274). To coat culture plates, dilute collagen solution 1:10 with sterile H_2O. Add 0.1 mL of this solution to each 100-mm Primaria culture dish or 0.1 mL solution to each 35-mm plate. Incubate overnight at room temperature. When dry, store the coated plates wrapped with Parafilm at 4°C. The plates can be stored for up to 2 mo.
2. Cell culture medium: For PC12 cells, use Dulkecco's Modified Eagle's Medium (DMEM) without NaPyruvate (Gibco-BRL [Grand Island, NY], cat. no. 11965-092) plus 5% fetal bovine serum (Atlanta Biologicals [Atlanta, GA], cat. no. S11550), 10% horse serum (Atlanta Biologicals, cat. no. S12150), 60 mg/L penicillin (Sigma [St. Louis, MO], cat. no. P-3032), and 100 mg/L streptomycin (Sigma, cat. no. S-9137). For the other cell lines, use the medium recommended by American Type Culture Collection (ATCC, Manassas, VA).
3. LipofectAMINE™ Reagent (Gibco-BRL, cat. no. 18324-012).
4. Heat-inactivated horse serum: Incubate thawed horse serum at 55°C for 30 min, and then store 100-mL aliquots at –20°C. Before use, thaw at room temperature. Store at 4°C after use.
5. G418 and hygromycin solution: Dissolve G418 or hygromycin separately in 0.1 M HEPES (pH 7.2) at a concentration of 25 mg/mL. Filter-sterilize and store as 5-mL aliquots at –20°C until use. Once thawed, solution can be stored at 4°C.
6. 150 mM NaCl solution: Dissolve 0.87 g of NaCl in 100 mL of H_2O. Filter-sterilize and store at 4°C.
7. Transfection medium: DMEM without NaPyruvate (Gibco-BRL, cat. no. 11965-092). Before transfection, add 300 µL insulin stock (50 µM), 30 µL transferrin stock (100 mg/mL), 6 µL putrescine solution (30 mM), and 3 µL selenium salt solution (30 µM)/30 mL DMEM.
8. Plasmid DNA: The LacSwitch vector system was purchased from Stratagene (La Jolla, CA; cat. no. 217450) and the human α_{1A}-AR cDNA ligated into the multiple cloning site of the operator vector. DNA was prepared by alkali lysis

followed by differential precipitation with polyethylene glycol. This provides a relatively simple, but reproducible method for obtaining a high yield of pure plasmid DNA for transfection. A detailed protocol can be found in **ref. *13***.

3. Methods
3.1. Cell Culture

Cells are handled according to protocols recommended by ATCC. PC12 cells are maintained in DMEM without NaPyruvate plus 10% horse serum and 5% FBS in a 37°C incubator with 5% CO_2. Confluent cells are subcultured in a 1:3 ratio every other day.

3.2. Transfection

Exponentially growing cells are passaged 1 d before transfection into collagen-coated Primaria 100-mm plates. Cells should be about 60–70% confluent on the day of transfection (*see* **Note 1**). To transfect cells:

1. Mix 8 µg of the operator vector (pOphα_{1A} that constitutively expresses the cDNA for the human α_{1A}-AR in the absence of repressor vector) and 8 µg of the repressor vector (p3'SS) with 400 µL of 150 mM NaCl solution in a sterile propylene tube and wait 10 min at room temperature. Meanwhile, mix 400 µL of NaCl solution with 50 µL of Lipofectamine in another tube, and incubate for 10 min at room temperature.
2. Combine the contents of the two tubes, mix gently, and incubate at room temperature for 20 min.
3. Add 6.4 mL of transfection DMEM, and incubate at room temperature for 35 min.
4. At the end of the 35 min of incubation, wash cells to be transfected with transfection DMEM twice (Note: this medium does not contain any antibiotic or serum), and add 7 mL of the transfection mixture to one 100-mm plate of cells. Return the plate to incubator, and incubate at 37°C for 6 h (*see* **Note 2**).
5. At the end of the incubation, add an equal amount (7 mL) of regular DMEM with 20% horse serum and 10% FBS. Do not remove the transfection medium already in the plate. Incubate for an additional 48 h.
6. At the end of 48 h, change culture medium to regular DMEM supplemented with 250 µg/mL hygromycin and 500 µg/mL G418 to start selecting transfected cells. Replace medium with freshly added hygromycin and G418 every 3 d to remove dead cells.
7. After 2 wk in selection medium, switch to medium with 100 µg/mL hygromycin and 250 µg/mL G418. At this stage, almost all the dead cells are removed. Therefore medium can be changed every week with freshly added hygromycin and G418. In about a month, the resistant clones will appear under the microscope (*see* **Notes 3–6**).

3.3. Propagation of Stable Subclones

1. Mark the position of the clones on the bottom of the 100-mm plate. Once the clones are visible under the microscope, let them grow for an additional 3–4 d.
2. Pick clones: remove medium. Enclose each clone using a sterile cloning cylinder with vacuum grease on bottom.
3. Wash the cells off the plate with medium by pipeting several times with a sterile glass pipet. Transfer each subclone to separate wells in a 24-well culture plate. Pipet several times to separate cell clumps, and then allow the cells to attach and grow. When they reach confluency, cells are passed into 6-well culture dishes or 35-mm plates. When confluent again, pass the cells into three 35-mm plates. One plate is used for propagation of the clone, one for basal constitutive expression, and the last for induction with IPTG to induce receptor expression.

3.4. Screening Cell Lines for Endogenous Receptor Expression

To characterize the endogenous ARs expressed by particular cell lines, a variety of pharmacological and molecular approaches are necessary.

1. The first approach is usually to screen for receptor families using radioligand binding assays. For initial screening, expression of α_1-, α_2-, or β-ARs is estimated by measuring specific binding of a single K_D concentration of a radioligand specific for that subtype (**Table 2**) under three tissue dilutions. If specific binding is detected, the density of binding sites is determined by saturation analysis using a range of concentrations from 10 to 95% of the expected K_D value for the particular radioligand being utilized. Finally, specificity of the binding sites and the proportions of subtypes can be examined by competition curves using family-selective or subtype-selective antagonists. For example, α_1-ARs can be labeled with ^{125}IBE or ^{3}H-prazosin, specificity determined by competition with the α_1-selective antagonist prazosin and the α_2-selective antagonist yohimbine, and subtypes determined by competition with the α_{1A}-selective antagonist (+)niguldipine and the α_{1D}-selective antagonist BMY 7378 *(10)*.
2. The second approach is to characterize agonist-mediated second messenger responses. These would include norepinephrine- (100 μ*M*) stimulated ^{3}H-inositol phosphate formation or calcium mobilization for α_1-ARs *(10)*, UK 14,304- (1–10 μ*M*) mediated inhibition of forskolin-stimulated cyclic AMP formation for α_2-ARs *(15)*, and isoproterenol- (10 μ*M*) stimulated cyclic AMP accumulation for β-ARs *(6)*. If second messenger responses to a particular AR family are observed, then the subtype(s) involved can sometimes be identified by use of agonists and antagonists that distinguish between closely related subtypes *(1)*. Unfortunately, in some cases, such selective drugs are either not available or insufficiently selective for use in functional assays. In this case, it is necessary to use either the radioligand binding approaches discussed above or the molecular hybridization discussed below.

Table 2
Commonly Used Radioligands Selective for the Major AR Families

Radioligand	$\sim K_D$, pM^a	Known selectivity within family
α_1-ARs		
^{125}IBE (^{125}I-HEAT)	50–200	None
^{3}H-prazosin	200–500	None
α_2-ARs		
^{3}H-yohimbine	500–2000	$\alpha_{2A} > \alpha_{2B}$
^{3}H-rauwolscine	500–1000	None
β-ARs		
^{125}I-pindolol	100–500	$\beta_2 > \beta_1 \gg \beta_3$
^{125}I-cyanopindolol	20–50	$\beta_2 > \beta_1 \gg \beta_3$
^{3}H-dihydroalprenolol	500–1000	$\beta_2 > \beta_1 \gg \beta_3$

[a]Note the range of K_D values is typical for results in the literature. All radioligands listed are highly selective for the AR family that they label. The "known selectivity" refers to their selectivity between closely related subtypes within that family. Adapted from Esbenshade and Minneman *(24)*.

3. Molecular approaches can be utilized when sufficiently selective drugs are unavailable to distinguish adequately closely related subtypes. Isolation and electrophoresis of mRNA from the cells, transfer to membranes, and hybridization to subtype-specific probes on Northern blots will give information on which subtypes are being expressed. When species-specific sequences are available, more sensitive detection can be done by the use of the reverse transcriptase (RT) and the polymerase chain reaction (PCR) using appropriate primers. Finally, more quantitative information on RNA expression can be obtained with RNase protection assays. Note that these approaches can have greater specificity in identifying message expression, but provide no evidence for expression or function of the protein.

4. Finally, in some cases, subtype-specific antibodies are available. Cell lysates can be prepared, run on gels, and blotted with subtype-specific antibodies. For the ARs, however, because of the general availability of highly selective radioligands and subtype-specific antagonists, few highly specific antibodies have been developed. Since Western blots are less quantitative than radioligand binding assays, the latter are preferred when available.

3.5. Screening Transfected Subclones for Receptor Expression

Antibiotic-resistant subclones isolated following transfection must be screened to ensure that the expressed receptors show the correct pharmacological properties and are functional.

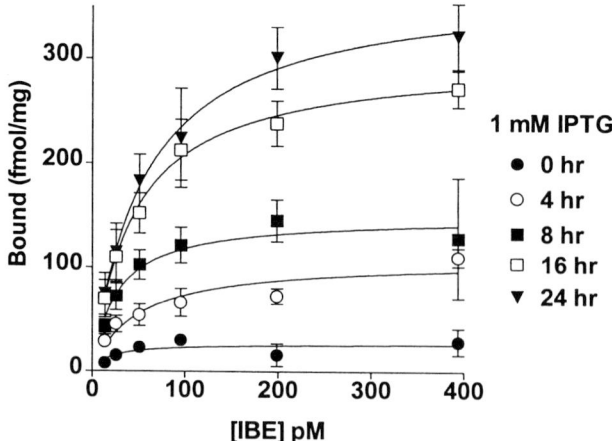

Fig. 1. Inducible expression of α_{1A}-ARs in PC12 cells. Stably transfected subclone α_{1A}-AR#28 was incubated with 1 mM IPTG for the indicated times and membranes prepared for radioligand binding. Saturation analysis of specific ^{125}IBE 2254 binding was performed as previously described *(10)*. Each value is the mean ± SEM of three experiments performed in duplicate.

1. Resistant subclones are first screened for expression by measuring receptor density with radioligand binding assays, with and without exposure to 1 mM IPTG for 48 h. For initial screening, receptor expression is estimated by radioligand binding as described above. Clones with low constitutive and highly inducible expression are then characterized further.
2. Saturation binding is used to determine receptor density (B_{max}) before and after induction (**Fig. 1**), and competition binding with subtype-selective antagonists is used to confirm the pharmacology of the expressed receptor.
3. Finally, NE-stimulated second messenger accumulation, such as increases in inositol phosphates (**Fig. 2**), and increases in intracellular Ca^{2+} (**Fig. 3**) are measured to be certain that the expressed receptors are functional.

4. Notes

1. The most important factor for high-efficiency transfection is the density of cells at the time of transfection. For most cell lines, a density of 50–70% confluency seems to work best. If the density is too high, cells will be overconfluent before starting selection, which tends to decrease transfection efficiency. If the density is too low, many cells will die during the incubation with DNA precipitates, which also decreases transfection efficiency. For PC12 cells, cultures that are about 60% confluent at the time of transfection give the best results. In this case, no passage is required before starting selection with antibiotics.
2. Another important factor in transfection is the duration of incubation of the DNA mixture with the cells. Different cell lines show different tolerances toward this

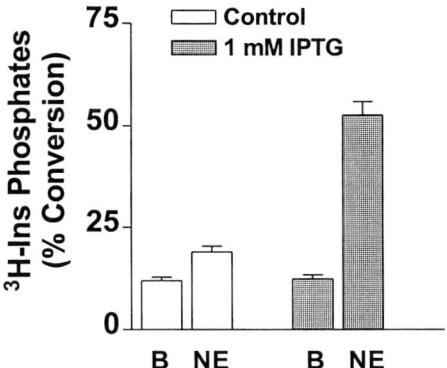

Fig. 2. Norepinephrine-stimulated ^{3}H-inositol phosphate formation in the α_{1A}-AR#28 PC12 subclone. Cells were exposed to 0 (control) or 1 mM IPTG for 48 h. Following prelabeling for 48 h with ^{3}H-inositol, basal- (**B**) and norepinephrine- (100 μM; **NE**) stimulated formation of ^{3}H-inositol (Ins) phosphates was determined as previously described *(10)*. Each value is the mean ± SEM of three experiments performed in duplicate.

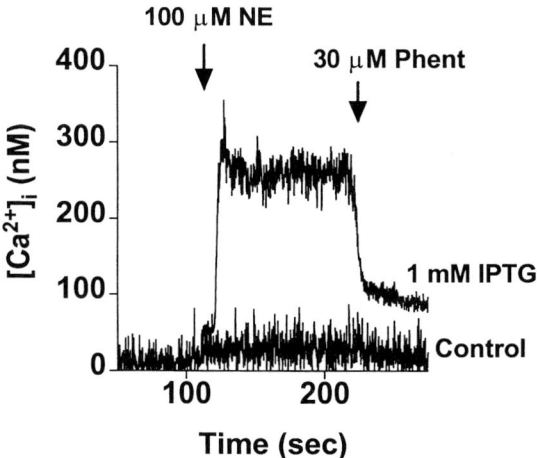

Fig. 3. Norepinephrine-stimulated Ca^{2+} mobilization in the α_{1A}-AR#28 PC12 subclone. Cells were exposed to 0 (control) or 1 mM IPTG for 48 h, loaded with fura-2, and intracellular Ca^{2+} levels determined as previously described *(10)*. Norepinephrine (NE) and phentolamine (Phent) were added at the indicated time-points. Data are from a single experiment representative of three similar experiments.

incubation. The optimal time can be determined empirically by incubating for various times and comparing the resulting transfection efficiency. For PC12 cells, 5–6 h of incubation has given us the best results.

3. During selection of stable transfectants, always keep a control plate that has not been transfected. Since spontaneous resistance may arise in the cells, it is important to compare the control plates with the transfected plates to ensure the transfection and selection have been successful.

4. For PC12 cells, obtaining a pool of transfected cells takes more time than obtaining single colonies. PC12 cells tend to grow in colonies, but fibroblast-like cells tend to spread out quickly to form a monolayer of mixed cells. Therefore, picking PC12 cell colonies directly from the plate using cloning cylinders has given rapid subclones for screening.

5. If stably transfected subclones are to be used for long-term studies of receptor function, it is necessary to include antibiotics in the culture medium intermittently to ensure that untransfected cells are killed and will not overgrow the plate.

6. Different cell lines show different sensitivities to hygromycin and G418. The optimal concentration of each antibiotic must be determined before transfection for each individual cell line. To do this, untransfected cells should be incubated with increasing concentrations of antibiotics, and the concentration that kills 90% of cells within 1 wk should be used for selection.

Acknowledgments

This work was supported by NS32706 and NS 21325. The technical assistance of Deborah Lee is gratefully acknowledged.

References

1. Bylund, D. B., Eikenberg, D. C., Hieble, J. P., Langer, S. Z., Lefkowitz, R. J., Minneman, K. P., et al. (1994) International Union of Pharmacology nomenclature of adrenoceptors. *Pharmacol. Rev.* **46,** 121–136.

2. Granneman, J. G., Lahners, K. N., and Rao, D. D. (1992) Rodent and human beta 3-adrenergic receptor genes contain an intron within the protein-coding block. *Mol. Pharmacol.* **42,** 964–970.

3. Hirasawa, A., Shibata, K., Horie, K., Takei, Y., Obika, K., Tanaka, T., et al. (1995) Cloning, functional expression and tissue distribution of human alpha 1c-adrenoceptor splice variants. *FEBS Lett.* **363,** 256–260.

4. Townsend-Nicholson, A. (1997) Approaches to the stable transfection of G protein-coupled receptors. *Methods Mol. Biol.* **83,** 45–54.

5. Fonseca, M. I. and Brown, R. D. (1997) Immunocytochemical methods for investigating receptor localization. *Methods Mol. Biol.* **83,** 91–106.

6. Zhong, H., Guerrero, S. W., Esbenshade, T. A., and Minneman, K. P. (1996) Inducible expression of β_1- and β_2-adrenergic receptors in rat C6 glioma cells: functional interactions between closely related subtypes. *Mol. Pharm.* **50,** 175–184.

7. DuCoeur, L. C., Wyborski, D. L., and Short, J. M. (1992) Control of gene expression in eukaryotic cells using the lac repressor system. *Strategies Mol. Biol.* **5,** 70–72.

8. Gossen, M. and Bujard, H. (1992) Tight control of gene expression in mammalian cells by tetracycline-responsive promoters. *Proc. Natl. Acad. Sci. USA* **89,** 5547–5551.

9. Esbenshade, T. A., Han, C., Theroux, T. L., Granneman, J. G., and Minneman, K. P. (1992) Coexisting β_1- and atypical β-adrenergic receptors cause redundant increases in cyclic AMP in human neuroblastoma cells. *Mol. Pharmacol.* **42,** 753–759.

10. Esbenshade, T. A., Han, C., Murphy, T. J., and Minneman, K. P. (1993) Comparison of α_1-adrenergic receptor subtypes and signal transduction in SK-N-MC and NB41A3 neuronal cell lines. *Mol. Pharmacol.* **44,** 76–86.

11. Esbenshade, T. A., Wang, X. F., Williams, N. G., and Minneman, K. P. (1995) Inducible expression of α_{1B}-adrenoceptors in DDT_1MF-2 cells: Comparison of receptor density and response. *Eur. J. Pharm.* **289,** 305–310.

12. Theroux, T. L., Esbenshade, T. A., Peavy, R. P., and Minneman, K. P. (1996) Coupling efficiencies of human α_1-adrenergic receptor subtypes: titration of receptor density and responsiveness with inducible and repressible expression vectors. *Mol. Pharm.* **50,** 1376–1387.

13. Sambrook, J., Fritsch, E. F., and Maniatis, T. (1989) *Molecular Cloning, A Laboratory Manual*, 2nd ed. Cold Spring Harbor Laboratory, Cold Spring Harbor, NY.

14. Schaak, S., Cayla, C., Blaise, R., Quinchon, F., and Paris, H. (1997) HepG2 and SK-N-MC: two human models to study alpha-2 adrenergic receptors of the alpha-2C subtype. *J. Pharmacol. Exp. Ther.* **281,** 983–991.

15. Bylund, D. B., Ray-Prenger, C., and Murphy, T. J. (1988) Alpha-2A and alpha-2B adrenergic receptor subtypes: antagonist binding in tissues and cell lines containing only one subtype. *J. Pharmacol. Exp. Ther.* **245,** 600–607.

16. Musgrave, I. F. and Seifert, R. (1997) Alpha 2A-adrenergic receptor stimulated calcium release is transduced by Gi-associated G(beta gamma)-mediated activation of phospholipase C. *Biochemistry* **36,** 6415–6423.

17. Wayman, G. A., Impey, S., Wu, Z., Kindsvogel, W., Prichard, L., and Storm, D. R. (1994) Synergistic activation of the type I adenylyl cyclase by Ca^{2+} and Gs-coupled receptors in vivo. *J. Biol. Chem.* **269,** 25,400–25,405.

18. Sager, G., Bang, B. E., Pedersen, M., and Aarbakke, J. (1988) The human promyelocytic leukemia cell (HL-60 cell) beta-adrenergic receptor. *J. Leukoc. Biol.* **44,** 41–45.

19. Duman, R. S., Fishman, P. H., and Tallman, J. F. (1994) Induction of beta 2-adrenergic receptor mRNA and ligand binding in HeLa cells. *J. Receptor Res.* **14,** 1–10.

20. Harden, T. K., Su, Y. F., and Perkins, J. P. (1979) Catecholamine-induced desensitization involves an uncoupling of beta-adrenergic receptors and adenylate cyclase. *J. Cyclic Nucleotide Res.* **5,** 99–106.

21. Shimura, H., Endo, T., Tsujimoto, G., Watanabe, K., Hashimoto, K., and Onaya, T. (1990) Characterization of alpha 1-adrenergic receptor subtypes linked to iodide efflux in rat FRTL cells. *J. Endocrinol.* **124,** 433–441.

22. Esbenshade, T. A., Theroux, T. L., and Minneman, K. P. (1994) Increased voltage-dependent calcium influx produced by alpha 1B-adrenergic receptor activation in rat medullary thyroid carcinoma 6-23 cells. *Mol. Pharmacol.* **45,** 591–598.

23. Homburger, V., Lucas, M., Rosenbaum, E., Vassent, G., and Bockaert, J. (1981) Presence of both beta1- and beta2-adrenergic receptors in a single cell type. *Mol. Pharmacol.* **20,** 463–469.

24. Esbenshade, T. A. and Minneman, K. P. (1994) Adrenergic receptor subtypes: pharmacological approaches. *Neuroprotocols* **4,** 2–13.

25. Takahashi, H. and Iizuka, H. (1991) Regulation of beta 2-adrenergic receptors in keratinocytes: glucocorticoids increase steady-state levels of receptor mRNA in foetal rat keratinizing epidermal cells (FRSK cells). *Br. J. Dermatol.* **124,** 341–347.

26. Pittman, R. N. and Molinoff, P. B. (1980) Interactions of agonists and antagonists with beta-adrenergic receptors on intact L6 muscle cells. *J. Cyclic Nucleotide Res.* **6,** 421–435.

27. Han, C., Esbenshade, T. A., and Minneman, K. P. (1992) Subtypes of alpha 1-adrenoceptors in DDT1 MF-2 and BC3H-1 clonal cell lines. *Eur. J. Pharmacol.* **226,** 141–148.

28. Gleason, M. M. and Hieble, J. P. (1991) Ability of SK&F 104078 and SK&F 104856 to identify alpha-2 adrenoceptor subtypes in NCB20 cells and guinea pig lung. *J. Pharmacol. Exp. Ther.* **259,** 1124–1132.

29. Stadel, J. M., Poksay, K. S., Nakada, M. T., and Crooke, S. T. (1987) Regulation of beta-adrenoceptor number and subtype in 3T3-L1 preadipocytes by sodium butyrate. *Eur. J. Pharmacol.* **143,** 35–44.

30. Feve, B., Emorine, L. J., Lasnier, F., Blin, N., Baude, B., Nahmias, C., et al. (1991) Atypical beta-adrenergic receptor in 3T3-F442A adipocytes. Pharmacological and molecular relationship with the human beta 3-adrenergic receptor. *J. Biol. Chem.* **266,** 20,329–20,336.

31. Shear, M., Insel, P. A., Melmon, K. L., and Coffino, P. (1976) Agonist-specific refractoriness induced by isoproterenol. Studies with mutant cells. *J. Biol. Chem.* **251,** 7572–7576.

32. Cotecchia, S., Schwinn, D. A., Randall, R. R., Lefkowitz, R. J., Caron, M. G., and Kobilka, B. K. (1988) Molecular cloning and expression of the cDNA for the hamster alpha 1-adrenergic receptor. *Proc. Natl. Acad. Sci.* **85,** 7159–7163.

33. Meier, K. E., Snavely, M. D., Brown, S. L., Brown, J. H., and Insel, P. A. (1983) alpha 1- and beta 2-adrenergic receptor expression in the Madin-Darby canine kidney epithelial cell line. *J. Cell Biol.* **97,** 405–415.

34. Crespo, P., Cachero, T. G., Xu, N., and Gutkind, J. S. (1995) Dual effect of beta-adrenergic receptors on mitogen-activated protein kinase. Evidence for a beta gamma-dependent activation and a G alpha s-cAMP-mediated inhibition. *J. Biol. Chem.* **270,** 25,259–25,265.

35. Murphy, T. J. and Bylund, D. B. (1988) Characterization of alpha-2 adrenergic receptors in the OK cell, an opossum kidney cell line. *J. Pharmacol. Exp. Ther.* **244,** 571–578.

16

Transient Transfection and Adrenergic Receptor Promoter Analysis

Yi-Tang Tseng and James F. Padbury

1. Introduction

There are several methods currently available for transfection of DNA into mammalian cells. These include transfection with calcium phosphate (*1*), diethylaminoethyl (DEAE)-dextran, cationic liposomes, nonliposomal lipid compounds, and electroporation (*2*). Each of these methods has advantages and disadvantages; selection of transfection method is based on the cell types used and personal preference. Transfection using calcium phosphate is highly efficient and is the most cost-effective method; therefore, this is the method of choice for larger-scale experiments. It is also the most widely utilized method and has been successful for DNA transfection into a variety of cell types. Our laboratory has used calcium phosphate-mediated transfection to introduce β_1-adrenergic receptor (β_1-AR) gene fragments into SK-N-MC, C6, and W1 cells (*3,4*). The protocol for transient transfection of adherent cells with calcium phosphate will be described in this chapter. As with other transfection methods, the efficiency of calcium phosphate-mediated transfection is largely determined by the recipient cell type. An optimized protocol for one cell type does not guarantee success in another cell type. From our experience, lipofectin transfection reagent works well for transfection of β_1-AR gene fragments into adherent cell lines, but is much less effective for transfection of DNA into primary cultures. As noted by other investigators (*5*), we have applied FuGENE™ 6 transfection reagent, a nonliposomal lipid compound, to transfect β_1-AR gene fragments into 3-d-old neonatal cardiac myocytes with great efficiency (**Fig. 1**). Therefore, the optimal transfection method is highly variable between cells. It is absolutely critical to optimize the transfection condition for each cell type.

From: *Methods in Molecular Biology, vol. 126: Adrenergic Receptor Protocols*
Edited by: C. A. Machida © Humana Press Inc., Totowa, NJ

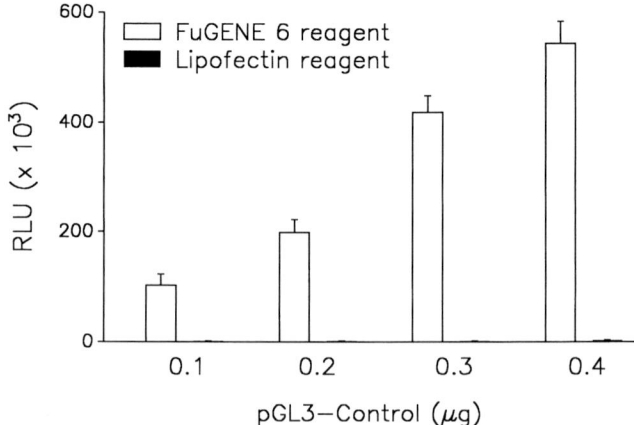

Fig. 1. The transfection efficiency of two different transfection reagents, FuGENE 6 (Boehringer Mannheim, Indianapolis, IN) and Lipofectin (Gibco-BRL, Grand Island, NY). Cardiac myocytes were prepared from 3-d-old neonatal rats. Cells were transfected with pGL3-control vector 2 d after plating in 24-well plates to test efficiency on primary culture. The DNA doses are indicated. Several doses of Lipofectin reagent were used, ranging from 0.8 to 1.5 μL/well. The ratio of DNA (μg)/FuGENE 6 reagent (μL) was 1/5. Cells were exposed to the transfection reagent for 20 h before harvesting for luciferase activity. Data are mean ± SEM of luciferase relative light unit from four separate experiments.

The system of transient transfection using calcium phosphate is comprised of two major components. Solution A is a mixture of DNA and calcium chloride. Solution B is a 2X HEPES-buffered saline (HBS) containing Na_2HPO_4. An equal volume of solution A is slowly added to Solution B to form the DNA/calcium phosphate precipitate. After a brief incubation, the mixture is applied onto the cells. The DNA enters the cell potentially by endocytosis. Depending on the recipient cell type, the transfected cells are washed 6–20 h later and allowed to recover. Cells can then be treated or lysed, target protein- or nucleic acid-extracted, and measured. For reporter genes expressing a measurable enzyme, such as luciferase or chloramphenicol acetyltransferase (CAT), the enzyme activity is usually measured 24–48 h after transfection.

2. Materials

1. DNA for transfection dissolved in water (*see* **Note 1**).
2. Eukaryotic cells for transfection, e.g., SK-N-MC cells.
3. All reagents and medium for culturing the cell chosen, e.g., for SK-N-MC cells MEM with glutamine, penicillin-streptomycin, and fetal bovine serum (EBS).
4. Cell culture equipment and instruments, including a humidified incubator with

controlled temperature setting and CO_2 supply, and 24-well plates or other types of tissue culture plates.

5. 2X HBS: 280 mM NaCl, 10 mM KCl, 1.5 mM Na_2HPO_4, 12 mM dextrose, 50 mM HEPES. If HEPES acid is used, use 5 N NaOH to adjust the pH to 7.05 (*see* **Note 2**).
6. 2 M $CaCl_2$ (*see* **Note 2**).
7. Autoclaved deionized water.
8. Phosphate-buffered saline (PBS).

3. Methods

1. Plate cells for transfection in 24-well plates 24–48 h before transfection. The cells should be approx 75% confluent at the time of transfection.
2. Replace the medium with fresh medium (0.8 mL/well) 3–4 h prior to the transfection.
3. Thaw all components. Ensure that all components for transfection are at room temperature at the time of transfection. Prepare solution A and solution B as follows (per group in 24-well plate (in quadruplicate, prepare 4.1X):

	DNA	RSV-CAT	2 M $CaCl_2$	2X HBS	Water
Solution A	4.1 µg	1.23 µg	38.45 µL	—	Add to 307.5 µL
Solution B	—	—	—	307.5 µL	—
Final /well	1 µg	0.3 µg	125 mM	1X	—

   (*see* **Note 3**)
4. Adjust vortex to the lowest setting, and gently vortex solution B. Avoid spillover. Very slowly add solution A dropwise to solution B. The solution will turn slightly opaque after addition. Incubate the mixture for 30 min at room temperature.
5. Gently and briefly vortex the mixture again. Apply 150 µL of the mixture evenly onto a well containing the cells while continuously rocking the plate. Repeat with the other three wells. After adding the mixture to all four wells, rock the plate again to ensure even distribution of DNA to cells (*see* **Note 4**).
6. Six to 24 h later, wash the wells twice with PBS, and replace with regular medium (*see* **Note 5**). The cells can then be left to recover overnight.
7. Change the medium with fresh medium or medium containing drug (*see* **Note 6**).
8. Twenty-four to 48 h after treatment, cells can be harvested for functional testing (*see* **Note 7**).

4. Notes

1. The quality of DNA used for transfection is critical for transfection efficiency. Plasmid DNA prepared by using *Qiagen*™ (Valencia, CA) column or CsCl density gradient centrifugation is recommended. The A_{260}/A_{280} of DNA should be 1.8 or greater. To control for transfection efficiency, a second plasmid expressing a different protein is usually cotransfected. For example, we subcloned all ovine β_1-AR promoter constructs (**Fig. 2**) into the pGL2 luciferase vector. A RSV-CAT plasmid is cotransfected with pGL-β_1-AR constructs during transfection to control for efficiency. The RSV-CAT plasmid is a useful vector to

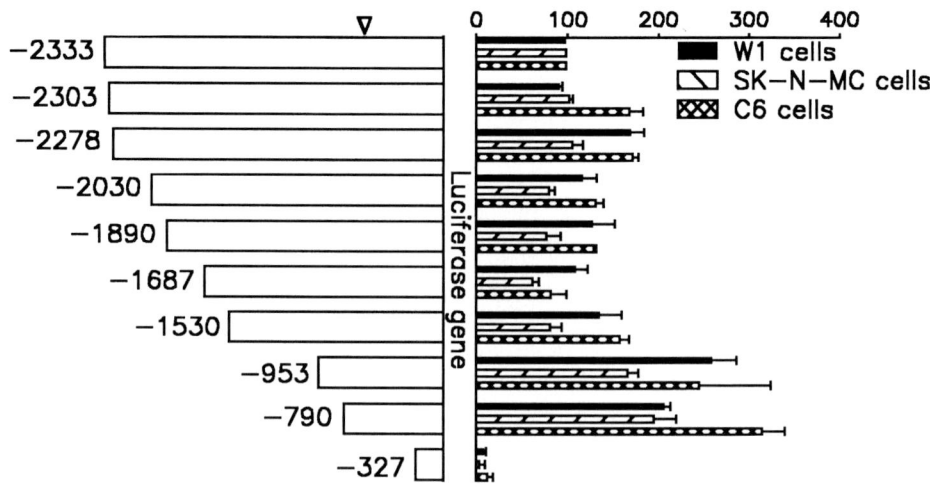

∇, major transcription start site

Fig. 2. A schematic of β_1-AR promoter constructs (left panel). Each construct was cut 151 bp upstream from the initiator methionine and ligated directly upstream from the luciferase gene of the pGL2-basic plasmid *(3)*. Numbers listed on the left side are relative to the initiator methionine. Cells were transiently transfected with individual β_1-AR promoter construct for 12 h using the calcium phosphate method. Cells were washed 12 h later and were harvested 48 h after transfection. Results of the basal transcription activities of these constructs in three cell lines are shown on the right panel. Data are mean ± SEM of luciferase relative light unit from three separate experiments, and are normalized to the cotransfected RSV-CAT activities. All activities are expressed as relative to the activity of the full-length promoter illustrated on top.

determine transfection efficiency. The RSV promoter is a strong, promiscuous promoter that provides high-level expression in nearly all eukaryotic cells.

2. Depending on the cell type chosen, the optimal pH of the 2X HBS used for the transfection cocktail will vary. The optimal pH should be determined (between pH 6.95 and 7.10) for each recipient cell type. Sterilize the final solution by filtration through a 0.22-μm filter. For transfections using the same cell type, consistency in transfection efficiency is primarily determined by consistency in the quality of 2X HBS. Transfection efficiency may fluctuate from one experiment to another, if different batches of 2X HBS are used. To achieve consistency over long periods of time, it is advisable to prepare a large batch of 2X HBS, test transfection efficiency, and store in 5- to 10-mL aliquots at –70°C. However, long-term storage will occasionally change the pH of 2X HBS. This will in turn affect the transfection efficiency. Therefore, the pH of a stored batch of 2X HBS needs to be monitored if a dramatic reduction in transfection efficiency is noted. The 2 M CaCl$_2$ must be sterilized in the same manner and stored in 1- to 2-mL aliquots at –70°C.

3. The protocol described here is for transfection of cells prepared in 24-well plates, which is ideal for treating cells in quadruplicate samples. If transfection is conducted in 12-well, 6-well, or 60-mm plates, the volumes of reagents should be increased accordingly.

4. Continuous swirling of the plate will cause DNA to concentrate in the center of the well, leaving large areas of untransfected cells. To avoid this complication, rock the plate only once in each direction (front, back, left, and right; lay the plate flat between each rocking). The plate should be rocked again in the same manner when returned to the incubator. After adding DNA precipitate to cells, the color of the medium will turn yellow-orange. Under phase-contrast microscopy, there should be a few visible precipitates immediately following DNA application to cells. Large numbers of coarse precipitates at this time are usually an indication that low transfection efficiency will occur. This usually indicates that the pH of the 2X HBS is not optimal or that the mixing of solution A and solution B is too vigorous.

5. Unlike liposome and FuGene 6 reagent, calcium phosphate reagents are more toxic to cells. A preliminary experiment can be performed to determine the optimal application time for the DNA/calcium phosphate precipitate for a particular recipient cell.

6. The DNA/calcium phosphate precipitates should be visible in a phase-contrast microscope at this time. A large quantity of fine precipitates should be observed settling on the cells; in the region of the plate without cells, there should be very few precipitates observed. If the purpose of the experiment is to test a regulatory element of a gene, treatment of cells after transfection is usually required. For example, to test the glucocorticoid response element in the ovine β_1-AR gene in transfected SK-N-MC cells, the cultures are treated with glucocorticoid.

7. If no treatment of cells is intended, the cells can be harvested in 48–60 h after transfection. Depending on the type of transfected DNA, the cells now provide a source of nucleic acid or protein, including enzymatic activity, for further analysis.

References

1. Graham, F. L. and van der Eb, A. J. (1973) A new technique for the assay of infectivity of human adenovirus 5 DNA. *Virology* **52,** 456–467.
2. Kingston, R. E. (1990) Transfection of DNA into eukaryotic cells, in *Current Protocols in Molecular Biology*. John Wiley, New York, pp. 4.8.1–4.8.3.
3. Padbury, J. F., Tseng, Y.-T., and Waschek, J. A. (1995) Transcription initiation is localized to a TATAless region in the ovine β_1 adrenergic receptor gene. *Biochem. Biophys. Res. Commun.* **211,** 254–261.
4. Tseng, Y. T., Waschek, J. A., and Padbury, J. F. (1995) Functional analysis of the 5′ flanking sequence in the ovine β_1-adrenergic receptor gene. *Biochem. Biophys. Res. Commun.* **215,** 606–612.
5. Noah, D. L., Blum, M. A., and Sherry, B. (1998) Transfection of primary cardiac myocyte culture with DNA and anti-sense oligonucleotides using FuGENE™ 6 transfection reagent. *Biochemica* **2,** 38–40.

17

Antisense RNA/DNA-Based Techniques to Probe Adrenergic Receptor Function

Hsien-Yu Wang, Fubao Lin, and Craig C. Malbon

1. Introduction

1.1. Oligodeoxynucleotides (ODNs) as Antisense Tools for Cells in Culture

Ablation of the mRNA of a targeted protein by the use of antisense DNA and RNA provides degrees of freedom not available in many other strategies to suppress or eliminate gene products *(1–3)*. Numerous examples exist demonstrating the utility of the antisense DNA/RNA strategy for study of signaling *(4–18)*. In the case of ODNs, preparation of reagents requires no additional skill other than knowing the commercial supplier for ODN synthesis and purification *(19)*. Expression of antisense RNA requires a scientific facility exhibiting simple techniques of molecular biology and can be accomplished by a variety of approaches, including constitutive expression by a strong promoter; this latter approach requires no regulation and assumes functional compatibility with the targeted cells *(4,9,11,12)*. Promoters that can be "induced" afford an additional capability; expression of antisense RNA being turned "on" and again "off" in response to molecular signals provide approaches to RNA induction or suppression. The inducibility of antisense RNA is of particular utility in the suppression of mRNAs that encode proteins necessary for viability in cells or in the whole animal. Traditional "knockout" of genes by homologous recombination that are crucial targets leads to lethality in the transgenic mice system and consequently no viable pups. Inducible antisense RNA transgenes are maintained "silently" *in utero* and later can be turned "on" at birth or thereafter, permitting production of viable transgenic pups. In the "tech-

From: *Methods in Molecular Biology, vol. 126: Adrenergic Receptor Protocols*
Edited by: C. A. Machida © Humana Press Inc., Totowa, NJ

nical knockouts" (TKOs) rendered by inducible antisense RNA vectors, the additional time and expense of breeding to homozygosity in traditional knockouts is avoided, the output of antisense RNA product from a single transgene being sufficient to silence the mRNA for most protein targets.

The antisense RNA approach provides a strategy for targeting multiple mRNAs at one time. Selecting an antisense sequence to the 5' untranslated region (UTR) of one member of a homologous family can create a vector that solely ablates the mRNA of that member, whereas selecting an antisense sequence to a common coding region provides a vector that can ablate the mRNA of all members. This extraordinary flexibility is as great as that of the Cre lox P system for multiple targets *(20)*, but with greatly reduced technical demand. The three approaches will be detailed below. These strategies represent an evolution in antisense DNA/RNA technology, a technology with many skeptics in its earlier years, but one that now enjoys widespread use in research applications and more recently in human therapeutics.

1.2. Vector-Expressed Antisense RNA In Vitro

Mammalian expression vectors have been created that enable the expression of RNA antisense to mRNAs of target proteins. The antisense RNA sequences are expressed most often in the context of a larger mRNA that acts as a carrier. Small fragments of RNA appear to suffer from instability. Incorporating the antisense RNA sequence into a larger fragment of RNA enables accumulation that is not possible for expression of the short RNA sequences.

1.3. Inducible, Tissue-Specific Expression of Antisense RNA In Vivo

Application of antisense RNA in tandem with transgenic mice offers an exciting new opportunity for study of G-protein-linked receptor signaling. The approach is far less technically demanding than homologous gene interruption. Katsuki et al. *(21)* demonstrated the proof of concept early, succeeding in the transfer of an antisense RNA-producing "minigene" against myelin basic proteins (MBP) to fertilized mice zygotes. The transgenic mice developed a shiverer phenotype. Antisense MBP mRNA was shown to be expressed in the mice, and endogenous MBP mRNA, the gene product, and central nervous system (CNS) myelination were found to be reduced in the mutant mice.

2. Materials

1. ODNs: May be prepared in-house or obtained commercially as cell-culture grade materials (preferred supplier: Operon).
2. Cationic carrier molecules: lipofectin, cell fectin, or (DOTPA) N-[1-(2, 3-dioleoyloxy)propyl]-N,N,N-trimethylammonium methylsulfate.

3. Cells: PC12 pheochromocytoma cells, F9 mouse teratocarinoma cells, or FTO-2B rat hepatoma cells.
4. Appropriate tissue-culture medium, such as Dulbecco's Modified Eagle Medium (DMEM), and other supplements (e.g., fetal bovine serum [FBS]).
5. pLNCX retroviral expression vector.
6. pCMV5 and pcDNA3 cytomegalovirus expression vectors.
7. PEP-CK-AS expression vector (transactivation by cAMP).
8. p-IND expression vector (ecdysome-inducible promoter; contains 5 E/GREs) (Invitrogen).
9. pVsRxR expression vector (expresses the nuclear receptor for the response element contained in p-IND).
10. Muristerone A, potent ecdysone agonist.
11. *Lac*Switch lac operon system (Stratagene), including inducer ligand isopropyl thiogalactose (IPTG).
12. 2X HBS: 6.75 mL 2 M NaCl, 1.0 mL 0.5 M KCl, 0.7 mL 100 mM Na_2HPO_4, 1.2 mL 0.5 M glucose, and 4.0 mL 0.5 M HEPES. Add water to 100 mL final volume. Adjust to pH 7.08 with NaOH. Filter and store at 4°C.
13. Calcium phosphate precipitation reagents: 20 µg/mL plasmid DNA, 20 µg/mL salmon sperm DNA, and 125 mM $CaCl_2$.
14. G418 (Geneticin, Gibco-BRL).

3. Methods

3.1. ODNs as Antisense Tools for Cells in Culture

3.1.1. Selection of Proper Sequences for Synthesis of ODNs

Application of ODNs to target the mRNA of a specific gene product requires judicious selection of the sequence to be targeted. Almost all early examples of ODN applications uniformly selected target sequences that included the ATG initiator codon. The theoretical basis for including the ATG is obvious, but empirical determinations suggest that the sequence to be targeted can be selected from virtually any region of a particular mRNA. Successful ablation of signaling elements, such as G-protein-linked receptors and G-proteins by ODNs, has been achieved through targeting sequences in the 5' and 3' UTR, as well as the open reading frame (ORF).

The "proper" length of ODNs for optimal use remains an overarching consideration for the researcher. ODNs <6–9 nucleotides (NT) in length do not appear capable of forming stable homoduplexes with their target RNAs, suggesting that 9 nt is the lower limit for this parameter *(22)*. Some controversy still exists concerning the upper limit of the length of ODNs that is optimal. Examples taken from the literature include a few in which antisense sequences representing the entire cDNA of a gene product have been employed. Although not to be considered a typical ODN, these inverted cDNAs represent the extreme case and also raise significant issues of specificity for targeted mRNAs. The issue of specificity of an inverted cDNA of a large protein relates to the simple empirical observation that within relatively long pieces of DNA will be found domains of homology/identity with other nontargeted mRNAs. A phenotype of interest is virtually guaranteed by use of a very long antisense RNA. The inherent lack of specificity, however, reduces the probability that the phenotype observed is related to the targeting of solely the "parent" sequence. For the selection of the proper sequence for antisense targeting that is not common, but rather unique with respect to target mRNA, the database of the GenBank seems the best source. A search of the GenBank is absolutely necessary to uncover other likely inadvertent targets. By sliding the antisense sequence to other upstream or downstream sequences to limit homology to unwanted, known targets, one can reduce the chances of nonspecific effects to acceptable limits. One must be cognizant that the search for homology of unwanted targets will be confined to "known" sequences in the database. Possible targets not yet sequenced are sure to exist, and this consideration demands careful attention in the interpretation of any study relying solely on antisense ODNs as a tool.

The best approach for the design of ODNs is minimalist, i.e., designing the shortest sequence with the greatest specificity is optimal. Analysis of the theoretical "optimal" size for ODNs confirms the empirically derived data of those in the field. ODNs ranging between 20 and 30 nt in length are optimal and a good starting point for the researchers naive to the strategy. The expense associated with ODN use for antisense approaches to cells in culture can be significant. Minimizing the length of the ODNs therefore can minimize unnecessary expense for the reagent.

Whether or not to utilize the unmodified ODNs or ODNs with chemically modified linkages that theoretically make the oligonucleotides resistant to metabolism is a common question. There are valid arguments to support the use of either modified or unmodified ODNs. From our perspective, the cost, availability, and need for modified ODNs is probably best driven by empirical data (22). Inability of an ODN to provoke the decline in the mRNA or expression of the gene product targeted may represent lack of stability of the particular ODN or many other parameters. We recommend that researchers first employ the unmodified ODNs and ascertain their effectiveness. Use of modified ODNs may be warranted if outcomes with the more simple, inexpensive oligonucleotides prove unsuccessful. We have found few instances in which the ODNs of the proper sequence and length were not capable of suppressing the expression of G-protein subunits. This experience is shared by other laboratories.

In spite of the common synthesis procedures for production of ODNs, all ODNs are not the same. Experience of many in this field make a compelling argument for the need to utilize ODNs of high purity in cell-culture work. Use of ODNs prepared "in-house" often results in a toxic effects on cells, in a manner unrelated to ODN sequences. Proprietary procedures have been developed by commercial suppliers that are capable of preparing "cell-culture grade" ODNs at approximately the same cost as ODNs prepared and purified in-house. Operon is a leading commercial supplier of cell-culture-grade ODNs. Over the course of many years, we have found such cell-culture-grade ODNs to display superior performance (*see* **Notes 1** and **2**).

3.1.2. Introduction of ODNs to Cells in Culture (see **Notes 3–5**)

The introduction of ODNs directly to the medium of cultured cells results in uptake via a poorly understood mechanism(s). According to this strategy, the cell-culture media is first substituted with serum-free media for a short period prior to the introduction of the ODNs. The ODNs are made soluble in serum-free media and at concentrations 20- to 50-fold higher than the final concentration selected. Successful use of ODNs introduced directly into the culture media is maintained typically at 10- to 30-μM final concentrations. The ODNs were introduced in the serum-free media and the cells maintained under standard culture conditions for at least 30 min. At the end of the 30-min incubation period, culture was supplemented with serum to complete the media to the typical composition and the cells then maintained in this manner until they are fed again *(10)*.

Improvements in the uptake of plasmid DNA for mammalian cell transfections created a derivative benefit to studies using direct addition of ODNs to cells in culture, i.e., cationic carrier molecules. A variety of small organic molecules of little known toxicity to most cells are available commercially for facilitating the uptake of plasmid DNA and work very well for facilitating the uptake of ODNs. Several cationic carrier molecules including LipoFectin®, CellFectin®, and DOTPA, are available currently, with new carriers being developed all the time. The use of a carrier can reduce the amount of ODNs required to achieve adequate uptake in mammalian cells in culture. The use of the carrier with smaller amounts of ODNs reduces the cost of antisense ODN-based experiments significantly. Whether or not these individual carriers are nontoxic and beneficial to a specific cell type in culture can be determined only empirically. The evolution to antisense RNA from ODNs was stimulated in our laboratory by our inability to use the technique efficiently in certain cell types, prominently PC12 pheochromocytoma cells and initially in mouse F9 teratocarcinoma cells *(12)*. The read-out for assay of the target must be miniaturized to a rather small mass of cells, since introduction of ODNs to mass culture of cells would be prohibitively expensive (*see* **Notes 6** and **7**).

3.1.2.1. PROTOCOL FOR INTRODUCTION OF ODNS INTO CELLS IN CULTURE

1. Grow cells in 6-, 12-, and 24-well plates or 8-well chamber slides to >50% confluent.
2. Change media on the day before the ODN treatment.
3. Remove media, and gently rinse cells with new media without serum twice.
4. Add minimum volume of medium (without serum) into chambers as shown in **Table 1**.
5. Add ODNs to final concentrations to 15–30 mM. Incubate cells at 37°C for 30 min to 2 h.
6. Add FBS into the media to final of 10% and continue to incubate cell for 48 h.

3.1.3. Direct Microinjection to Cell Nuclei

Microinjection directly into the nucleus of target cells remains the most efficient means of introducing ODNs into cells. Although once the domain of relatively few, highly trained researchers using manual manipulators to inject cell nuclei, nuclear microinjection has been brought within the grasp of most researchers by the advent of the new micromanipulator/microinjector systems, which require no more than average capabilities in manual dexterity. Microinjection has been used with great success for introducing plasmid DNA as well as ODNs into cells *(15–18,23,24)*. Some level of vagary is encountered in the use of ODNs microinjected into cell nuclei, since the amount of injectable material that is retained, its relative concentration, and lifetime are essentially not known. This criticism applies equally to the general approach of using ODNs in cells, since the addition of a fixed concentration of ODNs to a culture media, either alone or with a cationic carrier molecule, does not lend any greater level of knowledge about how much ODNs is taken up, how much ODNs is delivered to the nucleus, and the integrity of the ODNs within the cells. In spite of these limitations, antisense ODNs are proven, powerful tools in the analysis of the function of a protein through ablation of expression. A great deal of thought must be given to design of the proper read-outs to ensure that the ODNs are providing suppression of the target mRNA and thereby target protein (*see* **Notes 6** and **7**).

3.1.4. Constitutive Expression of Antisense RNA

The best approach to the expression of RNA antisense to targeted gene products takes advantage of mammalian expression vectors of demonstrated capability. These vectors can employ a variety of promoters, including that for early

Table 1
Minimum Volumes for Introduction of ODNs to Cells in Culture

	8-Well chamber	6-Well plate	12-Well plate	24-Well plate
Medium w/o serum	100 μL	1000 μL	200 μL	300 μL
ODNs, 300 mM in medium w/o serum	11 μL	110 μL	22 μL	33 μL
FBS	12 μL	120 μL	24 μL	36 μL

SV40, cytomegalovirus promoters, and retroviral long terminal repeats (LTRs), to provide constitutive expression of antisense RNA. By judicious selection of the cell type and vector employed, high-level, constitutive expression of antisense RNA can be obtained. For cells resistant to transfection by standard protocols, a retroviral construct can be used, which when properly packaged, can provide high-efficiency introduction into a wide variety of cells. A broad spectrum of mammalian vectors have been created to suit most needs, and these typically can be obtained from commercial suppliers.

3.1.5. Retroviral Expression Vectors

In an effort to transfect cell lines efficiently that are resistant to transfection with calcium-phosphate precipitates of vector plasmid DNA, the pLNCX retroviral vector was adapted to the task *(4,9,11,12)*. The pLNCXAS vector was first employed to infect mouse F9 teratocarcinoma cells with a construct harboring an antisense sequence to the G-protein $G_{i\alpha2}$. This vector has proven useful in retroviral infection of a many cell types. Improvements in transfection approaches and reagents over the interim period have reduced to a much smaller number of cell types for which retroviral infection is required.

3.1.6. Cytomegalovirus Promoter-Driven Constructs

Experience indicates that mammalian expression vectors in which antisense sequences are driven by the cytomegalovirus (CMV) promoter are of the greatest utility. Harbored in the pCMV5, pCDNA3, and similar expression vectors, antisense RNA sequences embedded in the context of a carrier RNA prove very effective at yielding high levels of expression of antisense RNA. Under selection with the NEO[r] resistance gene, stable transfectants can be created in which constitutive expression of antisense RNA is robust and the expression of the targeted protein is abolished.

3.1.7. Inducible Expression of Antisense RNA

The ideal for ablation of the production of a protein would be a vector that can be induced to express the antisense RNA. Stable transfectants can be isolated and propagated harboring an inducible construct targeting virtually any gene product. Constitutive expression of antisense RNA throughout the selection process with a selectable marker, such as NEO^r, may be accompanied by adaptive changes in the cells that act to rectify the loss of the targeted protein. To minimize adaptive changes in the signaling pathways, a number of inducible promoters have been engineered into expression vectors, each basically offering the ability to be activated selectively and to induce the expression of antisense RNA, but with differing capabilities. Three examples of inducible promoter-driven vectors for antisense RNA are discussed that have been evaluated in our laboratory for the study of G-protein-linked receptor signaling. The utility of this strategy is described below. The list of inducible promoter constructs adapted for the production of antisense RNA includes heat-shock, heavy metals, and hormone-(steroid) dependent activation of gene expression.

3.1.7.1. PEP-CK PROMOTER-DRIVEN CONSTRUCTS

The utility of the pPEP-CK-AS vector to suppress expression of a G-protein subunit was evaluated first for study of $G_{i\alpha2}$ following transfection into FTO-2B rat hepatoma cells (5,6). These liver-derived cells display cAMP-inducible PEPCK gene expression and express $G_{i\alpha2}$ and $G\beta$ subunits. RNA antisense to $G_{i\alpha2}$ was expressed in FTO-2B clones transfected with pPCK-AS$G_{i\alpha2}$ as detected by reverse transcription of total cellular RNA followed by polymerase chain reaction (PCR) amplification. FTO-2B clones stably transfected with pPCK-AS$G_{i\alpha2}$ display wild-type levels of $G_{i\alpha2}$ expression in the absence of cAMP, an inducer of the PEPCK gene expression. $G_{i\alpha2}$ expression declined >85% when these same cells were challenged with the cAMP analog, 8-(4-chlorphenylthio)-cAMP (CPT-cAMP) for 12 d. FTO-2B clones stably transfected with the vector lacking the antisense sequence to $G_{i\alpha2}$ displayed no change in $G_{i\alpha2}$ expression. The expression of $G_{s\alpha}$ and $G_{i\alpha3}$, in contrast, was not changed in cells expressing the RNA antisense to $G_{i\alpha2}$, demonstrating that the antisense RNA sequence was specific for $G_{i\alpha2}$. The time elapsing between the induction of pPCK-AS$G_{i\alpha2}$ by CPT-cAMP and the decline of steady-state levels of $G_{i\alpha2}$ likely reflects the half-life of this subunit.

$G_{i\alpha2}$ is the most prominent member of the G_i family implicated in mediating the inhibitory adenylylcyclase pathway. Suppression of $G_{i\alpha2}$ expression in FTO-2B cells largely attenuates receptor-mediated inhibition of adenylylcyclase (5,7). Inhibition of forskolin-stimulated cAMP accumulation by either somatostatin or the A1 purinergic agonist (–)-N^6-(R-phenylisopropyl)-adenosine (R-PIA) is nearly abolished in transfectant cells in which expression of RNA antisense to $G_{i\alpha2}$ was first induced by CPT-cAMP for 12 d. Cells transfected with the vector lacking the antisense sequence for $G_{i\alpha2}$ displayed a normal inhibitory adenylylcyclase response following a 12-d challenge with CPT-cAMP. These data demonstrate that $G_{i\alpha2}$ mediates the hormonal inhibition of liver adenylylcyclase. Thus, the pPEPCK-AS construct is shown to be a strong, inducible promoter that drives the production of an antisense RNA within the context of the PEPCK gene. The parent RNA molecule that contains the antisense sequence is stable, accumulates at high levels in response to induction for FTO-2B cells, and can be easily detected by reverse transcriptase (RT)-PCR using flanking primers within the antisense cassette.

3.1.7.2. p-IND PROMOTER-DRIVEN CONSTRUCTS

We have explored the use of the ecdysone-inducible promoter for this same purpose. The ecdysone-inducible promoter is derived from *Drosophila melanogaster* and makes use of the pIND expression vector (Invitrogen) harboring five E/GREs and of a companion plasmid pVgRXR, which expresses the nuclear receptor for the response element (*see* manufacturer's handbook). Muristerone A is a very potent ecdysone agonist, which is used as an inducer in the stably transfected mammalian cells. One must first determine that muristerone A itself stimulates or alters the pathways of target cells and that the promoter construct is not "leaky" in the absence of the inducer. The ecdysone agonist induces the expression of the antisense RNA to very high levels within 24 h. In preliminary experiments of the G-protein subunits Gβ1 and Gβ2 suppression, we have observed that expression of antisense RNA induced by the steroid provokes an inducible decline in the expression within 48 h postinduction. RNA levels of $G_{\alpha i2}$ remain unaffected. Thus, antisense RNA strategies can make use of either constitutively active or inducible constructs. The inducibility may require complex transactivation in the presence of cAMP (PEP-CK-based system) or simple addition of a unique steroid, such as muristerone A (pIND system).

3.1.7.3. *LAC* OPERON-DRIVEN CONSTRUCTS

Recent advances taking advantage of the *lac* operon to provide an inducible eukaryotic expression system are also amenable to adaptation in antisense RNA strategies. Similar to the advantages provided by the ecdysone-inducible promoter described above, the advantages of the *lac* operon system include rapid, robust expression of the gene or antisense construct and reliable suppression of the target gene product by antisense with reduced worry of adaptive changes in the signaling pathways under study. The LacSwitch® system of Stratagene offers a facile approach for the purposes of expression of antisense RNA under an inducible promoter. The system makes use of two vectors, one into which the antisense construct is engineered and the other expressing high levels of the *lac* repressor enabling silencing of the first expression vector in the absence of the inducer. The vector into which the target gene or antisense construct is inserted contains multiple restriction sites to facilitate directional insertion. The repressor activity is suppressed by addition of the inducer ligand IPTG into the medium of stable transfectants harboring both plasmids. The IPTG is added directly to the medium at 1–5 m*M*, and induction is measurable within 4–8 h. For use in driving antisense RNA constructs, the suppression of target gene product will reflect the relative half-life of the pre-existing protein, following the destruction of the target mRNA by the RNA duplex formation with the antisense RNA. In our experience, the suppression of G-protein subunit expression requires 24–48 h postinduction with IPTG. Although not creating an instantaneous, "step-like" disappearance of the target protein, the system does offer an additional dimension to the use of inducible antisense RNA constructs.

3.1.7.4. Transfection Protocol

1. Grow cells in one 100-mm cell-culture dish (log phrase: $1–2 \times 10^6$ cells).
2. To a 13-mL polypropylene tube, add the following (in order):

	Stock solution	Per dish
H$_2$O		413.5 µL
Salmon sperm DNA	5.0 mg/mL	4 µL
Neo$^+$ Plasmid	1.0 mg/mL	20 µL
CaCl$_2$	2 *M*	62.5 µL

3. To a tube containing 0.5 mL of 2X HBS, add the above mixture dropwise. At the same time, bubble in air for 30 s to mix.
4. Let the tube sit at room temperature for 30 min (the solution should become cloudy).
5. Gently add 1.0 mL of the above solution to the cell dish. Swirl gently to mix.
6. Incubate for 4 h.
7. Aspirate off the medium, and wash the plate with 5.0 mL of Dulbecco's Modified Eagle's Medium (DMEM). Then add 7.0 mL of fresh medium.
8. The following day, replace the old medium with new medium containing 400 µg/mL of G418 (Geneticin®).
9. Change the medium every other day.
   Formulation of the 2X HBS:

	Stock solution	Volume
NaCl	2 *M*	6.75 mL
KCl	0.5 *M*	1.0 mL
Na$_2$HPO$_4$	100 m*M*	0.7 mL
Glucose	0.5 *M*	1.2 mL
HEPES	0.5 *M*	4.0 mL

   H$_2$O to 100 mL
   Adjust to pH 7.08, with NaOH; filter and store at 4°C.

3.2. Inducible, Tissue-Specific Expression
of Antisense RNA In Vivo

3.2.1. Goals of Inducible, Tissue-Specific Ablation
of Target Gene Products

In creating a strategy for suppressing the expression of a target protein in vivo, we were confronted by several formidable obstacles. Although adapted first to the elimination of a G-protein α-subunit, the approach was designed for the broader use in targeting G-protein-linked receptors and other accessory proteins. The approach of employing $G_{i\alpha2}$-specific antisense RNA is far simpler than gene disruption by homologous recombination. To ensure accumulation of the antisense RNA in vivo, the target sequence was inserted in the first exon of the rat PEPCK mRNA. The PEPCK gene was selected for this work based on three considerations *(5)*. First, this 2.8-kb hybrid mRNA would be far more stable than a comparatively short-lived antisense RNA oligonucleotide. Second, expression of PEPCK is controlled by several hormones, including glucagon (acting via cAMP), glucocorticoids, thyroid hormone, and insulin. Insertion of the antisense sequence within the PEPCK gene confers regulated expression of the desired antisense RNA sequences. cAMP acts coordinately to increase the transcription rate of the gene as well as the stability of the mRNA, providing an additional benefit. Third, expression of the PEPCK gene is regulated developmentally, i.e., initial appearance of the mRNA occurs at birth. PEPCK mRNA is highly abundant at birth, and neonatal levels approach 0.5–1.0% of total cellular RNA. Targeted expression of the antisense RNA after birth obviates the problem of inducing a potentially lethal event by the suppression of a G-protein subunit *in utero* that might preclude viable transgenic pups.

3.2.2. Screening of PEP-CK Promoter-Driven Constructs In Vitro

The utility of the construct pPCK-AS to suppress expression of a signaling molecule was evaluated first for $G_{i\alpha2}$ after transfection into FTO-2B rat hepatoma cells *(5,7,25,26)*. FTO-2B cells display cAMP-inducible PEPCK gene expression and express $G_{i\alpha2}$ and Gβ-subunits. The RNA antisense to $G_{i\alpha2}$ was detected in FTO-2B clones transfected with pPCK-AS$G_{i\alpha2}$ after induction with cyclic AMP, as detected by reverse transcription of total cellular RNA followed by PCR amplification. FTO-2B clones transfected with pPCK-AS $G_{i\alpha2}$ display normal levels of $G_{i\alpha2}$ expression in the absence of the inducer of the PEPCK gene expression. $G_{i\alpha2}$ expression declined >85% when these same cells were challenged with the cyclic AMP analog, 8-(4-chlorophenylthio)-cAMP (CPT-cAMP) for 12 d. FTO-2B clones transfected with the vector lacking the antisense sequence to $G_{i\alpha2}$ displayed no change in $G_{i\alpha2}$ expression. In marked contrast to the suppression of $G_{i\alpha2}$, the expression of $G_{s\alpha}$ and $G_{i\alpha3}$ was unchanged in cells expressing the RNA antisense to $G_{i\alpha2}$, providing compelling data as to the specificity of action of the RNA antisense to $G_{i\alpha2}$. The time elapsing between the induction of pPCK-ASG$_{i\alpha2}$ by CPT-cAMP and the decline of steady-state levels of $G_{i\alpha2}$ reflects the half-life of this signaling protein. The controls required to ensure that it is the antisense RNA expressed from the vector that is responsible for the suppression of the target mRNA and thereby target protein are important considerations (*see* **Notes 8–10**).

3.2.3. Creation of Transgenic Mice Harboring Inducible, Tissue-Specific Vectors

Transgenic lines of mice can be created at any number of commercial sites. We utilize the Transgenic Mouse Facility at SUNY Stony Brook, which employs standard techniques in transgenic work.

1. The antisense vector pPCK-ASG$_{i\alpha2}$ (or an alternative) is excised free of vector sequences and purified prior to microinjection into the single-cell stage, preimplantation embryos of either BDF1 or FVB strains of mice.
2. The microinjected embryos then are transferred to pseudopregnant female mice.
3. The offspring harboring the transgene are identified by PCR amplification and subsequent Southern analysis using a pPCK-ASG$_{i\alpha2}$-specific probe.
4. At least four separate founder lines should be created, identified by Southern analysis and then used to breed an F1 generation for use in experimentation.
5. Features of the necropsy and histological evaluation of transgenic mice and their littermates require availability of a competent mouse histologist. The mice carrying the pPCK-ASG$_{i\alpha2}$ transgene, for example, display a runted phenotype that includes a marked reduction in total body mass (20–40%), a sharp reduction in liver mass, and frank insulin resistance. More detailed analysis of the phenotype of transgenic mice harboring the inducible, tissue-specific promoter system for TKOs of G-protein α-subunits has been published elsewhere.

3.2.4. Creation of Transgenic Mice Overexpressing Mutant Signaling Proteins

The same strategy used to create loss-of-function mutants can be adapted to create gain-of-function mutants in which a constitutively active form of a signaling protein (e.g., $G_{i\alpha2}$) is expressed in a tissue-specific, inducible manner. This approach was instrumental in probing further the relationship between $G_{i\alpha2}$ and insulin action. $G_{i\alpha2}$-deficiency results in blunted glucose tolerance, loss of insulin-stimulated hexose transport, inability of insulin to recruit GLUT4 transporters to the plasma membrane, blunted glycogen synthase activation in liver and skeletal muscle of the transgenic mice, and enhanced expression of phosphotyrosine phosphatase 1b, but not *Syp (7)*. We explored the linkage between insulin action and $G_{i\alpha2}$ in the opposite direction, seeking to investigate the effects of increased $G_{i\alpha2}$ activity on insulin action in transgenic mice. The cDNA encoding Q205L $G_{i\alpha2}$ was engineered in a position 3' to and under the control of the PEPCK promoter and upstream of the PEPCK 3' UTR to enhance mRNA stabilization.

Embarking on the creation of transgenic mice in which to take advantage of antisense RNA vectors is a major decision. The limitations, pitfalls, and resources required deserve serious attention well in advance of the actual commitment (*see* **Note 11**). However, it must be highlighted that advantages in speed and reduced cost associated with the use of inducible, antisense RNA transgenes as compared to homologous recombination to "knock out" a target are quite significant.

4. Notes

1. Antisense ODNs may have nonspecific effects on cellular function. An antisense ODN-induced change in a response must be carefully verified with the proper controls to ensure that the change is not artifactual. Lacking any basis to expect a toxic reaction to ODNs by cells in culture, declines in cell growth, and increased apoptosis may indicate a non-selective toxicity of the cells to either the ODNs and/or carrier molecules, if employed. Cell types of interest should be screened to uncover any unique toxicity to ODNs, in general, as well as to cationic carrier molecules, such as DOTAP. In addition, these studies of toxicity provide a margin of safety for the selection of a concentration of carrier that will be well below the level resulting in alterations in cell growth.

2. For the DNA oligonucleotide reagents, the standard set of controls includes the use of ODNs that are antisense, sense, as well as missense to the targeted sequence. Control sequences must be treated in the same manner as the antisense ODNs, i.e., introduced in the same manner and employed for the same periods as their antisense counterparts. Nonspecific effects of ODNs that do not differentiate among the three classes of controls are readily apparent. The determination of targeted gene product expression defines the efficacy of the approach and the extent to which it is truly specific. Although our laboratory has had nearly a decade in experience using antisense ODNs, our experience reinforces the absolute requirement to include these three controls on a routine basis.

3. Prominent among pitfalls to be avoided in antisense ODN approaches are the following: lack of proper in-depth analysis of the "selectivity" of a sequence to be targeted; failure to include analysis of the issue of species differences at the nucleotide level for conserved proteins; improper synthesis and purity of the ODNs; failure to characterize the toxicity of a carrier molecule if employed; failure to include proper sense and missense controls with the antisense ODNs; inability to assay the expression of the gene product reliably with high sensitivity and selectivity; lack of appreciation for the complexities of potential isoforms of

the target protein; and failure to explore the competing effects of antisense ODNs vs the half-life of the protein target itself. It must be emphasized that for antisense ODNs to be useful in suppressing the expression of their target proteins, the half-life of the target protein cannot extend beyond the exposure to the ODNs. Treatment for 48 h with antisense ODNs would not be expected to impact significantly a protein with a half-life of 90 h. If the evidence indicates little effect of antisense ODNs on the steady-state levels of expression of the target protein, one can prolong the exposure to the ODNs. Underpinning the antisense ODN technique is the requirement that the amount of oligonucleotide needed for stoichiometric destruction of target protein mRNA can be delivered, in fact, to the intracellular compartment. One can envision that the steady-state expression of the target mRNA may exceed the capacity of the cell to take up antisense ODNs.

4. In the event that treatment with ODNs antisense to a specific target does not yield a decline in or loss of the target, several steps may be taken *(22)*. The sequence selected for suppression may be erroneous, and a new sequence can be selected and tested as an alternative. The conditions for suppression may be inadequate in relationship to the half-life of the pre-existing protein target. The cells may display a generalized toxic reaction either to an ODN or to a carrier molecule, and some alternative cell type may prove more profitable for the studies. The antisense ODN may be targeting an essential gene product required for cell growth and viability. The phenotype of these ODN-treated cells may be poor cell growth or apoptosis. The availability of multiple readouts for the antisense ODN experiments can be invaluable. The small size of the cultures dictated by the cost of cell-culture-grade ODNs does not enable biochemical analyses of the treated cells. To circumvent this formidable problem, expression of antisense RNA rather than antisense ODNs has been achieved for in vitro applications.

5. Antisense ODNs have proven invaluable to the suppression of G-protein-linked receptors as well as the protein kinases, phosphatases, and accessory proteins involved in their regulation *(4,8,10,15–18,24)*.

6. The ability to ensure by biochemical means that the expression of the protein targeted by the ODNs is truly sensitive to antisense DNA is dependent on the proper readout. To analyze whole-cell extracts for the level of expression of the targeted gene product, sodium dodecyl sulfate-polyacrylamide gel electrophoresis (SDS-PAGE) is optimal. Staining of vanishingly small amounts of protein is not suitable, and immunoblotting of the proteins on transfer to nitrocellulose is usually required. With high-quality antibodies, one can stain immunoblots with great sensitivity, exploring the extent to which antisense ODNs successfully reduce expression of a target protein. The dimension of ODNs-based antisense experiments dictates use of small masses of cells. Large-format gels and SDS-PAGE are not ideal or even possible in many cases. The minigel format for SDS-PAGE is ideal, being very reliable and managing the separation and transfer of very small quantities of proteins typically encountered for G-protein-linked receptors in mammalian cells. The Pharmacia Phast® microgels provide a system in which submicroliter volumes of protein extracts can be rapidly resolved

on precast gels, carefully transferred to nitrocellulose and then blotted by standard methods.

7. Although functional readouts are possible, at best they are only confirmatory of the biochemical data. Single-cell electrophysiology of ion conductances lend themselves to studies in which ODNs are microinjected into the cell nuclei and then later impaled with micropipets enabling study of ion conductances. This approach has been used with great success and is a powerful strategy when employed to study receptors and signaling molecules involved in regulating ion channel activity. Some cells, such as the mouse embryonic fibroblast 3T3-L1 cells, benefit by accumulating easily stained lipid droplets in their cytoplasm as a possible readout in signaling. Mouse F9 stem cells can be assessed for their ability to secrete tissue plasminogen activator, which can be assayed on extremely small amounts of conditioned media. The advent of green fluorescent protein-based assays has created new opportunities for linking G-protein-linked receptor signaling to this powerful new readout, permitting more cells to come under scrutiny with the micropipet.

8. Controls for use of vector-expressed antisense RNA in vitro in stably transfected mammalian cells are the same as those detailed in **Note 2**. Data gathered from studies with antisense ODNs are a useful guide for the selection of the target sequence used in vector-driven antisense RNA. The benefits of using vectors capable of expression of antisense RNA over antisense ODNs are obvious, and substantial quantities of cells can be prepared in which antisense RNA has suppressed a targeted gene product. Biochemical analysis of the stably transfected clones is facile, avoiding miniaturization of assays for protein expression and the assays of signaling pathways confined to those with single-cell capability.

9. Prominent among pitfalls in this application is the variance among different clonal derivatives of a single transfection and the drift that sometimes occurs in clones expressing antisense RNA. At least 10 stable transfectant clones should be selected for expansion and characterization of both expression of the antisense RNA (utilizing unique primers and RT-PCR) and of the targeted protein (utilizing immunoblotting). The variance among clones from a single transfection can range from little suppression to virtual suppression of the targeted protein. The variance is encountered routinely, and multiple clones must be selected and characterized. Clones with optimal suppression are selected and used immediately for analysis of possible suppression of other related proteins, analysis of signaling pathways, and general characterization of the rate of growth.

Stable transfectants can display a "drift" in the amount of expression of antisense RNA with cell passage. The extent to which the clones can be frozen, thawed, and maintain the suppression phenotype is also quite variable. The drift in the cell population is rarely a general decline in the expression of the antisense RNA vector, but rather the emergence of subclones that have lost the expression of the vector and overgrow the cells in which the antisense RNA continues to suppress expression of the targeted protein. Repeating the transfection and selection protocol entirely is more prudent than is an attempt to "rescue" cells from the drifting population. In our experience, antisense RNA expression and target protein suppression of many clones remain quite stable over time and cell passage. The extent to which suppression of a given protein is "toxic" to the cell and reduces cell growth or viability may dictate the extent to which the population drifts. The ability to suppress the expression of certain targets proteins displays a limit that may well indicate an important role in cell viability that cannot be sustained if suppressed further.

10. Constitutively, expression of RNA antisense to protein kinase A, protein kinase C, and G-protein-coupled receptor kinases (GRKs), as well as protein phosphatases, has been used to probe the role of these proteins in the regulation of G-protein-linked receptors.

11. For transgenic mice, the controls rely heavily on proof of concept developed using cells in culture as a screen. It is not financially practical to create a series of transgenic mice in which missense and/or sense constructs are substituted for the antisense sequence. Our own experience of targeting several different G-protein subunits and exhaustively characterizing the derivative transgenic mice provides the proof-of-concept for the strategy.

A major limitation for studies of the transgenic mice is the cost incurred in the creation and maintenance of the mice colonies. The fee for creating transgenic mice depends on the facility and whether the activity is subsidized. For academic settings, the costs range between $2000 and $5000 fees, with the commercial rates not very different from the costs of unsubsidized academic facilities. Four or five founder lines must be generated and maintained for each antisense vector construct. The animal husbandry for adequate production of mice for experimental purposes is 1500 to 2000 mice/construct. The mice that have been created using the pPEP-CK transgene have been bred for more than 10 generations, are easily rederived from the single-cell embryos by microinjection, and are stable. The antisense RNA expression in vivo has surpassed our greatest expectation and remains restricted to the sites targeted in the initial design for the project now more than 8 yr ago.

The pPEPCK-AS vector system is patented and is licensed for use through the Research Foundation of SUNY/Stony Brook. For expression of antisense RNA in other target tissues, ample reports of analogous tissue-specific elements for the heart, regions of the CNS, and other tissues are easily retrieved from a search of the Medline.

References

1. Miller, P. S., Braiterman, L. T., and Ts'o, P. O. (1977) Effects of a trinucleotide ethyl phosphotriester, Gmp(Et)Gmp(Et)U, on mammalian cells in culture. *Biochemistry* **16,** 1988–1996.

2. Haseloff, J. and Gerlach, W. L. (1988) Simple RNA enzymes with new and highly specific endoribonuclease activities. *Nature* **334,** 585–591.

3. Goodchild, J. (1989) Oligodeoxynucleotides-antisense inhibitors of gene expression. (Cohen, C. H., ed.), CRC, Boca Raton, FL, pp. 53–77.

4. Bahouth, S. W., Park, E. A., Beauchamp, M., Cui, X., and Malbon, C. C. (1996) Identification of a glucocorticoid repressor domain in the rat beta 1-adrenergic receptor gene. *Receptors Signal Transduc.* **6,** 141–149.
5. Moxham, C. M., Hod, Y., and Malbon, C. C. (1993) Gi alpha 2 mediates the inhibitory regulation of adenylylcyclase in vivo: analysis in transgenic mice with Gi alpha 2 suppressed by inducible antisense RNA. *Dev. Genet.* **14,** 266–273.
6. Moxham, C. M., Hod, Y., and Malbon, C. C. (1993) Induction of G alpha i2-specific antisense RNA in vivo inhibits neonatal growth. *Science* **260,** 991–995.
7. Moxham, C. M. and Malbon, C. C. (1996) Insulin action impaired by deficiency of the G-protein subunit Gi alpha2. *Nature* **379,** 840–844.
8. Shih, M. and Malbon, C. C. (1994) Oligodeoxynucleotides antisense to mRNA encoding protein kinase A, protein kinase C, and beta-adrenergic receptor kinase reveal distinctive cell-type-specific roles in agonist-induced desensitization. *Proc. Nat. Acad. Sci. USA* **91,** 12,193–12,197.
9. Shih, M. and Malbon, C. C. (1996) Protein kinase C deficiency blocks recovery from agonist-induced desensitization. *J. Biol. Chem.* **271,** 21,478–21,483.
10. Wang, H. Y., Watkins, D. C., and Malbon, C. C. (1992) Antisense oligodeoxynucleotides to Gs protein alpha-subunit sequence accelerate differentiation of fibroblasts to adipocytes. *Nature* **358,** 334–337.
11. Watkins, D. C., Johnson, G. L., and Malbon, C. C. (1992) Regulation of the differentiation of teratocarcinoma cells into primitive endoderm by G alpha i2. *Science* **258,** 1373–1375.
12. Watkins, D. C., Moxham, C. M., Morris, A. J., and Malbon, C. C. (1994) Suppression of Gi alpha 2 enhances phospholipase C signalling. *Biochem. J.* **299,** 593–596.
13. Dean, N. and McKay, R. (1995) Inhibition of protein kinase C-alpha expression in mice after systemic administration of phosphorothioate antisense oligodeoxynucleotides. *Proc. Nat. Acad. Sci. USA* **91,** 11,762–11,766.
14. Shih, M., Lin, F., Scott, J. D., Wang, H.-Y., and Malben, C. C. (1999) Dynamic complexation of β2-adrenergic receptors with protein kinases and phosphatases. *J. Biol. Chem.* **274,** 1588–1595.
15. Kleuss, C., Hescheler, J., Ewel, C., Rosenthal, W., Schultz, G., and Wittig, B. (1991) Assignment of G-protein subtypes to specific receptors inducing inhibition of calcium currents. *Nature* **353,** 43–48.
16. Kleuss, C., Scherubl, H., Hescheler, J., Schultz, G., and Wittig, B. (1992) Different beta-subunits determine G-protein interaction with transmembrane receptors [*see* comments]. *Nature* **358,** 424–426.
17. Kleuss, C., Scherubl, H., Hescheler, J., Schultz, G., and Wittig, B. (1993) Selectivity in signal transduction determined by gamma subunits of heterotrimeric G proteins. *Science* **259,** 832–834.
18. Kleuss, C., Schultz, G., and Wittig, B. (1994) Microinjection of antisense oligonucleotides to assess G-protein subunit function. *Methods Enzymol.* **237,** 345–355.
19. Wagner, R. W. (1994) Gene inhibition using antisense oligodeoxynucleotides. (review) (53 refs). *Nature* **372,** 333–335.

20. Sauer, B. (1993) Manipulation of transgenes by site-specific recombination: use of Cre recombinase. *Methods Enzymol.* **225,** 890–900.
21. Katsuki, H., Kaneko, S., and Satoh, M. (1992) Involvement of postsynaptic G proteins in hippocampal long-term potentiation. *Brain Research* **581,** 108–114.
22. Agrawal, S. (1996) *Methods in Molecular Medicine, vol. 1: Antisense Therapeutics.* Humana, Totowa, NJ.
23. Gollasch, M., Kleuss, C., Hescheler, J., Wittig, B., and Schultz, G. (1993) Gi2 and protein kinase C are required for thyrotropin-releasing hormone-induced stimulation of voltage-dependent Ca2+ channels in rat pituitary GH3 cells. *Proc. Natl. Acad. Sci. USA* **90,** 6265–6269.
24. Kleuss, C., Raw, A. S., Lee, E., Sprang, S. R., and Gilman, A.G. (1994) Mechanism of GTP hydrolysis by G-protein alpha subunits. *Proc. Nat. Acad. Sci. USA* **91,** 9828–9831.
25. Galvin-Parton, P. A., Chen, X., Moxham, C. M., and Malbon, C. C. (1997) Induction of Galphaq-specific antisense RNA in vivo causes increased body mass and hyperadiposity. *J. Biol. Chem.* **272,** 4335–4341.
26. Guo, J. H., Wang, H. Y., and Malbon, C. C. (1998) Conditional, tissue-specific expression of Q205L G-alpha12 in vivo mimics insulin activation of Jun N-terminal kinase and P38 kinase. *J. Biol. Chem.* **273,** 16,487–16,493.

18

Targeted Disruption of Adrenergic Receptor Genes

Daniel K. Rohrer

1. Introduction

The use of targeted gene disruptions, or knockouts, has become common-place in many basic research laboratories. In its most common application, this technique enables the researcher to disrupt expression of a specific gene product selectively. This approach has been particularly useful for studies on adrenergic receptor (AR) function, having been used to effect disruption of at least six of the nine known AR genes *(1–6)*. As presented here, the knockout strategy is essentially a synthesis of two separate techniques. The first utilizes the phenomenon of homologous recombination, a process whereby foreign DNA (homologous at least in part to portions of the host genome) introduced into cells undergoes strand exchange and integration into the host genome at a specific locus. The second utilizes the pluripotent nature of embryonic stem (ES) cells, which can be cultured and manipulated ex vivo and reintroduced into host embryos. Engineered ES cells having undergone homologous recombination can be incorporated into such "chimeric" embryos, potentially giving rise to all cell types of the adult mouse, including germ cells. Through simple mating experiments, one can then transmit the ES cell-derived disrupted allele to progeny, and eventually intercross heterozygotes to generate mice homozygous for the disrupted allele.

In brief, AR gene disruption involves the following discrete steps:

1. Cloning of the AR genomic locus.
2. Design and construction of a targeting vector suitable for homologous recombination.
3. Transfection of the targeting vector into ES cells and selection of homologous recombinants.

From: *Methods in Molecular Biology, vol. 126: Adrenergic Receptor Protocols*
Edited by: C. A. Machida © Humana Press Inc., Totowa, NJ

4. Injection or aggregation of targeted ES cells with normal mouse embryos to create chimeras.
5. Genotyping, breeding, and phenotypic characterization of genetically modified mice.

This technique has emerged as a powerful tool for the analysis of adrenergic receptor gene function over the past several years. The high degree of sequence similarity observed among the nine currently identified AR genes is mirrored by many functional similarities in the corresponding receptor proteins, including their common activation by the endogenous catecholamines norepinephrine and epinephrine. Overlapping pharmacological profiles among each of the three α_1-, α_2-, or β-AR members can make functional assignments of specific subtypes difficult, and the knockout technique has been effectively used in conjunction with more classical pharmacological approaches to define subtype-specific function and mode of activation better. More recent iterations of the knockout technique can allow for simultaneous ablation of multiple AR members, site-specific mutagenesis, as well as tissue-specific and temporal control of gene expression.

The techniques described here combine disciplines from molecular, cellular, and whole animal biology. Microinjection is one aspect of knockout methodology that is considered to be a specialized technique, but apart from that, the other components are not overly difficult for the average bench scientist with a molecular biology background. As an alternative to microinjection, we describe here an aggregation technique that enables those without access to a microinjectionist the ability to make ES cell–embryo chimeras. Thus, given the proper resources and time commitment, targeted gene disruptions can be performed in most labs.

2. Materials

2.1. Screening of Mouse Genomic Libraries and Creation of Gene Targeting Vector

1. Gene-specific DNA probe, suitable for high-stringency washing and detection of gene sequences on Southern blots.
2. Mouse genomic library of 129 strain origin. These are available from a variety of sources, e.g., Stratagene (La Jolla, CA) or Genome Systems (St. Louis, MO). Screening of such libraries can even be contracted out (Genome Systems).
3. 50X Tris-acetate ethylenediamine tetra-acetic acid buffer (TAE; Gibco-BRL, Gaithersburg, MD).
4. 1% Agarose in 1X TAE buffer.
5. T4 DNA ligase and 5X buffer (1U/μL; Gibco-BRL).
6. 20% Sodium dodecyl sulfate (SDS).
7. 20X Standard sodium citrate (SSC): 3 M NaCl, 0.3 M sodium citrate, pH 7.0.

8. Phenol:chloroform:isoamyl alcohol (25:24:1; Gibco-BRL).
9. Competent bacteria (e.g., DH5α; Gibco-BRL).
10. [α-^{32}P]dCTP (3000 Ci/mmol; Amersham Life Science, Arlington Heights, IL).
11. Random primer labeling system (Rediprime; Amersham Life Science).
12. 50% Dextran sulfate (Oncor, Gaithersburg, MD).
13. Nylon supported blotting membranes (e.g., Nytran; Schleicher and Schuell, Keene, NH).
14. 2X hybridization solution: 12X SSC, 10X Denhardt's solution, 200 μg/mL sheared, denatured salmon sperm DNA (Gibco-BRL).
15. DNA separaration columns (Centri-sep; Princeton Separations, Adelphia, NJ).
16. HEPES-buffered normal saline: 20 mM HEPES, 137 mM NaCl, 5 mM KCl, 0.8 mM Na$_2$HPO$_4$, 5.6 mM *d*-glucose, pH 7.4.
17. DNA isolation buffer: 0.5% SDS, 50 mM Tris-HCl, pH 7.5, 10 mM EDTA, 100 mM NaCl, and 100 μg/mL proteinase K.
18. TE buffer: 10 mM Tris-HCl, pH 8.0, 1 mM EDTA.

2.2. Cell-Culture Supplies

1. Dulbecco's Modified Eagles Medium (DMEM—high glucose; Gibco-BRL).
2. Penicillin/streptomycin (100X; Gibco-BRL).
3. 10 mM nonessential amino acids (Gibco-BRL).
4. 100 mM sodium pyruvate (Gibco-BRL).
5. 55 mM 2-mercaptoethanol (Gibco-BRL).
6. 200 mM l-glutamine (Gibco-BRL).
7. ES-qualified fetal calf serum (FCS) (Gibco-BRL).
8. Leukemia inhibitory factor (LIF or ESGro; Gibco-BRL).
9. 50 mg/mL Geneticin (G418; Gibco-BRL).
10. 2 mM ganciclovir (Cytovene, available by prescription).
11. 1 mM mitomycin C (Sigma Chemicals, St. Louis, MO).
12. G418-Resistant mouse embryonic fibroblasts (MEFs) (Genome Systems, St. Louis, MO).
13. 129-Derived mouse embryonic stem cells (Genome Systems).
14. M2/M16 medium (Specialty Media, Philipsburg, NJ).
15. ES cell growth medium: DMEM plus 20% FCS, 0.1 mM nonessential amino acids, 2 mM l-glutamine, 1 mM sodium pyruvate, 1X penicillin/streptomycin, 0.1 mM 2-mercaptoethanol, 1000–2000 U/mL LIF.
16. MEF cell growth medium: DMEM plus 10% fetal calf serum, 1X penicillin/ streptomycin.

2.3. Mice

1. C57Bl6, CD-1, and 129Sv mouse strains (normal or vasectomized) can either be obtained through Jackson Labs (Bar Harbor, ME) or Taconic (Germantown, NY).
2. Pregnant Mare's Serum (PMS) and human chorionic gonadotropin (hCG) can be obtained through Sigma Chemicals.

3. Avertin anesthetic: 2.5% (w/v) 2,2,2, tribromoethanol, 3% (v/v) *tert*-amyl alcohol.

3. Methods
3.1. Cloning of the Adrenergic Receptor Locus

1. A suitable DNA probe specific for the gene of interest must first be obtained. The following criteria are important:
   a. The probe should be gene-specific under high-stringency washes of mouse Southern blots.
   b. Ideally, the probe will be >200 bp in length and <1000 bp, for ease of labeling.
   c. The probe will preferably be located within close proximity to the point of gene disruption.
2. Screening of a mouse 129 strain-derived genomic library is the first step in isolation of the target gene (*see* **Note 1**). Ten to 30 ng of purified DNA probe (≥0.2 ng/µL) are used to make a high specific activity [^{32}P]-labeled DNA probe, using [α-^{32}P]dCTP and the Redi-Prime kit. Labeled DNA fragments are separated from free [α-^{32}P]dCTP using Centri-Sept columns, and a small aliquot of the excluded volume is counted by liquid scintillation to determine incorporation efficiency. One should obtain radiolabeled probes of ~10^9 cpm/µg SA.
3. The mouse genomic library is then screened with the radiolabeled DNA probe. The appropriate number of recombinants are plated out and transferred to nylon membranes (*see* **Note 2**). Prehybridization of filters at 42°C is carried out for ≥4 h with hybridization buffer (available in 2X concentration; *see* **Subheading 2.**) plus 40% formamide, 0.5% SDS, and 5% dextran sulfate. Use enough hybridization buffer to cover membranes fully (~0.05–0.10 mL hybridization buffer/1-cm^2 filter surface area). Following prehybridization, radiolabeled probe (denatured at 100°C for 5 min prior) is added to the filters, at 10^5–10^6 cpm/mL hybridization fluid. Hybridization is then carried out overnight, and the following day, filters are washed and exposed to film. A typical washing regimen would include a initial room temperature wash in 2X SSC, 0.1% SDS for 15 min, followed by a 37°C wash in 0.1X SSC, 0.1% SDS for 30 min, and a final high-stringency wash at 65°C in 0.1X SSC and 0.1% SDS for 30 min. Membranes are then wrapped in Saran wrap and exposed to X-ray film (Kodak XAR-5 or the equivalent). Positive clones are purified to homogeneity, and processed for DNA as recommended by the manufacturer or by conventional means *(7)*, depending on the cloning vector.
4. Identification and verification of the mouse genomic clone can be carried out by a combination of restriction mapping and sequencing (*see* **Note 3**). Given the wide scope contained within the knockout methodology, it is assumed that certain techniques, such as DNA restriction mapping and sequencing, are familiar to the investigator. Restriction mapping should include common restriction enzymes frequently found within multiple cloning sites and flanking drug resistance

cassettes (e.g., *Apa*I, *Bam*HI, *Bgl*II, *Cla*I, *Hind*III, *Eco*RI, *Eco*RV, *Kpn*I, *Not*I, *Sac*I, *Sac*II, *Sal*I, *Sma*I, *Spe*I, *Xba*I, *Xho*I), as well as any others that would provide a convenient means to insert or shuttle drug resistance gene cassettes (*see* **Subheading 3.2.**). Sequencing can be contracted out to local core facilities or to commercial sequence facilities (e.g., Sequatech, Mountain View, CA).

3.2. Targeting Vector Design and Construction

1. Using **Fig. 1** as a guide, identify the potential 5′ and 3′ "arms" of homology, which will be used to flank the *neo* gene cassette, making sure to include sufficient sequence information (at least 1 kb on each arm; *see* **Note 5**), and a strategy that will effectively perturb AR gene expression (*see* **Notes 4** and **6**). This tripartite sequence is then flanked by the negative selectable marker thymidine kinase (*TK*), as well as irrelevant plasmid vector sequences.

2. Once the targeting vector has been constructed and the relevant junctions between target sequence and disrupting cassettes verified, the DNA is ready for linearization and electroporation into embryonic stem cells. Forty micrograms of supercoiled targeting vector DNA should be digested to completion with the chosen restriction enzyme, and a small aliquot (200–400 ng) removed and electrophoresed on a 0.8% agarose gel (1X TAE in gel and running buffer) to verify that the vector runs as a single species (i.e., a linear molecule that has been completely digested). When verified, the remaining sample is then extracted with an equal volume of phenol:chloroform:isoamyl alcohol (25:24:1), and the aqueous layer is ethanol-precipitated by adding ammonium acetate to a final concentration of 2.5 *M*, followed by addition of 2 vol of absolute ethanol. The resulting precipitate is centrifuged, washed with 70% ethanol, briefly air-dried, and resuspended in HEPES-buffered normal saline at ~1 mg/mL. The linearized DNA can be stored at 4°C until ES cells are ready for transfection.

3.3. ES Cell Culture, Transfection, and Homologous Recombination

3.3.1. Feeder Cell Culture

1. For most of their time in culture, ES cells are grown on a mitotically inactivated feeder cell layer. This protocol describes the use of mouse embryonic fibroblast (MEF) feeders, which are derived from midgestation mouse embryos. These in turn are frequently derived from mice that have an incorporated neo^r gene, in order not to interfere with the antibiotic selection process, which follows electroporation of ES cells. MEF cells derived from commercial sources (*see* **Subheading 2.**) should be grown in DMEM (high glucose) plus 10% FCS and penicillin/streptomycin. These cells have a limited mitotic life-span, and an approach we have found to be successful is to passage the cells twice (at a 1:5 dilution at both steps, using 100-mm tissue-culture dishes) and then freeze these cells for later use. Freezing these third-passage (P3) cells is performed following a standard 5-min trypsinization of a near-confluent monolayer, vigorous

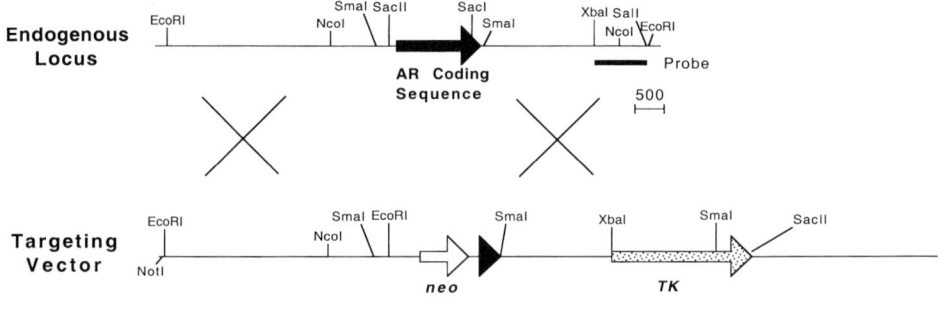

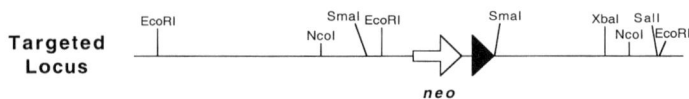

Fig. 1. Schematic of targeting vector design and homologous recombination event. The upper panel depicts a partial restriction map of an endogenous AR locus, with the single exon coding sequence indicated by the dark arrow. The middle panel depicts the targeting vector, which is derived from parts of the endogenous AR gene and the *neo* and *TK* resistance genes, respectively. The long arm of homology was derived from the ~3500 bp *Eco*RI/*Sac*II fragment of the AR gene. This was then ligated to the *neo* drug resistance cassette (open arrow), which contains the PGK enhancer to drive expression *(14)*. The short arm of homology was derived from the ~1500 bp *Sac*I/*Xba*I fragment of the AR gene. The short arm is then flanked by the *TK* gene (stippled arrow), also driven by the PGK enhancer *(14)*. This in turn is flanked by plasmid vector sequences. Note that this approach to constructing the targeting vector results in excision of a majority of AR coding sequence. Also note the acquisition of a new *Eco*RI site just upstream of the *neo* cassette in the targeting vector. The two crosses are meant to represent crossover events, which occur during homologous recombination. Such a double crossover would be expected to eliminate the *TK* gene from the integration site while incorporating the neo gene, rendering such cells resistant to both G418 (through neo acquisition) and ganciclovir (through *TK* elimination). The targeted locus is depicted at the bottom. In order to confirm this type of integration event, a probe is derived from a region outside of that used to make the targeting vector (top panel, right side). The newly acquired *Eco*RI site in the targeted locus changes what is normally an ~8-kb *Eco*RI fragment detected by the probe to an ~5-kb fragment. This strategy is adapted from disruption of the β_1-adrenergic receptor *(3)*.

disaggregation, and trypsin inactivation by addition of an equal volume of growth media. Cells are then centrifuged at 1000*g* for 5 min, the supernatant aspirated off, and the cell pellet resuspended in growth media plus 10% dimethyl sulfoxide (1 mL/100-mm dish). Vials are slowly frozen at –70°C in a cell freezing container

(Nalgene), and transferred the next day to liquid N_2. These P3 MEF cells can then be thawed out onto 100-mm dishes at any time, and passaged two to four times subsequently.

2. Mitotic inactivation of MEF cells can be performed by either γ-irradiation (3000 rad) or treatment with the mitotic inhibitor mitomycin C. Mitomycin C treatment is described here, since many laboratories may not have access to a γ-irradiation source. Near-confluent MEF cells are treated with mitomycin C at 10 μg/mL in growth media for 2 h in a 37°C incubator. Following treatment, cells are washed three times with phosphate-buffered saline (PBS) before any further treatment. Following PBS washes, mitotically inactivated feeder cells may be trypsinized and aliquoted to the desired tissue-culture plate format, and left in growth media until ES cells are ready to be added (generally not to exceed 5 d). The appropriate density of feeder cells to be plated out is ~5 × 10^4 cells/cm^2, which should form a continuous monolayer on the dish.

3.3.2. ES Cell Culture

1. In the studies described here, the R1 ES cell line has been used (gift from A. Nagy, University of Toronto). The cell culture, transfection, and selection conditions described here can also be applied to most other commonly and commercially available ES cells as well (*see* **Subheading 2.**). Growth medium for ES cells is listed in **Subheading 2.2., item 15**.
2. ES cells are thawed out onto a confluent, mitotically inactivated feeder cell monolayer in the above growth media (minus selection agents). ES cells tend to form tightly packed colonies, with a smooth or encapsulated outer appearance. ES cell cultures should be refed daily, and should not be allowed to grow too densely in order to avoid differentiation (*see* **Note 7**). ES cell cultures are split frequently at 1:3–1:5 dilutions. Using trypsin-EDTA, ES cells are vigorously disaggregated into a single-cell suspension at each passage step, or prior to electroporation. Trypsinization is as follows: for a 100-mm tissue-culture plate, media are aspirated off, and plate is washed with 10 mL PBS and then aspirated. One milliliter of trypsin–EDTA is added to the plate, and is returned to a 37°C incubator for 5 min. The plate is then tapped gently to ensure cell detachment, and 4 mL PBS are added. The cell suspension is then vigorously disaggregated (10–20 times with 5-mL pipet), and an equal volume of ES cell growth media is added to inactivate trypsin. The cell suspension is centrifuged at 1000*g* for 5 min, followed by media aspiration and resuspension of the pellet in growth media in an appropriate volume for a 1:3–1:5 dilution.

3.3.3. ES Cell Transfection

A 100-mm tissue-culture dish of exponentially growing ES cells is typically used for one electroporation. Trypsinized ES cells from the 100-mm dish in single-cell suspension are centrifuged at 1000*g* for 5 min, resuspended in 10 mL PBS, and recentrifuged. The resulting cell pellet is resuspended in

0.8 mL HEPES-buffered saline. Twenty to 30 μg of linearized targeting vector (*see* **Subheading 3.2.**) are added to the cell suspension, and the mixture transferred to a 0.4-cm electroporation cuvet (Bio-Rad). Electroporation of the cell suspension is then carried out. Conditions which we have optimized for the R1 cell line are as follows: 0.25 kV, 500 μF, using a Bio-Rad Gene Pulser with capacitance extender. Following electroporation, the cell suspension is allowed to sit in the cuvet at room temperature for 10 min, after which time cells are removed and ES cell growth medium is added. Typically, the contents of one electroporation will be split out equally among five 10-cm tissue-culture plates containing mitotically inactivated MEF cells. These transfected ES cells are left to recover in growth media minus selection agents for the first 24 h, after which time G418 is supplemented at 250 μg/mL. The following day (48 h posttransfection), ganciclovir is also added to a final concentration of 2 μ*M*. Under this positive/negative selection scheme, one can typically expect to recover 200–500 individual G418/ganciclovir-resistant colonies when cells are fed daily for 8–12 d. The addition of ganciclovir, which aids in reducing the number of nonhomologous recombinants recovered (**Fig. 1**; *see* **Note 4**), usually imparts a ~10-fold reduction in colony number over G418 alone.

3.3.4. Expansion and Freezing of Transfected ES Cells

1. Individual G418/ganciclovir-resistant colonies are then picked, expanded, and screened for homologous recombination events. Processing one 100-mm plate at a time, media are aspirated off, the plate washed once with PBS, and replaced with just enough PBS to cover the dish (~4–5 mL). Individual colonies are then picked using a micropipeter and sterile micropipet tips. Colonies are dislodged with the pipet tip under a dissecting microscope, picked up in as small a volume as possible (5–10 μL), and then transferred to the individual wells of a 96-well tissue-culture plate (U-shaped bottom) containing 25 μL of trypsin. After 96 colonies are picked, the plate is incubated at 37°C for 10 min, and cells vigorously disaggregated with a multipipeter. The trypsinized colonies are then transferred to a 96-well plate containing mitotically inactivated MEF cells plus ES cell growth media, and allowed to adhere overnight. Media are replaced daily until the majority of wells are near-confluent. Using a multichannel pipeter, media are removed, cells are washed once with PBS, and 100 μL trypsin are added to each well. After 5–10 min at 37°C, cells are disaggregated. Fifty microliters of this cell suspension are removed and transferred to 1 mL of growth media in an individual well of a 24-well tissue-culture plate precoated with 0.1% gelatin. To the remaining 50 μL of cells on the 96-well plate, add 50 μL of 2X freezing media (DMEM with 40% FCS and 20% dimethylsulfoxide [DMSO]), mix, and seal the plate with Parafilm. This plate is placed in a styrofoam box, which is then transferred to a –70°C freezer. Cells frozen in this way should remain viable for ~6 mo.

2. At this point, the 96-well plate (referred to as the "Master Plate") will remain frozen until positive clones are identified, at which point individual clones can be thawed out and expanded. The gelatin-coated 24-well plate serves as a source for genomic DNA; MEF cells are omitted in order not to contaminate genomic ES cell DNA samples with MEF cell DNA.

3.3.5. Genotype Analysis of ES Cell Clones

1. Once the cells on the 24-well plate approach confluency, they are ready for DNA isolation. Wells are washed once with PBS, replaced with 500 µL DNA isolation buffer (*see* **Subheading 2.1.**), and returned to 37°C for 3–12 h. Plates are then removed, 500 µL of isopropanol added to each well, and then the whole plate is transferred to an orbital shaker. Plates are swirled (150–200 rpm) for at least an hour, and during this time, DNA will precipitate out in the well. Precipitates can then be removed with a sealed glass capillary pipet, washed in 70% ethanol, and briefly air-dried. The DNA and pipet are then placed in 100–200 µL TE buffer, the glass pipet removed, and DNA left at 4°C overnight to dissolve. DNA (~10–20 µg, approx 1/4 of total harvested) can then be digested with restriction endonucleases to determine those ES colonies that have undergone the desired recombination event (*see* **Fig. 1** for typical strategy; *see* **Note 8**).
2. Once correctly targeted ES colonies have been identified, the appropriate cells can be thawed out from the master plate. The 96-well plate is removed from –70°C and placed in a shallow 37°C water bath (do not submerge plate). After cells are thawed out, the exterior of the plate is wiped down with 70% ethanol. The 50 µL of cell suspension are transferred to the individual well of a 48-well plate containing mitotically inactivated MEF cells in 250 µL of ES cell growth medium. Medium is changed after cells have adhered (~12 h). When thawed out cells approach confluency, they are ready to be passaged onto larger plates. The contents of one well from a 48-well plate can be passaged up to an MEF-containing 12-well plate, which, following growth to near-confluency, can then be passaged onto 60-mm dishes. At such point, a near-confluent 60-mm dish of a targeted ES cell clone can be trypsinized (1 mL of trypsin/EDTA for 5 min at 37°C, followed by vigorous disaggregation, then inactivation with equal volume of growth media), centrifuged at 1000*g* for 5 min, and then resuspended in three 0.5-mL aliquots of freezing media (DMEM plus 20% FCS and 10% DMSO). These aliquots are frozen slowly at –70°C for 24 h and then transferred to liquid nitrogen indefinitely. Frozen cells can be thawed out onto individual wells of a 24-well plate (containing mitotically inactivated MEFs), grown to near-confluency, and prepared for microinjection into blastocysts or aggregation with morula.

3.4. Creation of Chimeric Mice

By far the most frequent method for creation of chimeric mice with targeted ES cells is microinjection of blastocyst-stage mouse embryos. This technique, however, is highly specialized, requiring extensive training and a substantial investment in microinjection equipment to be consistently effective. Many institutions will either have a core microinjection facility, or in some cases, such work can be contracted out (e.g., University of Cincinnati Transgenic Mouse Service, Cincinnati, OH). Details of this technique are well described elsewhere *(8)*.

For those investigators without access to such microinjection expertise, the author provides here an alternative that, with practice, can also be an efficient means of creating chimeric mice. This technique has been pioneered by several research groups *(9)* and has been aptly demonstrated for the R1 ES cell line. The success of this technique with other ES cell lines has not been rigorously established, however.

In brief, the technique involves the following:

1. Aggregation of morula-stage mouse embryos (donors) with targeted ES cells.
2. Overnight culture of aggregates to the blastocyst stage.
3. Uterine transfer of chimeric blastocysts into pseudopregnant mice (hosts).

3.4.1. Preparation of Donor Embryos and Pseudopregnant Hosts

1. The preparation of donor embryos and pseudopregnant hosts is coordinated over an ~4-d period, and can be seen schematically in **Fig. 2**. Day 1: for donor embryos, one needs prepubescent females (CD-1 strain, 3–4 wk old). In the middle of their day cycle (usually at 1:00 PM), 10 females are injected ip with 5 U PMS each (*see* **Subheading 2.**), in normal saline. Day 3: approx 46 h later (11:00 AM), each donor mouse receives 5 U of hCG ip, also in saline. Matings (1:1 ratio of females:males) with proven male CD-1 studs are set up overnight. They can be set up directly following hCG injection, up to early evening. Day 4: the following morning, check females for vaginal plugs, segregating all females from males and the plugged females from nonplugged. The plugged females are considered to be 0.5 d pregnant at this point. An 80% plugging rate is considered good (*see* **Note 9**).
2. With respect to host mice, the same day schedule applies. Day 4: mature females (7 wk old minimum) are set up with vasectomized males at a 3:1 ratio of females:males. These are not superovulated, and therefore, one needs to set up many more females since only ~25% of them will be in estrus at any time. A routine setup would be 50–80 host females for every 10 donors, but this is quite variable, and may require less host females if the plugging rate is high, and more if they are not (it is better to err on the side of having excess hosts rather than excess donor embryos). Day 5: check hosts for vaginal plugs. Segregate all females from males and the plugged females from nonplugged females. If plugged hosts end up not being used, refrain from mating for ~3 wk (*see* **Note 9**).

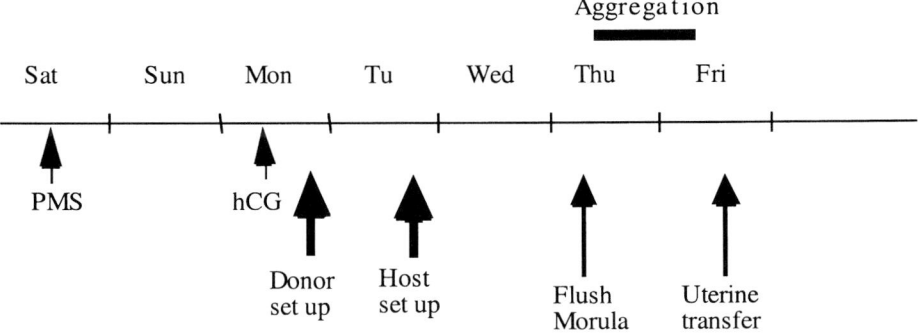

Fig. 2. Typical weekly schedule for chimera production by the aggregation method. On a cycle arbitrarily picked to begin on Saturday, mice are injected with PMS at 1:00 PM. Forty-six hours later, the same mice are injected with hCG, and that evening, set up to mate with male studs (donor setup). The following evening (Tu), host pseudopregnant females are created by mating with vasectomized males (host setup). On Thursday of this schedule, donor morula are flushed out in the morning, aggregated with ES cells during the day, and allowed to develop into blastocysts overnight. On Friday, developed blastocysts are transferred into pseudopregnant females. Chimeric pups can be expected 15–18 d following transfer.

3.4.2. ES Cell: Embryo Aggregation

On day 6 of the experiment (**Fig. 2**), donor embryos and targeted ES cells are harvested, and aggregated together. The sacrificed females are prepared as follows:

1. Wipe down belly with 70% ethyl alcohol (EtOH).
2. Open and expose peritoneal cavity as cleanly as possible.
3. Using a sterile pair of scissors and forceps, pick up one side of uterus, and clean off associated tissue. While holding the uterus with forceps, locate the junction between ovary and oviduct. Very carefully cut between the two at the infundibulum, and then remove ~1-cm section by cutting uterus distally. Place collected uterus-oviducts in small tissue culture dish containing M2 medium. Do the same for the other side.
4. In order to flush morula from oviducts, take a 30- or 31-gage needle, with the tip sanded blunt and smooth (200–400 grit). Attach to a 1-mL syringe filled with M2 media (clear out all bubbles!), and move one uterus to a dry 10-cm tissue-culture plate. Using a dissecting microscope, grab uterus with watchmaker's forceps (size 3). Though it is difficult at first, locate the infundibular opening and immobilize the oviduct near it with forceps. Carefully insert the needle, grasp needle surrounded by oviduct with forceps, and gently flush with ~0.5 mL M2. You

should see a swelling of the oviduct and fluid movement if this is done correctly. Remove uterus and search for morula. Collect these with a drawn-out Pasteur pipet (so-called handling pipette, ~140- to 220-μm fire polished opening), and transfer using mouthpiece and rubber tubing to an M2 droplet, which is overlayed with mineral oil (Sigma, embryo tested). Do the same for all uteri.

5. After all morula have been collected, prepare to strip off zona pellucida. On a 10-cm tissue-culture plate, place 5 droplets (50–100 μL) of acid Tyrodes (Sigma, embryo tested). Place five droplets of M2 parallel to Tyrodes droplets. Taking 10–20 morula at a time, transfer from M2 into acid Tyrodes. Clear out handling pipet, load with Tyrodes, collect morula, and transfer to next droplet. Continue moving through droplets until zona pellucida starts dissolving, at which point morula should be transferred out of the acidic solution into M2. Embryos will disaggregate if left too long in acid Tyrodes. Move morula through the successive droplets of M2. After last droplet wash, transfer morula to a droplet of M16 + 10% FCS and pen/strep under mineral oil. M16 should be equilibrated in a 5% CO_2 incubator before making the overlayed droplets. Place morula in 37°C CO_2 incubator.

6. While morula are in the CO_2 incubator, ES cells are prepared (these cells should be grown without selection agents for the previous 48 h in order not to interfere with the donor embryo on aggregation). Trypsinize a single well from a 6- or 24-well plate, but reduce trypsinization time to one-half or one-fourth of normal (1.5–2.5 min). The object is to recover small clumps of ES cells (3–12 cells), so the following trypsin treatment cells are not disaggregated as vigorously. Add the trypsin suspension to 10 mL growth media, and centrifuge cells for 5 min at 1000g. Resuspend cell pellet in 5 mL PBS, and recentrifuge. Aspirate off the supernatant, and resuspend cell pellet in a small volume of either growth media minus selection agents, or M16 + 10% FCS and pen/strep (both of these have been used with success). Plate the cell suspension on a 35- or 60-mm cell-culture dish, and return to the CO_2 incubator for approx 1 h to allow feeder cells to attach to dish. After 1 h, tip the dish gently, remove the media containing unattached ES cells, and transfer to another tissue-culture dish. These cells are then ready for aggregation, and can be left in a 37°C CO_2 incubator until morula are ready.

7. On a 60-mm dish, place five to six droplets of CO_2-equilibrated M16 + 10% FCS and pen/strep. Quickly overlay with mineral oil to maintain pH. Using a Hungarian darning needle (probably other suitable instruments can be used—the important thing is to have a very smooth point, which can be used to make small depressions in TC plastic. The point of our needle is approximately the diameter of a thumbtacks' point). Make 6–8 depressions/droplet using a smooth, circular motion and consistent, but light downward pressure. We have successfully used Falcon tissue-culture plates for this purpose. With practice, one can make depressions that do not crack the dish, but are deep enough to keep the morula from sloshing out during routine movement. It is also important that the sides of these depressions are smooth, and do not "snag" morula or ES cells.

8. Under the microscope, place 1 morula/depression. After this is done, place dispersed ES cells. This also takes some practice, but ideally, one wants to place 2–8 cells/morula into each depression. They will eventually "aggregate" in the bottom. Be careful when moving the dish around, since cells can easily slosh out. Place in 37°C CO_2 incubator overnight.

9. The next morning, observe the aggregates under the microscope. Hopefully, one sees a high percentage of morula that have gone on to develop into blastocysts, have a blastocoel, and have expanded considerably. Generally, it is advisable to segregate the nice-looking blastocysts from the bad or, alternatively, just to discard the less well-developed ones: few pregnancies usually result from the transfer of such undeveloped embryos. Transfer the blastocysts into mineral oil-covered M2; these are now ready for uterine transfer.

10. Uterine transfer is performed on d 7 (**Fig. 2**) on anesthetized hosts. A convenient anesthetic agent is Avertin, at a dose of ~300 µL/25–30 g mouse. After mouse is unconscious and unresponsive to a toe pinch, wipe the back down with 70% EtOH, and make a lateral incision just distal to the end of the ribs, approx 0.5–1 cm below the spine. Locate the fat pad attached to ovary, and gently pull out. Grab fat pad (avoid contact with ovary/uterus if possible) with bulldog clip, exposing proximal end of uterus.

11. Using a pulled and flamed glass transfer pipet (Drummond Scientific Pyrex, 150 mm long, 0.75 mm interior diameter. Pull and fire-polish to interior diameter of 110–150 µm). Suck up enough mineral oil to fill pipet halfway. Wipe off the end with a Kimwipe, wet some M2 into the tip of the pipet (some people like to put bubbles in here to visualize the interface between oil/medium when transferring), and suck up 10–15 blastocysts into the pipet.

12. Using forceps and a 27-gage needle, pierce the uterus, and make a passageway for the transfer pipet: make sure the needle is in the lumen of the uterus. Pull the needle out, and while still holding uterus (gently!) with forceps, insert transfer pipet plus blastocysts into uterus (approx 3–5 mm). Apply pressure until you see the bubbles at the interface reach the uterine opening, avoiding the transfer of bubbles. Remove transfer pipet, and gently place the fat pad back into the peritoneal cavity. Place one to two autoclips on the outer incision, and place mouse in warm environment until recovered from anesthesia.

3.5. Maintenance and Characterization of Knockout Mouse Lines

The transfer recipients should be expected to deliver pups anywhere from 15 to 18 d following transfer. The degree of chimerism, or contribution from ES cells, is grossly determined by coat color: in the example given here, where the R1 ES cell line is used together with CD-1 donor embryos, nonchimeric donors would be albino, whereas a fully ES-derived pup would be agouti (*see* **Fig. 3**). Most chimeric mice will display graded levels of chimerism, manifested as random coat spotting. With heavy contributions from ES cells in the

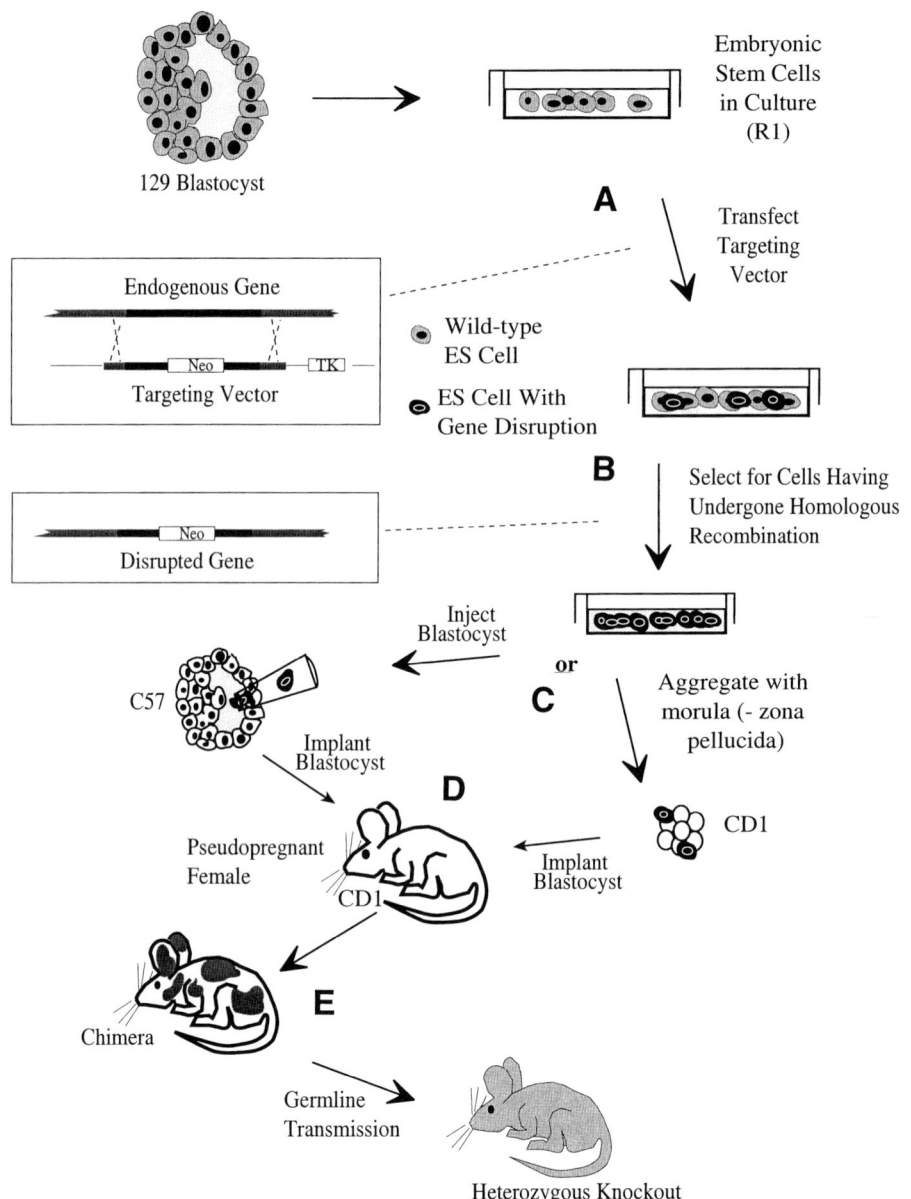

Fig. 3. General schematic for the development of knockout mice. Beginning at the top left, ES cells are derived from the inner cell mass of the 129 mouse blastocyst, and accommodated to tissue culture. ES cells are then transfected with the gene targeting vector (**A**). Schematic of crossover event seen at left. Cells having undergone homologous recombination are then selected for with G418 and ganciclovir (**B**), and characterized by Southern blotting. Correctly targeted cells are then used to create chimeras by blastocyst microinjection or morula aggregation (**C**), followed by uterine transfer of manipulated embryos (**D**). Host females then give birth to chimeric mice

chimeras, one should also see a skewing of sex ratios favoring males, since the R1 cell line is male-derived.

1. Once the chimeras reach sexual maturity (~6 wk for females, ~8 wk for males), they can be set up as a mating pair with another mouse. The strain of mouse chosen may be influenced by the eventual application for the knockout, since there are significant interstrain variations in many physiological parameters (*see* **Note 10**). A common choice is to breed chimeras to the C57Bl/6 strain, in which case transmission from ES-derived gametes is seen as agouti offspring, whereas transmission from CD-1-derived gametes is seen as a black coat color. Transmission of the disrupted allele is determined by taking tail biopsies from agouti offspring, and performing Southern blots in the same manner as for screening targeted ES cells. Approximately 1 cm of the tail is removed from a weanling mouse, placed in 400 μL DNA isolation buffer, and digested overnight with shaking at 60°C. The following morning, add 75 μL 8 *M* potassium acetate plus 500 μL chloroform, and briefly vortex. Place on ice for ~30 min. Samples can then be microfuged for 5–8 min, and the upper, aqueous layer carefully removed and placed in another tube. To the aqueous sample, add 2 vol of ethanol, and gently rock tube back and forth 5–10 times. DNA will precipitate out, and can be spooled on a glass capillary pipet. Spooled DNA is dipped in 70% ethanol, briefly air-dried, and resuspended in 100–200 μL TE buffer. Fifteen to 30 μL of this DNA sample can be used for Southern blot analysis.
2. Once germline transmission of the targeted allele is verified, these heterozygous mice can be intercrossed to generate homozygous knockout mice. The offspring expected from a heterozygote:heterozygote intercross (as long as the gene of interest is on an autosome) would be 25% wild type, 50% heterozygous knockout, and 25% homozygous knockouts.
3. A variety of techniques may be used to assess the physiological implications of adrenergic receptor gene knockout. To date, many of these have focused on the cardiovascular, metabolic, and behavioral effects *(1–6)*, although analysis should be tailored to the existing knowledge of that particular receptor subtype.

4. Notes

1. In the initial strategy of screening genomic libraries for the gene of interest, it is important to use a DNA source that is isogenic with the ES cell line, which will eventually be transfected. Nonisogenic DNA is a poorer substrate for homolo-

---

Fig. 3. *(continued)* (**E**), which can be crossed to a nonagouti mouse strain (e.g., C57Bl/6). Germline transmission is verified by the acquisition of the agouti coat color; transmission of the targeted gene (theoretically in 50% of all agouti mice) is further verified by genotyping such agouti mice. C57 and CD1 refer to the mouse strain of origin used as donor blastocysts and morula, respectively.

gous recombination than is isogenic DNA, owing to sequence polymorphisms. If ES cell lines from other mouse strains are used (e.g., C57Bl/6), the genomic libraries should also be derived from that source.

2. Prior to screening the mouse genomic library, one must first calculate the minimum number of library clones to be screened, which is based on the (1) average insert size within the library and (2) desired probability of finding the given sequence represented *(7)*. For a phage genomic library containing an average insert size of 10 kb and a 99% desired probability, one would need to screen ~1.5 × 10^6 recombinants. For bacterial artificial chromosome (BAC) libraries containing an average insert size of 100 kb and a 99% probability, one would need to screen ~1.5 × 10^5 recombinants. Both of these approximations assume a genome size of 3 × 10^9 bp.

3. Whereas full sequence information of the genomic locus is helpful for targeting vector construction, such a task is not always necessary and can be overly taxing on resources in some cases. Importantly, the junctions between the mouse genomic sequence and the disrupting gene cassettes should be known and verified by sequencing before such a vector is used for gene disruption. In many cases, a working knowledge of the pertinent exon–intron boundaries and a detailed restriction map are sufficient.

4. The design of the targeting vector can greatly influence the ultimate success of the gene disruption technique, and is pertinent to both the efficiency of homologous recombination and the ultimate functional effect on the disrupted gene. With regard to the efficiency of homologous recombination, there are minimal requirements for the length of homology between the targeting vector and the endogenous gene. The most common strategy for designing a targeting vector is to introduce a drug resistance gene cassette (such as *neo*^{r}; *see* **Fig. 1**) in a functionally important domain within the gene of interest. The drug resistance cassette is then flanked on either side by mouse genomic sequences. The flanking "arms" of homologous sequence then serve as the substrate for homologous recombination in ES cells: a crossover event on each arm (i.e., a double crossover) will result in incorporation of the neo^r gene within the gene of interest. Frequently, many investigators also utilize an added step of including a negative selectable marker, such as viral thymidine kinase (TK), at the end of either arm. Since a double crossover would eliminate incorporation of the negative selectable marker cassette (i.e. TK) into the genome, this allows for selection against random, nonhomologous insertion events by use of viral-specific nucleoside analogs, such as ganciclovir. Nonhomologous integrants that retain the TK cassette will be rendered sensitive to ganciclovir. A typical targeting vector and strategy are shown in **Fig. 1**.

5. With respect to the flanking arm lengths, it has been shown that targeting vectors containing short arms with <1 kb homology perform poorly, and that dramatic increases in targeting frequency occur with increasing arm length, up to a distance of ~6 kb *(10)*. Thus, it is prudent to design and incorporate the homologous arms such that >1 kb homology is present on both, if possible. Commercially

available neo and HSV tk gene cassettes are becoming available (Stratagene) and possess multiple flanking restriction sites, enabling one to insert flanking genomic mouse sequences with relative ease, once a detailed restriction map becomes available. It is also important to design the targeting vector such that a unique restriction endonuclease site exists between the homologous sequence and vector DNA, for linearization prior to electroporation (linearized targeting vector seen in **Fig. 1**). Furthermore, it is important that the targeting vector DNA be fully linearized at the correct restriction site: the strategy employed here, known as gene replacement, can be misapplied if DNA is either incompletely linearized or linearized at the wrong site. In such cases, one may favor gene insertional events over replacement, leading to improper gene targeting.

6. With respect to effectiveness of gene disruption, it is important to consider the functionally important domains of the gene to be targeted, and either delete these during construction of the targeting vector or ensure that the disrupting resistance cassette is positioned to interrupt such domains. There have been instances where despite successful gene disruption, functional gene products could be made by either alternatively spliced transcription products or even truncated proteins from the disrupted allele *(11)*. Such "leaky" knockouts may be useful for analysis of genetic hypomorphs, but in most cases are problematic in the clarification of gene function. In the example provided, the majority of coding sequence is actually omitted from the targeting vector, ensuring that homologous recombination results in a null allele (**Fig. 1**). AR genes are particularly amenable to such an approach owing to their small gene size and frequently intronless structure.

7. Embryonic stem cells are pluripotent cell lines derived from the inner cell mass of the developing blastocyst (**Fig. 3**). As such, their pluripotent state must be carefully maintained; cells grown too densely, infrequent media changes or incomplete media, or excessive passage in culture can cause these cells to lose their pluripotency. These cells must have daily media changes to guard against this propensity to differentiate, and are always grown in the presence of LIF. Many laboratories have found it useful to perform karyotype analysis on targeted ES cell clones. The extensive culture in vitro, passaging, and selection pressures frequently result in ES cell clones with an abnormal karyotype. These clones almost certainly have lost the ability to form viable chimeras, and can end up costing the investigator needless time and money. Karyotype analysis can be performed by a qualified lab or can be contracted out (e.g., Genome Systems).

8. It is frequently desirable to identify targeted clones using two different probes, both outside of the targeting vector sequence (in order not to detect any nonhomologous integrants), preferably flanking the 5' and 3' arms of homology. Such probes must be able to detect specifically only the targeted locus, and should be chosen in accordance with the restriction maps of the endogenous locus vs targeted locus so that restriction fragment-length differences between the two loci are readily apparent by Southern blot analysis (**Fig. 1**). Alternatively, one can genotype by polymerase chain reaction (PCR) if the assay has been proven reliable. A general strategy is to design PCR primer pairs such that one member

lies within the drug resistance cassette, but the other is specific to a region just outside of one the homologous arms. Frequently, however, such a strategy results in PCR products that are 2 kb or greater, which can be difficult to amplify reliably. Southern blotting of genomic DNA is slower, but more reliable.

9. Great care and vigilance must be practiced with the mice used as donors and hosts. A regular day/night cycle must be maintained, and mice must be handled as minimally as possible. This includes excessive noise. Males used for donor embryos should not be overworked; poor mating performance will result if they are used for mating on consecutive days. Also keep close tabs on the performance of vasectomized males, culling the poor performers and replacing with new ones. Female recipient mice should also be monitored in the same way.

10. Despite the fact that most ES cell lines are derived from the 129 mouse strain, few laboratories maintain their knockouts on this background. One hundred twenty-nine mice are notoriously difficult to breed, are more disease-prone, and possess significant behavioral deficits that may cloud the interpretation of gene knockouts *(12)*. Given these drawbacks, it is common to begin backcrossing chimeras to other mouse strains in an attempt to "dilute" the 129 genetic contribution. Approximately 20 backcross generations are accepted as the standard for creating a congenic mouse strain by such methods, although in practice, many labs are performing less backcrosses. One can accelerate the process of backcrossing by either superovulation of prepubescent females, or by so-called speed congenic methods *(13)*.

References

1. Link, R. E., Stevens, M. S., Kulatunga, M., Scheinin, M., Barsh, G. S., and Kobilka, B. K. (1995) Targeted inactivation of the gene encoding the mouse alpha 2c-adrenoceptor homolog. *Mol. Pharmacol.* **48,** 48–55.
2. Susulic, V. S., Frederich, R. C., Lawitts, J., Tozzo, E., Kahn, B. B., Harper, M. E., et al. (1995) Targeted disruption of the beta 3-adrenergic receptor gene. *J. Biol. Chem.* **270,** 29,483–29,492.
3. Rohrer, D. K., Desai, K. H., Jasper, J. R., Stevens, M. E., Regula, D. J., Barsh, G. S., et al. (1996) Targeted disruption of the mouse beta1-adrenergic receptor gene: developmental and cardiovascular effects. *Proc. Natl. Acad. Sci. USA* **93,** 7375–7380.
4. Link, R. E., Desai, K., Hein, L., Stevens, M. E., Chruscinski, A., Bernstein, D., et al. (1996) Cardiovascular regulation in mice lacking alpha2-adrenergic receptor subtypes b and c. *Science* **273,** 803–805.
5. MacMillan, L. B., Hein, L., Smith, M. S., Piascik, M. T., and Limbird, L. E. (1996) Central hypotensive effects of the alpha2a-adrenergic receptor subtype. *Science* **273,** 801–803.
6. Cavalli, A., Lattion, A., Hummler, E., Nenninger, M., Pedrazzini, T., Aubert, J., et al. (1997) Decreased blood pressure response in mice deficient of the alpha 1b-adrenergic receptor. *Proc. Natl. Acad. Sci. USA* **94,** 11,589–11,594.

7. Sambrook, J., Fritsch, E. F., and Maniatis, T. (1989) Analysis and cloning of eukaryotic genomic DNA, in *Molecular Cloning: A Laboratory Manual*. Cold Spring Harbor Laboratory, Cold Spring Harbor, NY.
8. Hogan, F. (1986) *Manipulating the Mouse Embryo*. (Hogan, B., Constantini, F., and Lacy, E., eds.), Cold Spring Harbor Laboratory, Cold Spring Harbor, NY.
9. Wood, S. A., Allen, N. D., Rossant, J., Auerbach, A., and Nagy, A. (1993) Non-injection methods for the production of embryonic stem cell-embryo chimeras. *Nature* **365,** 87–89.
10. Hasty, P., Rivera-Perez, J., and Bradley, A. (1991) The length of homology required for gene targeting in embryonic stem cells. *Mol. Cell. Biol.* **11,** 5586–5591.
11. van Deursen, J., Ruitenbeek, W., Heerschap, P., Jap, P., ter Laak, H., and Wieringa, B. (1994) Creatine kinase (CK) in skeletal muscle metabolism: a study of mouse mutants with graded reduction in muscle CK expression. *Proc. Natl. Acad. Sci. USA* **91,** 9091–9095.
12. Gerlai, R. (1996) Gene-targeting studies of mammalian behavior: is it the mutation or the background genotype? *Trends Neurosci.* **19,** 177–181.
13. Markel, P., Shu, P., Ebeling, C., Carlson, G. A., Nagle, D. L., Smutko, J. S., et al. (1997) Theoretical and empirical issues for marker-assisted breeding of congenic mouse strains. *Nat. Genet.* **17,** 280–284.
14. Soriano, P., Montgomery, C., Geske, R., and Bradley, A. (1991) Targeted disruption of the c-src proto-oncogene leads to osteopetrosis in mice. *Cell* **64,** 693–702.

IV

RECEPTOR AND SECOND MESSENGER ANALYSIS

19

Development of Antibodies to Adrenergic Receptors

Suleiman W. Bahouth

1. Introduction

Adrenergic and other G-protein-coupled receptors (GPCRs) are transmembrane molecules with complex secondary and tertiary organization intended to generate the binding pocket within the transmembrane loops *(1)*. Although there are numerous differences in the primary amino acid composition of the various receptor classes, these molecules consist of seven intermembranous hydrophobic cores encompassing half of the molecule. Within the connecting extracellular loops, there is extensive glycosylation in the amino terminus and in the second extracellular loop in some GPCRs, as well as extensive disulfide bridging between the loops *(2)*. The intracellular carboxy-terminus is palmitoylated to generate a fourth miniloop *(3)*. In addition, the carboxy-terminus and the third cytoplasmic loop are extensively phosphorylated as a consequence of agonist binding *(4)*.

The development of antibodies to GPCRs has been hampered by the constraints mentioned above and by difficulties associated with purifying low-abundance membrane proteins. Moreover, even when sufficient quantities of purified β_1- and β_2-adrenergic receptors (β_1-AR and β_2-AR) became available, these molecules were not sufficiently immunogenic owing to their high hydrophobicity *(5)*. Consequently, a strategy involving the development of antibodies to discrete peptides within these receptors proved very promising *(6)*. Antipeptide antibodies offered several advantages that were particularly useful for probing the structures of an ever-expanding family of adrenergic receptors. These advantages include ease of preparation, selectivity to a par-

From: *Methods in Molecular Biology, vol. 126: Adrenergic Receptor Protocols*
Edited by: C. A. Machida © Humana Press Inc., Totowa, NJ

ticular receptor subtype, and straightforward immunopurification and neutralization (7).

The first step in the production of a good antipeptide antibody is the choice of the peptide sequence for immunization. These domains are usually found within the amino- or carboxy-terminal regions or in the extra- or intracellular loops connecting the transmembranal domains. Hydrophilic strings of about 10–15 amino acids are chosen, and each is tested to ascertain that none of their amino acids are glycosylated, palmitoylated, or phosphorylated. If a subtype selective antibody is desired, the sequences that are unique to a given receptor subtype are selected for further analysis (8). The choice of a carboxy-terminal domain should be considered first, because a peptide derived from this domain is often more mobile than the rest of the molecule and can be coupled easily to the carrier. Peptides derived from C-termini are coupled at their amino ends, whereas peptides derived from the amino-terminus are coupled at their carboxy-termini, in order to approximate their configuration in the native antigen. To achieve this organization, it is preferable to add an N-terminal linker cysteine to carboxy-terminal-derived peptides and to add a C-terminal linker cysteine to amino-terminal-derived peptides. If a naturally occurring cysteine is present, it is preferable that it is positioned either in the first two or the last two amino acids of a given sequence. Selection of a sequence from one of the extracellular or intracellular loops requires the use of algorithms for determining the hydrophobicity of a given protein, such as the Hopp and Woods (9) or Kyte and Doolittle (10) algorithms in the Macvector™ computer program, for example. Generally, the peptide sequence containing the highest number of hydrophilic residues is usually the most immunogenic (11).

The optimal length of a peptide antigen is between 10 and 15 residues. Shorter peptides may not crossreact with the native antigen, and longer peptide sequences may be less soluble. Peptide synthesis is the most efficient and cost-effective method. Synthesis of at least 20 mg of high-performance liquid chromatography (HPLC) purified peptide with purity >90% will provide sufficient quantities for immunization, purification, and characterization of the antiserum. The author strongly recommends that you biotinylate at least 10 mg of the peptide, particularly if you desire to prepare peptide-specific affinity-purified IgGs. Request the biotinylation of a portion of the newly synthesized peptide when you place your order, because the synthesis facility is more experienced and it is less expensive to biotinylate at time of synthesis. In some instances, it is preferable to acetylate and block the α-amino group of a peptide that is derived from the N-terminal domain so that the peptide mimics the native antigen as closely as possible (**Fig. 1**).

In the event that a longer sequence is desired and synthesis is not feasible, polymerase chain reaction (PCR) amplification followed by molecular cloning

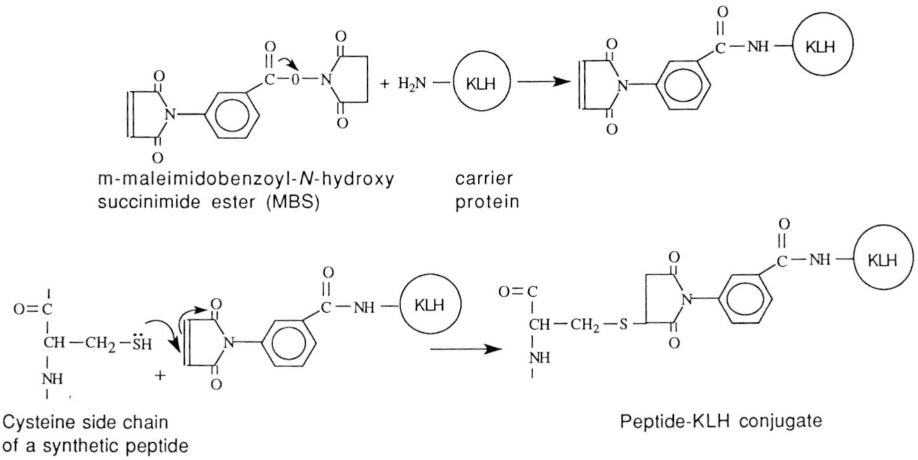

Fig. 1. Coupling of synthetic peptides via cysteines. The reaction on top shows the derivatization of KLH with the bifunctional reagent MBS. The reaction below shows the coupling of the cysteinyl residue in a peptide to the carrier protein KLH via MBS *(14)*.

into a suitable epitope containing vector (such as glutathione or others) is the easiest. The desired peptide sequence is amplified, purified by affinity chromatography, and isolated after cleavage of the epitope tag.

The most important requirement in a carrier for immunization is for the carrier molecule to remain soluble after coupling and to contain a high number of coupling lysines. The carrier most often used in coupling peptides derived from adrenergic receptor sequences has been keyhole limpet (*Megathura crenulata*) hemocyanin (KLH). This protein is quite soluble, more immunogenic than albumin, and is not expressed in mammals.

The choice of adjuvant depends on the rules and regulations that govern the use of animals in the research institution. If the institution forbids the use of Freund's adjuvant because of the distress and pain it causes in the animal, there are other commercially available, less-painful adjuvants that have been used successfully for the production of antibodies *(6)*. However, the cost of Freund's adjuvant is significantly lower than other adjuvants. This chapter will describe the use of Freund's adjuvant and the Ribi Adjuvant System™ *(12)*. Freund's adjuvant is a two-component system, composed of Complete Freund's Adjuvant (CFA), which contains heat-killed *Mycobacterium tuberculosis* bacilli and Incomplete Freund's Adjuvant (IFA). The Ribi Adjuvant System™ is a trivalent composition of monophosphoryl Lipid A, trehalose dicorynomycolate, and cell-wall skeleton *(13)*.

The choice of animal for production of antibodies depends on the amount of antiserum desired as well as the species from where the adrenergic receptor sequence is derived. In most instances, rabbits have been used for this purpose because they are genetically divergent from human, rat, and murine species. New Zealand white rabbits are generally the best strain because they have a life-span of about 6 yr, and between 30 and 50 mL of whole blood can be obtained at each bleed. At least two rabbits should be immunized with the antigen to have a reasonable chance of obtaining an animal with a good titer. These rabbits should be ordered early in order to prepare sufficient amounts of preimmune serum from each animal. Moreover, an identification tag must be affixed in their ears for identification.

To detect specific adrenergic receptor antibodies in rabbit serum, two types of assays have been used. The first assay is an indirect enzyme-linked immunosorbent assay (ELISA) to determine the presence of antipeptide antibodies in the serum and to determine their titer. This assay is rapid (<1 d) and can be used as an initial screen to determine the quality of the antiserum. The second assay is a Western blotting procedure to characterize the antibody specificity for denatured or native adrenergic receptor proteins. In this assay, membranes prepared from cells that either express or do not express the desired adrenergic receptor or its subtype are probed with various dilutions of preimmune and immune serum. The immunoreactive peptides are detected by the methodology outlined in Chapter 20. Based upon the signal-to-noise ratio of the immunoblots, a general idea of the crossimmunoreactivity of the antiserum will be known. Moreover, initial estimates of the size of the immunoreactive protein, the titer of the antiserum, and the potential selectivity and subtype specificity of the antiserum will be learned from this technique.

Once an antiserum has been characterized for specificity and selectivity by Western blotting, the antiserum is further purified by immunoaffinity chromatography to isolate peptide or epitope-specific immunoglobulins. The potential usefulness of these IgGs as probes for adrenergic receptor distribution in tissues by indirect immunofluorescence microscopy is determined by the methodology outlined in another chapter of this series. In addition, the usefulness of the IgG fraction or preferably the peptide-purified fraction in immunoprecipitating a particular adrenergic receptor subtype will be described in this chapter.

2. Materials

2.1. Production of Antisera to Adrenergic Receptors

2.1.1. Conjugation of Peptide to Hemocyanin

1. KLH (Sigma [St. Louis, MO], H-2133).
2. *m*-Maleimido-benzyl-*N*-hydroxysuccimide (MBS) (Sigma, M-8759).
3. Sephadex G-25 from Amersham Pharmacia Biotech (Piscataway, NJ).

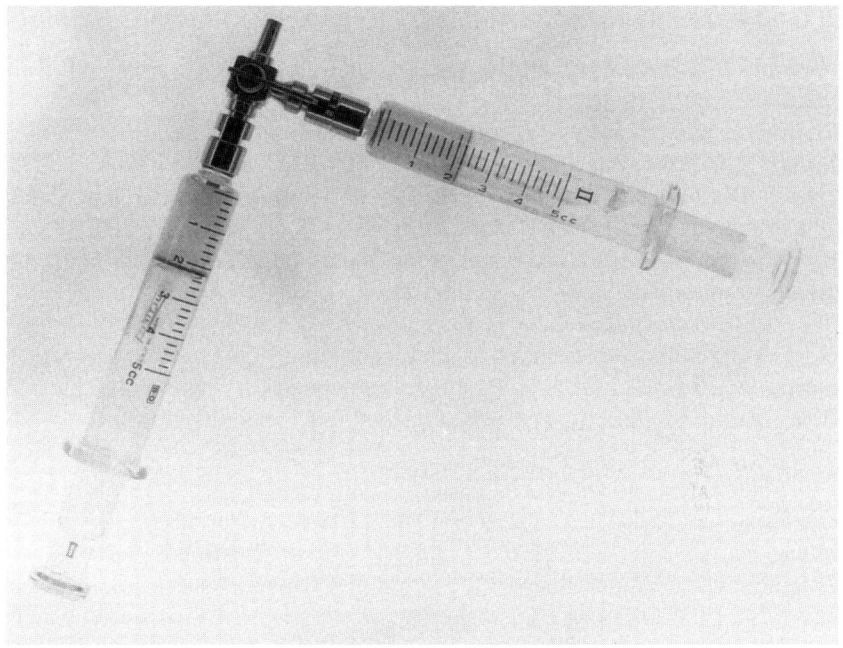

Fig. 2. Two 5-mL glass Luer lock syringes connected by means of a double-ended Luer lock are used to prepare the antigen-Freund's adjuvant emulsions.

4. 0.1 *M* sodium phosphate, pH 6.8 prepared as follows:
   a. Stock A: 1.42 g Na_2HPO_4 (anhydrous) in H_2O to 100 mL.
   b. Stock B: 1.56 g $NaH_2PO_4 \cdot 2_2O$ in H_2O to 100 mL.
      Add 51 mL of stock A to 49 mL of stock B.
   c. Disposable columns (Kontes [Vineland, NJ], 420160).
   d. 20 m*M* ethylenediamine tetra-acetic acid (EDTA) in 0.1 *M* sodium phosphate buffer, pH 6.8. Prepared by dissolving 0.74 g of $Na_2 \cdot EDTA \cdot 2H_2O$ in 100 mL 0.1 *M* sodium phosphate, pH 6.8.

2.1.2. Immunization of Rabbits and Collection of Serum

1. New Zealand White female rabbits, 2.7–3.1 kg, from J. R. H. Biosciences (Lenaxa, KS) or other vendors.
2. CFA and IFA (Sigma, F-4258 and F-5506, respectively).
3. Double-ended locking hub connector (luer lock) and 5 mL metal-luer glass syringes are from Poper and Son (**Fig. 2**).
4. Ribi Adjuvant System; Ribi Immunochem (Hamilton, MT), cat. no. R-730.
5. Serum separation tubes, Vacutainer with a wax-plug, Becton-Dickinson (Franklin Lakes, NJ).
6. Butterfly infusion set, 19 or 21 gage, Desert Medical (Sandy, VT), 38-5386.

2.1.3. Characterization of Antisera by ELISA and Western Blotting

1. Immulon 2 (Dynatech (Chantilly, VA) #011-010-3450) or equivalent 96-well titer plates or strips.
2. Microtiter plate reader equipped with a 405-nm filter.
3. Borate-buffered saline (BBS) composed of 0.015 M $Na_2B_4O_7$, 0.12 M NaCl, pH 8.5. Dissolve sodium borate (5.7 g) and NaCl (7 g) in 1 L of water, and then adjust to pH 8.5 with NaOH.
4. Blocking buffer: BBS containing 0.05% Tween 20, 1 mM EDTA, 0.4% bovine serum albumin (BSA), 0.02% sodium azide.
5. Sep-Pak C18 cartridge (Waters, Marlboro, MA).
6. Goat antirabbit IgG-conjugated to alkaline phosphatase (various sources).
7. p-nitrophenyl phosphate (pNPP) substrate; Sigma Fast pNPP (N1891).
8. Reagents for Western blotting are described in Chapter 20.

2.1.4. Affinity Purification of IgG Fractions

1. Diethylaminoethyl (DEAE)-Affi Gel Blue, Bio-Rad (Hercules, CA) 153-7307.
2. Econo-Pac 10 disposable chromatography column, Bio-Rad 732-10101.
3. Loading buffer: 20 mM Tris-HCl, 28 mM NaCl, 0.02% NaN_3, pH 8.0, Dissolve Tris (2.42 g), NaCl (1.64 g), and sodium azide 0.2 g in 1 L of water, then adjust the pH to 8.0 with NaOH.
4. Elution buffers: 20 mM Tris-HCl, 50 mM NaCl, 0.02% NaN_3pH 8.0, 20 mM Tris, and 100 mM NaCl, pH 8.0.
5. Streptavidin-agarose, Life Technologies (Grand Island, NY), 15942-014.
6. Dialysis buffer: 50% glycerol, 25 mM Tris-HCl, 0.1 M NaCl, pH 7.4. Prepared by dissolving NaCl (5.8 g), Tris-HCl (3 g), and 500 mL glycerol in water to 1 L and then adjust the pH to 7.4 with NaOH.

2.2. Immunoprecipitation of Adrenergic Receptors

1. Phosphate-free Dulbecco's Modified Eagles Medium (DMEM) Gibco-BRL.
2. $^{32}PiO_4$ radionucleotide, New England Nuclear (Boston, MA), NEX-053.
3. Lysis buffer: 20 mM Tris-HCl, pH 7.4, 150 mM NaCl, 5 mM EDTA, 50 mM sodium fluoride (NaF), 40 mM sodium pyrophosphate, 50 mM potassium phosphate, 10 mM sodium molybdate, 2 mM orthovanadate, 1% Triton X-100, 0.1% sodium dodecyl sulfate (SDS), leupeptin (5 μg/mL), aprotinin (5 μg/mL), benzamidine (0.1 mg/mL), bacitracin (0.1 mg/mL), 0.6 mM dithiothreitol (DTT), and 1 mM phenyl methyl sulfonyl fluoride.

    Lysis buffer is prepared as a two-component buffer. The first component is a 2X phosphatase inhibition buffer that is prepared by dissolving NaCl (35 g), Na_2·EDTA·$2H_2O$ NaF (4.2 g), sodium pyrophosphate (35.7 g), KH_2PO_4 (13.6 g), and sodium molybdate $2H_2O$ (4.8 g) per 1 L of solution. This solution is stable at 4°C for a month. The other components are added immediately before use from stock solutions to 5 mL of 2X phosphatase inhibition buffer, followed by adjusting the volume of the final solution to 10 mL with water. To 5 mL of 2X phosphatase add sodium orthovanadate 50 mM (0.4 mL), Triton X-100 20% (0.5 mL), SDS 10% (0.1 mL), leupeptin 10 mg/mL (5 µL), aprotinin 10 mg/mL (5 µL), benzamidine (1 mg), bacitracin (1 mg), DTT 1 M (6 µL), and Tris-HCl 1 M, pH 7.4, (0.2 ml) in 5 mL of water.

4. Immunoprecipitation (IPA) buffer with detergents: 20 mM Tris-HCl, pH 8.0, 150 mM NaCl, 5 mM EDTA, 1% (v/v) Triton X-100, and 0.2% (w/v) SDS. Dissolve Tris-HCl (2.4 g), NaCl (8.8 g), EDTA (1.9 g), Triton X-100 (10 mL), SDS (2 g), and adjust the pH to 8.0 with NaOH.
5. IPA buffer without detergents does not contain Triton X-100 and SDS.
6. Protein A-agarose, Life Technologies, 15918-014.
7. Microtube transfer pipe, Research Products International Corp. (Mt. Prospect, IL), 147500.
8. 2X Laemmli sample buffer: 125 mM Tris-HCl, pH 6.8, 4% SDS, 5% sucrose, 0.03% bromophenol blue. Add 20 µL of 1 M dithiothreitol/mL before use.

3. Methods

3.1. Production of Antisera to Adrenergic Receptors

3.1.1. Conjugation of MBS to KLH

The first step in this procedure is to determine the solubility of your synthetic peptide and to prepare the Sephadex G-25 column.

1. Test the solubility of the peptide in 20 mM EDTA, 0.1 M sodium phosphate, pH 6.8. If the peptide is not soluble in this buffer, try the buffers described in **Subheading 4.** (*see* **Notes 1** and **2**).
2. Swell the Sephadex G-25 in 0.1 M sodium phosphate, pH 6.8, and load into the disposable polyethylene column.
3. Prior to coupling the peptide to KLH, the lysine side chain in KLH should be derivatized with a bifunctional crosslinking reagent, such as MBS (**Fig. 1**). This step will introduce a bifunctional crosslinking reagent that will crosslink the thiol group of cysteine in the peptide with lysine in the carrier molecule *(14,15)* (*see* **Note 3**).
4. Dissolve 4 mg of KLH in 0.25 mL of 0.1 M sodium phosphate buffer, pH 6.8, in a 1.7-mL Eppendorf tube.
5. Dissolve 0.7 mg MBS in 70 µL of dimethyl formamide (DMF).
6. Add MBS in dimethyl formamide by pipeting 7- to 10-µL aliquots to the KLH solution, and vortex vigorously, to avoid increasing the local concentration of DMF >30%, since KLH is insoluble at these high DMF concentrations.
7. After adding all of the MBS, the solution must appear clear with no visible precipitate. Mix the MBS-KLH solution on a rotator for 30 min at room temperature.
8. Load the MBS-KLH onto the Sephadex G-25 column, elute with 0.1 M sodium phosphate, pH 6.8, and collect 5–6 drops/fraction into Eppendorf tubes (*see* **Note 4**).

9. Read A_{280} of each fraction, and collect the first protein peak that elutes (this peak will appear yellow from KLH). The first peak appears in the void volume and contains the MBS–KLH complex, and the second peak contains free MBS. Pool the first peak into a 13-mL polypropylene tube.

3.1.2. Conjugation of Peptide to Carrier Protein

1. Dissolve 5 mg of synthetic peptide in 1 mL of 20 mM EDTA, 0.1 M sodium phosphate, pH 6.8, or suitable buffer.
2. Add the peptide to the MBS–KLH solution and mix by inversion for 4 h at room temperature. **No precipitate should form during the coupling procedure**.
3. If you wish to determine the efficiency of peptide coupling to KLH, add between 10^4 and 10^5 cpm of ^{125}I-Bolton-Hunter coupled peptide to the mixture, prepared as described below.
4. After coupling, dialyze the peptide–KLH mixture at 4°C against 3 L of PBS overnight. In the morning, repeat the dialysis for an additional 4 h.
5. If **step 3** was performed, calculate the recovery by dividing recovered cpm by added cpm.
6. Aliquot about 1 mg of peptide/Eppendorf tube, lyophilize, and store at –70°C.

3.1.3. Measurement of Coupling Efficiency

The derivatization of a peptide with the ^{125}I-Bolton-Hunter reagent (*see* **Note 5**) is as follows:

1. Dissolve 0.1 mg of peptide in 1 mL 0.1 M sodium phosphate buffer, pH 8.5.
2. Evaporate 10 μL of ^{125}I-Bolton-Hunter reagent under N_2, and add 10 μL of peptide (~1 μg), agitate once, and incubate on ice for 1 h.
3. Hydrate a Sep-Pak C18 cartridge with water, and then pass the peptide sample through the cartridge.
4. Wash the cartridge with water to remove unincorporated Bolton-Hunter reagent.
5. Elute the ^{125}I-Bolton-Hunter-coupled peptide with 50% isopropanol in 0.5-mL aliquots *(16)*.
6. Count by Cerenkow to identify the ^{125}I peak, then lyophilize the pooled fractions, and rehydrate in 0.2 mL of 0.1 M sodium phosphate buffer, pH 8.5.
7. Count 1 μL of this solution in a γ-counter and add the desired cpm to the coupling reaction.

3.2. Immunization of Rabbits and Collection of Serum

3.2.1. Collection and Preparation of Preimmune Rabbit Serum

1. Prior to immunizing the rabbits, preimmune serum is collected from each rabbit to neutralize nonspecific immunoreactive proteins.
2. Blood is collected from the ear vein of the rabbit.

3. Dilate the ear vein by swiping the external region of each ear with a xylene-saturated gauze.
4. Remove the red top cap from the Vacutainer tube.
5. After dilation, insert the needle into the vein using the butterfly handle as a guide.
6. When blood begins to flow in the catheter, connect its end into the Vacutainer tube, and allow the blood to flow directly inside the tube.
7. Fill as many Vacutainer tubes as possible from each ear.
8. When blood flow from the ear vein ceases, place gauze around the ear and remove the needle. Press firmly on the ear to prevent blood loss and keep pressing for at least 2 min until blood flow from the site of injection ceases.
9. If you need more blood, use the other ear and follow **steps 3–8**.
10. Allow the collected blood from both ears to stand in the Vacutainer tube for 1–4 h at room temperature in order for the clot to form (*see* **Note 5**).
11. Spin the tube at 2000*g* for 15 min to separate clotted blood from the serum.
12. Collect the serum fraction from above the wax plug, and store in 2-mL aliquots at −70°C.

3.2.2. Preparation of Freund's Immunogen

1. For immunization with Freund's adjuvant (*see* **Note 6**), rehydrate 1 mg of peptide–KLH with 2 mL of PBS, and place into a 5-mL glass syringe equipped with a metal luer lock (**Fig. 2**).
2. Shake CFA, and draw 2 ml into a separate 5-mL glass syringe equipped with a Luerlock.
3. Connect the KLH–peptide-containing syringe to the CFA syringe by means of a double-ended Luer lock or a 3-way stopcock (**Fig. 2**).
4. Force the mixtures back and forth from one syringe to the other repeatedly for at least 40 times.
5. When the mixture is white and homogeneous, divide the solution equally into both syringes, and separate them from the Luer lock. Attach a 26-gage, 0.5-in. needle to each syringe.
6. Restrain the animal.
7. Inject 0.1 mL of the adjuvant-antigen emulsion into 20 intradermal sites.
8. Inject the contents of the other syringe into a second rabbit as in **step 7** above.
9. Return the rabbits to their cages, and monitor their skin reaction for 4 wk.
10. After 4–6 wk, rehydrate 1 mg of KLH–peptide conjugate in 2 mL of PBS. Emulsify the KLH-peptide conjugate with 2 mL IFA by the same method used to emulsify CFA.
11. Restrain the rabbits.
12. Administer the first booster immunization using IFA by injecting 0.1 mL of emulsion into 20 intradermal sites in the vicinity of the first injection sites.
13. After 7–10 d, collect a blood sample from the ear vein and prepare the serum as described in **Subheading 3.2.1.**
14. If further boosters are required, KLH–peptide–CFA emulsions may be administered at 2- to 3-wk intervals.

3.2.3. Preparation of Ribi-Adjuvant System™ Immunogen

1. Dissolve the entire contents of the RAS adjuvant vial into 2 mL of warm PBS by vigorous vortexing for at least 5 min.
2. Rehydrate 1 mg of the KLH–peptide antigen with 2 mL of reconstituted RAS adjuvant, and vortex vigorously for 5 min.
3. Divide into two tuberculin syringes (1 mL/syringe) equipped with a 23-gage 1 in. needle.
4. Inject the antigen into the popliteal lymph node first using this procedure:
   a. Restrain the rabbit, and sedate with an anesthetic approved by the Animal Care Facility. We inject 0.25 mL/kg of ketamine/xylazine (35/5) im for this purpose.
   b. After the animal is sedated, shave the hair from the inner portion of both hindlegs behind the knee joint with an electric clipper.
   c. The popliteal lymph node is a lentil-size lymph node behind the knee of the rabbit (*see* **ref. *17*** for details).
   d. Manually, squeeze the region behind the knee, and palpitate to feel the node in this region.
   e. Administer 0.1 mL of KLH–peptide RAS adjuvant into each popliteal region of rabbits.
5. Administer 0.3 mL intradermally (0.05 mL in each of six sites).
6. Administer 0.4 mL im (0.2 mL into each hindleg)
7. Finally, inject 0.1 mL sc in the neck region.
8. Repeat this procedure after 4, 8, and 12 wk.
9. After 3 d from the third injection, collect a blood sample from the ear vein and prepare the serum as described in **Subheading 3.2.1.**

3.3. Characterization of Antisera by ELISA and Western Blotting

3.3.1. Enzyme Linked Immunosorbent Assay (ELISA)

1. The first step in characterizing the serum is to determine if it contains antibodies directed against the peptide antigen by ELISA *(18)*.
2. Dissolve 0.5 mg of peptide/5 mL of water, and place 50 µL into each well of an Immulon™ 2 or equivalent 96-well titer plate (*see* **Note 7**).
3. Precipitate the peptide onto the well by adding to each well an equal volume of absolute ethanol *(19)*.
4. After 2 h, rinse each plate over the sink by filling the wells with deionized water. Flick the water into the sink, and rinse with water two more times.
5. Fill each well with blocking buffer, and incubate for 30 min at room temperature.
6. Rinse the plate three times with water.
7. Prepare appropriate dilutions of the immune serum in blocking buffer. Begin with a 1:20 dilution and increase the dilution by increments of 10 until 1:200,000 dilution is obtained.
8. Add 50 µL of diluted antiserum to each of the coated wells, and incubate for 3 h at room temperature.

9. Wash the plate by rinsing three times with deionized water, followed by a rinse in blocking buffer and subsequently three times with water.
10. Dissolve 20 µL of goat antirabbit IgG conjugated to alkaline phosphatase in 10 mL blocking buffer.
11. Using a multichannel pipet, add 50 µL/well for 2 h at room temperature.
12. Wash the plate three times with deionized water, once with blocking buffer, and three times with deionized water.
13. Prepare the *p*NPP reagent by dissolving one tablet of Sigma-fast *p*NPP in 5 mL of deionized water.
14. Using a multichannel pipet, add 75 µL of *p*NPP substrate, and incubate for 1 h at room temperature. Hydrolysis may be monitored visually by the presence of a yellow color.
15. Read *p*NPP hydrolysis using a spectrophotometer at A_{405} absorbance.

3.3.2. Western Blotting

1. The quality of the antiserum is initially determined by Western blotting.
2. Prepare membranes from cells that express the adrenergic receptor subtype and from those that do not express the receptor of interest. (These must be prepared and frozen at −70°C prior to starting this procedure) (*see* **Note 8**).
3. To prepare the membranes under denaturing conditions:
   a. Dissolve 20 µg of membrane protein in 30 µL of 2X Laemmli sample buffer containing 20 m*M* DTT for 30 min at 37°C.
   b. Add iodoacetamide to a final concentration of 40 m*M*, and incubate for 1 h at room temperature in the dark to alkylate the free sulfhydryl groups.
   c. In addition, electrophorese sample containing prestained marker to determine the approximate size of the immunoreactive band.
4. After the bromophenol dye exits from the gel, transfer the proteins in the gels electrophoretically to nitrocellulose and visualize the antigen–antibody complexes using the methodology described in Chapter 20.
5. A good antipeptide antibody reacts with a single protein band in the reduced and alkylated membranes that corresponds to the expected size of the adrenergic receptor under study. The immune complexes will be evident in membranes that express the antigen, but not in the nonexpressing membranes. An example of these analyses is shown in **Fig. 3**. The data in **Fig. 3** is an immunoblot analysis of sera from two rabbits immunized with a peptide derived from the sequence between 407 and 419 in the carboxy-terminal domain of the human β_1-AR. The sera were probed against 15 (lane 1) and 45 µg (lane 2) of rat fat membranes that express β_1-AR *(20)*. The serum from rabbit SB-03A crossreacted with multiple bands, whereas the serum of rabbit SB-03B crossreacted with a single band of 68 kDa, which corresponds to the size of the β_1-AR in these membranes. Therefore, serum SB-03A was not analyzed further. The intensity of the 68-kDa complex generated by SB-03B in lane 2 was three-fold higher than that in lane 1, indicating that the SB-03B antibody reacted quantitatively with increasing amounts of the antigen.

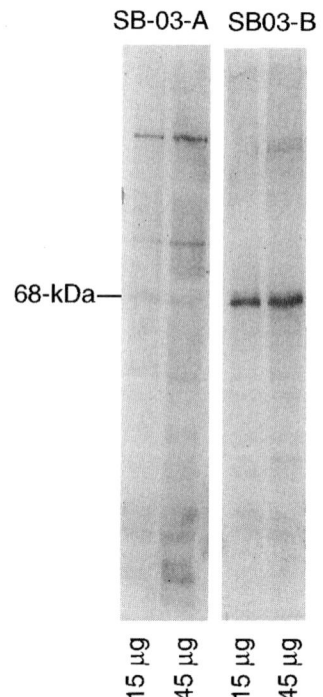

Fig. 3. Characterization of sera by immunoblotting. Rat fat cell membranes were subjected to SDS-PAGE and electrophoretically transferred to nitrocellulose. Lane 1 in each blot contained 15 μg of protein, and lane 2 contained 45 μg of membrane protein. Each blot was probed separately with a 1:100-dilution of rabbit serum that was obtained from two rabbits immunized with a β_1-AR-derived peptide-KLH emulsions prepared in the Ribi Adjuvant System™. The blots were developed with ^{125}I-goat antirabbit IgG and autoradiographed with a Kodak X-AR 5 film for 6 h. The migration of the 68-kDa mol-wt standard is shown on the right.

3.4. Affinity Purification of IgG Fractions

3.4.1. Preparation of IgG Fractions by DEAE Affi-Gel Blue

1. Dialyze the rabbit serum sample overnight at 4°C against column loading buffer composed of 20 mM Tris, 28 mM NaCl, pH 8.0.
2. Prepare a column of DEAE Affi-gel blue using 4 mL of bed volume/mL of initial antiserum volume (read the manufacturer's label for proper ratio of gel to serum) (*see* **Note 9**).
3. Load ~10 mL of the gel into an Econo-Pac 10 column, and if the gel has not been previously used, wash it as described by the manufacturer.
4. Wash the gel with at least 10 bed volumes of column loading buffer.

5. Apply the dialyzed serum to the column, close the column, and rock for 1 h.
6. Wash the column with 3 bed volumes of loading buffer at its own flow rate.
7. Elute the IgG sample first with 3 bed volumes of 20 m*M* Tris-HCl, 50 m*M* NaCl, pH 8.0, and then with 3 bed volumes of 20 m*M* Tris-HCl, 100 m*M* NaCl, pH 8.0.
8. Collect these samples into fractions of ~ 1/5 bed volumes.
9. Determine the presence of the IgG peak in each fraction as follows:
   a. Dissolve 10 μL of each fraction in 10 μL of 2X Laemmli buffer with 20 m*M* dithiothreitol.
   b. Boil the samples for 5 min.
   c. Load onto a 10% sodium dodecylsulphate-polyacrylamide gel electroporesis (SDS-PAGE) with an additional sample of SDS-PAGE mol-wt markers.
   d. Stain the gel with Coomassie blue, and then destain to visualize the protein.
   e. The IgG fraction will yield polypeptides of approx 25 and 55-kDa on SDS-polyacrylamide gels.
10. Combine the peak IgG fractions, and concentrate by ammonium sulfate precipitation (optional).
11. Saturated ammonium sulfate, pH 7.2, is added to the pooled fractions until the final concentration of ammonium sulfate is 40%.
12. The material is stirred overnight at 4°C and then centrifuged at 4000 rpm (2500*g*) in a swinging-bucket rotor for 30 min at 4°C.
13. Resuspend the precipitate in water to about 50% of its initial volume.
14. Dialyze this solution once against 2 L PBS, and then repeat against 5 L of PBS. This procedure concentrates the dialysate to about 1/5 its original volume.
15. Centrifuge and aliquot the supernatant into 0.1-mL aliquots and store at –70°C.

3.4.2. Affinity Purification of IgG Fractions

3.4.2.1. PROCEDURE FOR COUPLING OF BIOTINYLATED PEPTIDE TO STREPTAVIDIN-AGAROSE COLUMN (SEE **NOTE 10**)

1. Dissolve 5 mg of biotinylated peptide in a 5-mL column buffer composed of 25 m*M* Tris-HCl, 0.3 *M* NaCl, pH 7.4.
2. Load 2 mL of streptavidin–agarose into a disposable polypropylene column with a 10-mL reservoir, and wash the resin with 10 bed volumes of column buffer.
3. Add the peptide to the resin, cap the column at both ends, and mix by inversion for 1 h at room temperature.
4. Drain the column, and wash the resin with 5–10 bed volumes of column buffer. We routinely obtain coupling of 2 mg/mL of resin using this procedure.

3.4.2.2. PURIFICATION OF RABBIT IgG BY AFFINITY CHROMATOGRAPHY

1. Apply 10 mL of serum to the equilibrated packed resin.
2. Cap the column at both ends, and rock for 30 min to 1 h at room temperature.
3. Drain the column by gravity flow, and then wash it with 3 volumes of column buffer.

4. Antipeptide IgG is eluted with a buffer consisting of 0.2 *M* glycine-HCl, 0.3 *M* NaCl, pH 2.7, into collecting tubes containing solid Tris-HCl. We collect 8 drops into each tube containing 15 mg of solid Tris-HCl.
5. The fractions containing IgG, are determined by OD_{280}.
6. Immediately pool these fractions, and dialyze them against 50% glycerol, 25 m*M* Tris-HCl, 0.1 *M* NaCl, pH 7.4, and stored at −70°C.
7. The yield of affinity-purified IgG is between 100 and 200 µg/10 mL of immune serum.

3.5. Immunoprecipitation of β_1-AR

3.5.1. Labeling ATP Pools in Cells with $^{32}PiO_4$

1. All the procedures described in this section are to provide information concerning immunoprecipitation to researchers who have formal training in the use and handling of radioisotopes (*see* **Note 11**).
   a. These procedures must be performed behind adequate shielding, and appropriate protection for the researcher must be in force at all times.
   b. All experiments should be performed following the guidelines set forth by the Institution's Radiation Safety Office.
   c. After use, discard all the radioactivity in properly shielded and labeled containers for disposal.
2. Subconfluent cultures in the 10-cm plate are incubated for 1 h in DMEM-deficient phosphate at 37°C in a CO_2 incubator.
3. The medium is then supplemented with ^{32}P-orthophosphate (1 mCi/1 mL of medium) and incubation is continued for 4 h at 37°C in a CO_2 incubator.

3.5.2. Conjugation of Preimmune IgG to Protein A-Agarose

1. While the ATP pools are being labeled, conjugate protein A-agarose to preimmune IgG as follows.
2. Take 600 µL of protein A-agarose slurry, and centrifuge to pellet.
3. Resuspend the agarose in 800 µL of IPA buffer with detergents, and mix by vortexing.
4. Sediment the resin by centrifugation at 15,000*g* for 30 s, and carefully aspirate the aqueous phase with a microtransfer pipet ensuring than no resin is aspirated.
5. Repeat the washing procedure three times.
6. After the third wash, add 400 µL of IPA buffer and 120 µL of preimmune rabbit serum, and incubate for 2 h with full-rotation at 37°C.
7. Wash the resin with IPA buffer with detergents three times as described earlier.
8. This is a recipe for five immunoprecipitations and should be modified depending on the number of samples.

3.5.3. Cell Lysis and Removal of Nonspecific Immunoreactants

1. After labeling, wash the cells twice with ice-cold PBS and lyse the cells directly in lysis buffer. Use 1 mL of lysis buffer to lyse the cells on a 10-cm plate.

2. Cell lysates are transferred into a screw-cap Eppendorf tube and boiled for 5 min with intermittent vortexing.
3. After boiling, the tube is centrifuged in a microfuge equipped with an aerosol-guarded lid at 15,000g for 5 min.
4. The supernatant is obtained, and 4 μL are removed from each tube to measure ^{32}P by liquid scintillation counting.
5. You can remove an additional 4 μL and subject them to SDS-PAGE to ascertain the equal incorporation of ^{32}P.
6. Perform a Bradford assay on 5–10 μL in duplicate *(21)*. **Be careful. The sample is extremely radioactive**.
   a. Calculate the concentration of protein in μg/μL.
   b. Use the entire lysate in the tube with the lowest amount of protein.
   c. Use volume of lysate from the other tubes equivalent to the amount of protein contained in the tube with the lowest protein amount.
7. Incubate the appropriate volume of supernatant with protein A-agarose-coupled to preimmune serum IgG (80 μL of resin/mL of lysate) at room temperature on a rotating platform for 2 h.
8. Centrifuge the resin, and place the supernatant into a new tube.

3.5.4. Conjugation of β_1-AR Antiserum with Protein A-Agarose

1. During the preimmune serum incubation period, couple the antiadrenergic receptor IgG to protein A-agarose.
2. Using 300 μL of protein A-agarose slurry, centrifuge, and wash three times with 600 μL IPA buffer.
3. After the third wash, add 200 μL IPA buffer with 25 μg of antireceptor IgG (for β_1-AR immunoprecipitation, we use SB-03 IgG *[20]*) and incubate for 2 h with full rotation at 37°C.
4. After conjugation, wash the resin with IPA buffer with detergents three times as described earlier.
5. This is a recipe for five immunoprecipitations and should be modified depending on the number of samples.

3.5.5. Immunoprecipitation of Adrenergic Receptors

1. Incubate ~40 μL of antiadrenergic receptor IgG-coupled to protein A-agarose slurry with the lysate overnight in a rotator at 4°C.
2. During the interim period, prepare a 10% SDS-polyacrylamide gel.
3. In the morning, pour the stack and then proceed to wash the resins as follows:
   a. Spin the tube at 15,000g for 1 min and carefully aspirate the supernatant.
   b. This supernatant may be used for another immunoprecipitation if desired.
   c. Wash the resin with 1 ml ice-cold IPA buffer with detergents by vortexing for 1 min, and then spin in microfuge at 15,000g for 1 min.
   d. Carefully discard the supernatant making sure that you do not aspirate any resin.
   e. Repeat this procedure four times (total is five washes).

4. Add 1 mL ice-cold IPA buffer without detergents to the resin, vortex for 1 min, centrifuge for 1 min, and then discard supernatant.

5. Add 100 μL of 2X Laemmli sample buffer containing 20 mM DTT, vortex for 1 min, and then boil for 5 min.

6. Prepare SDS protein markers (5 μL + 95 μL 2X Laemmli + 20 mM DTT), and boil for 5 min.

7. Vortex the tubes with resin vigorously, and concentrate the resin by centrifugation at 15,000g for 2 min.

8. Using a protein loading pipet, aspirate all of the aqueous phase without removing any of the resin, and load into electrophoresis wells.

9. After loading all the samples, electrophorese until the bromophenol dye exits from the gel. Stain and destain, and then dry the gel.

10. Expose the dried gel to a Kodak XAR-5 film at −70°C.

4. Notes

1. Determine first if your peptide is soluble in 20 mM EDTA, 0.1 M sodium phosphate, pH 6.8. If not, try to dissolve in PBS, **or** in 0.1 M sodium borate, pH 9.0, and finally test its solubility in 1 M sodium acetate, pH 4.0.

2. The procedure described in **Subheading 3.1.1.** for conjugating MBS to KLH also works well for conjugating albumin to MBS.

3. If your peptide does not contain a lysine, you may substitute glutaraldehyde for MBS as the bifunctional coupling reagent. In this case, do not add a cysteine to the amino acid sequence. Test the solubility of your peptide in 0.15 M NaCl, 2.5 mM Na$_2$HPO$_4$, pH 7.4, or in 0.1 M sodium borate, pH 9.0. Conjugate your peptide directly to KLH in a 13-mL glass tube as follows: Dissolve 4 mg of KLH in 2 mL of suitable buffer, then add 5 mg of peptide, and vortex gently. Slowly add 1 mL of 0.3% glutaraldehyde in buffer, and mix rigorously at room temperature for 2 h. The solution will turn yellow, and a precipitate will form. Collect the precipitate by centrifugation at 10,000g for 5 min, and wash it once in the same buffer without glutaraldehyde. Homogenize the pellet in 1 mL buffer, and emulsify with an equal volume of Freund's adjuvant.

4. Identify the first peak of KLH-MBS by its opaque yellow color. The second peak, which contains unreacted MBS, is colorless.

5. Because of its cost, ^{125}I-Bolton-Hunter reagent is used to estimate the coupling efficiency when you have more than one peptide to conjugate. An alternative method, if the peptide contains a tyrosine, is to iodinate it by the chloramine-T method with ^{125}I, as described *(22)*. To access the dilated ear vein, hold the wings of the butterfly catheter between the thumb and index finger. Guide the needle in a plane parallel to the ear surface into the vein and collect blood from the end of the catheter into a tube. If you are not using a Vacutainer tube with a wax plug, allow the collected blood to coagulate for 4 h so that the clot will form and retract. Then spin and collect the yellow-like aqueous serum on top.

6. Obtain approval from the Institutional Animal Care and Use Committee before any animals are injected with Freund's adjuvant. The sites of the 20 intradermal

injections of immunogen in Freund's adjuvant run parallel to the spinal cord in the back of the rabbit. Inject at 10 sites on each side at about 1.5 cm from the spinal cord. To inject the Ribi adjuvant into the popliteal lymph node, guide the needle about 1 cm into the back of the knee in between the muscles, and inject 0.1 mL in the general vicinity of the lymph node. The lymphatic popliteal node will absorb the antigen adjuvant complex from the site of injection. Although two rabbits are the minimum to immunize, in our experience, some antipeptide antibodies require immunization in multiple rabbits in order to obtain a good antiserum *(6)*.

7. The surfaces of Immunlon 2 plates are positively charged and will avidly bind to peptides. By adding ethanol, you will significantly increase the coupling efficiency of the peptide (and proteins) significantly.

8. The cell membranes used in the immunoblotting procedure in **Subheading 3.3.2.** may be obtained from a variety of sources. Whenever possible, express the cDNA of interest by transient expression in HEK 293 cells, and prepare membranes from cells expressing your cDNA and from cells transfected with the empty mammalian expression vector. This procedure results in membranes with a high concentration of your adrenergic receptor of interest. Denatured membranes are first reduced with DTT to break the cysteinyl disulfide bridges, and then alkylated with iodoacetamide to prevent refolding and to sharpen the individual protein bands.

9. The DEAE-Affi gel blue method will rarely produce a pure IgG unless used with another method, such as the ammonium sulfate precipitation or immunoaffinity purification on protein A-agarose. The ammonium sulfate method is used when a pure IgG is not crucial. Otherwise, protein A-agarose is preferred. The binding of IgG (or serum) to protein A-agarose is identical to the methodology used to bind the biotinylated peptide to streptavidin agarose. Similarly, the washing and elution of the IgG from the resin (with 0.2 M glycine) is as described in **Subheading 3.4.2.**

10. If your peptide is not biotinylated, it may be coupled to an Affigel 10 resin (Bio-Rad) or to a cyanogen bromide column by the methodology described in Chapter 19 of *Methods in Molecular Biology, vol. 59, Protein Purification Protocols (23)*. The use of Affigel is preferred, because the resin has a 10 residue spacer, arm which improves the presentation of the immobilized peptide to the antibody. The flowthrough fraction that does not bind to the peptide should not be immunoreactive on the Western blot. Read the OD_{280} of the affinity-purified IgG fractions in a cuvet with a narrow path (volume 0.4 mL). Prepare a neutralized glycine solution in Tris-HCl to zero the instrument, then load the entire fraction into the cuvet and read its absorbance. Clean the cuvet thoroughly, and read the OD_{280} for subsequent fractions. Dialyze the fractions with the highest absorbance, and store in small aliquots at $-70°C$.

11. Use screw-cap tubes with O-rings in every immunoprecipitation procedure. These tubes come in a variety of sizes from Sarstedt Inc. (Newton, NC) or as Cryo vials from Nalge (Hudson, OH) and other vendors. Since the cpm of ^{32}P-adrenergic

receptor bound to antireceptor IgG-protein A-agarose conjugates is low, it is possible to use click-seal Eppendorf tubes to wash the resin. We prefer to use colored Eppendorf tubes for resin washing to improve the contrast between the white resin and the tube.

References

1. Strader, C. D., Sigal, I. S., and Dixon, R. A. F. (1989) Structural basis of β-adrenergic receptor function. *FASEB J.* **3,** 1825–1832.
2. Fraser, C. M. (1989) Site-directed mutagenesis of β-adrenergic receptors: Identification of conserved cysteine residues that independently affect ligand binding and receptor activation. *J. Biol. Chem.* **264,** 9266–9270.
3. O'Dowd, B. F., Hantowich, M., Caron, M. G., Lefkowitz, R. J., and Bouvier, M. (1989) Palmitoylation of the human β$_2$-adrenergic receptor: Mutation of CYS341 in the carboxyl tail leads to an uncoupled nonpalmitoylated form of the receptor. *J. Biol. Chem.* **264,** 7564–7569.
4. Sibley, D. R., Benovic, J. L., Caron, M. G., and Lefkowitz, R. J. (1987) Regulation of transmembrane signaling by receptor phosphorylation. *Cell* **48,** 913–922.
5. Malbon, C. M., Moxham, C. P., and Brandwein, H. J. (1991) Antibodies to β-adrenergic receptors, in *The Beta-Adrenergic Receptors* (Perkins, J. P., ed.), Humana, Totowa, NJ, pp. 181–261.
6. Wang, H. Y., Lipfert, L., Malbon, C. C., and Bahouth, S. W. (1989) Site-directed anti-peptide antibodies define the topography of the β-adrenergic receptor. *J. Biol. Chem.* **264,** 14,424–14,431.
7. Bahouth, S. W., Wang, H. S., and Malbon, C. C. (1991) Immunological approaches for probing receptor structure and function. *Trends Pharmacol. Sci.* **12,** 338–343.
8. Getzoff, E. D., Tainer, J. A., Lerner, R. A., and Geysen, H. M. (1988) The chemistry and mechanism of antibody binding to protein antigens. *Adv. Immunol.* **43,** 1–98.
9. Hoop, T. P. and Woods, K. R. (1983) A computer program for predicting protein antigenic determinants. *Mol. Immunol.* **20,** 483–489.
10. Kyte, J., and Doolittle, R. F. (1982) A simple method for displaying the hydropathic character of a protein. *J. Mol. Biol.* **157,** 105–132.
11. Hoop, T. P. and Woods, K. R. (1981) Prediction of protein antigenic determinants from amino acid sequences. *Proc. Natl. Acad. Sci. USA* **78,** 3824–3828.
12. Deeb, B. J., DiGiacomo, R. F., Kunz, L. L., and Stewart, J. L. (1992) Comparison of Freund's and RIBI adjuvants for inducing antibodies to the synthetic antigen (TG)-AL in rabbits. *J. Immunol. Methods* **152,** 105–113.
13. Smith, D. E., O'Brien, M. E., Palmer, V. J., and Sudowski, J. A. (1992) The selection of an adjuvant emulsion for polyclonal antibody production using a low-molecular weight antigen in rabbits. *Lab. Anim. Sci.* **42,** 599–601.
14. Green, N., Alexander, H., Olson, A., Alexander, S., Shinnick, T. M., Sutcliffe, J. G., et al. (1982) Immunogenic structure of the influenza virus hemagglutinin. *Cell* **28,** 477–487.

15. Lui, F.-T., Zinnecker, M., Hamask, T., and Katz, D. H. (1979) New procedures for preparation and isolation of conjugates of proteins and a synthetic copolymer of D-amino acids and immunochemical characterization of conjugates. *Biochemistry* **1,** 690–697.

16. Miller, J. J., Schultz, G. S., and Levy, R. S. (1984) Rapid purification of radioiodinated peptides with Sep-Pak™ reverse phase cartridges and HPLC. *Int. J. Pep. Protein Res.* **24,** 112–122.

17. Sigel, M. B., Sinha, Y. N., and Vanderlaan, W. P. (1983) Production of antibodies by inoculation into lymph nodes. *Methods Enzymol.* **93,** 3–12.

18. Engvall, E., and Pelman, P. (1971) Enzyme-linked immunosorbant assay (ELISA):Quantitative assay of immunoglobulin G. *Immunochemistry* **8,** 871–879.

19. Biesiegel, U., Schneider, W. J., Goldstein, J. L., Anderson, R. G. W., and Brown, M. S. (1981) Monoclonal antibodies to the low density lipoprotein receptor as probes for study of receptor-mediated endocytosis and genetics of familial hypercholestrolemia. *J. Biol. Chem.* **256,** 11,923–11,931.

20. Bahouth, S. W., Gokmen-Polar, Y., Coronel, E. C., and Fain J. N. (1996) Enhanced desensitization and phosphorylation of the β_1-adrenergic receptor in rat adipocytes by peroxovanadate. *Mol. Pharmacol.* **49,** 1049–1057.

21. Greenwood, F. C. Hunter, W. M., and Glover, J. S. (1963) The preparation of ^{125}I-labelled human growth hormone of high specific activity. *Biochem. J.* **89,** 114–123.

22. Bradford, M. M. (1976) A rapid method for quantitation of microgram quantities of protein utilizing the principle of protein-dye binding. *Anal. Biochem.* **72,** 248–254.

23. Jack, G. W. and Beer, D. J. (1996) Immunoaffinity chromatography, in *Methods in Molecular Biology, vol. 59: Protein Purification Protocols* (Doonan, S., ed.), Humana Press, Totowa, NJ, pp. 187–196.

20

Western Blot Detection of Adrenergic Receptors

Suleiman W. Bahouth

1. Introduction

The power of polyacrylamide gel electrophoresis in resolving individual proteins in complex mixtures permitted the application of follow-up techniques to identify the separated components. A popular follow-up technique is immunoblotting (often referred to as Western blotting), which is used to identify specific antigens recognized by antibodies *(1)*. Membrane-bound or purified adrenergic receptors (ARs) are solubilized with sodium dodecyl sulfate (SDS) and reduced with dithiothreitol (DTT). After solubilization, proteins are separated by electrophoresis on SDS-polyacrylamide gels *(2)*. The separated proteins are electrotransferred "blotted" from the gel onto the surface of an inert membrane, such as nitrocellulose, polyvinylidene difluoride (PVDF), or nylon membrane. The proteins may be transferred in a tank or by a semidry apparatus to the membrane support *(1,3–5)*. The proteins are immobilized on the surface to become accessible to interaction with immunodetection reagents. To probe for specific antibody–antigen reactions, excess binding sites are blocked by immersing the membrane in a blocking solution containing albumin, nonfat milk, or detergent, such as Tween-20 *(6)*. The membrane is probed with an appropriate dilution of the primary antibody, washed, and the antibody–antigen complexes are tagged with a secondary anti-IgG antibody coupled either to horseradish peroxidase (HRP) or alkaline phosphatase (AP) *(1,7)*. Chromogenic or luminescent substrates are then used for visualization *(1,7,8)*. This technique provides qualitative information concerning the expression of adrenergic receptors in a given cell or tissue and the size of the immunoreactive species. If the purified antigen is available, quantitative information, such as the concentration of the adrenergic receptor in membranes, may be

From: *Methods in Molecular Biology, vol. 126: Adrenergic Receptor Protocols*
Edited by: C. A. Machida © Humana Press Inc., Totowa, NJ

calculated from a standard curve. In this case, an ^{125}I-iodinated secondary anti-IgG antibody is used to generate a standard curve of cpm incorporated/ng of protein, which is then used to deduce the concentration of antigen in the unknown samples subjected to a parallel analysis *(9)*.

2. Materials

2.1. Electrophoretic Transfer of Proteins from Gels to Nitrocellulose

2.1.1. Electrophoresis Reagents

1. 2X-Laemmli sample buffer: Dissolve Tris base (1.51 g), SDS (4 g), and sucrose (5 g) in ~80 mL of water, and adjust the pH to 6.8 by careful addition of 1 *M* HCl. Weigh 30 mg of bromophenol blue, add, and stir for 30 min. Adjust the volume to 100 mL using a graduated cylinder. Filter solution through a 0.22-µm bottle top filter. Divide the solution into 1 mL aliquots and store at –20°C for up to 1 yr. On the day of the experiment, prepare a fresh solution of 1 *M* DTT by dissolving 0.154 g DTT in 1 mL of deionized water. Add 20 µL of 1 *M* DTT to 1 mL of 2X Laemmli before use.
2. Prestained markers monochrome (Life Technologies, Grand Island, NY), Kaleidoscope (Bio-Rad, Hercules, CA): If the prestained marker is lyophilized, add 0.5 mL deionized water (with 1 m*M* DTT) and boil for 5 min. Divide into 5-µL aliquots in an Eppendorf tube, and store at –70°C.

2.1.2. Protein Blotting in a Tank System

1. Transfer buffer: This buffer is usually prepared as a 10X stock by dissolving Tris base (30.3 g), glycine (144 g), and SDS (10 g) in a liter of water. There is no need to adjust the pH of this solution. To prepare a liter of transfer buffer, mix 100 mL of 10X transfer buffer with 700 mL of cold deionized water in a beaker. Then slowly add 200 mL of methanol, mix, and store cold until use. Prepare 1 L for each miniblot.
2. 0.5 *M* 2-iodoacetamide (Sigma, St. Louis, MO): Weigh 92.5 mg, and add to 1 mL of deionized water.
3. Electroblotting apparatus: Mini Trans-Blot™ electrophoretic transfer cell, for 8 × 10 cm gels; Trans-blot™ for 14 × 14 cm gels, from Bio-Rad or Hoefer, Pharmacia Biotech, San Francisco, CA.
4. Transfer membrane: Nitrocellulose BA 85, 0.45 µm from Schleicher and Schuell (Keene, NH). On the day of the experiment, cut an 8 × 10 cm or 14 × 14 cm piece, place in a pyrex tray, add ~500 mL of deionized water, and boil for 1 min. Nitrocellulose will explode if heated to dryness. Therefore, never leave the blot unattended while it is being boiled. Remove the tray from the heating plate when finished. Then remove the nitrocellulose from the tray, and place in cold transfer buffer before use. PVDF membranes must be prewetted in absolute ethanol to activate the membrane, then rinsed in water and placed in cold transfer buffer before use.

5. Fiber pads: Scotch-Brite (3M) or equivalent.
6. Whatmann 3MM filter paper precut to 8 × 10 cm or 14 × 14 cm sheets. You need six sheets for each transfer.
7. Cooling reservoir or coil: Store the coil at −20°C. If the Trans-blot system comes with a cooling reservoir, fill this reservoir with water and freeze.
8. Power supply: Model 200/2.0 from Bio-Rad or equivalent.
9. Latex gloves: These must be worn throughout this procedure.
10. Indelible marker.

2.1.3. Protein Blotting with Semidry Systems

Use a semidry transfer unit (Hoefer, Bio-Rad, or Sartorius, Edgewood, NY). Additional supplies are described in **Subheading 2.1.2.**

2.2. Immunoprobing with Conjugated Secondary Antibody

1. Tris-buffered saline-Tween 20 (TBST): Dissolve Tris base (12.1 g), NaCl (17.5 g), and Tween 20 (2 g) in ~1.8 L of water, and adjust the pH to 8.0 by careful addition of 6 M HCl in a fume hood. Prepare 2 L with water.
2. Blocking solution (2% milk in TBST): To 500 mL of TBST, add 10 g of nonfat dry milk (Carnation, Nestle, Inc., Solon, OH). Dissolve the milk by stirring/warming (setting #2 in heating plate) for about 4 h. Allow the solution to equilibrate to room temperature before storing at 4°C. Discard after 2 mo.
3. Primary antibody specific for protein of interest.
4. Horseradish peroxidase- (HRP) linked F(ab')$_2$ fragment (from donkey), Amersham Pharmacia Biotech (Piscataway, NJ) cat. # NA9340. This solution is stable for 2 yr at 4°C.
5. Enhanced chemluminescence (ECL) detection reagents, ECL™-kit (Amersham Pharmacia Biotech, RPN 2106) Renaissance™ enhanced luminol (DuPont-NEN, Boston, MA; NEL-102), LumiGLO™ (Kirkegaard & Perry, Gaithersburg, MD), and others.
6. Thermo Impulse Sealer Model 210B-1, National Instrument Company (Austin, TX).
7. Heat-sealable plastic bag (Dazey Corp. [New Century, KS] model 50358).
8. Hematology/chemistry mixer, Fisher Scientific, Pittsburgh, PA 14-059-346.
9. Calf AP–anti-Ig conjugate (Organon Teknika/Cappel [Durham, NC], Kirkegaard and Perry): Prepare as a 0.5 mg/mL solution, divide into 10-µL aliquots, and store at −70°C. Dilute as directed by manufacturer in TBST before use.
10. 50 mM glycine, pH 9.6. Dissolve 3.755 g glycine in 1 L of water, pH to 9.6 with 10 N NaOH.
11. *p*-Nitroblue tetrazolium chloride (NBT), various sources. Prepare 1 mL of 1 mg/mL solution in glycine buffer immediately before use.
12. 5-Bromo-4-chloro-3-indoyl phosphate (BCIP), various sources. Prepare 0.2 mL of 5 mg/mL solution in dimethyl formamide.
13. 1 M MgCl$_2$. Dissolve 203.3 MgCl$_2$·6H$_2$O in 1 L of water.

14. BCIP/NBT Phosphatase Substrate™, Kirkegaard and Perry, product no. 50-81-08.
15. [^{125}I]protein A (NEX-146) or [^{125}I]goat antirabbit IgG (NEX-155) from DuPont-NEN Nuclear. For monoclonal antibodies (MAbs), use [^{125}I]protein G.
16. Phosphate-buffered saline (PBS): Dissolve anhydrous disodium hydrogen ortho-phosphate (11.5 g), sodium dihydrogen phosphate (2.96 g), and sodium chloride (5.84 g) in a liter of water and adjust the pH to 7.5.
17. 10% Albumin in PBS: Dissolve 10 g of bovine serum albumin (BSA) (IgG-free, fraction V) in 100 mL of PBS and readjust the pH to 7.5 with 1 *N* NaOH. Divide into 10-mL aliquots and store at –20°C for up to 1 yr.

3. Methods

3.1. Electroblotting from Gels to Nitrocellulose

3.1.1. Electrophoresis of the Protein and Prestained Marker Samples

1. Dissolve the membrane sample (10–50 µg) in an equal volume of 2X Laemmli sample buffer containing 20 m*M* DTT by vortexing for 1 min.
2. Incubate the solubilized protein at 37°C for 30 min to reduce the disulfide bridges in the protein with DTT (*see* **Note 1**).
3. Add iodoacetamide to a final concentration of 40 m*M* (1.6 µL/20 µL of 2X Laemmli) to alkylate the free sulfhydryl groups, and incubate in the dark for 45 min.
4. In the interim period, using 5 µL of prestained marker, add 15 µL of 2X Laemmli buffer with DTT, and boil for 5 min.
5. After the iodoacetamide step, your samples may appear yellow. Add to each sample 1–2 µL of saturated Tris base solution to restore the blue color of the tracking dye.
6. Load the samples on a 10% polyacrylamide (8 cm × 10 cm × 0.75 mm) minigel and electrophorese at 100 V until the bromophenol blue dye exits from the gel. For 14 × 14 cm gels, electrophorese at 30 mA/gel (constant current) until the tracking dye enters the running gel. Then increase the current to 35 mA/gel.

3.1.2. Electrophoretic Transfer in a Tank System

1. During gel electrophoresis, boil the nitrocellulose, and prepare the cold transfer buffer as described in **Subheading 2.**
2. Transfer the proteins from gel to nitrocellulose immediately on completion of SDS-PAGE to avoid leaching of protein from the gel.
3. Place the black side (cathode) of the plastic sandwich transfer apparatus on the tabletop (use **Fig. 1** as a guide).
4. Wet the Scotch-Brite pad in transfer buffer, and place over the sandwich. Wet three pieces of 3MM filter papers, and place individually over the Scotch-Brite pad.
5. Carefully disassemble the gel, and remove the shorter glass plate. Cut the stacking gel with a razor blade and discard. Place the gel face down over the wet filter

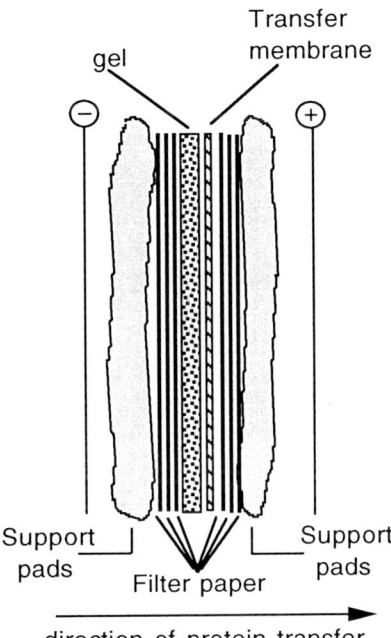

Fig. 1. Immunoblotting with a tank unit. On the cathode side, overlay the gel on three pieces of filter paper. Then the transfer membrane is overlaid on top of the gel, followed by three additional pieces of filter paper. The filter paper containing the gel and transfer membrane is sandwiched between Scotch-Brite pads, and the entire assembly is placed in a plastic support. The support is placed within the tank to transfer the proteins from the cathode to anode.

      paper, and remove the remaining glass plate using a spatula to tease the gel from the plate.

6. Fill a 10-mL plastic pipet with cold transfer buffer, and drip over the gel. Then smooth the gel gently by rolling a pipet over the surface of the gel. Add another 10 mL of transfer buffer to wet the gel completely.

7. Place the piece of nitrocellulose or PVDF on top of the wet gel. Carefully remove the air bubbles between the gel and nitrocellulose by gently rolling the pipet over the nitrocellulose using the palm of both hands to slide the pipet. Removal of the air bubbles is most crucial, because air bubbles prevent transfer (*see* **Notes 2** and **3**).

8. Place 3 pieces of wet 3MM paper over the nitrocellulose, and then the wet Scotch-Brite pad.

9. Fold and close the sandwich, and then lock the top half of the transfer cassette into place.

10. Place the transfer cassette inside the chamber in the proper orientation (**Fig. 1**), and fill the chamber with transfer buffer. The protein will be transferred from the

cathode (negatively charged) to the anode. Therefore, ensure that the nitrocellulose faces the anode side of the tank. You may place up to two transfer cassettes per tank.

11. Place a stirring bar and the frozen cooling reservoir or coil inside the tank. To transport the filled tank to the cold room, lift the cassettes by about 2 in. to avoid spillage. Then proceed to the cold room.

12. In the cold room, place the tank over a stirrer and initiate stirring. Place the electrode lid in the proper orientation over the tank. Connect the leads in the power supply to the corresponding anode and cathode sides of the electroblotting apparatus.

13. Electrophoretically transfer the proteins from the gel to the membrane for 1 h at 100 V (constant voltage) for 0.75-mm thick gels or for 3 h for 1.5-mm thick gels, with cooling and stirring. Make sure that the Bio-Ice cooling unit is submerged inside the tank to prevent the transfer buffer from overheating. If you desire to transfer overnight, transfer at 15 V (constant voltage) in the cold room.

14. Turn off the power supply, remove, and disassemble the sandwich. Remove the membrane, and note the orientation by cutting a corner. If you have used prestained markers, pinpoint their position with an indelible ink marker.

15. Place the membrane in deionized water to wash out the transfer buffer, then place membrane over a sheet of 3MM paper, and air-dry for 30 min to fix the proteins.

16. If you are using a PVDF membrane, avoid drying the membrane. Simply wash the membrane with water and proceed with immunostaining.

17. During the interim period, you may stain the gel for total protein with Coomassie blue to verify transfer efficiency (*see* **Note 4**).

3.1.3. Protein Blotting with Semidry Systems

1. Assemble the transfer stack over the anode side of apparatus as follows (**Fig. 2**):
   a. Mylar mask.
   b. Three sheets of buffer-soaked precut 3MM filter paper.
   c. Equilibrated transfer membrane.
   d. Gel.
   e. Three sheets of buffer-soaked precut 3MM filter paper.

2. Roll out the bubbles as each component is added to the stack.

3. Multiple gels may be transferred in the semidry system. Instead of placing three sheets of filter paper over the gel, simply place a sheet of porous cellophane (Hoefer) equilibrated in transfer buffer over the gel, and assemble the stack as described in **Fig. 2**.

4. Place top cathode electrode onto transfer stack.

5. Carefully connect the high-voltage leads to the power supply.

6. Apply a constant current of 60 mA for small (8 cm × 10 cm × 0.75 mm) or 200 mA for large (14 cm × 14 cm × 1.5 mm) gel for 1 h.

7. After transfer, turn off power supply, and disassemble the unit.

8. Remove the transfer membrane, and mark its orientation as described above.

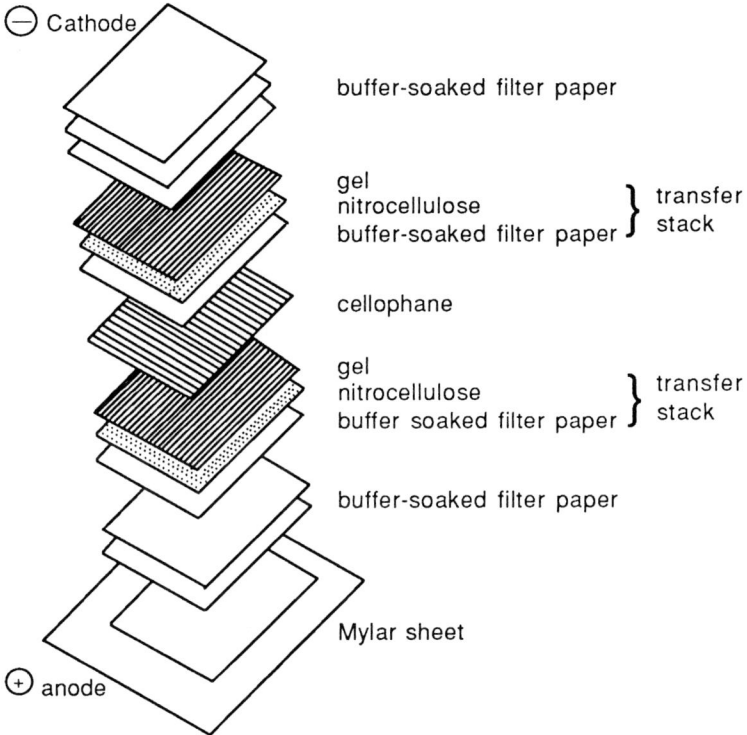

Fig. 2. Immunoblotting with a semidry transfer unit. In these units, check if the anode is the lower electrode in the unit. Place the mylar mask (optional in some units) properly on the anode. To transfer a single gel, overlay the mylar sheet with three sheets of transfer buffer-soaked filter paper, the transfer membrane, the gel, and three sheets of transfer buffer-soaked filter paper. To transfer multiple gels, construct the transfer stacks as illustrated.

3.2. Immunoprobing with Antibodies

3.2.1. Immunoprobing with Primary Antireceptor Antibody

1. Place membrane (use gloves) in a heat-sealable bag with 5 mL blocking buffer (2% milk in TBST), tease out the air bubbles, and seal the bag (*see* **Notes 5** and **6**).
2. Paste the bag on the surface of the mixer with scotch tape, and incubate for 1 h at room temperature with complete rotation.
3. Remove the nitrocellulose from the bag (use gloves), and place in a small tray. Wash the blot for 15 min in 80 mL of TBST. Discard the TBST, and repeat with three additional washes of TBST.

4. Dilute the primary antiadrenergic receptor antibody in 5 mL TBST (for a 1:100 dilution, add 50 µL antiserum/5 mL of TBST) depending on the antibody and the detection method (*see* **Note 7**).
5. Place the membrane (use gloves) into a new heat-sealable bag with 5 mL of diluted primary antibody, and seal the bag. Paste the bag onto the surface of the mixer, and incubate for 1 h at 37°C with full rotation in a warm room or incubator.
6. Remove the membrane from the bag, and place in a tray. Wash the blot four times for 15 min each in 80 mL of TBST.

3.2.2. Immunoprobing with Conjugated Secondary Antibody

1. Dilute HRP (1:5000; 1 µL/5 mL) or AP (1:1000; 5 µL/5 mL) in 5 mL of TBST (*see* **Notes 8** and **9**).
2. Place the membrane into a new heat-sealable bag with 5 mL of diluted secondary antibody, and seal the bag. Paste the bag onto the surface of the mixer, and incubate for 20 min at room temperature with full rotation for HRP or for 1 h for AP.
3. Remove the membrane from the bag (use gloves), and place in a tray. Wash the blot four times for 15 min each using 80 mL of TBST.

3.2.3. Visualization with ECL Detection System

1. Remove the ECL-kit™ (*see* **Notes 10**, **13**, and **14**) from the refrigerator and pipet 6 mL of detection solution 1 into a 13-mL polypropylene tube. Close tube with a stopper. Then use a clean pipet, and transfer 6 mL of detection solution 2 into another tube, and close using stopper.
2. Place the following items on a tray:
   a. X-ray film cassette (8 × 10 in.).
   b. Small tray wrapped with Saran Wrap. Using gloved hands, spread the Saran wrap to cover the bottom and the sides of the tray.
   c. Timer.
   d. X-ray film.
   e. The tubes containing the ECL solutions.
   f. 5-mL transfer pipet *(2)*.
   g. Gloves.
   h. Paper towels.
   i. Forceps.
   j. Saran wrap.
   k. Washed blot in container with TBST.
3. Take the tray to the darkroom.
4. Remove the membrane from the tray (use gloves).
5. Drain the excess TBST from the membrane. Touch the edge of the membrane against paper towels and place into the Saran-covered tray, **protein side up**.

6. Mix 6 mL of detection dilution 1 with an equal volume of detection solution 2, and mix well.
7. Quickly pour the detection solution on the protein side of the membrane, completely covering the entire membrane. Do not allow the membrane to become uncovered. If so, use a transfer pipet to cover with solution again.
8. Incubate for precisely 1 min at room temperature without agitation.
9. Drain the excess detection reagent (use paper towels as above), and wrap the membrane in Saran wrap. Gently smooth out the air bubbles.
10. Tape the covered blot on a larger filter paper with scotch tape. Make sure that the protein side of the membrane is face up. Then punch several holes with a handheld puncher.
11. Switch off the lights in the darkroom. With only the safety light on, carefully place a sheet of autoradiography film in the cassette.
12. Place the nitrocellulose, **protein side facing the film**, in the film cassette.
13. Mark the position of the holes onto the film with an indelible marker. Work as quickly as possible.
14. Close the cassette, and expose for 1 min.
15. Develop the film immediately.
16. If necessary, place another piece of film onto the membrane, repeat the alignment procedure, and incubate for longer or shorter time depending on the amount of luminescence on the membrane.
17. Align the film with the blot, and use the position of the prestained markers to determine the order of the lanes and the size of your immunoreactive species.

3.2.4. Visualization with Chromogenic Substrates (see **Notes 11**, **13**, and **14**)

1. Using gloves, place the washed membrane protein side up in a suitable tray.
2. Prepare developing reagent for staining immediately before use by mixing the following reagents in a tube:
   a. 5 mL of 50 mM glycine, pH 9.6.
   b. 1 mL of 1 mg/mL NBT in glycine buffer.
   c. 40 μL of 1 M MgCl$_2$.
   d. 100 μL of 5 mg/mL BCIP in DMF.
3. Wash the blot with 20 mL of 50 mM glycine, pH 9.6, for 30 s.
4. Drain the glycine, and add the developing substrate ensuring that the substrate covers the membrane.
5. If using the K&P BCIP/NBT Phosphatase Substrate™, simply add 5–10 mL of the substrate over the washed membrane and wait for color development.
6. Bands should appear in 10–30 min.
7. Terminate the reaction by removing the blot with forceps, and place the blot in a large tray full of distilled water.
8. Air-dry and photograph (optional) for a permanent record.

3.2.5. Visualization with Radioactive Probes
(see **Notes 12–14***)*

1. Using gloves, place the membrane into a heat-sealable bag.
2. Use 5 mL of 10% albumin in PBS to block the membrane for 1 h at room temperature.
3. Wash the membrane inside the bag twice with 20 mL of deionized water, then add 5 mL of the primary antibody in TBST, and incubate as described above.
4. Remove the membrane from the bag (using gloves). Wash in 80 mL TBST for 15 min, and repeat with three additional washes.
5. Place the membrane into a new heat-sealable bag.
6. All the remaining steps must be performed behind a lead shield.
7. To the membrane, add 5 mL of TBST containing 1 million cpm/mL of [^{125}I]protein A or [^{125}I]goat antirabbit IgG (if your primary antibody is raised in rabbits). Seal the bag.
8. Paste the bag onto the surface of the mixer, and incubate for 1 h at 37°C with full rotation in a warm room or incubator.
9. Remove the membrane from the bag with gloved hand and place in a tray. Discard the radioactive solution appropriately.
10. Wash the blot four times for 15 min each in 80 mL of TBST.
11. Place the blot on a filter paper of equal size, and cover it with Saran wrap.
12. Tape the covered blot on a larger filter paper with scotch tape. Then punch several holes with a handheld puncher.
13. In the darkroom, put the blot over an X-ray film, and mark the position of the holes onto the film with an indelible marker. Then expose the blot to an X-ray film until bands are evident.
14. Re-expose as necessary until you obtain satisfactory signal-to-noise ratio.
15. To determine the cpm per band, align the film over the blot using the holes in the filter paper and the markings on the film. Cut the desired bands, and place them in a counting vial. Cut an equal size strip from anywhere in the blot (this will serve as background). Count the cpm in each band using a γ-counter. Subtract the cpm in each band from the background counts.

3.3. Stripping and Reprobing of Membranes (10,11)

1. Prepare 0.2 *M* NaOH by dissolving 8 g of NaOH in a liter of water.
2. After probing, wet the blot with water, and then place in 0.2 *M* NaOH for 5 min.
3. Wash the blot with water for 5 min.
4. Proceed with the immunoblotting procedure.

4. Notes

1. The use of DTT and iodoacetamide in the solubilized protein sample prevents the refolding of the reduced membrane protein and sharpens the immunostained bands.

2. Elimination of air bubbles between the transfer membrane and the gel is a critical parameter. Air bubbles among the filter paper, gel, and membrane will block current flow and prevent protein transfer. These air pockets will cause sharply defined blank areas that are devoid of proteins. Therefore, during the assembly of the cassette, place the transfer membrane over a thoroughly wet gel and tease out the air bubbles. Air bubbles are difficult to displace from a dry gel.

3. Nitrocellulose is the most popular support for electroblotting, since it is compatible with most immunodetection systems and stains, is relatively inexpensive, and has a high protein binding capacity (249 μg/cm^2 [8,9]). Dry nitrocellulose sheets are very brittle and break easily. Therefore, they must be handled carefully and cut with a sharp razorblade to avoid breakage. Boiling of nitrocellulose removes the air that gets trapped in these sheets. PVDF has a high mechanical strength and a binding capacity similar to nitrocellulose (172 μg/cm^2 [12]). This support can be reversibly stained much more easily than nitrocellulose and is compatible with most immunoblotting protocols.

4. A variety of methods have been described for the detection of total protein patterns on nitrocellulose and PVDF membranes following immunoblotting (2). However, most of these methods (including Coomassie brilliant blue R-250, amido black 10 B, india ink, colloidal gold) are incompatible {with subsequent} immunoblotting (4). A 2% Ponceau S (3-hydroxy-4-[2-sulfo-4-sulfo-phenylazo phenylazo]-2,7-naphthaline disulfonic acid) stock solution is prepared by dissolving 2 g Ponceau S in 100 mL of 30% trichloroacetic acid and 30% sulfosalicylic acid in water. The concentrated dye solution is diluted (1:10) in 10 mL of water to prepare the working solution. This method provides a rapid and sensitive (15 ng/band) method for visualization of protein patterns on nitrocellulose or PVDF membranes. To probe the membrane, place in 0.2% Ponceau S solution for 5 min at room temperature, and then destain for 5 min in water. Photograph the blot to obtain a permanent record. Then completely destain the membrane by soaking for an additional 10 min in water, and proceed with blocking.

5. After transfer, the membrane may be stored in a plastic bag at 4°C for up to 1 yr. Wet the nitrocellulose membranes in TBST, and proceed with blocking as described. For PVDF membranes, wet the membrane in methanol for 2 s and then in distilled water to remove the methanol.

6. Plastic incubation trays (the cover of a pipet-tip rack, for example) may be used in place of heat-sealable bags, and are especially useful if probing multiple blots in different primary antibody solutions. Remember to increase the volumes so that the blot is completely covered by the solution.

7. The primary antibody solution may be stored at 4°C for up to a week.

8. AP-based luminescent reagents include substituted 1,2 dioxane phosphates such as AMPPD (Tropix, Bedford, MA) or lumigen-PPD (disodium 3-(4-methoxy-spiro[1,2-dioxetane-3,2′-tricyclo3.3.1.13,7decan]-4-yl)phenyl phosphate) and Lumi-Light Plus (Roche Molecular Biochemicals, Indianapolis, IN). These systems give reasonable sensitivity on all types of membranes.

9. HRP-based chromogenic systems utilize 4-chloro-1-naphthol (4CN) or 3,3′-diamino-benzidine (DAB/NiCl$_2$). 4CN is relatively insensitive, but provides an intense blue-black reaction that fades on storage. DAB is an exceptionally sensitive substrate that yields a brown reaction product. Nickel further enhances the sensitivity of DAB. The drawback of DAB is its fast reaction rate, making high backgrounds unavoidable. A 3,3′, 5,5′-tetramethylbenzidine (TMB) kit made by Kirkegaard and Perry laboratories appears to be as sensitive as the BCIP/NBT reagent, but is more expensive.

10. Enhanced HRP-based chemluminescent systems are most convenient, very sensitive, but are more expensive. In luminescent detection, H$_2$O$_2$-oxidized luminol emits blue light. Enhanced chemluminescence is achieved by oxidizing luminol in the presence of *p*-iodophenol. This increases the intensity of the light output by several orders of magnitude as well as its duration. Consequently, this is a very convenient and sensitive system that provides a permanent record on the film. The chemluminescent reaction is detected within a few seconds to an hour and works well on all types of membranes.

11. AP-based chromogenic detection involves BCIP and NBT. BCIP hydrolysis produces indigo precipitate after oxidation with NBT. Reduced NBT precipitates as a dark blue-gray stain over the immunoreactive bands. This system is the least expensive method for chromogenic detection and is as sensitive as more expensive methods.

12. Radioactive probes are used in quantitative immunoblotting to determine the relative intensity of the immune complexes. If an Instaimager™ (Packard, Meriden, CT) is available, the blot may be analyzed by electronic autoradiography without ruining the membrane in the process. The 20 × 24 cm counting surface of this instrument will count up to 1.2 million cpm with high spacial readout precision. The total number under each peak will be automatically determined after subtraction of the background counts.

13. A frequently encountered problem in immunoblots is the high background consisting either of additional bands or diffuse signals covering the entire membrane. Extra bands arise by contaminating antibodies present in polyclonal sera or by crossreactions between different antigens. Diffuse background results either from insufficient blocking of the membrane or from incompatibilities between the detection reagent and the blocking buffer.

14. If background bands appear, first determine whether the secondary detecting agent is the culprit. Omit the primary antibody and observe the background generated by the secondary reagent alone. If the primary antibody is involved, try modifying the blocking solution. For example, increase the nonfat milk to 5% or use 10% bovine serum albumin (BSA), reduce the incubation time of primary and secondary antibodies, titrate the dilution of primary and secondary antibodies, increase the duration of washing, add 1% of the nonionic detergent NP-40 to the incubation media of primary and secondary antibodies, or finally try using another antibody. A case in point is the detection of the popular hemagglutinin (HA) epitope tag. The HA sequence (YPYDVPYA) was cloned at the amino-

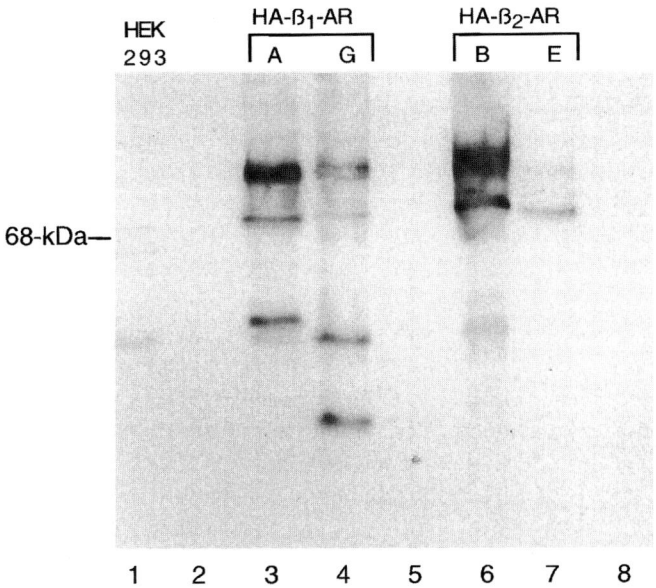

Fig. 3. Western blotting of HA-tagged β-AR. An HA epitope was added to the amino-terminus of the rat β₁-AR and β₂-AR using PCR. The cloned cDNA for each β-AR subtype was stably expressed in HEK-293 cells. Clonal cells expressing either β₁-AR or β₂-AR were selected and expanded. Membranes were prepared from two β₁-AR and two β₂-AR clones. The density of the β₁-AR in the membranes of the A and G clones was 7.8 and 1.2 pmol/mg protein, respectively. The density of the β₂-AR in the membranes of the B and E clones was 8.2 and 2.2 pmol/mg protein, respectively. Lane 1 contains 5 µg of untransfected HEK-293 membranes. Lanes 2, 5, and 8 are blank. Lanes 3 and 4 contains 5 µg of A and G clones of the β₁-AR, respectively. Lanes 6 and 7 contain 5 µg of B and E clones of the β₂-AR, respectively. The proteins were subjected to SDS-PAGE and transferred to nitrocellulose. The blot was blocked in 2% milk in TBST, washed in TBST, and then incubated with a 1:400 dilution of a rabbit polyclonal anti-HA IgG for 1.5 h at 37°C. The blot was washed in TBST and incubated with 1:5000 dilution of donkey antirabbit IgG-coupled to HRP for 20 min at room temperature. After washing, the blot was developed enhanced chemluminescence (ECL™-kit from Amersham Pharmacia Biotech). The blot was exposed to film for 5 s.

terminus of rat β₁- and β₂-ARs using polymerase chain reaction (PCR), and the resulting cDNAs were expressed in human embryonal kidney 293 cells. Clonal cell lines from each subtype of the β-AR were selected and expanded, and membranes were prepared from individual clones expressing β₁- and β₂-ARs (**Fig. 3**). Western blotting was used to correlate the intensity of the immune reaction with the density of β₁- and β₂-ARs in these membranes, as determined by radioligand binding. The mouse monoclonal anti-HA antibody generated multiple specific and nonspecific bands that could not be blocked (data not shown). To reduce the background, we switched to a rabbit polyclonal anti-HA antibody (Y-11, Santa Cruz Biotechnology, Santa Cruz, CA). Immunoblot analysis revealed that the HA-tagged β-AR migrated as 75-kDa proteins, because the HA-epitope increased the apparent size of the receptor (lanes 3, 4, 6, and 7). HEK-293 membranes (lane 1) that do not contain the HA tag did not crossreact with the antibody. The intensity of the anti-HA immune complexes for β_1-AR (lanes 3 and 4) and β_2-AR (lanes 6 and 7) were correlated with the density of these receptors in the membranes (**Fig. 3**).

References

1. Towbin, H., Staehelin, T., and Gordon, G. (1979) Electrophoretic transfer of proteins from polyacrylamide gels to nitrocellulose sheets: procedures and some applications. *Proc. Natl. Acad. Sci. USA* **76,** 4350–4354.
2. Dunn, M. J. (1993) *Gel Electrophoresis: Proteins.* Bios Scientific Publishers, Oxford, UK.
3. Gooderham, K. (1984) Transfer techniques in protein blotting, in *Methods in Molecular Biology, vol. 1: Proteins* (Walker, J. M., ed.), Humana, Totowa, NJ, pp. 165–178.
4. Dunn, M. J. (1996) Electroblotting of proteins from polyacrylamide gels, in *Methods in Molecular Biology, vol. 59: Protein Purification Protocols* (Doonan, S., ed.), Humana, Totowa, NJ, pp. 363–370.
5. Peferoen, M. (1988) Blotting with plate electrodes, in *Methods in Molecular Biology, vol. 3: New Protein Techniques* (Walker, J. M., ed.), Humana, Totowa, NJ, pp. 395–402.
6. Craig, W. Y., Pollin, S. E., Collins, M. F., Ledure, T. B., and Richie, R. F. (1993) Background staining in immunoblot assays. Reduction of signal caused by cross-reactivity with blocking agents. *J. Immunol. Methods* **158,** 67–76.
7. Blake, M. S., Johnston, K. H., Russel-Jones, G. J., and Gotsclich, E. C. (1984) A rapid, sensitive method for detection of alkaline phosphatase-conjugated anti-antibody on western blots. *Anal. Biochem.* **136,** 175–179.
8. Amersham (1995) ECL™ Western blotting protocols. Buckinghamshire, UK.
9. Bahouth, S. W. (1994) Effects of chemical and surgical sympathectomy on expression of β-adrenergic receptors and guanine nucleotide-binding proteins in rat submandibular glands. *Mol. Pharmacol.* **42,** 971–981.
10. Kaufmann, S. H., Ewing, C. M., and Shaper, J. H. (1987) The reusable western blot. *Anal. Biochem.* **161,** 89–95.
11. Tesfaigzi, J., Smith-Harrison, W., and Carlson, D. M. (1994) A simple method for reusing western blots on PVDF membranes. *BioTechniques* **17,** 268–269.
12. Pluskal, M. G., Prezekop, M. B., Kavonian, M. R., Vecoli, C., and Hicks, D. A. (1986) Immobilon PVDF transfer membrane: a new membrane substrate for western blotting of proteins. *BioTechniques* **4,** 272–283.

21

UV Crosslinking of Adrenergic Receptors and Ligands

Detection by SDS-PAGE

Michael K. Sievert, John F. Resek, Zhongren Wu, Yajing Rong, and Arnold E. Ruoho

1. Introduction

Most membrane functions are mediated by enzymes and specific receptors that bind small molecules: substrates, inhibitors, modulators, hormones, and neurotransmitters. At the molecular level, polypeptide chains that contribute to the binding site of a given receptor can be identified by sodium dodecyl sulfate-polyacrylamide gel electrophoresis (SDS-PAGE) if they are specifically labeled with a radioactive marker. Purification and subsequent N-terminal sequencing of these labeled peptides identify their location in the primary sequence of the receptor. Photoaffinity labeling has emerged as an effective tool to tag amino acid residues covalently that are in or near ligand binding sites *(1)*. The specificity of this method arises from the following characteristics.

The products of photolysis of phenyl azides, fluorenone, and diazo compounds react in the binding site before dissociation of the ligand occurs. The lifetimes of phenyl nitrenes or azepines, which are derived from phenylazides, or carbenes, which are derived from diazo-and α-ketophenyl-(benzophenone and fluorenone) containing compounds, are extremely short (nanosecond and millisecond). This means that even ligands with equilibrium dissociation constants between 10^{-6} and 10^{-4} M can react before dissociation can occur from the binding site. Higher-affinity photolabels ($K_D \leq 10^{-10}$ M), however, increase the ease of the photolabeling experiment, since excess photolabel can be

From: *Methods in Molecular Biology, vol. 126: Adrenergic Receptor Protocols*
Edited by: C. A. Machida © Humana Press Inc., Totowa, NJ

removed from the receptor preparation prior to photolysis either by centrifugation or dilution. This fact enhances the signal-to-noise ratio and has been the reason that this technique is so effective for identification of most ligand binding sites, where it is common to have equilibrium dissociation constants in the subnanomolar range.

In cases where removal of excess ligand prior to photolysis is not possible, the addition of "scavenger" molecules, such as mercaptans (β–mercapto-ethanol, glutathione) or amines (*para*-aminobenzoic acid), can be utilized *(2)*. Scavengers will react with the photolyzed molecules, which are not bound in the receptor site. This allows an enhancement of signal-to-noise ratio, which cannot be achieved in traditional covalent bond-forming affinity-labeling experiments in which the reaction rate in the binding site is slowly relative to the off-rate of the affinity label.

Reaction rates of the photo products of most photolabels are not affected by temperatures between 0 and 37°C. Since this temperature range is well tolerated in most biological symptoms, one can incubate the photolabel and receptor preparation in the dark to achieve equilibrium and then dilute into ice-cold buffer for photolysis. Since the dissociation rates of receptor–ligand complexes are usually much slower at 4°C than at 37°C, one can, in many cases, "trap" the photolabel–receptor complex during photolysis, whereas nonspecific binding sites, which are much lower affinity, will dissociate. This treatment will enhance the signal-to-noise ratio. This technique cannot usually be applied to affinity labels, since covalent reaction rates are highly dependent on temperature.

Several affinity labels have been synthesized and reported that derivatize the binding site of the β-adrenergic receptor (β-AR). These include antagonist bromoacetyl affinity labels *(3–6)*, which have been reported to derivatize the pure receptor *(4)*, and an associated lipid component *(5,6)*. Agonist affinity labels have been synthesized and reported to react covalently with the receptor *(7,8)*. Our laboratory has synthesized several antagonist and agonist carrier-free radioiodinated β-AR photolabels, which are effective in derivatization of the binding site of the β-AR *(9–13)* (**Fig. 3**). It is beyond the scope of this chapter to present the synthesis of the photoaffinity labels. The actual practice and use of this technique to label the β-AR in our laboratory is therefore presented. What is described below is the use of both agonists and antagonists for the β_2-AR from varying sources with two different photoreactive groups, a phenylazide and a novel flourenone moiety.

2. Materials

2.1. Reagents

1. Radiolabeled β-AR probe containing photoactive group (*see* **Note 1**).
2. Nonlabeled β-AR ligand, for example, alprenolol.

3. Source of active β-AR: This could be membranes isolated from lung tissue, cell membranes expressing recombinant β-AR, or purified receptor.

4. I$_3$ buffer: 10 m*M* Tris-HCl, pH 7.4, 5 m*M* ethyleneglycol-bis-(β-aminoethylether) -*N,N,N',N'*,-tetra-acetic acid (EGTA), 100 m*M* benzamidine, 100 μ*M* phenyl-methylsulfonyl fluoride (PMSF), 5 mg/mL soybean trypsin inhibitor, and 20 mg/mL leupeptin.

5. β–Mercaptoethanol.

6. 4X SDS-PAGE sample buffer: 20% β–mercaptoethanol, 40% glycerol, 10% SDS, 75 m*M* Tris-HCl, pH 6.8, 0.05% bromophenol blue.

7. LLD buffer: 0.02% digitonin, 150 m*M* NaCl, 10 m*M* Tris-HCl, pH 7.4.

8. HME buffer: 20 m*M* HEPES, pH 8.0, 2 m*M* MgCl$_2$, 1 m*M* EDTA, 1 m*M* benzamidine, 2 m*M* tetrasodium pyrophosphate, 10 μg/mL trypsin inhibitor, and 0.1 mg/mL bovine serum albumin (BSA).

9. HE buffer: 2 m*M* HEPES, pH 8.0, and 1 m*M* ethylenediamine tetra-acetic acid (EDTA).

10. Incubation buffer for photolabeling with [^{125}I]NAIN: 50 m*M* Tris-HCl, pH 7.4, 4 m*M* MgCl$_2$, 10^{-4} *M* ascorbate, and 1 m*M* glutathione.

11. Dilution buffer for photolabeling with [^{125}I]IAS: 10 m*M* Tris-HCl, pH 7.4, 100 m*M* NaCl, 2 m*M* EDTA.

12. Resuspension buffer used for photolabeling with [^{125}I]IAmF: 10 m*M* Tris-HCl, pH 7.4, 150 m*M* NaCl, 2 m*M* EDTA.

2.2. Equipment

1. Photolysis chamber (*see* **Note 2** and **Fig. 1**).

2. Rayonet reactor containing 16 lamps with a wavelength of 350 nm (*see* **Note 2** and **Fig. 2**).

3. Sorvall glass centrifuge tubes (DuPont catalog no. 00119, 2-mm thick Pyrex, 18 × 100).

4. SDS-PAGE apparatus.

5. X-ray film (Kodak X-Omat) and cassette.

6. Homogenizer, for making membranes.

3. Methods

3.1. Preparation of Membranes

3.1.1. Preparation of Guinea Pig Lung Membranes

1. Decapitate guinea pigs (300–500 g), and remove the lungs. Stored in liquid nitrogen.

2. Membranes are prepared by mincing the frozen lungs with a pair of tweezers and a spatula in the presence of I$_3$ buffer.

3. Resuspend approx 1 g of the preparation in 20 mL of I$_3$ buffer, and homogenize with a polytron homogenizer (full speed for 10 s).

4. Centrifuge the homogenate at 2000 rpm for 10 min using a Sorvall SS-34 rotor (4°C).

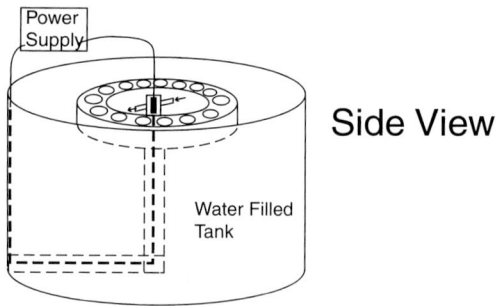

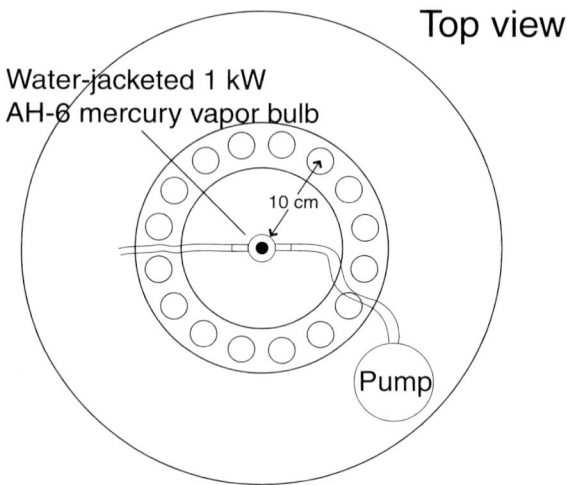

Fig. 1. Schematic representation of the photolysis apparatus that is used in our laboratory to activate phenylazide-containing photoaffinity labels, shown from the side
and top views. The apparatus has a stainless-steel tank designed to hold ice water (for
cooling both the lamps and the samples). The lamp is a 1-kW high-pressure mercury
vapor lamp (type AH-6, Advanced Radiation Corp.). The tubes are placed in an acrylic
rack that is 10 cm from the light source. The water level is such that it covers the
samples as well as the bulb. A pump circulates water through the custom-made water
jacket of the bulb, also to maintain cooling of the bulb.

5. Discard the pellet, and recentrifuge the supernatant at 15,000 rpm for 45 min
   using a Sorvall SS-34 rotor (4°C).
6. The pellet is resuspended in 10 mL of I_3 buffer, and store in aliquots at −80°C.

3.1.2. Sf9 Membrane Preparation

1. Collect Sf9 cells expressing recombinant β_2-AR (*see* **Note 3**) by centrifugation
   (1000g for 10 min), and resuspend in 1.5 mL I_3 buffer.

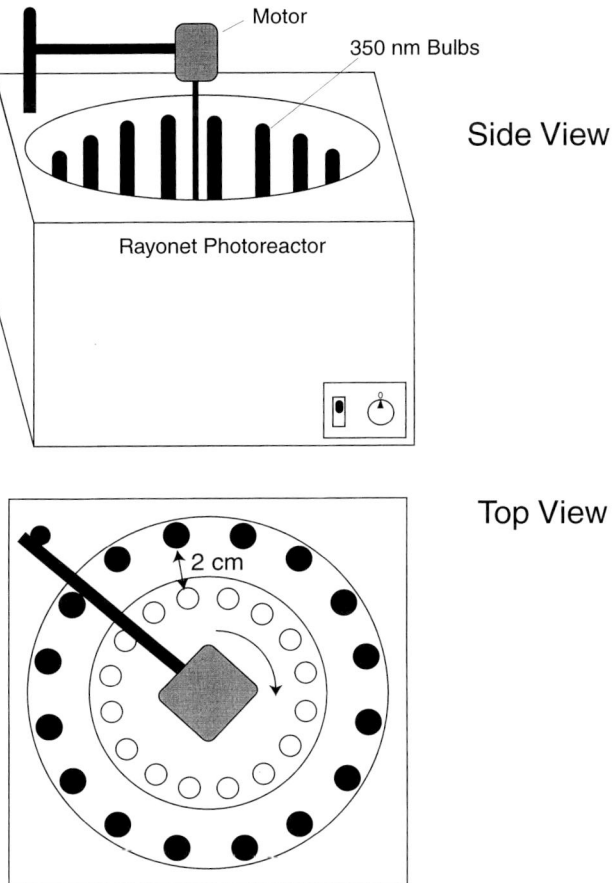

Fig. 2. The Rayonet Photochemical Reactor that is used to active benzophenone- and fluorenone-containing photoaffinity labels (purchased from The Southern New England Ultraviolet Co.). We outfitted the apparatus with 16- to 350-nm bulbs (Model RPR-100, The Southern New England Ultraviolet Co.). A motor that rotated the samples in a rack at a distance of approx 2 cm from the bulbs was added to the apparatus. The motor rotates at approx 10 rpm, and is used to average the exposure of the tubes to the bulbs. The photoactivation was performed in the cold room to prevent heating of the samples.

2. Shear the cells by rapidly passing them through a 27-gage needle three times.
3. Centrifuge the sheared cells at 15,000g for 30 min.
4. Resuspend the pellet in 1.5 mL I$_3$ buffer at a concentration of 2.0 mg/mL. The membranes are aliquoted and stored at −80°C.

3.1.3. HEK 293 Membrane

1. HEK 293 cells stably transfected with the human recombinant β_2-AR were scraped from plates using HME buffer (8 mL of buffer for 150-mm dish).
2. Homogenize the cells on ice with seven strokes in a Dounce homogenizer.
3. Prepare 15 mL SW 28.1 rotor tubes with 2.5 mL of 23% sucrose in HE buffer layered carefully over 2.5 mL 43% sucrose in HE buffer.
4. Layer lysates carefully on top of sucrose.
5. Centrifuge at 4°C for 40 min at 25,000 rpm in a Beckman SW 28.1 rotor.
6. Membranes are collected at the interface of the two sucrose layers. Separate the membranes by centrifugation over a sucrose gradient.
7. Quantitate β_2-AR content in the membranes by [^{125}I]Iodocyanopindolol ([^{125}I]ICYP) binding and adenylyl cyclase assays.

3.2. Preparation of Purified β-Adrenergic Receptor

3.2.1. Purification of Recombinant Hamster Lung β_2-AR

1. Collect infected Sf9 cells (0.5 L suspension culture) expressing recombinant hamster lung β_2-AR by centrifugation at 300g for 15 min at 4°C.
2. Resuspend the cell pellet in 100 mL of I$_3$ buffer containing 1% digitonin and 100 mM NaCl.
3. Centrifuge the solution at 48,000g for 20 min at 4°C. The supernatant was stored on ice overnight, and assayed for receptor using 10 nM [^{125}I]ICYP and Sephadex G-50 columns to separate free [^{125}I]ICYP from the [^{125}I]ICYP–β-AR complex.
4. Load the digitonin extract on an (–)-alprenolol-Sepharose column.
5. Wash the column extensively LLD buffer.
6. Elute the bound proteins using an alprenolol gradient (0–400 μM).
7. Fractions should be assayed for β-AR using [^{125}I]ICYP.
8. Pool the fraction containing the purified receptor, and concentrate to approx 1 mL using an amicon concentrator.
9. Wash Con A-agarose (~150 μL; 6 mg lectin/mL, Vector Laboratories) twice with 1 mL of LLD buffer in a 1.5-mL Eppendorf tube.
10. Add the affinity-purified receptor concentrate to the Con A-agarose and tumble at 4°C for 1 h.
11. Collect the agarose beads by centrifugation for 5 min in a Beckman microfuge (5000g).
12. Wash the beads three times with 1-mL aliquots of LLD buffer, collecting by centrifugation.
13. To elute the bound proteins, add 0.5 mL of 200 mM α-D-methyl mannoside in LLD buffer and tumble at 4°C for 30 min.
14. Pellet the beads as before, saving the supernatant. Repeat the elution with an additional 0.5 mL of 200 mM α-D-methyl mannoside in LLD buffer at 4°C for 30 min.
15. Pool this supernatant with the first α-D-methyl mannoside fraction.
16. Assay the receptor using [^{125}I]ICYP binding assays.

3.3. Antagonist Photoaffinity Labeling Using Probes Containing the Phenylazide Photoreactive Group

3.3.1. Photoaffinity Labeling of Guinea Pig Lung Membranes with [^{125}I]iodoazidobenzylpindolol ([^{125}I]IABP)

1. Thaw guinea pig lung membranes prepared as above.
2. Incubate the membrane preparation with 0.5 nM of [^{125}I]IABP in the dark for 45 min at 30°C in the thick-walled Pyrex tubes. To one of the samples, add 10 µM of nonlabeled alprenolol in I$_3$ buffer (*see* **Note 4**).
3. Place samples in the photolysis apparatus (*see* **Note 2** and **Fig. 1**). The samples are partially submerged in ice water at a distance of 10 cm from the 1-kW lamp.
4. Dilute the samples with 5 mL of I$_3$ buffer (*see* **Notes 3–8**).
5. Turn on lamp for 5 s.
6. Immediately after photolysis, add 50 µL of β-mercaptoethanol (final concentration of 1%; *see* **Note 9**).
7. Transfer samples to polycarbonate ultracentrifuge tubes.
8. Collect the membranes by ultracentrifugation for 30 min at 100,000g in a Beckman type 65 rotor.
9. Resuspend the membranes in and electrophorese on a 10% Laemmli gel *(14)*.
10. After staining (silver or Coomassie), destaining, and drying, the gel is placed in a X-ray cassette with Kodak X-Omat film and a Dupont Cronex Quanta III intensifier.
11. Develop film after 12–24 h (*see* **Fig. 3**).

3.3.2. Photoaffinity Labeling of Purified Hamster Lung β$_2$-AR (r-β$_2$-AR) with [^{125}I]IABP

1. Thaw purified, recombinant hamster lung β$_2$-AR (r-β$_2$-AR).
2. Remove free (–)-alprenolol using Sephadex G-50 chromatography columns (Bio-Rad Econo-Columns 0.7 × 10 cm, containing 3.5 ml of Sephadex G-50).
3. Aliquot the sample into the thick-walled Pyrex tubes, and incubate with 0.5 nM of [^{125}I]IABP in the dark for 45 min at 30°C in the presence and absence of excess (10 µM) of nonlabeled alprenolol.
4. Place samples in the photolysis apparatus (do not dilute; *see* **Note 10**).
5. Turn on lamp for 5 s.
6. Immediately after photolysis, add β-mercaptoethanol to a final concentration of 1%.
7. Solubilize the samples in electrophoresis sample buffer, and run on a 10% SDS-polyacrylamide gel.
8. Prepare the gel for autoradiography as described above (*see* **Fig. 3**).

3.4. Agonist Phenylazide Labeling

3.4.1. Photolabeling of Guinea Pig Lung Membranes with [^{125}I]-(–)-N-(p-azido-m-iodophenethylamidoisobutyl)norepinephrine ([^{125}I]NAIN)

1. Thaw guinea pig membranes.
2. Dilute guinea pig lung membranes (prepared as above) threefold in degassed

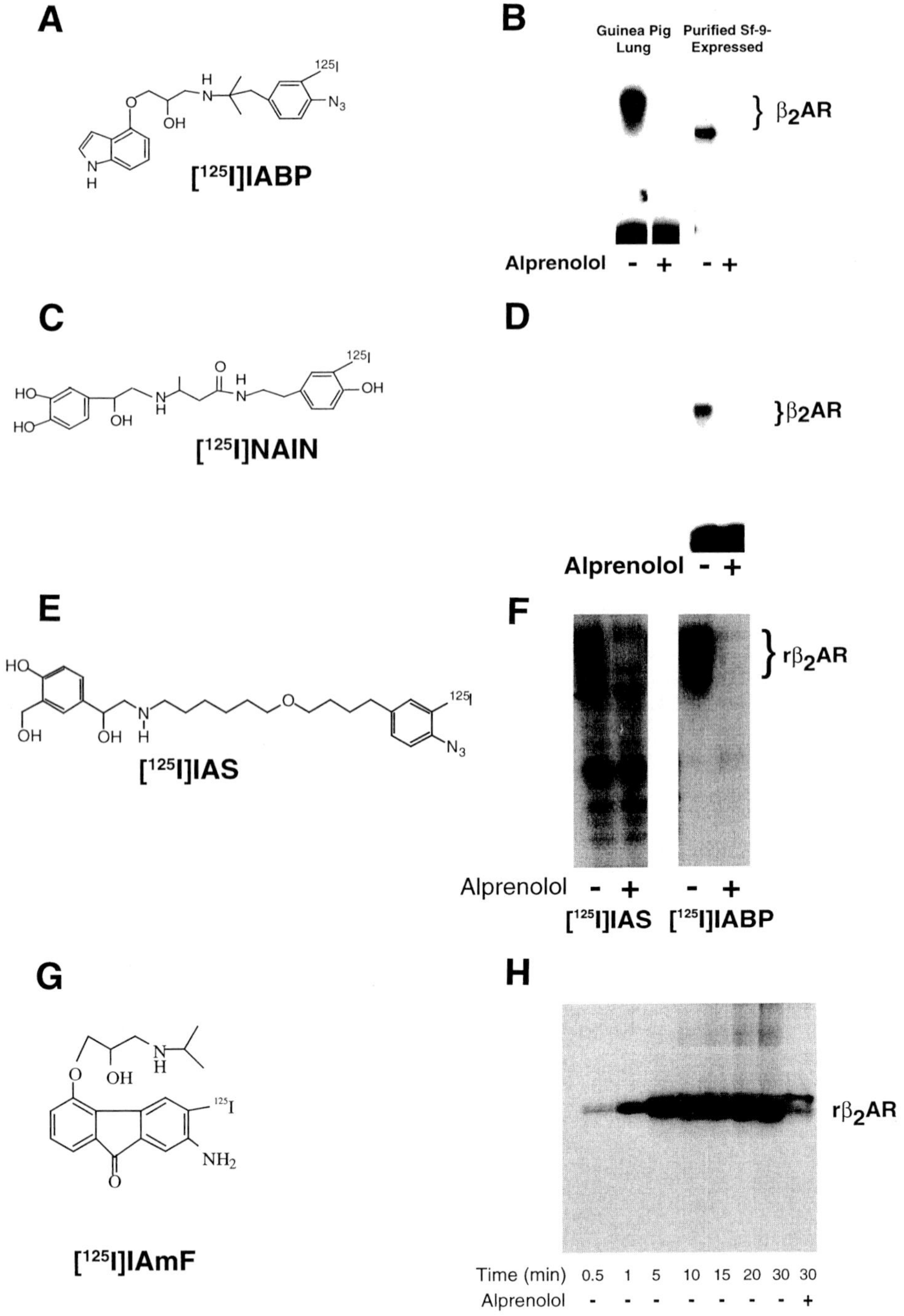

A

[125I]IABP

B

Guinea Pig Lung   Purified Sf-9-Expressed

} β_2AR

Alprenolol - +   - +

C

[125I]NAIN

D

}β_2AR

Alprenolol - +

E

[125I]IAS

F

} rβ_2AR

Alprenolol - +   - +
         [125I]IAS   [125I]IABP

G

[125I]IAmF

H

rβ_2AR

Time (min) 0.5 1 5 10 15 20 30 30
Alprenolol   -   - -   -   -   -   -   +

incubation buffer containing 50 mM Tris-HCl, 4 mM MgCl$_2$, pH 7.4, catechol (10^{-4} M), ascorbate (10^{-4} M), and 1 mM glutathione (*see* **Note 11**).

3. Incubate guinea pig lung membranes at 22°C for 1 h in the dark with 2 nM [^{125}I]NAIN in the presence and absence of 10 μM alprenolol.
4. Following incubation, dilute the membranes with 5 mL of ice-cold degassed incubation buffer, and immediately photolyze for 5 s at a distance of 10 cm from the 1-kW mercury lamp.
5. Transfer samples to polycarbonate ultracentrifuge tubes.
6. Collect the membranes by ultracentrifugation at 300,000g for 30 min in a Beckman type 65 rotor.
7. Resuspend the membranes in SDS-PAGE sample bufffer and electrophorese using a 10% Laemli gel *(14)*.

3.4.2. Photolabeling of HEK Cell Membranes with [^{125}I]iodoazidosalmeterol ([^{125}I]IAS

1. Thaw HEK 293 membranes (approx 3 mg/mL).
2. Transfer 100 μL of membranes (in 20 mM HEPES, pH 8.0, 1 mM EDTA) into the appropriate number of thick-walled Pyrex tubes.
3. Add [^{125}I]IAS to a final concentration of 0.7–1.0 nM. To one of the tubes, add 10 μM nonlabeled aplrenolol.
4. Incubate samples for 30 min at 30°C.
5. Place tubes in photolysis apparatus.
6. Dilute with 5 mL of ice-cold degassed dilution buffer, and turn on lamp for 5 s.
7. Add 50 μL of β-mercaptoethanol.

Fig. 3. *(opposite page)* Photoaffinity labels that are used in our laboratory and results obtained with them. (**A**) The antagonist photoaffinity label, [^{125}I]IABP. (**B**) An autoradiogram of [^{125}I]IABP photoaffinity labeling of guinea pig lung membranes and purified β$_2$-AR in the absence (–) and presence (+) of 10 μM alprenolol. The β$_2$-AR migrates as a diffuse polypeptide with an apparent molecular mass of 70–90 kDa. The diffuse nature of the labeling is owing to the glycosylation of the β$_2$-AR in the membrane. This diffuseness is common among other glycosylated proteins when they are run on an SDS-polyacrylamide gel. The Sf9-expressed β$_2$-AR migrates with an apparent molecular mass of 50–55 kDa. The difference in size is most likely owing to differences between glycosylation levels in guinea pig lung membranes and the Sf9 cells. (**C**) The agonist photoaffinity label, [^{125}I]NAIN. (**D**) An autoradiogram of the [^{125}I]NAIN-labeled β$_2$-AR from guinea pig lung membranes. Again, note the diffuseness of the labeling. (**E**) The agonist photoaffinity label, [^{125}I]iodoazidosalmeterol, a long-acting β$_2$-AR agonist. (**F**) An autoradiogram of [^{125}I]IAS labeling of the β$_2$-AR expressed in HEK 293 cells; [^{125}I]IABP labeling of the same membranes is shown for comparison. (**G**) The novel antagonist photoaffinity label, [^{125}I]Iodoaminoflisopolol. (**H**) An autoradiogram of the [^{125}I]IAmF labeling of the β$_2$-AR expressed in Sf9 cells.

8. Pellet the membranes by ultracentrifugation at 100,000*g* for 30 min in a Beckman type 70.1 Ti rotor.
9. Solubilize the pellet in SDS-PAGE sample buffer.
10. Electrophorese on a 12.5% SDS-polyacrylamide gel.
11. Prepare gel for autoradiography as described earlier. Results are usually obtained with a 48-h autoradiogram using a Quanta III intensifier screen (*see* **Fig. 3**).

3.5. Antagonist Photolabeling of Sf9-Expressed r-β_2-AR Using the Novel Fluorenone-Based Photoaffinity Label [^{125}I]iodoaminoflisopolol ([^{125}I]IAmF)

1. Incubate Sf9 membranes (approx 20 μg protein) containing r-β_2-AR (prepared as described above) at 30°C for 30 min with 1 n*M* [^{125}I] IAmF, in the absence or presence of 10 μ*M* (–)-alprenolol in a polypropylene test tube. Bring the total volume to 100 μL with resuspension buffer.
2. Load the tubes into the Rayonet photoreactor.
3. Turn on the overhead motor and the lamps. Photolyze samples for 15 min at 4°C. The distance between the lamps and the sample tubes should be approx 2 cm.
4. Solubilize the samples in SDS-PAGE sample buffer, and electrophorese using a 10% polyacrylamide gel.
5. Prepare the gel for autoradiography as above (*see* **Fig. 3**).

4. Notes

1. Our laboratory prepares our own photoaffinity labels for the β-AR receptor. Pierce Chemical Co. sells a variety of reagents with photoactive groups that can be used to derivatize active groups chemically on molecules. Some of these can be radioactively derivatized with ^{125}I. Our laboratory uses ^{125}I because of its high specific activity (2200 Ci/mmol) and relative ease of detection (γ-counters and autoradiography). Because it is a γ-emitter, necessary precautions must be utilized (lead shielding should be used for protection).
2. We use several methods of activating the photoaffinity labels, depending on the photoactive group. For photoaffinity labels containing phenylazide moieties, we use a custom-made apparatus (*see* **Fig. 1**). There are reports in the literature of using handheld mineral lamps. In these reports, the lamp (Ultra-violet Products, Inc. model UVG-11, 254 nm) is held at a distance of approx 2–5 cm from the samples on ice. The samples can be placed in wells created on parafilms. Photoactivation requires approx 5 min. Although this method works, we believe the exposure of proteins to 254 nm light for extended time inactivates them. This is why we use the short times with the intense light (1 kW). The bulb was purchased from Advanced Radiation Corporation (Santa Clara, CA) and fitted with a water jacket by Norman Erway (Madison, WI).

   For activating photoaffinity labels containing α-ketophenyl groups (fluorenone, benzophenone), we use a Rayonet Photoreactor (*see* **Fig. 2**). This was purchased from The Southern New England Ultraviolet Co. (Branford, CT). We outfitted the apparatus with 16–350-nm bulbs (Model RPR-100, The Southern New England Ultraviolet Co.). The apparatus is placed in a cold room to avoid heat build-up. It also has a motor connected to a carousel, so that the samples can be rotated (at approx 10 rpm) to obtain even light exposure. Time of photoreaction is generally 15 min (*see* **Fig. 3** for time-course of [^{125}I]IamF labeling).

3. Culture and infection of Sf9 cells: Our laboratory uses Ex-Cell 401-defined medium supplemented with 2.5 µg/mL of gentamicin, 2.5 mg/mL of streptomycin, and 2.5 U/mL of penicillin G (media from JRH Biosciences, antibiotics from Gibco Life Sciences). The cells are maintained in suspension cultures grown in Erlenmeyer flasks (filled to 20% of their volume) at 25–28°C at 110 rpm. The cells are split 1:5 when they reach a density of 5×10^6 cells/mL. For the infection of cells, the cells are grown to a density of $4–5 \times 10^6$ cells/mL and infected with the recombinant baculovirus at an MOI of approx 3.0. After 1 h of incubation with the virus at room temperature, the cells are diluted with fresh medium to a density of approx 1×10^6 cells/mL and grown for 2–4 d, at which time the cells are harvested by centrifugation and prepared as described.

4. The incubations with the photoaffinity labels are performed in low-light conditions. It is doubtful that there is a significant activation of the photoaffinity labels under these conditions. The benzophenone- and fluorenone-containing compounds are more resistant to ambient light, but these incubations are performed in low light as a precaution.

5. The photoaffinity labels are usually stored in methanol. We have found no deleterious effects of methanol at concentrations up to 5%.

6. A washing procedure is occasionally used to remove ligand from nonspecific binding sites prior to photolysis from membranes containing receptor. This procedure cannot be used for soluble proteins or photoaffinity labels with low affinity. Following the incubation, membranes are diluted with 5 mL of ice-cold degassed incubation buffer. The membranes are then collected by centrifugation at 300,000g for 15 min, resuspended in 200 µL of ice-cold incubation buffer, and diluted with 5 mL of ice-cold degassed incubation buffer. Immediately following dilution, the membrane preparation is photolyzed (*see* **Fig. 4** for comparison of washing). The main drawback of this procedure is the potential for loss of samples.

7. Glutathione can be added to the dilution buffer if nonspecific labeling is a problem. The glutathione is added at a conentration of approx 10 mM. The dilution buffer must be degassed before addition of the glutathione to prevent oxidation. Additionally, the addition of excess unlabeled protector (1–10 µM) can be added to the dilution buffer to lower nonspecific photoreaction.

8. We have constructed a simple apparatus to help us dilute many samples (up to 12) at the same time. It consists of 12 test tubes (13 × 100 mm) in a rack. Each tube has a length of Tygon tubing attached to it (approx 30 cm long). The end of the tubing has a 5-mL pipet tip (with the tip cut off). Prior to photolysis, the ends of the tubing are placed in each sample. The rack is simply turned upside down to dilute the samples simultaneously.

9. The addition of β-mercaptoethanol is helpful after photolysis of the photo-label,

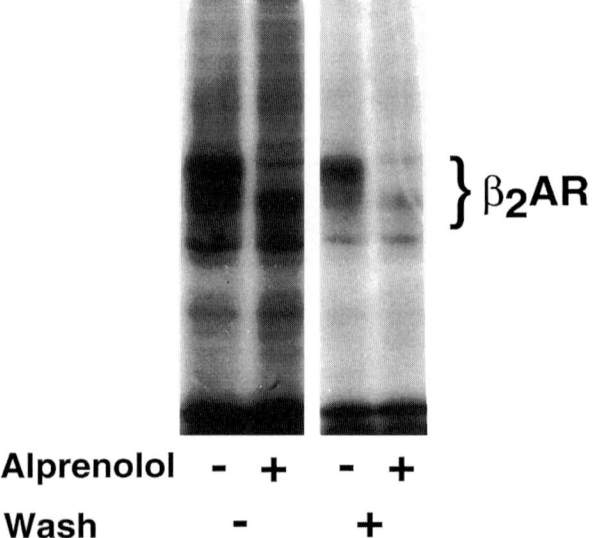

$\left.\right\}\beta_2\mathbf{AR}$

Alprenolol - + - +

Wash   -   +

Fig. 4. The effect of washing prior to photolysis on the specificity of [^{125}I]NAIN labeling of the β_2-AR from guinea pig membranes. Lanes 1 and 2 were photolabeled with [^{125}I]NAIN after dilution, but were not washed by centrifugation. Lanes 3 and 4 were incubated with [^{125}I]NAIN, but were diluted with incubation buffer and the membranes were collected by a rapid centrifugation step. The membranes were resuspended in incubation buffer and diluted prior to photolysis. Shown is an autoradiogram of an SDS-polyacrylamide gel.

  since there is a long-lived reactive species that is generated. The β-mercaptoethanol quenches this reactive species, thus reducing potential non-specific reactions.

10. The purified receptor cannot be diluted prior to photolysis owing to the fact that it is soluble. It is difficult to recover the labeled, soluble proteins for electrophoresis when they are in a dilute state. In this case, since the receptor has been highly purified, there typically is not much nonspecific labeling with which to contend. If nonspecific labeling is a problem, scavengers, such as β-mercaptoethanol or glutathione, at a concentration of 10 mM can be added to the sample prior to photolysis.

11. The incubation buffer for the agonist photoaffinity label, NAIN, needs to be degassed, because the catechol hydroxyls of the photoaffinity label are readily oxidized (which is why ascorbate, catechol, and glutathione are present as well).

References

1. Ruoho, A. E., Rashidbaigi, A., and Roeder, P. E. (1984) Approaches to the identification of receptors utilizing photoaffinity labeling, in *Membranes, Detergents, and Receptor Solubilization*, Liss, New York, pp. 119–160.

2. Ruoho, A. E., Kiefer, H., Roeder, P. E., and Singer, S. J. (1973) The mechanism of photoaffinity labeling. *Proc. Natl. Acad. Sci. USA* **70,** 2567–2571.
3. Atlas, D. and Levitski, A. (1978) Tentative identification of beta-adrenoreceptor subunits. *Nature (Lond.)* **272,** 370–373.
4. Dickinson, K. E. J., Heald, S. L., Jeffs, P. W., Lefkowitz, R. J., and Caron, M. G. (1985) Covalent labeling of the beta-adrenergic ligand-binding site with *para*-(bromoacetamidyl)benzylcarazolol. A highly potent beta-adrenergic affinity label. *Mol. Pharmacol.* **27,** 499–506.
5. Chorev, M., Fergenbaum, A., Keenan, A. K., Gilon, C., and Levitski, A. (1985) *N*-Bromoacetyl-amino-cyanopindolol: a highly potent beta-adrenergic affinity label blocks irreversibly a non-protein component tightly associated with the receptor. *J. Biochem.* **146,** 9–14.
6. Bar-Sinai, A., Aldouby, Y., Chorev, M., and Levitski, A. (1986) Association of turkey erythrocyte beta-adrenoceptors with a specific lipid component. *EMBO J.* **5,** 1175–1180.
7. Baker, S. P., Liptak, A., and Pitha, J. (1985) Irreversible inactivation of the beta-adrenoreceptor by a partial agonist. Evidence for selective loss of the agonist high affinity binding sites. *J. Biol. Chem.* **260,** 15,820–15,828.
8. Milecki, J., Baker, S. P., Standifer, K. M., Ishizu, T., Chida, Y., Kusiak, J. W., et al. (1987) Carbostyril derivatives having potent beta-adrenergic agonist properties. *J. Med. Chem.* **36,** 1563–1566.
9. Rashidbaigi, A. and Ruoho, A. E. (1981) Iodoazidobenzylpindolol, a photoaffinity probe for the beta-adrenergic receptor. *Proc. Natl. Acad. Sci. USA* **78,** 1609–1613.
10. Rashidbaigi, A. and Ruoho, A. E. (1982) Synthesis and characterization of iodoazidobenzylpindolol. *J. Pharm. Sci.* **71,** 305–307.
11. Rashidbaigi, A. and Ruoho, A. E. (1982) Photoaffinity labeling of beta-adrenergic receptors: identification of the beta-receptor binding site(s) from turkey, pigeon, and frog erythrocyte. *Biochem. Biophys. Res. Commun.* **106,** 139–148.
12. Resek, J. F. and Ruoho, A. E. (1988) Photoaffinity labeling the beta-adrenergic receptor with an iodoazido derivative of norepinephrine. *J. Biol. Chem.* **263,** 14,410–14,416.
13. Ruoho, A. E., Clark, R., Feldman, R. D., and Rashidbaigi, A. (1987) Photoaffinity labeling in the study of lymphoid cell beta-adrenergic receptors, in *Methods in Enzymology, vol. 150: Immunochemical Techniques* (Di Sabato, G., ed.), pp. 492–502.
14. Laemmli, U. K. (1970) Cleavage of structural proteins during the assembly of the head of bacteriophage T4. *Nature* **227,** 680–685.

22

Detection of β-Adrenergic Receptors by Radioligand Binding

Cheryl D. Dunigan, Patricia K. Curran, and Peter H. Fishman

1. Introduction

The catecholamines epinephrine and norepinephrine produce a wide variety of physiological responses in target cells. Some of these are mediated by β-adrenergic receptors (β-ARs) through the stimulation of adenylyl cyclase and the production of cyclic AMP as first demonstrated by Sutherland and his colleagues *(1)*. Although β-ARs were initially divided into β_1-AR (norepinephrine ≥ epinephrine) and β_2-AR (epinephrine > norepinephrine) subtypes by Lands et al. *(2)*, a third subtype, the β_3-AR, has now been recognized. The latter was first identified in adipocytes based on atypical responses to antagonists and later to novel agonists (reviewed in **ref. 3**). The three subtypes, from both human and rat, have been cloned, and although they share limited homology in amino acid sequences, they all are members of the seven membrane-spanning rhodopsin superfamily, and couple to the stimulatory G-protein and activate adenylyl cyclase.

Much of the progress in our identification and characterization of β-AR has occurred through the development of the radioligand binding assay (reviewed in **ref. 4**). This in turn has depended on the synthesis of radioligands that have high specificity and affinity as well as unlabeled antagonists with similar characteristics. The latter are used to define nonspecific binding. In addition, the development of highly selective subtype-specific antagonists has simplified the ability to detect the different subtypes in the same cells and tissues. Currently, the most useful radioligands are: (–)-[^{3}H]dihydroalpenolol, (–)-[^{125}I]iodocyanopindolol (^{125}ICYP), (–)-[^{125}I]iodopindolol, and (–)-[^{3}H]CGP 12177. Whereas the first three are hydrophobic and more suitable for

From: *Methods in Molecular Biology, vol. 126: Adrenergic Receptor Protocols*
Edited by: C. A. Machida © Humana Press Inc., Totowa, NJ

measuring receptors in broken cell and membrane preparations, [³H]CGP
12177 is hydrophilic and the radioligand of choice for whole-cell binding
assays. To determine nonspecific binding, the antagonists (–)-propranolol and
(–)-alprenolol are most useful, although a high concentration of the agonist
isoproterenol is occasionally used. CGP 20712A is the most selective β_1-AR
antagonist, whereas ICI 118,551 is a potent β_2-AR antagonist.

The most important criteria for a radioligand binding assay include both the
theoretical and the practical *(4,5)*. For the former, it is essential that the binding
sites identified by the assay are physiologically relevant receptors, and thus,
both recognize agonists and antagonists with the appropriate specificity and
affinity, and mediate a biological response. Thus, in order to develop a
radioligand binding assay for β-AR, one should use cells or tissues that display
a β-adrenergic response. Next, one does kinetic and equilibrium binding assays.
The kinetic data provide the rates of association and dissociation that can be
used to calculate independently the dissociation constant for the radioligand
(K_d). The association rate will also determine the minimum incubation time for
the assay. The dissociation data will establish that the binding is reversible,
one of the key criteria expected of a receptor. The equilibrium binding data
will provide the other key criteria: saturability and selectivity. For saturation
binding, one incubates a constant amount of receptors with increasing concentrations of radioligand (**Fig. 1A**). Often the binding data are analyzed after a
linear transformation, referred to as a Scatchard analysis, which will produce a
straight line when the binding is to a single class of sites (**Fig. 1A**, inset). From
the data, one can calculate both the K_d and the total number of receptors (B_{max}).
For theoretical reasons *(4)*, it is better to use a nonlinear curve-fitting computer
program for data analysis (*see* **Note 1**). To characterize β-AR further, one also
can do competition binding where a single concentration of radioligand is used
in the presence of increasing concentrations of unlabeled antagonists and agonists (**Fig. 1B**). Whereas the competition curves for antagonists are relatively
steep and yield a single class of binding sites, those for agonist are shallow and
indicative of two classes of binding sites *(4)*.

Regarding the practical aspects, it is important that the signal-to-noise ratio
(specific to nonspecific binding) be sufficiently strong in order to ensure that
the assay is reliable and reproducible. The reagents and other materials should
be commercially available or, in the case of some radioligands, can be synthesized using established protocols. The assay should be flexible enough to
handle a few as well as many samples, and easy to set up and conduct. Because
one is dealing with radioactivity, it is essential the procedures minimize any
radiation hazards, such as exposure, spills, or contamination, as well as the
generation of radioactive waste.

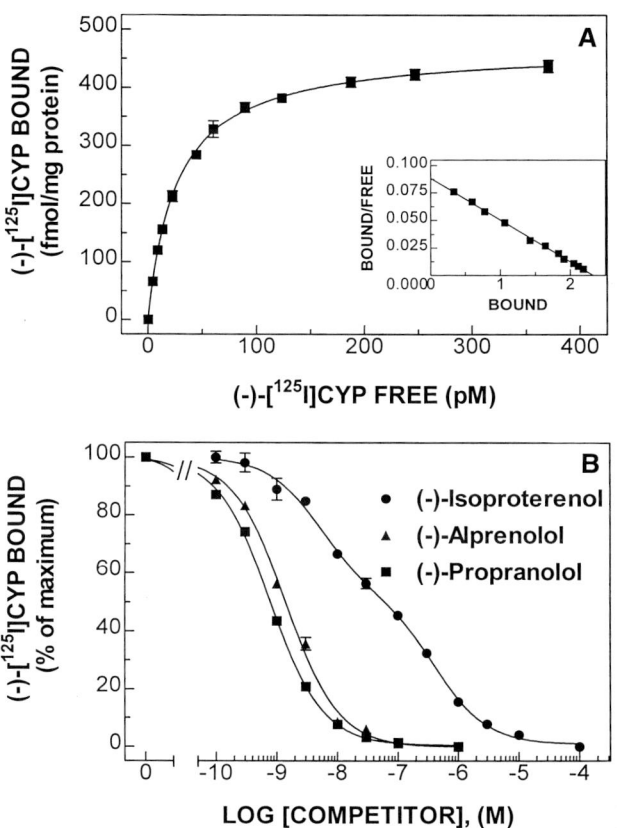

Fig. 1. Saturation and competition binding for ^{125}ICYP to membranes from CHW-h-β_2-AR cells. Membranes prepared from Chinese hamster fibroblast (CHW) cells stably expressing hβ_2-AR were incubated either with (**A**) increasing concentrations of ^{125}ICYP or (**B**) 30 p*M* ^{125}ICYP and increasing concentrations of competitor for 3 h at 30°C as described under **Subheading 3.** In (A), the binding data have been fitted to a one-site hyperbolic binding curve by nonlinear regression analysis. The inset shows a linear transformation of the data as a Scatchard plot. In (B), the binding data have been fitted to a one-site competition curve for the antagonists (▲,■) or a two-site competition curve for the agonist (●).

In this chapter, we will described experimental protocols established in our laboratory for assaying radioligand binding to β-AR that address three different aspects. In the first, we use ^{125}ICYP for measuring β-AR in cell lysates and membrane preparations. Because of its high specific radioactivity (2200 Ci/mmol), high affinity (~10 p*M* for β_1-AR and β_2-AR), and low nonspecific binding (usually ≤10% of total binding), it is the radioligand of choice for detecting low levels of receptors. Since ^{125}ICYP is also hydrophobic, it will detect β-AR that has become sequestered in endocytic vesicles and other subcellular membrane fractions. Thus, the assay detects the total population of β-AR and, therefore, can be used to follow receptor downregulation *(6–9)*. In the second, we use [^{3}H]CGP 12177 to measure surface β-AR on intact cells *in situ*. This assay is very useful to follow β-AR internalization *(6–9)*. In the third, we describe our protocols for identifying and quantifying β-AR subtypes in cells that express more than one subtype *(6–8)*. For these assays, the highly selective β$_1$-AR antagonist CGP 20712A has proven invaluable.

2. Materials
2.1. Apparatus

1. We use the Brandel M-24R harvester (Gaithersburg, MD) for separating bound from free radioligand by filtration (*see* **Note 2**). Although models that handle 48 or more samples at a time are available, the 24-sample harvester offers more flexibility, since small assays can be done with minimum of waste, and it only takes a few minutes for a skilled operator to filter 96 samples. We have found that #32 glass fiber sheets (2.25 × 12.25 in., cat. # 46690 from Schleicher & Schuell, Keene, NH) work as well as more expensive filters from other sources. The wash buffers are kept in 2-L polyethylene carboys with a tubular outlet at the bottom (Nalgene [Rochester, NY] #2302-0005). The outlet is fitted with a short length of Tygon tubing and a hose clamp, and is connected to the inlet side of the harvester wash pump using a barbed quick disconnect fitting (*see* **Note 3**). The waste is carried through Tygon tubing to a 4-L glass flask with side arm and a one-hole rubber stopper. The tubing is connected to the top of the stopper by a barbed elbow. A piece of plastic pipe (made by cutting the tip off of a 5-mL pipet) is fitted to the bottom of the stopper to ensure that the liquid waste is not aspirated into the side arm. The latter is connected through Tygon tubing and quick disconnect fittings to a 1-L glass flask with side arm that is used to catch any overflow. The smaller flask is connected to the house vacuum with Tygon tubing. A charcoal filter is placed in this line to trap any volatile radioactivity, and the waste flasks are placed in a large polypropylene pan to contain any leaks, spills, or drips. The quick disconnect fittings allow the large waste flask to be removed and emptied easily.
2. For washing cells in 24-well clusters, we use a 9-in. glass Pasteur pipet connected by Tygon tubing to a vacuum flask. A separate flask is used to collect any radioactive solutions. The clusters are tilted up at a 45° angle on a long edge, and the pipet tip is placed at the bottom edge of the well. We use either an Eppendorf Repeater (Brinkman Instruments, Westbury, NY) or an EasyStep electronic dispenser (Continental Lab Products, San Diego, CA) fitted with an Eppendorf Combitip to add media, wash buffer, or radioactive solutions to the wells. It is important to place the tip of the pipeting device against the side of the well when adding liquid to avoid pipeting directly on the cells and possibly detaching them.

3. To count filters containing bound ^{125}ICYP, we use an ICN Micromedics Apex/ Plus (Huntsville, AL) 10/300 γ-counter that has 10 detectors and >80% counting efficiency. Although any γ- or liquid scintillation counter can be used, it is important to determine the counting efficiency for ^{125}I in order to accurately calculate K_d and B_{max} values. We use an ^{125}I standard to calibrate our γ-counter. To count samples containing bound [^{3}H]CGP 12177, we use Beckman liquid scintillation counters with H-number quench correction and a quench correction curve generated by a set of quenched standards to give the values in dpm (*see* **Note 4**).

2.2. Reagents and Solutions

1. Radioligands: (−)-[^{125}I]iodocyanopindolol (2200 Ci/mmol) is obtained from Dupont-NEN (Boston, MA), is made on the first Monday of each month (*see* **Note 5**), and is stored at 4°C. (−)-[^{3}H]CGP 12177 (45–51 Ci/mmol) is obtained from Amersham (Arlington Heights, IL), can be used for at least 5–6 yr, and is stored at −25°C after diluting with ethanol to 5 μ*M*.
2. Other reagents: (−)-alprenolol, (−)-propranolol, (−)-isoproterenol, and bovine serum albumin (BSA) are obtained from Sigma (St. Louis, MO); (±)-CGP 12177A HCl, CGP 20712A methanesulfonate, and ICI 118, 551 HCl are from Research Biochemicals International (Natick, MA). The stock solutions (10 m*M*) of the different antagonists are prepared by dissolving each in 5 mL of ethanol and are kept at −20°C in screw-capped glass tubes with Teflon liners. For agonists, such as isoproterenol, 4 μL of concentrated HCl are added (10 m*M* HCl final). The stocks solutions of antagonists are stable for years, whereas isoproterenol is prepared every 6 mo.
3. Solutions for ^{125}ICYP binding:
    a. A 5X binding assay buffer is prepared from 1 *M* HEPES, pH 7.5, 1 *M* MgCl$_2$, and 10% bovine serum albumin (BSA): concentrations are 250 m*M*, 20 m*M* and 0.2%, respectively.
    b. A 5X solution of ^{125}ICYP is prepared by diluting the stock solution with water so that 50 μL contains 250,000 cpm (*see* **Note 6**).
    c. A 50 μ*M* solution of (−)-propranolol is prepared by diluting the stock solution with water.
    d. The wash buffer consists of 20 m*M* Tris-HCl, pH 7.4, 2 m*M* MgCl$_2$ at room temperature.
    e. The cell lysis buffer consists of 1 m*M* Tris-HCl, 2 m*M* ethylenediamine tetraacetic acid (EDTA), pH 7.4. When membranes are prepared for adenylyl cyclase assays, we include protease inhibitors in the lysis buffer (10 μg/mL each of leupeptin, soybean trypsin inhibitor, and benzamidine, and 1 m*M* Pefabloc (from Boehringer Mannheim, Indianapolis, IN). The protease inhibitors are kept at −20°C as a 50X mixed stock in 1-mL portions.
4. Solutions for [^{3}H]CGP 12177 binding to intact cells:
    a. For binding assays at 37°C, the stock solution of radioligand is diluted in serum-free medium buffered with 25 m*M* HEPES and containing 30 μg/mL

BSA. We use the same type of medium used to grow the cells. Thus, for cells grown in Dulbecco's Modified Eagle's Medium (DMEM), we use DMEM/ HEPES and for cells grown in Eagle's Minimal Essential Medium (EMEM), we use EMEM/HEPES. For binding assays at 4°C, we use EMEM/HEPES to minimize changes in pH at the lower temperature. Normally, 5 nM [^{3}H]CGP 12177 is used to determine maximum binding, and nonspecific binding is determined in the presence of 10 μM (−)-propranolol.

b. For washing the cells, we use ice-cold Dulbecco's phosphate-buffered saline (DPBS).
c. To dissolve the cells, we use 0.5 M NaOH (*see* **Note 7**).
d. For liquid scintillation counting, we use 3a70B cocktail (Research Products International, Mount Prospect, IL).

3. Methods

3.1. Binding to Cell Lysates and Membranes

3.1.1. Preparation of Cell Lysates and Membranes

1. Routinely, we use cells that grow in monolayer culture; the culture medium is removed by aspiration, the monolayers are rinsed twice with ice-cold DPBS without calcium and magnesium, and once with lysis buffer without protease inhibitors. Then a small volume of lysis buffer (with or without protease inhibitors depending on the type of assay) is added to the culture vessel. For a 175-cm^2 flask, 8 mL are added, and the flask is placed at 4°C for ~10 min with occasional shaking. After the cells have detached and begun to lyse, the solution is transferred to a disposable capped, conical polypropylene centrifuge tube and vortexed vigorously. The lysate can be assayed for binding and protein or used to prepare membranes.

2. For the latter, the lysate is centrifuged at ~500g for 5 min, and the supernatant is transferred to a 10-mL centrifuge tube. The pellet (nuclei and cell debris) is suspended in 2 mL of lysis buffer, centrifuged as before, and the supernatant combined with the first. This postnuclei supernatant is centrifuged at 43,000g for 20 min. The resulting supernatant is discarded, and the membrane pellet is suspended in a small volume of lysis buffer. The recovery of β-AR binding activity is usually 65–75% with a fourfold enrichment over the cell lysate. The lysates and membranes can be frozen at −80°C and retain their binding activity for many months.

3. It may be necessary to dilute the lysates and membranes for the binding assay. As a rule, we try to limit the amount of receptors in the assay to no more than 10 fmol (*see* **Subheading 3.1.3.**). Since we find it convenient to add 50-μL portions to the assay, we usually dilute the samples in lysis buffer, so 50 μL ≈ 5–10 fmol of binding activity.

3.1.2. Binding Assay

1. The assay is set up in 12 × 75 mm polypropylene tubes placed in a 6 × 12 hole test tube rack (PGC Scientific [Frederick, MD] #282-001) to which are added in the

Table 1
Protocol for Setting Up β-AR Binding Assay[a]

Tube #	5X Buffer	Water	Propranolol	^{125}ICYP	Lysate	Sample
1–3	50	50	—	50	100	1
4–6	50	—	50	50	100	1
7–9	50	50	—	50	100	2
10–12	50	—	50	50	100	2
13–15	50	50	—	50	100	3
16–18	50	—	50	50	100	3
19–21	50	50	—	50	100	4
22–24	50	—	50	50	100	4

[a]All volumes are in μL.

following order: 50 μL of 5X binding buffer; water to provide a final volume of 250 μL; 50 μL of ^{125}ICYP; 50 μL of propranolol as indicated; and up to a 100 μL of cell lysate or membranes. The protocol for assaying four different samples is shown in **Table 1**. When we have to assay many samples, we set up all the reactions without adding the lysates or membranes and store them at 4°C. Then when the samples are ready, we add them to the tubes, mix well, and incubate.

2. After everything has been added to the reaction tubes, they are mixed by placing the rack on a vortex-type mixer. Then the rack is placed in a 30°C water bath with an orbiting platform (New Brunswick Scientific, Edison, NJ) for 1 h.

3. The rack is then removed from the water bath, and the reactions rapidly filtered using 4 × 4 mL of the wash buffer. The filters are transferred to 12 × 75 mm polystyrene tubes and counted in a γ-counter.

3.1.3. Optimizing the Binding Assay

There is an important relationship between the amount of receptors, the concentration of radioligand, and the incubation time. One wants to add a sufficient amount of binding activity to obtain an adequate signal, but not an excess so that the concentration of free ligand is depleted *(5)*. When free ligand is depleted, binding is no longer proportional to the amount of receptors present. This can easily occur when binding studies are on transfected cells expressing β-AR. **Figure 2A** shows data for binding to increasing amounts of lysate from CHW-hβ$_1$ cells that express ~1 pmol/mg protein. Binding increased proportionally until 20 μg of protein, at which point the free radioligand was reduced to <150 pM. On the other hand, endogenous β-AR are usually expressed at low levels (~100 fmol/mg protein), and increasing the signal-to-noise ratio becomes important. One way to accomplish this is to lower the concentration of radioligand. As seen in **Fig. 2B**, reducing the concentration of ^{125}ICYP from 250 to 150 pM had only a minimal effect on binding. The ratio of specific to

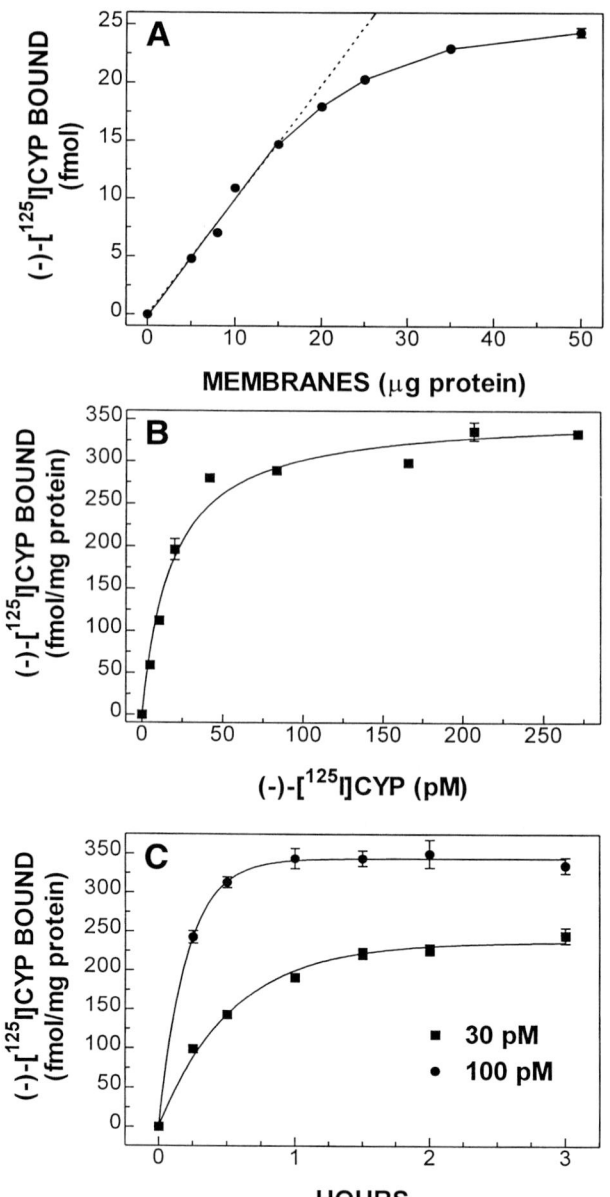

Fig. 2. Effects of receptor density, radioligand concentration, and incubation time on binding of [125]ICYP to hβ_1-AR. (**A**) Increasing amounts of membranes from CHW-hβ_1-AR cells were incubated with 220 p*M* [125]ICYP for 1 h at 30°C. (**B**) Membranes from human SK-N-MC cells were incubated with increasing concentrations of [125]ICYP for 3 h at 30°C. (**C**) Same as in (**B**), except the membranes were incubated for increasing times with the indicated concentrations of [125]ICYP.

nonspecific binding, however, increased from 9.0 to 13.6. When the radioligand concentration is reduced, it is important that the incubation time be long enough for equilibrium to be obtained as shown in **Fig. 2C**. Thus, 1 h is sufficient for equilibrium to be reach with 100 pM [125]ICYP, whereas at least 2 h are required with 30 pM. For this reason, we routinely use a 3-h incubation time for both saturation and competition binding assays. Furthermore, we increase the assay volume to 1 mL to ensure that the radioligand is not depleted at the lower concentrations. Under these conditions, we have obtained K_d values of 8–22 pM for [125]ICYP using a wide variety of cells, such as rat C6 glioma (endogenous $β_1$-AR and $β_2$-AR), human HeLa and A431 (endogenous $β_2$-AR), human SK-N-MC (endogenous $β_1$-AR), and stably transfected hamster cell lines (CHW, Chinese hamster ovary [CHO], baby hamster kidney [BHK]) expressing rat $β_1$-AR and human $β_1$-AR and $β_2$-AR (*see* **Note 8**).

3.2. Binding to Intact Cells

1. We routinely grow the cells in 24-well cluster dishes (Corning Costar [Cambridge, MA] #3524), but have used 12-well clusters (#3512) and 48-well clusters (#3548) for cells expressing low and high levels of β-AR, respectively. The cells are used when they become confluent.
2. For determining saturation binding at 37°C, the culture medium is removed by aspiration, the cells rinsed once with DPBS (1 mL/well for a 24-well cluster), and 500-μL portions of the solution containing increasing concentrations of [³H]CGP 12177 ± 10 μM propranolol are added. For each concentration, six wells are used, three for total and three for nonspecific binding. The clusters of cells are incubated at 37°C for 45 min, placed on a bed of ice, and rapidly washed with 3 × 0.5 mL portions of ice-cold DPBS. The cells then are dissolved by adding 1 mL of 0.5 M NaOH and warming at 37°C for ~2–3 h or leaving overnight at room temperature. After the cells have dissolved, a 100-μL portion of each sample is assayed for protein, and the rest is transferred to a glass scintillation vial, neutralized with 30 μL of glacial acetic acid, and diluted with 10 mL of liquid scintillation counting cocktail (*see* **Note 9**). The samples then are counted for ³H in dpm.
3. For determining the internalization of cell-surface receptors, the culture medium is removed and replaced with serum-free medium buffered with 25 mM HEPES (1 mL/well for a 24-well cluster). After the cells are equilibrated at 37°C for 10 min, they are exposed to agonist for increasing times. The clusters are placed on a bed of ice, and rapidly washed with 3 × 1 mL portions of ice-cold DPBS. Then 250-μL portions of the ice-cold solution containing 5 nM [³H]CGP 12177 ± 10 μM propranolol are added. The clusters are placed in a refrigerator at 4°C for 90 min, then returned to a bed of ice, and rapidly washed with 3 × 0.5 mL portions of ice-cold DPBS. The rest of the assay is carried out as described above.
4. Because (–)-[³H]CGP 12177 has a lower affinity for β-AR than (–)-[125I]CYP with K_d values of 100–250 pM, higher concentrations are required to reach saturation. In turn, the incubation times can be reduced. Thus, binding to intact cells

at 37°C requires only 45 min and at 4°C, equilibrium is reach in <90 min with 5 nM (–)-[^{3}H]CGP 12177.

3.3. Detection of Different β-Adrenergic Receptor Subtypes

3.3.1. Competition Binding

To determine the proportion of β$_1$-AR and β$_2$-AR in cells, such as rat C6 glioma, we use a competition binding assay and a subtype-selective antagonist. The most selective β$_1$-AR antagonist is CGP 20712A *(10)*. As seen in **Fig. 3A**, CGP 20712A is a thousand-fold more potent at inhibiting ^{125}ICYP binding to β$_1$-AR than to β$_2$-AR. To assay samples containing both subtypes, we use our standard assay with the following modifications: the volume is increased to 1 mL, the ^{125}ICYP concentration is reduced to 25–30 pM, and increasing concentrations of CGP 20712A from 10^{-10} to 10^{-4} M are used. The tubes are set up in an ice bath, and 100-μL portions of the sample are added last. The tubes are incubated for 3 h at 30°C and then filtered as described above. The binding data are analyzed by computer-assisted nonlinear regression analysis to fit a two-site competition curve (*see* **Note 1**). Typical results for rat C6 glioma cells are shown in **Fig. 3B** in which the ratio of β$_1$-AR to β$_2$-AR is ~2:1 *(6,7)*. We have used a similar approach to determine the distribution of the two subtypes on the surface of intact C6 cells. Briefly, the medium is changed to 400 μL/well of DMEM/HEPES/BSA, and 50-μL portions of increasing concentrations of CGP 20712A are added followed by 50-μL portions of 6 nM [^{3}H]CGP 12177 all in the same medium. The cells are incubated at 37°C for 45 min, washed, dissolved, and assayed for protein and bound ^{3}H as described in **Subheading 3.2.**, **step 1**.

3.3.2. Saturation Binding

Subtype analysis by competition binding is suitable as long as the radioligand itself is not highly subtype-selective. Whereas both ^{125}ICYP and [^{3}H]CGP-12177 exhibit little selectivity for β$_1$-AR and β$_2$-AR, they both have a much lower affinity for β$_3$-AR as do the antagonists alprenolol and propranolol. To discriminate between β$_3$-AR and β$_1$-AR, we use a saturation binding assay that has been modified from our standard assay as follows: the volume is kept at 250 μL, the concentration of ^{125}ICYP is varied from 10 to 2500 pM, and 100 μM alprenolol is used to define nonspecific binding. The membranes are added last, and the tubes are incubated for 2.5 h at 30°C. Ice-cold wash buffer is used for the filtration to minimize any dissociation of the radioligand from β$_3$-AR. The binding data are analyzed by nonlinear regression to fit a two-site hyperbolic binding curve (*see* **Note 1**). Such a plot is shown in **Fig. 4A** for SK-N-MC cells that express both β$_1$-AR and β$_3$-AR *(8)*. The broken lines represent the theoretical one-site binding curves for each subtype.

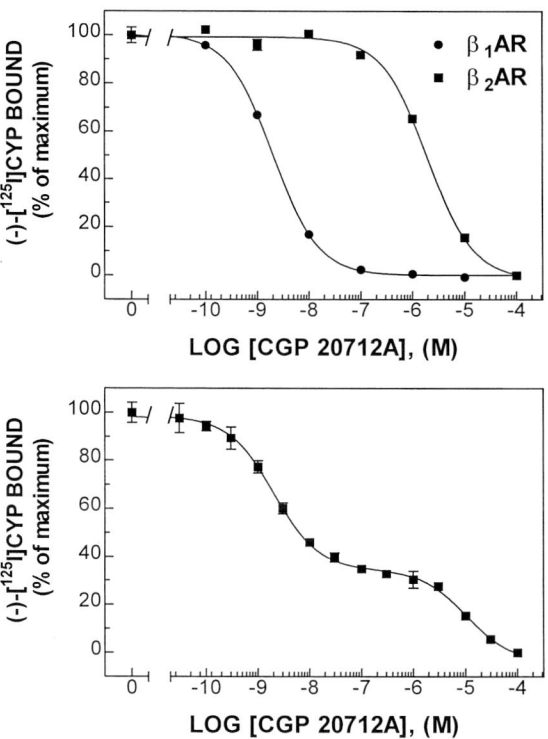

Fig. 3. Determination of β_1-AR and β_2-AR by competition binding with CGP 20712A. **(Top)** Lysates from CHW-hβ_1 (●) or CHW-hβ_2 (■) cells were incubated with 27.5 pM ^{125}ICYP and increasing concentrations of CGP 20712A for 3 h at 30°C as described under **Subheading 3.** Nonlinear regression analyses of the binding data gave IC$_{50}$ values of 2.07 nM and 1.92 μM for hβ_1-AR and hβ_2-AR, respectively. **(Bottom)** Same as in the top panel except lysates from rat C6 glioma cells were used. The respective IC$_{50}$ values were 2.18 nM and 10.7 μM, and the percentages of β_1-AR and β_2-AR were 62.3 and 37.7.

We also can discriminate between the two by using a single concentration of CGP-20712A (0.3 μM), which will occupy most of the β_1-AR, but very few of the β_3-AR *(8)*. In this procedure, we vary the concentration of radioligand and use 0.3 μM CGP-20712A to block binding to β_1-AR and 100 μM alprenolol to block all specific binding. The difference between total binding and binding in the presence of CGP-20712A represents β_1-AR, whereas the difference between the latter binding and nonspecific binding represents β_3-AR **(Fig. 4B)**. The same approach can be used to discriminated between β_1-AR and β_2-AR in rat C6 glioma cells *(6,7)*. If [^{3}H]CGP-12177 is used, then the distribution of subtypes on the cell surface can be determined.

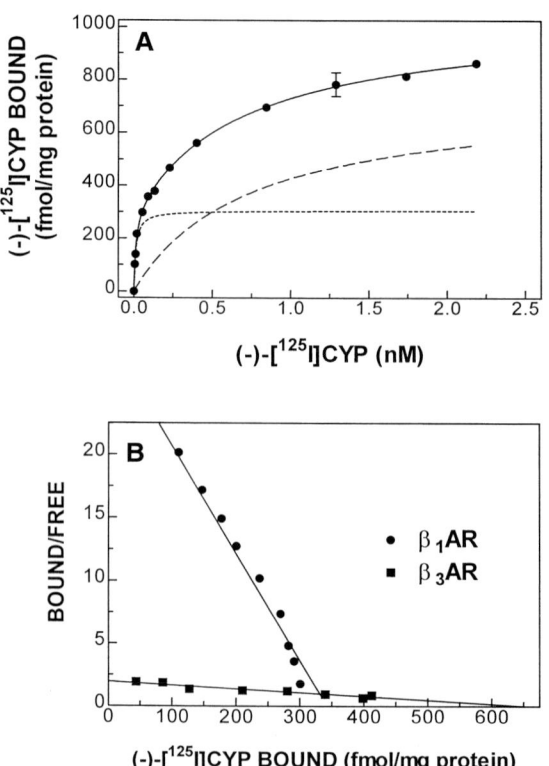

Fig. 4. Saturation binding of ^{125}ICYP to β_1-AR and β_3-AR in SK-N-MC cells. (**A**) Membranes from SK-N-MC cells were incubated with increasing concentrations of ^{125}ICYP ± 100 μ*M* alprenolol for 2.5 h at 37°C. The binding data have been fitted to a two-site hyperbolic binding curve by nonlinear regression analysis. The respective K_d values are 12.1 and 735 p*M*. The broken lines represent the theoretical one-site binding curves. (**B**) Same as in (A), except binding also was done in the presence of 0.3 μ*M* CGP 20712A. Binding to β_1-AR was defined as total binding minus binding in the presence of CGP 20712A, and binding to β_3-AR as the latter binding minus nonspecific binding. The binding data are shown as Scatchard plots with K_d values of 9.5 and 1280 p*M*.

Finally, we use the above assays with a single high concentration of radioligand to follow the internalization and downregulation of β-AR subtypes in rat C6 glioma and human SK-N-MC neurotumor cells *(6–8)*. The assay conditions are summarized as follows: (1) for internalization of β-AR, intact cells are exposed to agonist for different times, washed, and incubated with 0.5 mL/ well of 1 n*M* [^{3}H]CGP 12177 at 4°C for 4 h with no inhibitor, 0.3 μ*M* CGP 20712A, or 1 μ*M* propranolol to detect β_1-AR and β_2-AR; or with 100 n*M*

[³H]CGP 12177 at 4°C for 1 h with 0.3 μM CGP 20712A, or 100 μM CGP 12177A to detect β_3-AR; (2) for downregulation, lysates from control and treated cells are incubated with 100 pM ¹²⁵ICYP at 30°C for 75–90 min with no inhibitor, 0.3 μM CGP 20712A, or 1 μM propranolol to detect β_1-AR and β_2-AR; or 1 nM ¹²⁵ICYP at 30°C for 60 min with 0.3 μM CGP 20712A, or 100 μM alprenolol to detect β_3-AR.

4. Notes

1. We use the program Prism 2 from GraphPad Software (San Diego, CA) to analyze the binding data by nonlinear least-squares regression analysis. In addition to analyzing and plotting one- and two-site hyperbolic binding and competition curves, the program will do association and dissociation curves as well as concentration–response curves.
2. The Brandel is very reliable, but does need routine maintenance. We flush the system with 30% ethanol or 10% methanol about once a week and with a 30% bleach solution once a month to clean out the lines. If the harvester is being used often, we squirt a 70% ethanol solution on the stainless-steel filter screens and the o-rings to remove any buildup of glass fibers. Trouble signs to look for are: low vacuum; uneven delivery of wash solution to the binding assay tubes; uneven scoring of the filters by the o-rings.
3. It is important that the inner diameter of the tubing connecting to the wash pump not be restricted since this will reduce the flow of the wash buffer. We once made the mistake of placing a stopcock in the line.
4. To convert cpm of bound ¹²⁵ICYP to fmol, we use the following calculations:
   cpm/counting efficiency × 100 = dpm.
   dpm/2220 = nCi (1 μCi = 2.22 × 10⁶ dpm).
   nCi/2.2 = fmol (SA of ¹²⁵ICYP = 2200 Ci/mmol).
   With a counting efficiency of 82%, 4000 cpm = 1 fmol [4000 × (100/82)/ (2220 × 2.2)].
   To convert dpm of bound [³H]CGP 12177 to fmol, we use the following calculations:
   dpm/2.22 = pCi (1 μCi = 2.22 × 10⁶ dpm).
   pCi/45 = fmol (SA of our current lot of [³H]CGP 12177 = 45 Ci/mmol).
5. Although ¹²⁵ICYP undergoes radioactive decay with a 60-d half-life, the SA remains constant with time owing to decay catastrophe. We have used ¹²⁵ICYP over 120 d old and obtained the same binding activity as that observed with freshly made radioligand as shown in **Table 2**.
6. The working solution of ¹²⁵ICYP should be prepared in either a glass or polypropylene tube as the radioligand will absorb to polystyrene tubes. We often add BSA (100 μg/mL) to ensure that it will not be absorbed.
7. Some cells, such as Chinese hamster fibroblasts (CHW), do not readily dissolve in 0.5 M NaOH, so we add half the normal volume of 1 M NaOH. After the cells have dissolved, we add an equal volume of water.

Table 2
Effect of Radioactivity Decay on ^{125}ICYP Binding Activity[a]

Days old	Fraction remaining	cpm Added	Total cpm bound	nonspecific cpm bound	fmol of β-AR
4	0.955	200,313	36,795 ± 1120	648 ± 70	9.04 ± 0.28
123	0.2415	208,946	38,117 ± 903	627 ± 53	9.37 ± 0.23

[a]Membranes from CHW-hβ$_2$ cells were incubated with the indicated amounts of ^{125}ICYP from new and old lots using standard assay conditions.

8. The rate of binding also is inversely affected by temperature; lowering the temperature to 25°C will slow binding whereas at 37°C, equilibrium will be reached sooner. If because of incubation time or temperature, equilibrium is not reached, then one will be measuring apparent and not true K_d values.

9. More recently, we have been dissolving the cells in 250 μL of 1 *M* NaOH in order to reduce the amount of radioactive waste. After a 20-μL portion is removed for protein determination (we use a microtiter plate assay based on the Pierce BCA Protein Assay Reagent, Rockford, IL), the rest of the sample is transferred to a 7-mL glass minivial. Each well is rinsed with 250 μL of water that is added to the vial. Then 15 μL of glacial acetic acid and 4 mL of counting cocktail are added to each vial.

References

1. Robison, G. A., Butcher, R. W., and Sutherland, E. W. (1968) Cyclic AMP. *Annu. Rev. Biochem.* **37,** 149–174.
2. Lands, A. M., Arnold, A., McAuliff, J. P., Luduena, F. P., and Braun, T. G. (1967) Differentiation of receptor systems activated by sympathomimetic amines. *Nature* **214,** 597,598.
3. Arch, J. R. S. and Kaumann, A. J. (1993) β$_3$ and atypical β-adrenoceptors. *Med. Res. Rev.* **13,** 663–729.
4. Stadel, J. M. and Lefkowitz, R. J. (1991) Beta-adrenergic receptors: identification and characterization by radioligand binding studies, in *The Beta-Adrenergic Receptors* (Perkins, J. P., ed.), Humana, Clifton, NJ, pp. 1–40.
5. Keen, M. (1995) The problems and pitfalls of radioligand binding. *Methods Mol. Biol.* **41,** 1–16.
6. Hosada, K., Feussner, G. K., Rydelek-Fitzgerald, L., Fishman, P. H., and Duman, R. S. (1994) Agonist and cyclic AMP-mediated regulation of β$_1$-adrenergic receptor mRNA and gene transcription in rat C6 glioma cells. *J. Neurochem.* **63,** 1635–1645.
7. Fishman, P. H., Miller, T., Curran, P. K., and Feussner, G. K. (1994) Independent and coordinate regulation of β$_1$- and β$_2$-adrenergic receptors in rat C6 glioma cells. *J. Receptor Res.* **14,** 281–296.

8. Curran, P. K. and Fishman, P. H. (1996) Endogenous β_3- but not β_1-adrenergic receptors are resistant to agonist-mediated regulation in human SK-N-MC neurotumor cells. *Cell. Signal* **8,** 355–364.
9. Zhou, X.-M., Pak, M., Wang, Z., and Fishman, P. H. (1995) Differences in desensitization between human β_1- and β_2-adrenergic receptors stably expressed in transfected hamster cells. *Cell. Signal* **7,** 207–217.
10. Dooley, D. J., Bittiger, H., and Reymann, N. C. (1986) CGP 20712A: a useful tool for quantitating β_1- and β_2-adrenoceptors. *Eur. J. Pharmacol.* **130,** 137–139.

23

Assessing Adrenergic Receptor Conformation Using Chemically Reactive Fluorescent Probes

Anne Dam Jensen and Ulrik Gether

1. Introduction

It is believed that binding of agonist to a G-protein coupled receptor (GPCR) induces a set of structural changes in the tertiary structure of the receptor that can be recognized by the associated G-protein α-subunit. Many different methodological approaches have been applied over the years in the attempt to understand these conformational changes and establish the critical link between agonist binding and G-protein coupling *(1)*. However, most models for how GPCRs are activated have been based on indirect evidence; therefore, the receptor conformation has been inferred from the ability of the receptor to activate second messenger systems or from computational simulations *(2–6)*. Only during the last couple of years has the use of spectroscopic techniques on purified receptors preparations allowed direct insight into the structural changes underlying activation of GPCRs *(7–12)*. Thus far, a majority of the studies have been carried out in the photoreceptor, rhodopsin. There are abundant natural sources of rhodopsin, and its inherent stability makes it possible to produce and purify relatively large quantities of recombinant protein. In particular, the elegant use of electron paramagnetic resonance (EPR) spectroscopy by Hubbell, Khorana and coworkers has provided substantial insight into the conformational changes associated with photoactivation of rhodopsin *(9)*. Notably, spectroscopic approaches have also been used to examine the structural basis for the interaction between the α-subunit / $\beta\gamma_T$-subunit complex of transducin and rhodopsin *(13–15)*.

Efficient purification procedures for hormone-activated GPCRs, such as the β_2-adrenergic receptor (β_2-AR), have been established only very recently.

From: *Methods in Molecular Biology, vol. 126: Adrenergic Receptor Protocols*
Edited by: C. A. Machida © Humana Press Inc., Totowa, NJ

As described in this chapter, expression of an epitope-tagged version of the β_2-AR in Sf9 insect cells has allowed purification of this receptor in sufficient quantities for direct spectroscopic analysis. It has thus become possible to probe conformational changes in the receptor both using endogenous tryptophan fluorescence and by incorporating chemically reactive fluorescence probes *(10–12,16)*. Since endogenous tryptophan fluorescence is sensitive to polarity of the environment and the β_2-AR contains eight tryptophan residues dispersed throughout the receptor structure, tryptophan fluorescence may be used to monitor conformational changes in the receptor molecule. By utilizing this technique, it was found that dithiothreitol (DTT) induces a reversible confor-mational change of the β_2-AR and that antagonists were able to slow the rate of this DTT-induced conformational change *(16)*. Unfortunately, it was not pos-sible to assess ligand-induced conformational changes in the receptor owing to overlap between the absorption spectra for tryptophan and the adrenergic ligands. However, the use of a sulfhydryl-reactive environmentally fluorescent probe has permitted characterization and mapping of ligand-induced confor-mational changes in the β_2-AR *(10–12,16)*. The sulfhydryl-reactive fluores-cent probe, IANBD, was incorporated into the purified β_2-AR and used as a molecular reporter of structural changes that takes place on agonist binding to the receptor *(10–12)*. Utilizing a series of mutant receptors containing one, two, or three of the naturally occurring cysteines, it was possible to map ago-nist-induced structural changes to helix III and VI. These data supported the EPR spectroscopy studies in rhodopsin, suggesting that a counterclockwise rotation of helix VI is an essential part of the activation mechanism *(9,12)*. The EPR spectroscopy studies in rhodopsin also indicated that helix III and VI move away from each other during activation *(9)*. Moreover, the fluorescence spec-troscopy data indicated that helix III most likely undergoes a counterclockwise rotation *(12)*.

It is the aim of this chapter to describe the background and methods involved in using chemically reactive fluorescence probes to measure conformational changes in the β_2-AR in response to binding different classes of ligands. Since the purity of the protein used for the spectroscopic analysis is critical, we have included a description of the expression and purification procedure.

2. Materials

2.1. Culturing and Infection Sf9 Insect Cells

1. SF900-II medium.
2. Fetal calf serum (FCS), Gibco (Grand Island, NY).
3. Gentamycin.
4. Polypropylene Erlenmeyer flasks, Corning Costar (Acton, MA).
5. 2800-mL triple-baffled Fernbach flask, Bellco Glass Inc. (Vineland, NJ).

2.2. Purification of the β₂-AR

1. Chelating fast-flow Sepharose resin, Pharmacia (Uppsala, Sweden).
2. M1 immunoaffinity resin, Eastmann Kodak (Rochester, NY).
3. [^{3}H]dihydroalprenolol, Amersham (Arlington Heights, IL).
4. *n*-Dodecyl-β-D-maltoside (DβM), Anatrace Inc. (Maumee, OH).
5. Lysis buffer: 10 mM Tris-HCl, pH 7.5, 10 μg/mL, 10 benzaminde (Sigma), μg/mL leupeptin (Sigma), 200 μM PMSF (Sigma).
6. Solubilization buffer: 20 mM Tris-HCl, pH 7.5, 500 mM NaCl, 0.8% DβM, 1 μM alprenolol, 10 μg/mL benzamidine, 10 μg/mL leupeptin, 200 μM phenylmethylsulfonyl fluoride (PMSF).
7. High-salt buffer: 20 mM Tris-HCl, pH 7.5, 500 mM NaCl, 0.08% DβM.
8. Low-salt buffer: 20 mM Tris-HCl, pH 7.5, 100 mM NaCl, 0.08% DβM.
9. Binding buffer: 75 mM Tris-HCl, pH 7.4, 12.5 mM (for membranes) MgCl₂, 1 mM ethylenediamine tetra-acetic acid (EDTA).

2.3. Binding Assay on Purified β₂-AR

1. Sephadex G50 resin, Pharmacia.
2. Poly prep columns, Bio-Rad (Hercules, CA).

2.4. Fluorescence Spectroscopy Analysis

1. Fluorescence spectroscopy was done using a SPEX Fluoromax-2 spectrofluo-rometer connected to a PC equipped with the Datamax software package.
2. Nitrobenzdioxazol iodoacetamide (IANBD) Molecular Probes (Eugene, OR).
3. The 5 × 5 mm quartz cuvet (HELLMA [type no. 1001.016-QS]).
4. 2 × 2 mm magnetic stir bar Bel-Art Products (Pequannock, NJ).

3. Methods
3.1. Expression and Purification of the β₂-AR

Spectroscopic analyses require a reliable expression system and preferably an easy purification procedure. We have used the baculovirus/Sf9 cell system to express the β₂-AR. Sf9 cells are straightforward to grow, do not require CO_2, and can easily be adapted to suspension cultures in a standard shaker using either glass or polyethylene Erlenmeyer/Fernbach flasks. The most prominent problem of the insect cell expression system may be the varying fraction of improperly folded and thus nonfunctional protein *(17,18)*. In case of the β₂-AR, approximately half of the synthesized receptor is nonfunctional *(18)*. The fraction can vary dramatically from protein to protein and may hinder the use of the system. However, high levels of expression of several GPCRs in Sf9 insect cells have been reported *(19,20)*.

The generation of recombinant baculovirus encoding the modified receptor and the procedures involved in amplification of the virus are outside the scope of the present chapter and are described in Chapter 14. However, the procedures we use for growing and infecting large-scale Sf9 cell cultures are important for achieving optimal expression levels and will therefore be described.

3.1.1. Culturing and Infection of Sf9 Insect Cells for Purification of the β_2-AR

Sf9 insect cells are maintained in SF900-II medium supplemented with 5% heat-inactivated FCS and 0.1 mg/mL gentamycin (all products purchased from Gibco) (*see* **Note 1**). The cell stock is normally kept in 250-mL polypropylene Erlenmeyer flasks (Corning Costar, Acton, MA) at 27°C in a shaker set at 125 rpm. Each flasks contains 70–100 mL medium and the cells are kept at a density varying from 0.5 to 6×10^6 cells/mL.

1. For infections, 500–1000 mL of cells are grown in a 2800-mL triple-baffled Fernbach flask (Bellco Glass Inc.) until they reach a density of $5–7 \times 10^6$ cells/mL. The cells are split into two flask, and fresh medium is added 1:1.
2. The cells are grown until the next day (usually they will have reached a density of $5–6 \times 10^6$ cells/mL). The cells are infected by inoculating virus (1:20 to 1:200 dilution of virus stock). The optimal inoculum for each virus stock is determined by small-scale infections of 20 mL suspension cultures (*see* **Note 2**). To stabilize the receptor during infection, 1 μM of the antagonist alprenolol is added (*see* **Note 3**).
3. Cells are incubated for 48 h at 27°C under constant shaking, pelleted by centrifugation for 10 min at 2500g, and stored at −70°C until purification (*see* **Note 4**).

3.1.2. Purification of the β_2-AR for Spectroscopic Analysis

For purification of the β_2-AR, we express a modified form of the receptor in the baculovirus/Sf9 cell system. A cleavable influenza-hemaglutinin signal sequence followed by the M1 antibody "FLAG"-epitope is inserted at the amino-terminus *(17,18)*. It has previously been shown that the introduction of this signal-sequence results in an approximate twofold increase in expression *(17)*. At the carboxy-terminus, the receptor is tagged with six histidines allowing nickel affinity chromatography as a step in the purification procedure. The following purification procedure was developed to ensure the purification of full-length and active receptor by utilizing an initial nickel-chromatography step, followed by anti-FLAG immunoaffinity chromatography and alprenolol affinity chromatography *(18)*.

1. Cells from 500–1000 mL of infected culture are lysed in 100–200 mL of lysis buffer.
2. Lysed cells are pelleted by centrifugation at 35,000g for 30 min at 4°C and the supernatant is discarded.
3. To solubilize the receptor, the pellet is resuspended in 100–200 mL solubilization buffer. The resuspended cells are solubilized by douncing (20 strokes with

the tight pestle) followed by incubation at 4°C for 1.5–2 h under constant stirring using a magnetic stir bar.

4. The solubilized protein is separated from the particulate by centrifugation at 35,000g for 30 min at 4°C.
5. Binding of the solubilized receptor to chelating fast-flow Sepharose, charged with nickel and equilibrated in high-salt buffer, is done by batch absorption for 2–3 h at 4°C with gentle rotation. Before the resin is mixed with the solubilisate, imidazole is added to a final concentration of 50 m*M* to lower nonspecific adsorption of protein. In general, we use 1 mL of resin/100 mL cell culture.
6. The nickel resin is pelleted by centrifugation for 5 min at 2000g at 4°C, washed once in 4 vol high-salt buffer, and loaded onto a column. Additional washing is done with 5 vol high-salt buffer and 2 vol high-salt buffer containing 25 m*M* imidazol.
7. Elution of the receptor is done in 1/2 column volume fractions with high-salt buffer containing 200 m*M* imidazol. Elution of the protein from the column is followed by a simplified Bradford protein assay (*see* **Note 5**). Protein-containing fractions are screened for receptor binding (*see* **Subheading 2.3.**) and peak fractions are pooled.
8. CaCl$_2$ is added to the pooled fractions at a final concentration of 2.5 m*M*. The pooled fractions are loaded onto an M1 antibody column (0.2 mL/nmol receptor) equilibrated in low-salt buffer, and recycled four times by gravity flow. The column is washed with 4 vol low-salt buffer containing 2.5 m*M* CaCl$_2$ and eluted using low-salt buffer containing 1 m*M* EDTA in 1/4 column volume fractions. Fractions are analyzed for receptor binding and peak fractions pooled.

These two purification steps can produce almost pure protein (SA around 5 nmol/mg of protein). However, around half is nonfunctional *(18)*. To separate the nonfunctional receptor from functional receptors, we use alprenolol affinity chromatography, which is a standard procedure for purification of the β$_2$-AR *(21–23)*. It is important to note that we have been able to exclude the M1 immunoaffinity chromatography in some applications. This results in a specific activity of the purified receptor around 5–10 nmol/mg as compared to 10–15 nmol/mg for the three-steps purified. Approximately 5 nmol of purified protein generally can be obtained from a 1000-mL culture. Protein is determined using the detergent-insensitive Bio-Rad *DC* protein assay kit (Bio-Rad). Purified β$_2$-AR receptors are analyzed by classical 10% sodium dodecyl sulfate-polyacrylamide gel electrophoresis (SDS-PAGE). Notably, samples should not be boiled before loading on to the gel to prevent receptor aggregation. The receptor is visualized by standard Coomassie staining.

3.1.3. Binding Assay on Purified β$_2$-AR

The amount of active β$_2$-AR is determined in a binding assay using [^{3}H]dihydroalprenolol (^{3}H-DHA) as radioligand. Our standard procedure is as follows:

1. Purified β_2-AR (10 μL appropriately diluted) is incubated with 10 n*M* ^{3}H-DHA in a total volume of 100 μL in low-salt buffer for 1 h. Nonspecific binding is determined in the presence of 10 μ*M* alprenolol.

2. Free ^{3}H-DHA is separated from bound by rapid desalting on 2-mL Sephadex G50 (Pharmacia) columns (Poly Prep columns, Bio-Rad) (*see* **Note 6**). Columns are eluted with 1 mL of ice-cold low-salt buffer directly into scintillation vials. Scintillation fluid is added followed by counting in a scintillation counter.

3.2. Derivatization of the Purified β_2-AR with Chemically Reactive Fluorescence Probes

The emission from many fluorescent molecules is strongly dependent on the polarity of the environment in which they are located. Incorporation of fluorescent labels into proteins can therefore be used as sensitive indicators of conformational changes and protein–protein interactions that cause changes in polarity of the environment surrounding the probe *(15,24–26)*. Nitrobenzdioxazol IANBD is a highly fluorescent, cysteine-selective reagent *(25,26)*. As illustrated in **Fig. 1**, the fluorescence from IANBD increases as the polarity of the solvent decreases. Furthermore, decreasing the polarity of the solvent causes a blueshift in the emission maximum from 543 nm in aqueous buffer to 536 nm in 80% dioxane. We have used IANBD to probe conformational changes in the β_2-AR. To incorporate IANBD (or other cysteine reactive labels) into the purified β_2-AR we use the following procedure:

1. Purified receptor (usually a whole badge of 2–5 nmol) is incubated with 0.5 m*M* IANBD (*see* **Note 7**) for 20 min in high-salt buffer at room temperature in the dark.

2. To remove excess unreacted IANBD, the labeled receptor is bound to a 150 μL nickel column (chelating fast-flow Sepharose resin from Pharmacia equilibrated in high-salt buffer). Binding to the column is achieved by recycling the labeling mix five times over the nickel column.

3. Excess dye is removed by extensive washing of the column with approx 50 column volumes of high-salt buffer. Labeled receptor is eluted in 50-μL fractions with 200 m*M* imidazol in high-salt buffer. Fractions are assayed for protein content, and peak fractions are pooled. The labeled receptor can be used directly for the fluorescence spectroscopy analysis or stored on ice at 4°C. Under these conditions, the protein is stable for several days.

This labeling procedure generally results in incorporation of 1.2–2 mol IANBD/mol receptor, as determined by measuring absorption at 481 nm and using an extinction coefficient of 21,000^{-1} cm for IANBD and a mol wt of 50 kDa for the receptor. We should note that the covalent modification of the receptor can be confirmed by SDS-PAGE of the labeled receptor followed by visualization of the labeled receptor band under UV light. The specificity

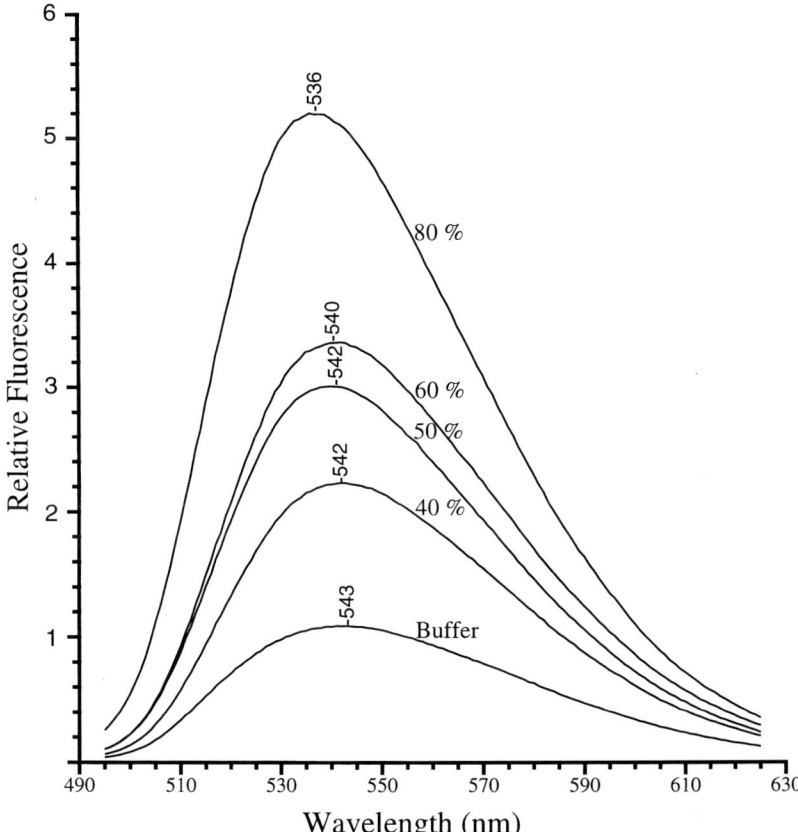

Fig. 1. Fluorescence properties of IANBD. Emission spectra of 1 μ*M* cysteine-reacted IANBD (Cys-NBD) in indicated concentrations of 1,4-dioxan in low-salt buffer (excitation was set at 481 nm).

of the labeling can be verified by blocking the incorporation of IANBD with the cysteine-specific, nonfluorescent reagents, iodoacetamide and *N*-ethylmaleimide *(10)* (*see* **Note 8**).

3.3. Fluorescence Spectroscopy Analysis of the IANBD-Labeled β_2-AR

Fluorescence spectroscopy analysis of the purified IANBD-labeled β_2-AR is performed at room temperature on a SPEX Fluoromax-2 spectrofluorometer connected to a PC equipped with the Datamax software package. During the experiments, we use the photon counting mode and an excitation and emission bandpass of 4.2 nm.

3.3.1. Emission Scan Experiments

In emission scan experiments, we use 15–30 pmol of IANBD-labeled β_2-AR in 400 µL of low-salt buffer in a 5 × 5 mm quartz cuvet (purchased from HELLMA, type no. 1001.016-QS). In a typical experiment, excitation is 481 nm, and emission is measured from 490 to 625 nm with an integration time at 0.3 s/nm. Emission scan analysis of the IANBD-labeled β_2-AR shows a strong fluorescence signal with an emission maximum at 523 nm *(10)*. The blue-shift in emission maximum, as compared to cysteine-reacted IANBD in aqueous buffer (540 nm), indicates that the modified cysteine(s) is located in an environment that on the average, is of lower polarity than 80% dioxane (**Fig. 1**). This would likely involve labeling of one or more of the five cysteine residues that are located in the transmembrane, hydrophobic core of the receptor.

3.3.2. Time-Course Experiments

Time-course experiments, in which the excitation and emission wavelength are fixed and fluorescence is measured over time, can be used as a highly sensitive method to identify ligand-induced conformational changes that lead to changes in the polarity of the environment surrounding the incorporated fluorophore *(10)*. As illustrated in **Fig. 2**, binding of the full agonist, isoproterenol, to the IANBD-labeled β_2-AR causes a decrease in fluorescence reaching a maximum amplitude below the extrapolated baseline after 10 min. The response to isoproterenol can be readily reversed by the active (–)-isomer of the antagonist propranolol, but not by the less active (+)-isomer (**Fig. 2**). The response to isoproterenol can also be reversed by several other antagonists, including alprenolol, ICI 118,551, pindolol, and dichloroisoproterenol *(10)*. Moreover, the isoproterenol response is dose-dependent and stereospecific *(10)*. Prior to adding ligand, we normally observe a slight, but constant decline in baseline fluorescence. This loss of fluorescence over time is likely caused by bleaching of the fluorophore combined with some loss of protein possibly owing to sticking of the protein to the inside of the cuvet (**Fig. 2** and *see* **Note 9**). The experiments are carried out as follows:

1. Usually 10 µL of IANBD-labeled β_2-AR receptor (15–30 pmol) are added to 490 µL low-salt buffer in a 5 × 5 mm quartz cuvet. To stabilize the baseline, the mixture is preincubated for at least 15 min in the cuvet before the experiment is started (*see* **Note 9**). Both during this period and during the time scan experiment, the mixture is kept under constant stirring using a 2 × 2 mm magnetic stir bar (Bel-Art Products).
2. The time-course experiment is routinely performed over 30 min, and the first addition of ligand is usually done after 5 min (*see* **Note 10**). During the experi-

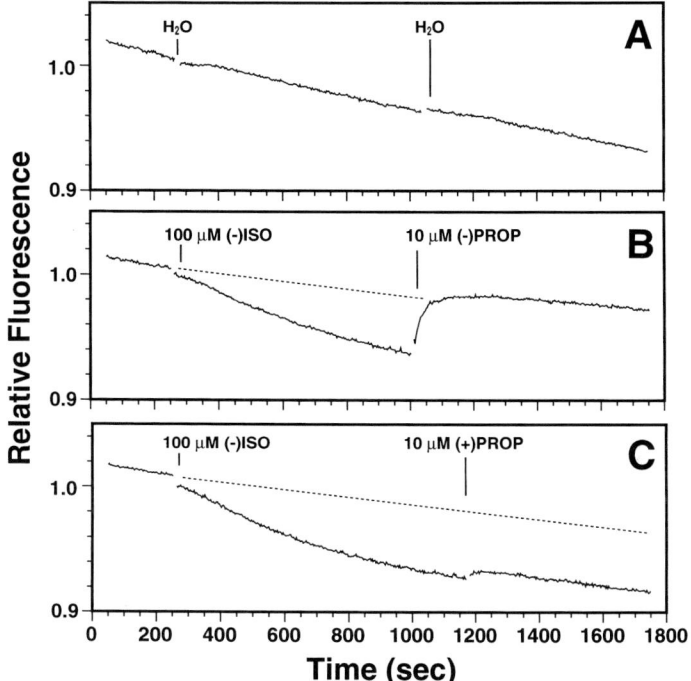

Fig. 2. Time-course and isoproterenol induced decreased fluorescence from IANBD-labeled β_2-AR (**A**) Control addition of water (H$_2$O). (**B, C**) reversal of the response to isoproterenol (ISO) by the active (–)-isomer of the antagonist propranolol, (–)-PROP, but not by the less active (+)-isomer, (+)PROP (**C**). Dotted lines indicate extrapolated baseline. Exitation was 481 nm and emission measured at 523 nm. Fluorescence in all the individual traces shown was normalized the fluorescence observed immediately after addition of ligand. All traces shown are representative of at least three identical experiments (reproduced with permission from **ref. *10***).

ment, the excitation wavelength is fixed at 481 nm, and emission measured at a wavelength of 525 nm. The volume of the added ligands is 1% of the total volume, and fluorescence is corrected for this dilution. The compounds tested in our fluorescence experiments have an absorbance of <0.01 at 481 nm and 525 nm in the concentrations used, excluding inner filter effects.

3.4. Site-Selective Fluorescent Labeling of the β_2-AR

The methods described above have dealt with the wild-type β_2-AR, which contains 13 cysteines. Notably, five of these cysteines are not expected to be available for chemical derivatization. In the extracellular loops, four cysteines ([106]Cys, [184] Cys, [190]Cys, and [191]Cys) form two disulfide bridges (**Fig. 1**)

(27–29), and in the intracellular carboxy-terminal tail, [341]Cys, has been shown to be palmitoylated *(30,31)*. To establish a system that would allow site-selective incorporation of chemically reactive fluorescent probes, such as IANBD, we mutated cysteines in the receptor and generated a series of mutant receptors with only one, two, or three of the natural cysteines available for chemical derivatization *(12)* (*see* **Note 11**). Importantly, all these mutants displayed minimal changes in pharmacological properties as compared to the wild-type both with respect to ligand binding and functional coupling to adenylate cyclase *(12)*.

Initially, we have used the series of mutants to identify the cysteine residue(s) responsible for the agonist-induced changes in fluorescence observed in the IANBD-labled β_2-AR *(12)*. The mutant receptors containing one, two, or three of the naturally occurring cysteines were all purified and labeled with the IANBD-fluorophore. The spectroscopic analyses of the mutants revealed that the agonist-induced changes in fluorescence require the presence of [285]Cys in transmembrane segment (TM) 6 and [125]Cys in TM 3 *(12)*. The data suggest that agonist binding to the β_2-AR promotes a conformational change in the receptor that exposes NBD attached to [125]Cys in TM 3 and to [285]Cys in TM 6 to a more polar environment. We attempted to predict the actual structural changes using molecular modeling and computational simulations *(12)*. According to this, our data are mostly consistent with a counterclockwise rotation of helix III and VI in response to agonist binding (*see* **Note 12**). This is consistent with the suggested rigid body movements of the corresponding helices in rhodopsin *(9)*.

The experimental procedures used for the mutants are identical to the methods described for the wild-type β_2-AR. The mutants were generated by standard molecular biology techniques.

3.5. Probing the Accessibility of the Fluorophore by Collisional Quenching

Collisional quenching involves a bimolecular collision between the fluorophore and the quencher leading to shortening of the fluorescence lifetime and, thus, a decrease in the quantum yield. The most ubiquitous quenchers are paramagnetic species, including O_2 and heavy atoms. Importantly, a hydrophilic quencher, such as I^-, can be used to evaluate the accessibility of a fluorophore to the solvent. Therefore, if the hydrophilic quencher is able to reduce the quantum yield, the fluorophore must be exposed to the solvent and accessible to the quencher. On the other hand, if the fluorophore is located in the interior of the receptor protein or pointing toward the lipid bilayer, it would be expected that the hydrophilic quencher would have less accessibility and therefore would not be able to reduce the fluorescence quantum yield. As shown in **Fig. 3**, a strong quenching of fluorescence from cysteine-reacted IANBD (Cys-NBD) is observed following addition of increasing concentrations of potassium iodine. Potassium iodine (KI) quenched the IANBD-labeled mutant receptors, Cys285 and Cys116,125, which only contained [285]Cys and [116]Cys/[125]Cys, respectively, significantly less than Cys-NBD, indicating that the labeled cysteines are only partly accessible to the solvent (**Fig. 3**). These quenching data suggest that NBD attached to cysteines at position 116, 125, or 285 are located in the interior of the protein or partly pointing toward the lipid bilayer and thereby less exposed to the solvent. The following protocol is used to probe the accessibility of the fluorophore by collisional quenching.

1. Usually 10 μL of IANBD-labeled β_2-AR (15–30 pmol) is added to 390 μL low-salt buffer in a 5 × 5 mm quartz cuvet (HELLMA, type no. 1001.016-QS). No stirring is required.
2. The data are collected using the Datamax Constant Wavelength program for the Fluoremax-2 with excitation set at 481 nm and emission set at 523 nm for the IANBD-labeled β_2-AR.
3. Following the first measurement in absence of quencher, 10 μL of 1 M KI containing 10 mM Na$_2$S$_2$O$_3$ (*see* **Note 13**) is added, and the content of the cuvet is mixed by pipeting up and down 10 times. Wait 30 s before making the measurement.
4. Sequentially repeat **step 3** by adding 10 μL of 1 M KI/10 mM Na$_2$S$_2$O$_3$ until a final concentration of 250 mM KI is reached.
5. To correct for dilution, **step 3** and **step 4** are repeated with a new sample of receptor using 1 M KCl instead of KI.
6. The corrected data are plotted according to Stern-Volmer equation *(32)* $F_o/F = 1 + K_{SV}[KI]$, where F_o/F is the ratio of fluorescence intensity in the absence and presence of quencher, and K_{SV} is the Stern-Volmer quenching constant.

3.6. Concluding Remarks

In this chapter, we have described the methods required for analyzing adrenergic receptor conformation using chemically reactive fluorescent probes. Specifically, we have been able to use a sulfhydryl-reactive and environmentally sensitive fluorescent probe, IANBD, as a molecular reporter to characterize and map ligand-induced conformational changes in the β_2-AR. To carry out these experiments, a convenient expression system and a purification procedure are required. The purification procedure should be reliable, relatively easy, and result in highly pure and active protein. Moreover, to achieve site-selective incorporation of the chemically reactive flurophores, it should be possible to mutate most endogenous cysteines without perturbing receptor function. The results described and discussed in the text have focused on the endogenous cysteines in the β_2-AR. Currently, we are systematically introducing cysteines

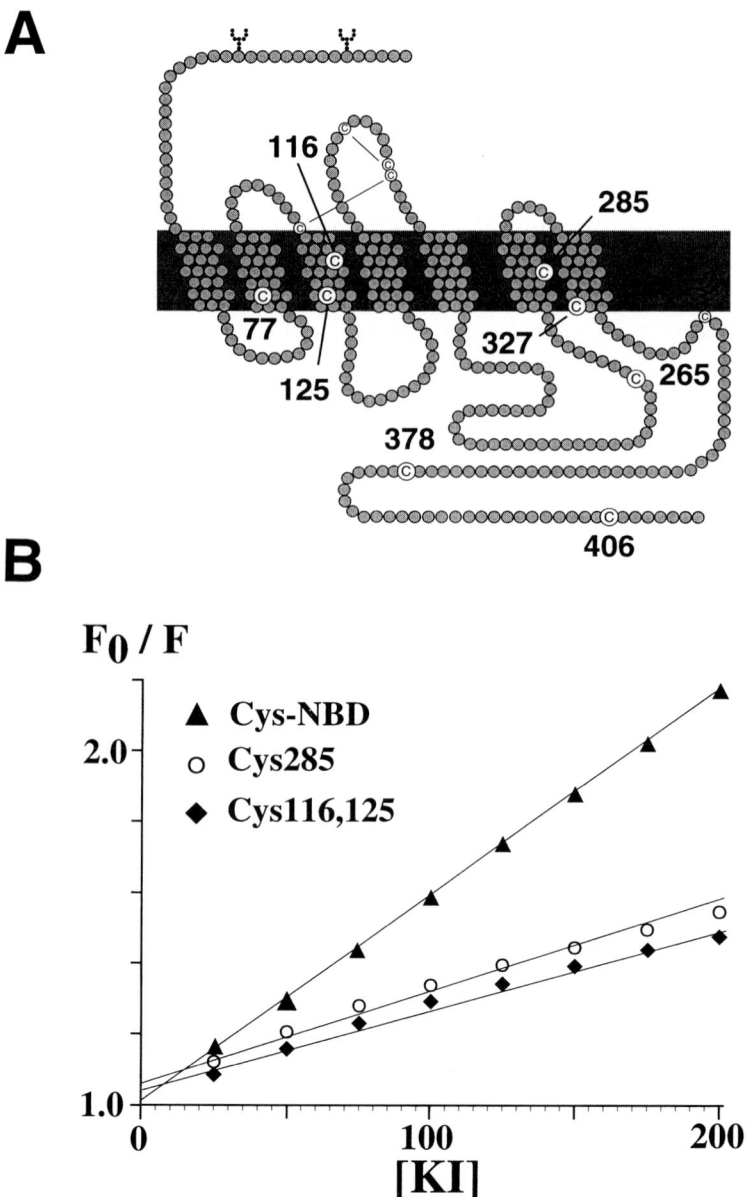

Fig. 3. (**A**) "Snake" diagram of the β_2-AR: The receptor contains 13 Cys residues, of which five ([77]Cys, [116]Cys, [125]Cys, [285]Cys, and [327]Cys) are predicted to be in the transmembrane domains. Three Cys residues are predicted to be in the cytoplasmic regions ([265]Cys, [378]Cys, and [406]Cys). Five Cys residues are not expected to be available for chemical derivatization: four residues ([106]Cys, [184]Cys, [190]Cys and [191]Cys) form two disulfide bridges *(27,28,30)* and in the intracellular carboxyterminal tail, [341]Cys, has been shown to be palmitoylated *(29,31)*. (**B**) Stern-Volmer plots of quenching of IANBD-labeled wild-type β_2-AR and mutant receptors: Increasing concentrations of

in new positions to map the conformational changes accompanying agonist-induced activation of GPCRs further. These new cysteines are introduced in a mutant β_2-AR still containing three endogenous cysteines to ensure a high level of expression. We now have evidence from these new mutants for significant movement at the cytoplasmic site of TM VI (Jensen and Gether, manuscript in preparation). These changes occur with the same kinetics as the changes discussed in this chapter, indicating that both responses reflects the same conformational change. The data underline that site-selective labeling with chemically reactive fluorescent probes may be used as a general approach for mapping ligand-induced conformational changes in a purified receptor protein.

4. Notes

1. The optional addition of serum allows the cells to grow at higher densities (up to 7 million cells/mL). Furthermore, it is our general experience that if the cell stock is grown in the presence of serum, the cells can be used for an extended period of time (usually 3–6 mo) before it is necessary to thaw a new sample from the nitrogen tank.

2. To achieve the highest level of expression, it is important to assess the optimal inoculum of each virus stock. Instead of time-consuming plaques assays to determine the number of PFU/mL of virus stock, we find it faster and more reliable to perform small-scale infections with different dilution of the virus stock. In a typical assay, we infect four 20-mL Sf9 cell suspension cultures (3×10^6 cells/mL) in 125-mL disposable Erlenmeyer flasks with 1, 10, 100, and 1000 µL, respectively, of the virus stock. After 48 h, 1.0-mL samples of the cultures are collected and centrifuged for 10 min at 16,000g in a microcentrifuge. The pellets are frozen at $-70°C$, thawed, and resuspended in 1000 µL of lysis buffer. The lysed cells are centrifuged (10 min at 16,000g) in a microcentrifuge, and the membrane pellet is resuspended in 1000 µL binding buffer. The amount of active receptor protein is determined in a binding assay using 50 µL membrane suspension incubated in binding buffer in a total volume of 500 µL mL with 10 nM ^{3}H-dihydroalprenolol (Amersham) for 2 h according to previously described methods *(10–12)*.

3. Adding a saturating concentration of antagonist or agonist during infection of the Sf9 cell culture may improve the apparent expression of active receptor protein up to 40% (**ref. *12*** and unpublished observations). In case of an inherently unstable receptor protein, such as the constitutively activated mutant of the β_2-AR (CAM), apparent expression may improve 200–250% *(12)*. Since both

---

Fig. 3. *(continued)* KI were added sequentially to cysteine-reacted IANBD (Cys-NBD, ▲), IANBD-labeled wild-type β_2-AR (●), Cys285 β-2AR (○), or Cys116, 125 β_2-AR (♦). The quenching constant K_{SV} was 5.88×10^{-3} ($n = 3$) for cysteine-reacted IANBD, 2.21×10^{-3} ($n = 3$) for Cys285, and 2.04×10^{-3} ($n = 3$) for Cys116, 125.

full agonists and antagonists can increase apparent expression, the effect is most likely owing to stabilization of already synthesized receptor protein *(12)*.

4. To keep track of receptor expression, it is important to take a 1-mL sample of each infection for determination of binding activity. Crude membranes are prepared of the sample and binding measured as described in **Note 2**. In general, we obtain an expression of 10–20 pmol receptor/mL culture for the wild-type receptor.

5. To assess the protein content, 1 μL of each fraction is added to 50 μL Bradford protein dye (Bio-Rad). After vortexing, the amount of protein present is roughly estimated visually by the intensity of the blue color.

6. It is important to block the Sephadex G50 columns with serum before the initial use to minimize nonspecific adsorption of receptor protein. Normally, we wash each column with 2 column volumes of serum followed by extensive washing with water. Once prepared, the columns can be reused many times. Columns can be regenerated by washing with 10 mL of water followed by equilibration in low-salt buffer.

7. IANBD is highly hydrophobic and should be dissolved in dimethyl sulfoxide (DMSO). Routinely, we make a 10 mM solution of IANBD in DMSO and add 1/20 vol of this solution directly to the receptor solution. The presence of 5% DMSO during the 20-min labeling reaction does not affect receptor stability.

8. It should be ensured that the fluorescent probe, when incorporated into the receptor, does not interfere with binding of the ligands. In the β_2-AR, this is highly unlikely. Labeling of the receptor with IANBD does not alter agonist or antagonist binding properties *(10)*, as would be expected if the bound NBD was positioned within the ligand binding pocket.

9. We usually observe a slight, but constant decline in baseline fluorescence over time prior to addition of ligand. This loss of fluorescence over time is likely caused by bleaching of the fluorophore combined with some loss of protein possibly owing to sticking of the protein to the inside of the cuvet. The decrease over time is unaffected by addition of 0.1% BSA, 10% glycerol, or phospholipids to the cuvet *(10)*. It is unlikely that the decline in fluorescence is owing to denaturation of the protein, since a similar loss of fluorescence also was observed with labeled receptor that was intentionally denatured in guanidinium chloride *(10)*. Preincubation of the receptor in the cuvet for 15 min or more under constant stirring before performing the experiment minimizes (but cannot eliminate) the constant decline in baseline fluorescence. It should be noted that even very small dust particles can cause significant noise in the fluorescence readout. It is recommended to centrifuge the receptor sample (e.g., 15,000g for 2 min in a microcentrifuge) and also the buffer before performing the experiments to precipitate dust particles.

10. Ligands are added using a gel-loading pipet tip (Corning-Costar, cat. # 4853). The pipet tip should be slid along the side of the cuvet to the bottom before the ligand is added. It is important to avoid touching the magnetic stir bar.

11. Mutation of several cysteine residues in the β_2-AR led to a reduction in receptor expression *(12)*. A mutant receptor with all free cysteines substituted expressed

so poorly that purification in sufficient quantities for fluorescence spectroscopy analysis was impossible *(12)*. Ideally, it should be possible to remove all endogenous cysteines and either reintroduce them one by one or introduce single cysteines in new positions. Unfortunately, this was not possible in the β_2-AR. Nevertheless, as described in Gether et al. *(12)*, it is possible to work with a system where not all cysteines can be removed. It just requires inclusion of proper control mutants.

12. It is important to note that the amplitude of the fluorescent change is only a rough indicator of the magnitude of conformational change. For example, we cannot assume that there is a linear correlation between change in fluorescence and magnitude of movements. Therefore, the movement of TM 3, for example, may not be of the same magnitude as movement of TM 6.

13. To avoid formation of I_2, it is advisable to prepare the 1 *M* KI solution fresh and add $Na_2S_2O_3$.

Acknowledgment

Part of the work described in the present chapter was carried out in Brian Kobilka's laboratory at Howard Hughes Medical Institute, Stanford University, CA.

References

1. Gether, U. and Kobilka, B. K. (1998) G-protein-coupled receptors. II. Mechanism of activation. *J. Biol. Chem.* **273,** 17,979–17,982.
2. Samama, P., Cotecchia, S., Costa, T., and Lefkowitz, R. J. (1993) A mutation-induced activated state of the beta 2-adrenergic receptor. Extending the ternary complex model. *J. Biol. Chem.* **268,** 4625–4636.
3. Lou, X., Zhang, D., and Weinstein, H. (1994) Ligand-induced domain motion in the activation mechanism of a G-protein-coupled receptor. *Protein Eng.* **7,** 1441–1448.
4. Ballesteros, J. A. and Weinstein, H. (1995) Integrated methods for the construction of three-dimensional models and computational probing of structure function relations in G-protein-coupled receptors. *Methods Neurosci.* **25,** 366–428.
5. Fanelli, F., Menziani, M. C., and Benedetti, P. G. (1995) Molecular dynamics simulations of m3-muscarinic receptor activation and QSAR analysis. *Bioorg. Med. Chem.* **3,** 1465–1477.
6. Scheer, A., Fanelli, F., Costa, T., De Benedetti, P. G., and Cotecchia, S. (1996) Constitutively active mutants of the alpha 1B-adrenergic receptor. A role of highly conserved polar amino acids in receptor activation. *EMBO J.* **15,** 3566–3578.
7. Altenbach, C., Yang, K., Farrens, D. L., Farahbakhsh, Z. T., Khorana, H. G., and Hubbell, W. L. (1996) Structal features and light-dependent changes in the cytoplasmic interhelical E-F loop region of rhodopsin: a site-directed labeling study. *Biochemistry* **35,** 12,470–12,478.
8. Farahbakhsh, Z. T., Ridge, K. D., Khorana, H. G., and Hubbell, W. L. (1995) Mapping light-dependent structal changes in the cytoplasmic loop connecting

helices C and D in rhodopsin: a site-directed spin labeling study. *Biochemistry* **34,** 8812–8819.

9. Farrens, D. L., Altenbach, C., Yang, K., Hubbell, W. L., and Khorana, H. G. (1996) Requirement of rigid-body motion of transmembrane helices for light activation of rhodopsin. *Science* **274,** 768–770.

10. Gether, U., Lin, S., and Kobilka, B. K. (1995) Fluorescent labelling of purified beta-2-adrenergic receptor: Evidence for ligand-specific conformational changes. *J. Biol. Chem.* **270,** 28,268–28,275.

11. Gether, U., Ballesteros, J. A., Seifert, R., Sanders-Bush, E., Weinstein, H., and Koblika, B. K. (1997) Structural instability of a constitutively active G-protein-coupled receptor. Agonist-independent activation due to conformational flexibility. *J. Biol. Chem.* **272,** 2587–2590.

12. Gether, U., Lin, S., Ghanouni, P., Ballesteros, J. A., Weinstein, H., and Kobilka, B. K. (1997) Agonists induce conformational changes in transmembrane domains III and VI of the beta-2-adrenergic receptor. *EMBO J.* **16,** 6737–6747.

13. Phillips, W. J. and Cerione, R. A. (1988) The intrinsic fluorescence of the α subunit of transducin. Measurement of receptor-dependent guanine nucleotide exchange. *J. Biol. Chem.* **263,** 15,498–15,505.

14. Phillips, W. J., Wong, S. C., and Cerione, R. A. (1992) Rhodopsin/Transducin interactions. I: Characterization of the binding of the transducin-$\beta\gamma$ subunit complex to rhodopsin using fluorescence spectroscopy. *J. Biol. Chem.* **267,** 17,032–17,039.

15. Phillips, W. J. and Cerione R. A. (1992) Rhodopsin/transducin interactions. II: Influence of the transducin-$\beta\gamma$ subunit complex on the coupling of the transducin-α subunit to rhodopsin. *J. Biol. Chem.* **267,** 17,040–17,046.

16. Lin, S., Gether, U., and Kobilka, B. K. (1996) Ligand stabilization of the β_2-adrenergic receptor: Effect of DTT on receptor conformation monitored by circular dichroism and fluorescence spectroscopy. *Biochemstry* **35,** 14,445–14,451.

17. Guan, X. M., Kobilka, T. S., and Kobilka, B. K. (1992) Enhancement of membrane insertion and function in a type IIIb membrane protein following introduction of a cleavable signal peptide. *J. Biol. Chem.* **267,** 21,995–21,998.

18. Kobilka, B. K. (1995) Amino and carboxyterminal modifications to facilitate the production and purification of a G-protein-coupled receptor. *Anal. Biochem.* **231,** 269–271.

19. Grisshammer, R. and Tate, C. G. (1995) Overexpression of integral membrane proteins for structural studies. *Q. Rev. Biophys.* **3,** 315–422.

20. Tate, C. G. and Grisshammer, R. (1996) Heterologous expression of G-protein-coupled receptors. *Trends Biotechnol.* **14,** 426–430.

21. Caron, M. G., Srinivasan, Y., Pitha, J., Kociotek, K., and Lefkowitz. R. J. (1979) Affinity chromatography of the beta 2-adrenergic receptor. *J. Biol. Chem.* **254,** 2923–2927.

22. Benovic, J. L., Show, R. G., Caron, M. G., and Lefkowitz, R. J. (1984) The mammalian beta 2-adrenergic receptor: Purification and characterization. *Biochemistry* **20,** 4510–4518.

23. Parker, E. M., Karneyama, K., Higashijima, T., and Ross, E. M. (1991) Reconstitutively active G-protein-coupled receptors purified from baculovirus infected cells. *J. Biol. Chem.* **266,** 519–522.

24. Cerione, R. A. (1994) Fuorescence assays for G-protein interactions. *Methods Enzymol.* **237,** 409–423.

25. Dunn, S. M. and Raftery, M. A. (1993) Cholinergic binding sites on the pentameric acetylcholine receptor of torpedo california. *Biochemistry* **33,** 8608–8615.

26. Gettins, P. G. W., Fan, B., Crews, B. C., Turko, I. V., Olson, S. T., and Streusand, V. J. (1993) Transmission of conformational change from the heparin binding site to the reactive center of antithrombin. *Biochemistry* **32,** 8385–8389.

27. Dohlman, H. G., Caron, M. G., DeBlasi, A., Frielle, T., and Lefkowitz, R. J. (1990) Role of extracellular disulfide-bonded cysteines in the ligand binding function of the beta 2-adrenergic receptor. *Biochemistry* **9,** 2335–2342.

28. Fraser, C. M. (1989) Site-directed mutagenesis of beta 2-adrenergic receptors. Identification of conserved cysteine residues that independently affect ligand binding and receptor activation. *J. Biol. Chem.* **264,** 9266–9270.

29. Mouillac, B., Caron, M. G., Bonin, H., Dennis, M., and Bouvier, M. (1992) Agonist-modulated palmitoylation of the beta 2-adrenergic receptor in Sf9 cells. *J. Biol. Chem.* **267,** 21,733–21,737.

30. Noda, K., Saad, Y., Graham, R. M., and Karnik, S. S. (1994) The high affinity state of the beta 2-adrenergic receptor requires unique interaction between conserved and non-conserved extracellular loop cysteines. *J. Biol. Chem.* **269,** 6742–6752.

31. O'Dowd, B. F., Hnatowich, M., Caron, M. G., Lefkowitz, R. J., and Bouvier, M. (1989) Palmitoylation of the human beta 2-adrenergic receptor. Mutation of cys341 in the carboxyl tail leads to an uncoupled non-palmitoylated form of the receptor. *J. Biol. Chem.* **264,** 7564–7569.

32. Lakowicz, J. R. (1983) *Principles of Fluorescence Spectroscopy.* Plenum, New York, pp. 258–262.

24

Biochemical Methods for Detection and Measurement of Cyclic AMP and Adenylyl Cyclase Activity

Steven R. Post, Rennolds S. Ostrom, and Paul A. Insel

1. Introduction

Many drugs and hormones interact with plasma membrane receptors to induce changes in the production of intracellular second messengers. The first such messenger to be identified was adenosine-3′,5′ cyclic monophosphate (cAMP), discovered by Sutherland and Rall in the late 1950s *(1)*. Changes in cellular cAMP levels can occur from increases or decreases in its biosynthesis (which results from the catalytic conversion of ATP to cAMP by the enzyme adenylyl cyclase) or by altering its degradation (which results from hydrolysis of the cyclic phosphodiester bond by cyclic nucleotide or cAMP phosphodiesterases).

Adenylyl cyclase is an example of an effector enzyme whose activity can be stimulated and inhibited by different GTP binding proteins. To date, 10 different isoforms of adenylyl cyclase have been identified, which differ in their primary sequence, tissue distribution, and regulation *(2,3)*. The α-subunit of the stimulatory G-protein, $G\alpha_s$, enhances the activity of every adenylyl cyclase isoform, whereas the α-subunit of the inhibitory G-protein, $G\alpha_i$, has been shown to inhibit types I, V, and VI of this enzyme directly *(4,5)*. G-protein activators include various hormone receptors, GTP analogs (in particular, nonhydrolyzable analogs, such as GTPγS and GppNHP), the exotoxin of *Vibrio cholerae*, a heat-labile toxin from *Escherichia coli*, and AlF_3 (more typically added to cell homogenates as NaF). In addition, the diterpene forskolin can be used to activate adenylyl cyclase directly to increase cAMP production. The different types of adenylyl cyclase share a common structure. Each has two domains, consisting of six putative membrane-spanning regions that are

From: *Methods in Molecular Biology, vol. 126: Adrenergic Receptor Protocols*
Edited by: C. A. Machida © Humana Press Inc., Totowa, NJ

separated by a large cytoplasmic loop, as well as a large cytoplasmic loop at the carboxyl-terminus. For a review of adenylyl cyclase isoforms, *see* **refs.** *(2,3,6)*.

Activation of many types of plasma membrane receptors promotes alterations in the amount of intracellular cAMP. Perhaps the best-studied receptors that increase cAMP are the β-adrenergic receptors (β-ARs). Activation of β-ARs via endogenous catecholamines (e.g., norepinephrine, epinephrine) or synthetic agonists (e.g., isoproterenol) promotes adenylyl cyclase activation via the stimulatory G-protein, G_s, and an increase in intracellular cAMP. The increase in cAMP activates cAMP-dependent protein kinase A, which phosphorylates specific substrates resulting in a variety of cell type-dependent effects. For example, activation of β-ARs in hepatocytes increases glycolysis and lipolysis, whereas in smooth muscle cells, increasing cAMP promotes relaxation. In contrast to β-ARs, $α_2$-ARs are coupled to inhibition of adenylyl cyclase in a variety of cell types (e.g., adipocytes, platelets, and so on) via activation of G-proteins (particularly G_i), which are inactivated by an exotoxin of *Bordetella pertussis* (pertussis toxin).

Many methods are available for determining the concentration of cAMP in cells. Such methods include protein kinase A activation, competitive binding assays, and immunoassays. In general, cells are treated with an agonist, and intracellular cAMP is extracted. The mass of cAMP present in the cellular extracts is then determined by one of the above methods. Alternatively, it may be desirable to assess adenylyl cyclase activity rather than cAMP accumulation. This can be accomplished in two ways. The adenine nucleotides (including ATP) of whole cells can be radiolabeled by incubating with [^{3}H]adenine and determining the amount of radiolabeled cAMP present in extracts from treated cells. Alternatively, adenylyl cyclase assays can be conducted using cell membranes or homogenates to assess the conversion of added radiolabeled ATP (commonly [^{32}P]) to cAMP. Both of these methods employ column chromatography to isolate the radiolabeled cAMP product *(7)*.

2. Materials

1. Dulbecco's Modified Eagles Medium (DMEM): Gibco-BRL Life Technologies (Grand Island, NY).
2. Adenosine 3′,5′-cyclic phosphoric acid 2′-*O*-succinyl-3-[^{125}I]iodotyrosine methyl ester ([^{125}I]cAMP): Used for radioimmunoassays. Can be purchased from NEN Life Science (Boston, MA; 2200 Ci/mmol).
3. [2,8-^{3}H]cAMP: Used for monitoring sample recovery following column chromatography. Can be purchase from Amersham Life Science Products (Piscataway, NJ; 30–50 Ci/mmol).
4. Durapore™ 96-well multiscreen filtration plates (0.65 μM, DV): Millipore (Bedford, MA).

5. Goat antirabbit IgG coupled to magnetic beads: Used to isolate cAMP–antibody complexes in immunoassays. Can be purchased from Polysciences, Inc. (Warrington, PA).
6. [α-^{32}P]ATP: Used as substrate in adenylyl cyclase activity assays. Purchased from Amersham Life Science Products (30 Ci/mmol).
7. [^{3}H]adenine: Used to label cells prior to isolation of [^{3}H]cAMP. Purchased from NEN Life Science (20–40 Ci/mmol).
8. Rabbit anti-cAMP antibody: Calbiochem (La Jolla, CA). Dilute in γ-globulin buffer for use in immunoassay.
9. γ-Globulin buffer: Dissolve human γ-globulin (Cohn Fraction II, III) in 50 mM sodium acetate buffer (pH 4.75) at 1 mg/mL. Use this buffer in radioimmunoassays (RIAs) for diluting anti-cAMP antibody and [^{125}I]cAMP.
10. Adenylyl cyclase activity assay buffer: Buffer can be stored at –4°C if made in separate solutions and mixed in equal parts for the assay. Solution A contains 30 mM Na-HEPES, pH 7.5, 400 mM NaCl, 2 mM, ethyleneglycol-bis-(β-aminoethylether)-$N,N,N,'N,'$-tetra-acetic acid (EGTA), 20 mM MgCl$_2$, 4 mM cAMP, 2 mM isobutylmethylxanthine, 0.04 mM ATP, and 20 mM phosphocreatine. Solution B contains 30 mM Na-HEPES, 120 U/mL creatine phosphokinase, and 0.2% bovine serum albumin (BSA). [α-^{32}P]ATP 20 μCi/mL) should be added to the cyclase assay buffer just prior to use. Omit [α-^{32}P]ATP and cAMP from buffer if assaying cAMP by RIA instead of separation by column chromatography.
11. 12% Polyethylene glycol (PEG): Dissolve 120 g of PEG (average mol wt 8000 from Sigma) in 1 L of 50 mM sodium acetate buffer (pH 6.2). Store at 4°C.
12. Dowex AG 50W-X4: Prepare slurry of Dowex AG 50 W-X4 (Bio-Rad) resin by mixing the resin with degassed dH$_2$O. Pour the slurry into disposable columns (e.g., Bio-Rad Econo-Pac columns) to obtain a resin bed volume of approx 1.0 mL. Prior to use, regenerate columns by washing with 3 mL 1 M NaOH, then 3 mL 1 M HCl, followed by 10 mL of dH$_2$O. Columns should also be regenerated in the same manner following each use.
13. Neutral alumina: Prepare alumina columns by placing 0.6 g neutral alumina (Sigma, St. Louis, MO) into disposable columns and equilibrating with 8 mL of 0.1 M imidazole (pH 7.5) followed by 8 mL dH$_2$O. Columns should be regenerated after use in the same manner.

3. Methods

3.1. Cell Treatment and Lysis

3.1.1. Adherent Cells

1. Cells are grown to suitable density in serum-supplemented DMEM (*see* **Note 1**).
2. Wash and equilibrate cells (typically 30 min, 37°C) in HEPES- (20 mM) buffered DMEM (DMEH, pH 7.4) to allow cells to be incubated outside a gassed tissue-culture incubator. Alternatively, DMEM or other medium that depends on

equilibration of CO_2 can be used for experiments conducted in a gassed tissue-culture incubator.

3. Replace medium with DMEH. If applicable, preincubate cells in the presence of a cAMP phosphodiesterase inhibitor (e.g., 1 mM isobutylmethylxanthine). If a receptor antagonist is used in any experimental conditions, the cells should also be preincubated with that agent.

4. Add agonist and incubate cells for the desired time (usually 1–15 min) at the desired temperature (typically 37°C). The total volume should be at least 0.5 mL for a 24-well plate.

5. Stop incubation by placing plates on ice and removing assay medium. Add 70% ethanol to plate, and incubate at room temperature for 15 min. Transfer ethanol extract to glass tube. Repeat extraction once more and dry under N_2. Dissolve dried sample in cAMP assay buffer (*see* **Subheading 3.2.**), and store frozen (–20°C) until assayed for cAMP. Alternative methods of cell lysis may be utilized, depending on the assay method to be employed (*see* **Note 2**).

6. If desired, dissolve the cellular protein in NaOH (0.5 M), and assay for protein content.

3.1.2. Cells in Suspension

1. Equilibrate cells in DMEH (typically 30 min, 37°C).
2. Wash and resuspend cells at a suitable density in DMEH (*see* **Note 1**).
3. If applicable, incubate cells in the presence of a cAMP phosphodiesterase inhibitor.
4. Add agonist and incubate cells for the desired time (usually 1–15 min) at the desired temperature (typically 37°C).
5. Stop incubations by addition of ice-cold ethanol (70% v/v final). Pellet cells, and transfer extract to glass tube. Wash cell pellet with 70% ethanol, and transfer to tube. Dry ethanol extract under N_2. Dissolve dried sample in cAMP assay buffer (*see* **Subheading 3.2.**), and store frozen (–20°C) until assayed for cAMP. Alternative methods of cell lysis may be utilized, depending on the assay method to be employed (*see* **Note 2**).

3.1.3. Tissues

Measurements of cAMP accumulation and adenylyl cyclase activity can be readily performed with tissues. The following is a general protocol often used for whole-cell assays that has been adapted for use with specific tissues.

1. Dissect tissue and immediately place in ice-cold physiologic buffer (such as Kreb's, Ringer's, and so forth), which is well gassed with O_2/CO_2 (95:5).
2. Slice tissue with a McIlwain tissue chopper two times at 350 µm, rotating the stage 90° between each pass. The number of passes and the size of the slices should be increased if the tissue is tough and resistant to chopping.
3. Wash tissue slices three times with oxygenated buffer to remove debris from broken cells, and equilibrate at 37°C for 15 min. If using [³H]adenine labeling

method for determining adenylyl cyclase activity (*see* **Subheading 3.3.2.**), label tissue with [^{3}H]adenine, typically for 30–90 min, under O_2/CO_2 (95:5), and wash extensively to remove unincorporated [^{3}H]adenine.

4. Wash tissue slices, replacing buffer with assay buffer (containing a phosphodiesterase inhibitor, if appropriate). Transfer tissue to a large glass test tube, and allow slices to settle to the bottom.
5. Add 70- to 100-µL aliquots of settled tissue slices to tubes containing assay buffer (as used in **step 1**, above) and various agonists and antagonists of interest (final volume 0.7 mL). Gas tubes with O_2/CO_2 (95:5), cap, and incubate for the desired time (1–15 min) at the desired temperature (typically 37°C).
6. Stop reactions with the addition of 0.3 mL trichloroacetic acid (30% w/v), and centrifuge tubes to pellet cells.
7. Assay supernatant for determination of cAMP. The supernatant can be used directly to isolate cAMP by column chromatography (*see* **Subheading 3.3.2.**) or can be neutralized for RIA (*see* **Note 2**). If desired, dissolve the cellular protein in NaOH (0.5 *M*) and assay for protein content.

3.2. Assessment of cAMP Formation

3.2.1. Radioreceptor Binding Assay

The concentration of cAMP in samples can be assessed by competitive binding assays using a cAMP binding protein *(8)*. The procedure described here is an adaptation of a method that was published by Nordstedt and Fredholm *(9)*. Briefly, this method involves the competition between cAMP present in samples or standards and [^{3}H]cAMP for association with a cAMP binding protein (usually the regulatory subunit of protein kinase A). Results of this assay compare favorably to those obtained using a sensitive RIA *(9)*. Many different methods can be used to separate bound from free cAMP (e.g., filtration, precipitation, and so forth). By using filtration to separate bound from free nucleotide, the time required for this assay is greatly diminished. This method can be used with samples containing ≥3 n*M* concentrations of cAMP.

3.2.1.1. Preparation of Binding Protein

1. Obtain five bovine adrenals from a slaughterhouse. Dissect the cortex free from the capsule and medullary tissue (*see* **Note 3**).
2. Homogenize the cortex in 10 vol of homogenization buffer consisting of 100 m*M* Tris-HCl (pH 7.4), 10 m*M* ethylenediamine tetra-acetic acid (EDTA), 250 m*M* NaCl, 0.1% 2-mercapoethanol, and 0.25 *M* sucrose.
3. Filter the homogenate through cheesecloth.
4. Centrifuge the homogenate at 30,000*g* for 60 min at 4°C.
5. Adjust the protein concentration of the supernatant to 16 mg/mL with homogenization buffer. Store aliquots frozen at <–20°C. This crude protein preparation can be used without further processing, and the activity should be stable for at least 2 yr.

3.2.1.2. PROTEIN BINDING ASSAY

1. Dispense 50 μL aliquots of standard (0–8 pmol cAMP diluted in 50 m*M* Tris-HCl buffer, pH 7.4, plus 4 mM EDTA) or sample into 96-well microtiter plate (*see* **Note 4**).
2. Add 50 μL of [^{3}H]cAMP (0.14 pmol, about 11,000 cpm diluted in dH$_2$O) to each well.
3. Initiate incubations by the addition of 200 μL binding protein (diluted to 40 μg/ 200 μL in homogenization buffer without sucrose).
4. Incubate for 150 min at 4°C.
5. Stop incubations by rapidly filtering through glass-fiber filters (Whatman GF/B). Wash filters 2 × 3 mL with ice-cold 50 m*M* Tris-HCl buffer.
6. Dry filters, and place in scintillation vials with 5 ml scintillation cocktail.
7. Alternatively, incubations can be stopped with the addition of 0.2 mL Tris-HCl buffer containing 1% activated charcoal and 0.1% BSA. Samples should then be centrifuged for 5 min at 9000*g* to pellet unbound cAMP (which adsorbs to charcoal) and the supernatant counted for protein-bound [^{3}H]cAMP.

3.2.2. Radioimmunoassay (RIA)

Another approach for determining cAMP concentration in cell extracts utilizes a cAMP-specific antibody in an immunoassay *(10,11)*. This procedure relies on a highly specific antibody generated against 2'-*O*-monosuccinyl cAMP for detection of femtomolar amounts of cAMP *(12)*. The antibody can be used either in RIA or enzyme immunoassays (EIA). The concept behind this assay is analogous to that of the radioreceptor assay. The difference is that competition between radiolabeled cAMP (RIA) or enzyme-conjugated cAMP (EIA) and cAMP from cell extracts occurs for binding to an antibody rather than a binding protein (for detailed review of RIA, *see* **ref. 13**). Sensitivity of these assays is enhanced by converting cAMP to an acetylated derivative that binds the antibody with higher affinity. Although more sensitive than the radioreceptor method, immunoassays require more sample manipulation (to remove potentially interfering compounds), which may increase variability and assay time. Several commercially available kits are available that provide detailed protocols and all the reagents necessary for completing the assay, thus making this assay convenient for the user who only occasionally needs to measure cAMP levels. However, the cost of these kits makes frequent determinations quite expensive. A less expensive alternative, which we describe herein, is to purchase individual assay reagents separately. We have modified such an RIA (Calbiochem), to quantify cAMP levels in extracts from several types of cells. In our assay, the use of centrifugation to isolate cAMP–antibody immunocomplexes has been replaced by a rapid filtration method using 96-well filtration plates (e.g., Millipore Multiscreen™ plates). Thus, the amount of time required for sample manipulation and the amount of radioactive waste generated are substantially decreased, whereas the number of samples that can be processed is increased. In addition, this approach should be readily adaptable for high-throughput screening.

1. Dilute stock solution of cAMP (5 µM in dH$_2$O) into 10 mM sodium acetate buffer (pH 4.75) to obtain standard solutions containing 5, 2.5, 1.25, 0.625, 0.313, 0.156, 0.078, 0.039, 0.019, and 0.009 nM cAMP. Also prepare a buffer only solution for determination of total binding. The final volume of all solutions should be 1 mL.
2. Acetylate standards and 1 mL of sample extracts (diluted in 10 mM sodium acetate buffer, pH 4.75) by the rapid sequential addition of 20 µL of triethylamine and 10 µL of acetic anhydride to each test tube (*see* **Note 5**). Vortex immediately. Samples should be assayed within 4–6 h of acetylation.
3. Prepare the 96-well filter plates by prewetting the plate with 100 µL of 10 mM sodium acetate buffer. Vacuum buffer through filters on the plate using vacuum manifold. Do not allow the filter membrane to dry under the vacuum.
4. Pipet 50 µL acetylated cAMP standards or samples into individual wells. Add 25 µL of [^{125}I]cAMP (10–12,000 cpm in γ-globulin buffer) and 25 µL of cAMP antiserum (diluted 1:3000 in γ-globulin buffer) to each well, except the well designated for nonspecific binding (which receives only [^{125}I]cAMP).
5. Cover plates, and incubate 16–24 h at 4°C.
6. Add 50 µL of antirabbit antibody coupled to solid support (e.g., magnetic beads or agarose) to each well, and vortex the plate. Incubate at 4°C for 1 h on rotating platform.
7. Add 100 µL of ice-cold PEG (12%, w/v) to each well, and filter with vacuum manifold. Wash filters twice with 200 µL of PEG.
8. Radioactivity in the immunoprecipitates is determined by removing the underdrain of the plate and punching filters into 12 × 75 mm glass tubes (*see* **Note 4**). Count samples on a γ-counter.

3.2.3. Data Analysis

For both the RIA and radioreceptor binding assay, the amount of cAMP in cell extracts is determined by comparison of the amount of radioactivity (RIA) or enzyme activity (EIA) present in samples with a standard curve generated using solutions containing known amounts of cAMP.

3.3. Assessment of Adenylyl Cyclase Activity

Measurements of adenylyl cyclase activity involve assessing the conversion of ATP to cAMP. Two such methods are widely used. The first utilizes [α-^{32}P]ATP, which is converted by adenylyl cyclase to [^{32}P]cAMP, to assess enzyme activity in cell or tissue homogenates or membranes. The second method involves incubating cells or tissues with [³H]adenine to label intracellular pools of adenine nucleotides (e.g., ATP). Accumulation of [³H]cAMP is then used as an index of adenylyl cyclase activity. Both of these methods require the separation of radioactively labeled cAMP from other components of the reaction mixture. This is efficiently accomplished by a method described by Salomon et al. utilizing sequential chromatography on Dowex cation-exchange and alumina columns *(7)*.

Another approach that can be utilized for determining adenylyl cyclase activity involves using an anti-cAMP antibody (described above) to assess cAMP formation following the incubation of cell membranes with ATP. This approach avoids the use of radioactive substrate and can also be used for assessing cAMP accumulation in intact cells. An important consideration of this assay is that only a small fraction of substrate (ATP) is converted to cAMP. Thus, there is the potential that components of the reaction mixture may interfere with the antibody–cAMP interaction. For this reason, it is necessary to confirm that the buffers used for cAMP generation do not interfere with the detection of cAMP in a solution of known concentration.

3.3.1. ATP Conversion Assay

As noted above, the principle of this assay is the conversion of $[\alpha\text{-}^{32}P]$ ATP to cAMP. Radioactively labeled cAMP is then isolated by sequential column chromatography. It is important to note that adenylyl cyclase activity can be influenced by several factors. First, degradation of ATP by enzymes (nucleotidases and hydrolases) present in cell homogenates can result in depletion of the substrate. To prevent this problem, incubations are performed in the presence of an ATP regeneration system typically consisting of creatinine phosphate and creatinine phosphokinase (or phospho[enol]pyruvate and pyruvate kinase). Second, adenylyl cyclase activity is dependent on the presence of divalent cation (Mg^{2+} or Mn^{2+}), which must be present during the incubation (the preferred substrate of the enzyme is MgATP). Third, phosphodiesterases present in cell membranes can hydrolyze cAMP to AMP. Therefore, incubations typically contain one or more phosphodiesterase inhibitors (e.g., isobutylmethylxanthine, rolipram, RO 20-1724) to prevent cAMP breakdown as well as unlabeled cAMP to dilute and "trap" the labeled cAMP product. Fourth, the regulation of adenylyl cyclase by G-protein-coupled receptors (e.g., adrenergic receptors) requires that GTP (or a nonhydrolyzable GTP analog) be included in the incubations. Fifth, it is important that cAMP generation is linear with respect to time and protein content. This may require modification of the amount of cellular protein, length of incubation, or incubation temperature. A factor that contributes to nonlinear (and loss) of enzyme activity is the relative instability of adenylyl cyclase owing to denaturation, especially at temperatures typically used for enzyme assays.

1. Into labeled assay tubes, aliquot 50 μL adenylyl cyclase activity assay buffer (*see* **Subheading 2., item 10**) and receptor agonist and/or antagonist (total volume of 70 μL).
2. Initiate reactions by the addition of 30 μL cell or tissue homogenate/membrane (containing 25–200 mg protein diluted into 30 mM Na-HEPES, 5 mM MgCl$_2$, and 2 mM, dithiothreitol [DTT], pH 7.5) to the assay tubes. The total volume in each tube should be 100 μL.
3. Incubate for appropriate amount of time (usually 5–15 min at 30–37°C).
4. Incubations are terminated by the addition of 100 μL stop solution (2% w/v sodium dodecyl sulfate [SDS] containing 40 mM ATP and 1.4 mM cAMP, pH 7.5).
5. Add about 5000 cpm of [^{3}H]cAMP (NEN Life Science) as an internal standard to monitor sample recovery during column chromatography. Be sure to assess the exact amount of internal standard cpm by counting several aliquots directly.
6. Bring sample volume to 1 mL with dH$_2$O, and carefully apply samples to Dowex columns. Allow loaded volume to pass through resin, and wash column two times with 1.25 mL dH$_2$O each. Discard the eluate.
7. Arrange columns such that Dowex columns elute into the alumina columns.
8. Pipet 5 mL dH$_2$O onto each Dowex column. This will elute the cAMP from the Dowex columns and onto the alumina columns. Discard eluate.
9. Remove Dowex columns and position the alumina columns such that that eluate will drip into 20-mL scintillation vials.
10. Add 4 mL of 0.1 M imidazole-HCl, pH 7.5, to the alumina columns to elute cAMP into the vials.
11. Add scintillation fluid and count the samples using dual-channel counting, whereby one channel monitors [^{3}H] for recovery of internal standard and the other channel monitors [^{32}P]-labeled product.

3.3.2. [^{3}H]cAMP Generation in Cells

1. Incubate cells for 1–18 h with 2 μCi/mL [^{3}H]adenine (26.3 Ci/mmol; NEN Life Science) in DMEM with serum. The time required for effective labeling of the ATP pool needs to be optimized for individual cell types.
2. Wash cells to remove extracellular [^{3}H]adenine with two changes of DMEM, and equilibrate cells in DMEH (typically 30 min, 37°C).
3. Replace medium with fresh DMEH. If appropriate, incubate cells in the presence of a phosphodiesterase inhibitor (e.g., isobutylmethylxanthine).
4. Add agonist, and incubate cells for the desired time (usually 1–15 min) at 37°C. The total volume should be at least 0.5 mL for a 24-well plate.
5. Reactions are stopped by aspirating media and adding 0.25 mL trichloroacetic acid (7.5% w/v). Approximately 2000 cpm of [^{32}P]cyclic AMP (*see* **Note 6**) is

added to each sample as an internal standard, and samples are brought up to 1-mL vol with water.

6. [^{3}H]cAMP is resolved from [^{3}H]ATP and other nucleotides by sequential chromatography on Dowex and alumina columns essentially as described in **Subheading 3.3.1.** However, eluates of sample and dH$_2$O washes from the Dowex columns must be collected, and the radioactivity measured to determine the incorporation of [^{3}H]adenine into [^{3}H]ATP.

7. Add scintillation fluid to the eluate from the alumina columns, and count the samples using dual-channel counting, whereby one channel monitors [^{32}P] for recovery of internal standard and the other channel monitors [^{3}H]-labeled product.

8. The [^{3}H]cAMP values are corrected for recovery of the internal standard and are expressed as percent conversion of the amount of total incorporated [^{3}H]adenine nucleotides ([^{3}H]cAMP cpm + [^{3}H]ATP cpm). This correction allows for variation in the number of cells or the extent of [^{3}H]adenine uptake and incorporation into [^{3}H]ATP in each assay.

3.3.3. Immunoassay

The conversion of ATP to cAMP in an adenylyl cyclase assay can also be determined by immunoassay *(13)*. This method has several advantages relative to the conversion assay described above (*see* **Subheading 3.3.1.**). First, it does not require the use of large amounts of ^{32}P. Second, cAMP does not need to be isolated by column chromatography. Third, the same assay can be used for both cell homogenates and intact cells. The procedure for the incubation of membranes is essentially the same as described above for the conversion assay. We prefer to stop reactions by boiling for 5 min. The determination of cAMP present in the boiled extract is identical to that described for cell lysates (*see* **Subheading 3.2.2.**). Because the amount of cAMP in the boiled extract represents only a small fraction of nucleotide, it is important to include a "blank" tube, which receives boiled membrane, to control for any interference of antibody–cAMP interaction by components of the assay mixture.

4. Notes

1. Other growth media (e.g., Roswell Park Memorial Institute medium [RPMI], Hank's, and so on) may be used as necessary for growth and incubation of cells. In addition, it should be noted that in some cell types, cell density could affect basal levels of cAMP. It may therefore be necessary to examine the effect of cell confluency on agonist response. For both cells in suspension and adherent cells, the number of cells can be varied such that basal levels are easily detectable in the cAMP assay.

2. Several methods have been described for terminating incubations and extracting cAMP from samples. The use of ethyl alcohol (EtOH) and drying permits resuspension of samples at appropriate concentrations in assay buffer. Other

methods of extracting cAMP include the use of trichloroacetic acid, perchloric acid, or HCl. Such agents may interfere with immunoassay and may therefore require neutralization. For cell suspensions or tissues, incubations can be terminated by placing samples in a boiling water bath for 5–10 min.

3. The cAMP binding protein can be isolated from other sources, such as rabbit or bovine skeletal muscle, or can be purchased in purified form (cAMP-dependent protein kinase, Sigma). The method described here for isolation from adrenal cortex followed by filtration to separate bound from free cAMP reportedly permits the use of a crude protein extract. Other methods of separation (e.g., precipitation) may require a more highly purified binding protein preparation.

4. The use of 96-well microtiter plates is convenient when a filtration apparatus for such a format is available (e.g., Millipore Multiscreen™ System). Alternatively, the assay can be conducted in tubes appropriate for use with Brandel cell harvesters.

5. Acetylation increases sensitivity of the assay >10-fold. If samples contain sufficient amounts of cAMP, this step can be omitted. If this is done, standard concentrations should be increased accordingly (e.g., 10-fold).

6. When using the sequential column method to separate [^{3}H]cAMP from [^{3}H]ATP, recovery of sample from each column can be corrected by including [^{32}P]cAMP and utilizing dual-channel counting. [^{32}P]cAMP can be purchased at considerable cost (NEN Life Science) or can be synthesized by incubating [α-^{32}P]ATP with adenylyl cyclase toxin from *B. pertussis* (List Biological, Campbell, CA). Specifically, 5 µg adenylyl cyclase toxin is incubated with 10 µM calmodulin, 10 mM ATP, 100 mM MgCl$_2$ and 250 µCi [α-^{32}P]ATP in 60 mM tricine buffer, pH 8.0, at room temperature for 4–12 h. [^{32}P]cAMP is then purified by column chromatography as described above and stored at 4°C. Alternatively, [^{14}C]cAMP or [^{14}C]ATP (used to produce [^{14}C]cAMP) can be purchased commercially (NEN Life Science) and used to correct for column recovery.

References

1. Sutherland, E. W. and Rall, T. W. (1958) Fractionation and characterization of a cyclic adenine ribonucleotide formed by tissue particles. *J. Biol. Chem.* **232,** 1077–1091.
2. Taussig, R. and Gilman, A. G. (1995) Mammalian membrane-bound adenylyl cyclases. *J. Biol. Chem.* **270,** 1–4.
3. Tang, W. J. and Hurley, J. H. (1998) Catalytic mechanism and regulation of mammalian adenylyl cyclases. *Mol. Pharmacol.* **54(2),** 231–240.
4. Taussig, R., Iniguez, L. J., and Gilman, A. G. (1993) Inhibition of adenylyl cyclase by Gi alpha. *Science* **261,** 218–221.
5. Taussig, R., Tang, W. J., Hepler, J. R., and Gilman, A. G. (1994) Distinct patterns of bidirectional regulation of mammalian adenylyl cyclases. *J. Biol. Chem.* **269,** 6093–6100.
6. Sunahara, R. K., Dessauer, C. W., and Gilman, A. G. (1996) Complexity and diversity of mammalian adenylyl cyclases. *Annu. Rev. Pharmacol. Toxicol.* **36,** 461–480.

7. Salomon, Y., Londos, C., and Rodbell, M. (1974) A highly sensitive adenylate cyclase assay. *Anal. Biochem.* **58,** 541–548.

8. Gilman, A. G. (1970) A protein binding assay for adenosine 3':5'-cyclic monophosphate. *Proc. Natl. Acad. Sci. USA* **67,** 305–312.

9. Nordstedt, C. and Fredholm, B. B. (1990) A modification of a protein-binding method for rapid quantification of cAMP in cell-culture supernatants and body fluid. *Anal. Biochem.* **189,** 231–244.

10. Steiner, A. L., Kipnis, D. M., Utiger, R., and Parker C. (1969) Radioimmunoassay for the measurement of adenosine 3',5'-cyclic phosphate. *Proc. Natl. Acad. Sci. USA* **64,** 367–373.

11. Steiner, A. L., Parker, C. W., and Kipnis, D. M. (1972) Radioimmunoassay for cyclic nucleotides. I. Preparation of antibodies and iodinated cyclic nucleotides. *J. Biol. Chem.* **247,** 1106–1113.

12. Harper, J. F. and Brooker, G. (1975) Femtomole sensitive radioimmunoassay for cyclic AMP and cyclic GMP after 2'0 acetylation by acetic anhydride in aqueous solution. *J. Cyclic Nucleotide Res* **1,** 207–218.

13. Brooker, G., Harper, J. F., Terasaki, W. L., and Moylan, R. D. (1979) Radioimmunoassay of cyclic AMP and cyclic GMP. *Adv. Cyclic Nucleotide Res.* **10,** 1–33.

25

Assay of Arachidonic Acid Release Coupled to α_1- and α_2-Adrenergic Receptors

Lincoln Edwards and Paul Ernsberger

1. Introduction

1.1. Biochemical Pathways for Generation of Arachidonic Acid (AA) from Phospholipids

AA is the precursor for prostaglandins, eicosanoids, and the endogenous cannabinoid ligands, such as anandamide *(1)*. The major pathways that lead to the liberation of free AA from phospholipids are shown in **Fig. 1**. Phospholipase A_2 (PLA_2) provides a direct pathway for AA release and is the most common source of cellular AA liberation. PLA_2 is the primary source of free AA in most cell systems, including neurons. Physiological regulation of PLA_2 occurs primarily through the action of calcium, which binds to several PLA_2 isoforms and is required for activity *(2,3)*. The low-molecular-weight secretory PLA_2 isoforms require millimolar calcium, whereas the cytosolic classes of PLA_2 need only micromolar concentrations. Calcium is required for translocation to membranes, where the cytosolic form of the enzyme can have access to its phospholipid substrate. Other isoforms are completely independent of calcium, but these variants may be primarily involved in membrane remodeling rather than signal transduction *(4)*. The calcium-independent forms of PLA_2 play an important role in the rapid incorporation of free AA into phospholipids, which is taken advantage of in the AA release assay described here.

Besides calcium, the other important physiological stimulus regulating PLA_2 is phosphorylation. In particular, PLA_2 is a substrate for mitogen-activated protein kinase (MAPK; p42 ERK-1 and p44 ERK-2). Phosphorylation of PLA_2 by MAPK leads to rapid activation *(5)*. Although stimulation of protein kinase C stimulates PLA_2, this action is apparently indirect and mediated through the

From: *Methods in Molecular Biology, vol. 126: Adrenergic Receptor Protocols*
Edited by: C. A. Machida © Humana Press Inc., Totowa, NJ

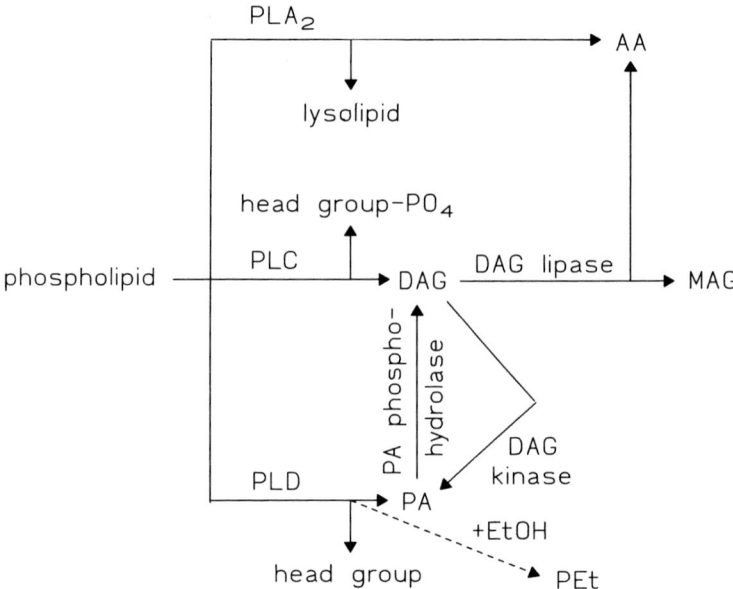

Fig. 1. Possible pathways for generation of AA from phospholipids. Hypothetical model of the biochemical pathways contributing to the liberation of AA. AA = arachidonic acid, PLA$_2$ = phospholipase A$_2$, PLC = phospholipase C, PLD = phospholipase D, PA = phosphatidic acid, DAG = diacylglyceride, MAG = monoacylglyceride, EtOH = ethanol, PEt = phosphatidylethanol.

downstream activation of MAPK by protein kinase C. Calcium and MAPK may act synergistically to activate PLA$_2$ *(2)*. Thus, receptors that stimulate both pathways, such as the α_1-adrenergic receptors (ARs), may be particularly effective in activating PLA$_2$ *(5)*. However, in smooth muscle cells, it has been proposed that the activation of PLA$_2$ elicited by α-1 adrenergic receptor stimulation is not mediated by MAPK, but rather through Jun kinase (JNK) *(6)*.

Specific inhibitors of different classes of PLA$_2$ are not yet available. The secretory classes of PLA$_2$ enzyme have as many as 16 cysteine resides *(2)* and, thus, can be inactivated in a relatively straightforward manner by the addition of reducing agents, such as dithiothreitol (typically at about 1.0 m*M*). A relatively selective inhibitor of PLA$_2$ is aristolochic acid *(7)*, typically used at a concentration of 100 µ*M*, which is typical of PLA$_2$ inhibitors. Arachidonyl trifluromethyl ketone was developed as selective inhibitor of the cytosolic isoforms of PLA$_2$, but has equal potency as an inhibitor of the calcium-insensitive isoforms *(4)*. Bromoenol lactone selectively inhibits the latter calcium-insensitive forms, but also has the additional action of reducing the formation of diacylglycerol (DAG) from phosphatidic acid (PA) by acting as an inhibitor of phosphatidic acid hydrolase. Classic PLA$_2$ inhibitors, such as mepacrine, have a variety of nonspecific and even toxic effects. Both secretory and cytosolic forms of PLA$_2$ exist as multiple isozymes, and they may not be equally affected by all inhibitors.

Activation of phospholipase C (PLC), either phosphatidylinositol-selective phospholipase C (PI-PLC) or phosphatidylcholine-selective phospholipase C (PC-PLC), can result in AA release through an indirect pathway. PLC generates DAGs, which can then be acted on by DAG lipase to produce mono-acylglycerols (MAGs), which are further broken down by MAG lipase. AA can be released either by DAG lipase or by MAG lipase, with DAG lipase being the rate-limiting step *(8)*. DAG lipase is particularly active in brain *(8)* and in adrenal chromaffin cells *(9,10)*. AA release mediated by DAG lipase is required for the secretion of catecholamines from chromaffin cells *(9)*. In brown adipocytes, stimulation of α$_1$-ARs leads to AA release via PLC and DAG lipase, without involvement of PLA$_2$ *(11)*, in contrast to renal cells where AA release in response to α$_1$-AR stimulation appears entirely mediated by PLA$_2$ *(5)*. As this example illustrates, even when the receptor subtype is identical and the cellular response is the same, the intermediate signaling steps may be different.

A third pathway with the potential to cause AA release is illustrated at the bottom of **Fig. 1**. Hydrolysis of phospholipids by phospholipase D (PLD) generates phosphatidic acid, which can be dephosphorylated by phosphatidic acid hydrolase to yield DAG. Also shown in **Fig. 1** is the common method for assaying PLD activity, which is to monitor the accumulation of phosphatidylethanol (PEt) in the presence of ethanol. Generation of DAGs through the sequential actions of PLD and phosphatidic acid hydrolase has been shown in endothelial cells in response to β-AR stimulation *(12)*.

1.2. Adrenergic Receptor Signal Transduction Pathways and AA Release

Adrenergic receptors are members of the large family of integral membrane proteins that are coupled to guanine nucleotide binding regulatory proteins or G-proteins. G-protein-coupled receptors are important physiological regulators, because they are the targets for a large number of hormones and neurotransmitters, as well as autocrine and paracrine factors. Adrenergic receptors form the interface between the sympathetic nervous system and the cardiovascular system as well as endocrine and parenchymal tissue.

The primary initial signaling pathways for the adrenergic receptor subtypes have been identified. Thus far, genomic or cDNA clones for nine subtypes of adrenergic receptors have been identified in mammalian species: three α$_1$-AR subtypes, three α$_2$-AR subtypes, and three β-AR subtypes *(13)*.

α_1-ARs are primarily coupled to stimulation of PI-PLC *(14)*. Additional activation of PLD may be independent of the PI-PLC pathway *(15)*. The activation of AA release in response to stimulation of α_1-AR is usually thought to be mediated by the influx of calcium triggered through inositol phosphates produced by PI-PLC activation *(16)*. However, the PI-PLC activation and AA release triggered by stimulation of α_1-ARs may result from separate processes. For example, the PI-PLC inhibitor prevents the rise in calcium induced by stimulation of either α_{1A}-AR or α_{1B}-AR, but the AA release response is unaffected *(17)*. Furthermore, a specific single amino acid substitution in the α_{1B}-AR eliminates the induction of AA release by adrenergic agonists, whereas PI-PLC activation and the accumulation of inositol phosphates are preserved or even facilitated *(18)*. The specific steps coupling α_1-ARs to AA release may be complex and may differ between cell types.

Activation of α_2-ARs results mainly in inhibition of adenylyl cyclase, but in some systems, such as smooth muscle and kidney epithelial cells, it results in stimulation of PI-PLC, thus triggering calcium entry, which in turn leads to PLA_2 activation and the accumulation of AA *(19)*. Chinese hamster ovary (CHO) cells stably transfected with α_{2A}-AR show release of [^{3}H]AA in response to epinephrine and other α_2-AR agonists *(20)*. In contrast, Axelrod's group found that α_{2A}-AR stably expressed in CHO cells were not able to elicit AA release unless endogenous purinergic receptors were simultaneously stimulated with adenosine triphosphate (ATP) *(21)*. This is reminiscent of other receptor–receptor interactions reported for the α_2-AR. Stimulation of α_{2A}-ARs can also lead to the activation of PLD, but this signaling pathway is apparently only activated when cells are artificially induced to express very high densities of receptor through gene transfection *(22)*. Subtypes of α_2-ARs other than the α_{2A}-AR have not been studied as extensively, but thus far, major differences in cell signaling have yet to be found among the α_2-AR subtypes.

An important action of α_2-ARs is the activation of potassium channels. The classic PLA_2 inhibitors mepacrine and bromophenacyl bromide interfere with action, which might seem to implicate a signaling pathway through PLA_2 *(23)*. However, the action of α_2-AR agonists was not blocked by more specific PLA_2 inhibitors, nor did AA mimic it. Thus, the PLA_2 pathway does not participate in the modulation of neuronal potassium channels by α_2-ARs, although it does subserve other physiological responses.

Information on the activation of AA release by β-ARs is more limited. These receptors are primarily coupled to the activation of adenylyl cyclase, leading to the accumulation of cAMP. It has been reported that in tracheal epithelium β_2-AR agonists trigger the release of AA and the production of prostaglandins *(24)*. The intermediate steps between the accumulation of cAMP and the release of AA are not yet known.

AA release is most commonly assayed in a simple static incubation procedure, which has been described in dozens of reports *(16,17,19,21,24–26)*. Cells are incubated with [^{3}H]AA until equilibrium incorporation into phospholipids is reached at 18–24 h. Then fresh medium-containing test agent is added. Fatty acid-free bovine serum albumin (BSA) is added as a carrier to absorb released AA. The cumulative release of [^{3}H]AA is determined by collecting the cell-free medium, typically after 30 min. Some of the limitations of this simple method are as follows:

1. [^{3}H]AA can be rapidly metabolized and taken back up by the cells, especially after prolonged incubation periods, although this can be minimized by adding sufficient BSA to absorb free AA as it is formed.
2. Rapid changes of medium with a pipet are sufficient in themselves to perturb cell membranes and evoke AA release (data not shown).
3. The response is highly variable, allowing detection only of large changes (two-fold or more over baseline).
4. No information is obtained on the time-course of the response.

A further complication of the static incubation method is that prostaglandins formed from AA can in term elicit further release of AA, potentially creating positive feedback. For example, in the uterus, prostaglandin $F_{2\alpha}$ elicits AA release *(27)*.

The regulation of AA release from neuronal cells *(28–32)* differs in several important respects from other cell types in which AA release has been closely studied, including CHO cells, lymphocytes, and epithelial and smooth muscle cells *(33,34)*. Calcium ionophore was only a modestly effective stimulus for the release of AA in PC12 cells, suggesting that elevated cell calcium may not be the primary stimulus for AA release in this neuronal cell type. Consistent with the present finding, we found that calcium ionophore induced an increase in the generation of prostaglandin E_2 by differentiated PC12 cells by only about 50% *(35)*. Similarly, bradykinin increases AA release from PC12 cells by only 33%, despite induction of a large spike of intracellular calcium *(31)*. This is in contrast to the majority of cell types, wherein the activation of phospholipase A_2 by calcium influx leads to AA release *(19,20)*.

1.3. Superfusion Assay of [^{3}H]AA Release

The method described here measures the release of [^{3}H]AA following incorporation into cellular phospholipids by using a superfusion system with BSA as a carrier. The use of a superfusion system eliminates each of the limitations of the static incubation method described above. The continuous flow of medium minimizes reuptake or metabolism of liberated AA. The cells are sealed in a Teflon chamber, removing the influence of pipeting and medium changes. Responses are still variable, but with standardized superfusion appa-ratuses allowing 6- to 12-cell aliquots to be tested in parallel, the opportunity to average data across replicates is improved. Perhaps most importantly, one can obtain a profile of the time-course of AA release with a resolution of 1 min or less. Our timecourse experiments, for example, indicate that most AA release responses are concluded within 10 min, so the 30-min collection period used in most static incubation protocols is probably not necessary. One drawback of the superfusion method is that it is more time-consuming, particularly from the standpoint of analyzing the large number of data points generated by the collection of multiple fractions from superfused cells. The method is described below.

1.4. Example: [^{3}H]AA Release from Superfused PC12 Pheochromocytoma Cells

In preliminary experiments, we characterized the release of AA from PC12 pheochromocytoma cells as a model neuronal cell. Specific [^{3}H]AA release in response to stimuli, such as phorbol ester, potassium depolarization, or chemi-cal hypoxia induced with sodium cyanide, was not detectable after only 6 h of labeling. There was no further increase in released radioactivity after more than 24 h. Thus, a 24-h labeling period was selected. In all cases, more than 90% of the radioactivity added to the medium was incorporated into the cells. Basal release of [^{3}H]AA or its metabolites from unstimulated PC12 cells was $0.22 \pm 0.01\%/min$ ($N = 36$), expressed as a fraction of total tritium incorpora-tion. This value is in close agreement with that reported for brain synapto-somes under similar conditions (0.19%/min) *(30)*. Typically, $1.2–1.8 \times 10^6$ dpm were incorporated into each of six aliquots derived from one 75-cm^2 flask, indicating that one-third to one-half of the total radioactivity added to the cul-ture medium remained in the cells after a 30-min washout period.

Assay of [^{3}H]AA release from superfused PC12 cells is shown in **Fig. 2**. Moxonidine administered in the superfusion medium at a concentration of 100 nM for 2 min elicited a rapid twofold rise in [^{3}H]AA release, followed by a return nearly to baseline 4 min later. The potent guanadinium α_2-AR agonist guanabenz was inactive at a 10-fold higher concentration, since only a gradual depletion of radioactivity was observed. The threshold concentration of moxonidine appears to be above 1 nM, since this concentration elicited only a slight response, which failed to reach statistical significance.

In experiments with CHO cells stably transfected with α_{2A}-ARs (Ernsberger, unpublished data), moxonidine at 100 nM had no effect on [^{3}H]AA release, consistent with its status as a weak α_2-AR agonist. In contrast, guanabenz induced a doubling of [^{3}H]AA release, which persisted for 5 min. This con-firms previous reports of [^{3}H]AA release induced by stimulation of α_2-ARs in

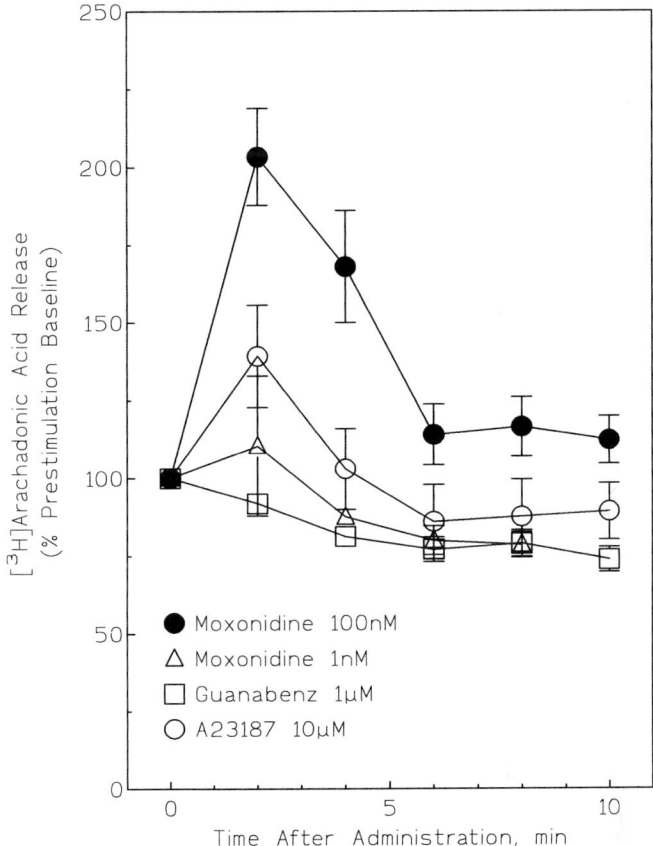

Fig. 2. Assay of [^{3}H]AA release from superfused PC12 cells. Shown are the mean results ± SE from 10–16 different superfused cell suspensions. Data were collected as dpm/2-min fraction, and are expressed as a percentage of prestimulation baseline period immediately preceding exposure to the test compound. Baseline release was 2160 ± 140 dpm/2-min fraction ($N = 16$). Reproduced with permission from *J. Auton. Nerv. Sys. (28)*.

stably transfected CHO cells *(20)*. In the superinfusion assay, the release of [^{3}H]AA was greater in magnitude than previously reported using static incubation.

2. Materials

2.1. Description of the Six-Channel Superfusion System

The methods were developed using the SF-06, a six-channel superfusion system from Brandel (Gaithersburg, MD). Identical methods can be used with the SF-12 or 12-channel system. The system is illustrated in **Fig. 3** using a

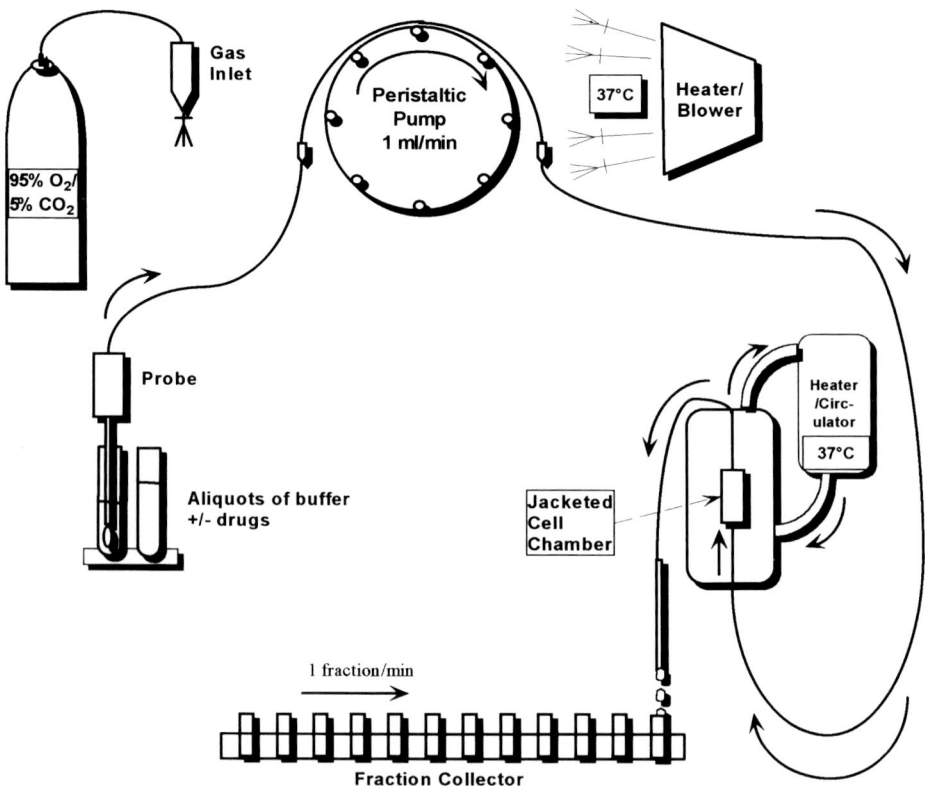

Fig. 3. Schematic of the Brandel superfusion system. For the sake of simplicity, only a single channel is depicted.

single channel for the sake of simplicity. A peristaltic pump pulls superfusion medium up into a probe initially from a trough (not shown). After the washout and baseline periods, agonists and antagonists can be delivered in the superfusion medium from test tubes, as shown in **Fig. 3**.

Medium flows through a stainless-steel tube into a water-jacketed block. In the center of the block is a Teflon chamber that holds the radioactively labeled cells. The transit time from the test tube to the chamber is approx 2 min at a pump speed of 1 mL/min. The chamber is sealed at both the bottom and top ends by glass fiber filters (GF/B, Whatman) held in place by O-rings. The cells are suspended in the chamber by the upward flow of medium, which just balances the downward pull of gravity. Superfusion medium passes out of the chamber through a stainless-steel tube and then descends through an outflow tube to the fraction collector. At regular intervals (usually 1 min), the fraction collector advances one rank, positioning a fresh miniscintillation vial (Beckman) under the outflow tube. A typical 60-min experiment will generate 60 samples/cell aliquot, or 360 samples for the six-channel system.

The apparatus is surrounded by an environmental chamber, which is maintained at 37°C by thermostatically-controlled heater/blower (*see* **Fig. 3**). The cell chamber is kept at 37°C by slowly circulating jacket water. An atmosphere of 95% O2/5% CO_2 is maintained by continuous flow from a gas cylinder. Also, superfusion medium in the trough or in test tubes is periodically gassed through a set of gas probes (not shown).

2.2. Cell Labeling Prior to Superfusion

1. Cells of interest, usually CHO cells stably transfected with an α_1- or α_2-AR subtype.
2. [5, 6, 8, 9, 11, 12, 14, 15-^{3}H]AA (200 Ci/mmol) is obtained from American Radiolabeled Chemicals (St. Louis, MO). [^{3}H]AA is aliquoted into 2.5-mL sterile microcentrifuge tubes, each containing 10 µCi of [^{3}H]AA in 50 µL of ethanol, which are then stored under nitrogen gas at –70°C[1].
3. Phorbol myristate acetate and A23187 are prepared as 1:1000 stock solutions in 100% dimethyl sulfoxide (DMSO) and stored at –70°C. Compounds are diluted in Krebs' buffer immediately prior to use.
4. Krebs' buffer: 124 mM NaCl, 25 mM NaHCO$_3$, 20 mM N-(2-hydroxyethyl) piperazine-N-(2-ethanesulfonic acid), pH 7.4 (HEPES), 11.1 mM glucose, 10 mM NaH$_2$PO$_4$; 4.7 mM KCl, 1.2 mM KH$_2$PO$_4$, 2.5 mM CaCl$_2$, and 1.2 mM MgSO$_4$. Buffer is aerated with 95% O$_2$/5% CO$_2$ at room temperature prior to adjusting pH to 7.4 with NaOH or HCl. Aerated and pH-adjusted Krebs' was then sealed in an air-tight container (Corning Fleaker) and warmed to 37°C.
5. Rinse buffer: Krebs' buffer without BSA.
6. Superfusion medium: Krebs' buffer with 0.01% BSA.
7. Lysis solution: 5.0 mM ethylenediamine tetra-acetic acid (EDTA) in 10% methanol, with 0.01% BSA, and 0.1% Triton X-100. Pilot studies show that this solution is optimal for removal of the greatest proportion of [^{3}H]AA in the minimum number of fractions at the end of the experiment.
8. Superfusion apparatus and GF/B filter circles used to seal the cell chambers are purchased from Brandel.

3. Methods
3.1. Superfusion Cell Labeling and Preparation

1. Seed CHO cells at low density onto uncoated 75-cm^2 flasks. Then 1–3 d later perform prelabeling with [^{3}H]AA (10 µCi in 10 mL of culture medium) for about 24 h.
2. Cell labeling should be done 24 h prior to the experiment. Take a microcentrifuge tube containing 10 µCi of [^{3}H]AA (*see* **Note 1**), and add it to the flask containing the cells. As a safety precaution, be certain to place a radioactive sticker on each flask containing [^{3}H]AA.

3. On the day of the superfusion experiment, turn the superfusion machine on, and allow the air in the enclosure to warm up to 37°C. It will take at least 30 min for the whole apparatus to equilibrate at this temperature. As the device warms up, wash the peristaltic tubing by running through reagent-grade water for 20 min at a high pump rate, approx 2.5 mL/min.

4. Warm the plastic trough that will hold the superfusion medium up to 37°C. Make up 1 L of Krebs' buffer, aerate, and then pH to 7.4 with NaOH.

5. Place 50 mL of Krebs' buffer without BSA into a 100-mL beaker and cover. Leave at room temperature for use as rinse buffer.

6. Warm the remaining 950 mL of Kreb's buffer to 37°C, and then add 95 mg BSA to yield a 0.01% solution of BSA in Krebs'. Stir sealed container slowly with a stir bar until the BSA dissolves. (Vigorous mixing will denature the BSA and induce foaming.)

7. Place superfusion medium (Krebs' buffer with BSA) into the trough, and wash the machine for 5 min at the experimental flow rate (usually 1 mL/min).

8. Transfer the cells on ice from the incubator to the workstation. Add 50 μL of 1 M EGTA to the flask containing the cells, and tap gently to remove adhering cells. (Scraping cells may cause cell membrane damage and evoke premature AA release). Transfer the cells to a 15-mL centrifuge tube containing 20 μL of 2.5 M $CaCl_2$ (to quench the ethyleneglycol-bis-(β-aminoethylether)-$N,N,N,'N,'$-tetra-acetic acid [EGTA]), and spin at 200g for 3 min at room temperature.

9. Remove the supernatant (see **Note 2**), and gently resuspend the cell pellet by rocking it with 10 mL of Krebs' (without BSA). Prior to this step, use the 10 mL of Krebs' to rinse the culture flask and recover additional cells.

10. Resuspend the final pellet in 3.2 mL of aerated Krebs' buffer without BSA (rinse buffer) by gentle rocking.

11. Remove the chambers from the superfusion machine, and dry with laboratory wipes. Add mesh disk and filter to the bottom of each chamber. Place on top of the O-ring on the lower probe.

12. Aliquot the final cell suspension into the six chambers (500 μL/chamber) of the superfusion device, and cover the aliquot in the chamber with a second filter, which will be held in place by the O-ring on the upper probe.

13. Replace the chambers into the superfusion device, and attach the plastic covers to create a partial air seal. Start flow of 95% O_2/5% CO_2 into the machine.

14. Start superfusion with fresh Krebs equilibrated with 95% O_2/5% CO_2 and containing 0.01% fatty acid-free BSA, at a rate of 1.0 mL/min for 30 min to achieve a stable baseline release, and remove any remaining unincorporated label (see **Note 3**). The rate of 1.0 mL/min provides for 2 medium changes/min in the cell chamber, which has a volume of 0.5 mL.

15. Continue superfusion with standard medium collecting 1-min fractions and collect 5–7 baseline fractions. Then deliver agonist ± antagonist (see **Note 4**). Usually 5 min of treatment with an adrenergic receptor agonist are sufficient to provoke a maximal response. Recovery to baseline may require an additional 5–10 fractions with normal superfusion medium.

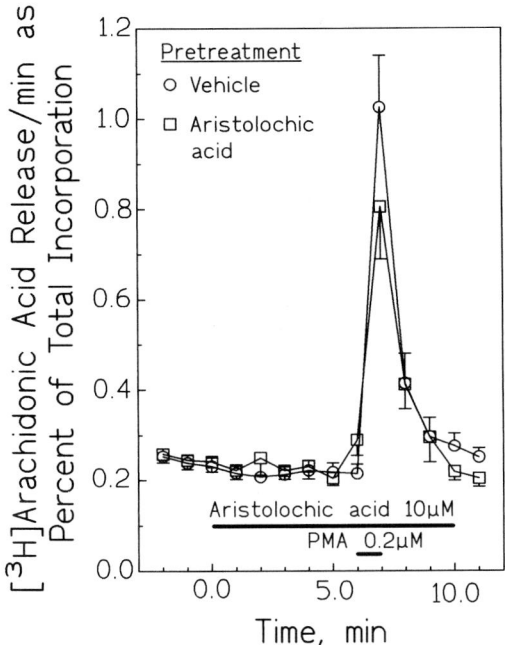

Fig. 4. Effect of the PLA$_2$ inhibitor aristolochic acid on [^{3}H]AA release elicited by phorbol myristate acetate (PMA). Shown are the mean results ± SE from 6 superfused cell suspensions. Data are radioactivity/1-min fraction as a percentage of the total incorporated radioactivity, corrected for preceding release (*see* **Fig. 3**). PMA (0.2 μ*M*) elicits a rapid increase in the release of [^{3}H]AA and its metabolites. Aristolochic acid (10 μ*M*) given as a pretreatment for 5 min failed to affect the basal rate of [^{3}H]AA release or its stimulation with phorbol myristate acetate. These data imply that resting and phorbol-stimulated AA release in PC12 cells does not involve PLA$_2$. Reproduced with permission from *J. Auton. Nerv. Sys. (28)*.

16. Depending on experimental design, a positive control or standard stimulus may be required. Phorbol myristate acetate at 0.2 μ*M* is very effective (*see* **Fig. 4**). Depending on cell type, the response is sustained for several minutes even after 2 min of exposure to this protein kinase C activator.

17. At the end of the experiment, the cells are lysed with 10% methanol/0.01% BSA/ 0.1% Triton X-100, and five additional 2-min fractions are collected until radioactivity returns to baseline. The lysis step allows determination of total radioactivity incorporated into the cells (*see* **Note 5**).

3.2. Data Analysis

1. The sum of all the radioactivity released is defined as the total [^{3}H]AA incorporation into cellular phospholipids. Data are expressed as a percentage of total incorpo-

rated radioactivity, corrected for release. In other words, the data represent the dpm released as a percentage of total remaining radioactivity. To perform this calculation, we enter the dpm value for each fraction (including the lysis step) into an Excel spreadsheet. Using the summation of total dpm released, we determine the dpm release/min as a percentage of total incorporation. This partially corrects for the rundown of [³H]AA release with time during the experiment.

2. Data can be analyzed by paired t-test to compare [³H]AA release before and after treatment. Dose–response curves can be analyzed by two-way ANOVA for time and treatment with Newman-Keuls multiple comparisons test (*see* **Note 6**).

3. Responses can also be compared using an approximation of the area under the curve. Total radioactivity released during stimulation is corrected for baseline release, with both expressed as a percentage of total incorporation. By this method, vehicle controls yield a slight negative value for net release, probably owing to selective depletion of an active pool of [³H]AA. For an example, see our recent reports *(28,29)*.

4. In the experimental design, it may be appropriate to use a positive control. Phorbol esters can be used as positive controls, for example, phorbol myristate acetate at 0.2 μM. This protein kinase C activator has been reported to elicit AA release from PC12 cells in static incubation assays *(31,32)*. As shown in **Fig. 4**, a marked spike of [³H]AA release is triggered by phorbol ester. Aristolochic acid had no effect on the large spike of [³H]AA release elicited by phorbol myristate acetate. This finding implies that phospholipase A_2 does not mediate the AA release mediated by activation of protein kinase C. Mepacrine (0.1 mM), another phospholipase A_2 inhibitor, actually increased baseline [³H]AA release (not shown), possibly through a nonspecific effect. These data imply that phospholipase A_2 does not play an important role in the liberation of free AA from PC12 cells under the current experimental conditions, thereby implicating an indirect pathway mediated by lipases. Further experiments (not shown) demonstrated that inhibition of DAG lipase with RHC 80267 (100 μM) inhibits baseline release of [³H]AA and prevents the response to phorbol ester treatment. Thus, it appears that the indirect pathway to AA release passing through DAG lipase may be of particular importance in this neuronal cell line, as documented for other neuronal cells *(36,37)* (*see* **Note 5**).

4. Notes

1. The supernatant is radioactive and should be discarded in accordance with radioactive waste disposal procedures.

2. AA is highly sensitive to oxidation across its multiple double bonds. It is also light-sensitive. Store under nitrogen or argon in an opaque container, and avoid direct exposure to sun or fluorescent light.

3. Aerate Krebs' buffer with 95% O_2/5% CO_2 at least every 10 min during washout and experiment. Also, aerate test tubes 1 min prior to placing uptake probes into them. Airflow should be set at a minimum. Dilute stock solutions of drugs and all serial dilutions during the 30-min washout period.

4. Place test tubes with Krebs' buffer in the environmental enclosure of the superfusion apparatus 5–10 min before usage, then place tubes into rack, add the appropriate amount of drug to each tube, cover with parafilm to prevent spillage, and mix.

5. The radiolabeled chemical species bound to BSA in the superfusion fluid is not identified, which is a limitation of the technique. At the option of the investigator, additional analysis, for example by high-performance liquid chromatography (HPLC) separation, could be run to identify the tritiated products released. However, this may not be necessary for several reasons. After 24 h, [^{3}H]AA is quantitatively incorporated into cellular phospholipids. Even in static incubations of cells prelabeled with [^{3}H]AA, >90% of the radioactivity is associated with free arachidonate when BSA is present as a carrier *(33)*. In a superfusion assay where rapid flow of the medium removes free AA, this proportion should be even greater. Moreover, even if a fraction of the AA released from cellular phospholipids were to be metabolized to eicosanoids or other products, the level of released radioactivity would still be an accurate index of phospholipid hydrolysis. Thus, chromatographic identification of the released tritiated products may not be desirable or necessary.

6. It is important to note that the results of superfusion assays of AA release may not produce identical results to those obtained by standard static incubation techniques. An illustration is provided by the response to phorbol ester illustrated in **Fig. 4**. The response to phorbol ester in a superfusion assay is much greater than the 50% increase in AA release previously reported in PC12 cells by static incubation *(32)*. This discrepancy illustrates the differences between the static incubation and superfusion techniques. A short pulse of [^{3}H]AA release can be detected more readily in a series of 1-min fractions than in a single 30-min sample, which reflects the average rate of release over the 30-min period.

References

1. Katsuki, H. and Okuda, S. (1995) Arachidonic acid as a neurotoxic and neurotrophic substance. *Prog. Neurobiol.* **46,** 607–636.
2. Murakami, M., Nakatani, Y., Atsumi, G., Inoue, K., and Kudo, I. (1997) Regulatory functions of phospholipase A2. *Crit. Rev. Immunol.* **17,** 225–283.
3. Heal, D. J., Butler, S. A., Hurst, E. M., and Buckett, W. R. (1989) Antidepressant treatments, including sibutramine hydrochloride and electroconvulsive shock, decrease β_1- but not β_2-adrenoceptors in rat cortex. *J. Neurochem.* **53,** 1019–1025.
4. Balsinde, J. and Dennis, E. A. (1997) Function and inhibition of intracellular calcium-independent phospholipase A$_2$. *J. Biol. Chem.* **272,** 16,069–16,072.
5. Xing, M. B. and Insel, P. A. (1996) Protein kinase C-dependent activation of cytosolic phospholipase A$_2$ and mitogen-activated protein kinase by alpha$_1$-adrenergic receptors in Madin-Darby canine kidney cells. *J. Clin. Invest.* **97,** 1302–1310.

6. Nishio, E., Nakata, H., Arimura, S., and Watanabe, Y. (1996) α_1-adrenergic receptor stimulation causes arachidonic acid release through pertussis toxin-sensitive GTP-Binding protein and JNK activation in rabbit aortic smooth muscle cells. *Biochem. Biophys. Res. Commun.* **219**, 277–282.

7. Hou, W., Arita, Y., and Morisset, J. (1996) Caerulein-stimulated arachidonic acid release in rat pancreatic acini: a diacylglycerol lipase affair. *Am. J. Physiol.* **271**.

8. Farooqui, A. A., Rammohan, K. W., and Horrocks, L. A. (1989) Isolation, characterization, and regulation of diacylglycerol lipases from the bovine brain. *Ann. NY Acad. Sci.* **559**, 25–36.

9. Rindlisbacher, B., Sidler, M. A., Galatioto, L. E., and Zahler, P. (1990) Arachidonic acid liberated by diacylglycerol lipase is essential for the release mechanism in chromaffin cells from bovine adrenal medulla. *J. Neurochem.* **54**, 1247–1252.

10. Zahler, P., Reist, M., Pilarska, M., and Rosenheck, K. (1986) Phospholipase C and diacylglycerol lipase activities associated with plasma membranes of chromaffin cells isolated from bovine adrenal medulla. *Biochim. Biophys. Acta* **877**, 372–379.

11. Schimmel, R. J. (1988) The alpha 1-adrenergic transduction system in hamster brown adipocytes. Release of arachidonic acid accompanies activation of phospholipase C. *Biochem. J.* **253**, 93–102.

12. Ruan, Y., Kan, H., and Malik, K. U. (1997) Beta-adrenergic receptor stimulated prostacyclin synthesis in rabbit coronary endothelial cells is mediated by selective activation of phospholipase D: inhibition by adenosine 3'5'-cyclic monophosphate. *J. Pharmacol. Exp. Ther.* **281**, 1038–1046.

13. Bylund, D. B. (1992) Subtypes of α_1- and α_2-adrenergic receptors. *FASEB J.* **6**, 832–839.

14. Guarino, R. D., Perez, D. M., and Piascik, M. T. (1996) Recent advances in the molecular pharmacology of the α_1-adrenergic receptors. *Cell. Signal.* **8**, 323–333.

15. Llahi, S. and Fain, J. N. (1992) α_1-Adrenergic receptor-mediated activation of phospholipase D in rat cerebral cortex. *J. Biol. Chem.* **267**, 3679–3685.

16. Blue, D. R., Jr., Craig, D. A., Ransom, J. T., Camacho, J. A., Insel, P. A., and Clarke, D. E. (1994) Characterization of the α_1 adrenoceptor subtype mediating [^{3}H]-arachidonic acid release and calcium mobilization in Madin-Darby canine kidney cells. *J. Pharmacol. Exp. Ther.* **268**, 1588–1596.

17. Perez, D. M., DeYoung, M. B., and Graham, R. M. (1993) Coupling of expressed α_{1B}- and α_{1D}-adrenergic receptors to multiple signaling pathways is both G protein and cell type specific. *Mol. Pharmacol.* **44**, 784–795.

18. Perez, D. M., Hwa, J., Gaivin, R., Mathur, M., Brown, F., and Graham, R. M. (1996) Constitutive activation of a single effector pathway: evidence for multiple activation states of a G-protein-coupled receptor. *Mol. Pharmacol.* **49**, 112–122.

19. Nebigil, C. and Malik, K. U. (1993) α-adrenergic receptor subtypes involved in prostaglandin synthesis are coupled to Ca^{2++} channels through a pertussis

toxin-sensitive guanine nucleotide-binding protein. *J. Pharmacol. Exp. Ther.* **266,** 1113–1124.

20. Jones, S. B., Halenda, S. P., and Bylund, D. B. (1991) α_2-Adrenergic receptor stimulation of phospholipase A_2 and of adenylate cyclase in transfected Chinese hamster ovary cells is mediated by different mechanisms. *Mol. Pharmacol.* **39,** 239–245.

21. Felder, C. C., Williams, H. L., and Axelrod, J. (1991) A transduction pathway associated with receptors coupled to the inhibitory guanine nucleotide binding protein G_i that amplifies ATP-mediated arachidonic acid release. *Proc. Natl. Acad. Sci. USA* **88,** 6477–6480.

22. MacNulty, E. E., McClue, S. J., Carr, I. C., Jess, T., Wakelam, M. J. O., and Milligan, G. (1992) α_2-C10 adrenergic receptors expressed in rat 1 fibroblasts can regulate both adenylylcyclase and phospholipase D-mediated hydrolysis of phosphatidylcholine by interacting with pertussis toxin-sensitive guanine nucleotide-binding proteins. *J. Biol. Chem.* **267,** 2149–2156.

23. Evans, R. J. and Surprenant, A. (1993) Effects of phospholipase A_2 inhibitors on coupling of α_2-adrenoceptors to inwardly rectifying potassium currents in guinea-pig submucosal neurones. *Br. J. Pharmacol.* **110,** 591–596.

24. Lew, D. B., Nadel, G. L., and Malik, K. U. (1992) Prostaglandin E_2 synthesis elicited by adrenergic stimuli in guinea pig trachea is mediated primarily via activation of β_2 adrenergic receptors. *Prostaglandins* **44,** 399–412.

25. Kanterman, R. Y., Felder, C. C., Brenneman, D. E., Ma, A. L., Fitzgerald, S., and Axelrod, J. (1990) α_1-Adrenergic receptor mediates arachidonic acid release in spinal cord neurons independent of inositol phospholipid turnover. *J. Neurochem.* **54,** 1225–1232.

26. Weiss, B. A., Slivka, S. R., and Insel, P. A. (1989) Defining the role of protein kinase C in epinephrine- and bradykinin-stimulated arachidonic acid metabolism in Madin-Darby canine kidney cells. *Mol. Pharmacol.* **36,** 317–326.

27. Hertelendy, F., Molnar, M., and Rigo, J., Jr. (1995) Proposed signaling role of arachidonic acid in human myometrium. *Mol. Cell Endocrinol.* **110,** 113–118.

28. Ernsberger, P. (1998) Arachidonic acid release from PC12 pheochromocytoma cells is regulated by I_1-imidazoline receptors. *J. Auton. Nerv. Syst.* **72,** 147–154.

29. Ernsberger, P. (1997) Release of arachidonic acid from PC12 pheochromocytoma cells is regulated by I_1-imidazoline receptors. *J. Auton. Nerv. Syst.* **65,** 144.

30. Lazarewicz, J. W., Leu, V., Sun, S. Y., and Sun, A. Y. (1983) Arachidonic acid release from K^+-evoked depolarization of brain synaptosomes. *Neurochem. Int.* **5,** 471–478.

31. Fink, D. W., Jr. and Guroff, G. (1990) Nerve growth factor stimulation of arachidonic acid release from PC12 cells: independence from phosphoinositide turnover. *J. Neurochem.* **55,** 1716–1726.

32. Zheng, W. H., Fink, D. W., Jr., and Guroff, G. (1996) Role of protein kinase C in nerve growth factor-induced arachidonic acid release from PC12 cells. *J. Neurochem.* **66,** 1868–1875.

33. Traiffort, E., Ruat, M., Arrang, J.-M., Leurs, R., Piomelli, D., and Schwartz, J.-C. (1992) Expression of a cloned rat histamine H_2 receptor mediating inhibition of arachidonate release and activation of cAMP accumulation. *Proc. Natl. Acad. Sci. USA* **89,** 2649–2653.

34. Xing, M. and Mattera, R. (1992) Phosphorylation-dependent regulation of phospholipase A_2 by G-proteins and Ca^{2+} in HL60 granulocytes. *J. Biol. Chem.* **267,** 25,967–25,975.

35. Ernsberger, P., Graves, M. E., Graff, L. M., Zakieh, N., Nguyen, P., Collins, L. A., et al. (1995) I_1-Imidazoline receptors: Definition, characterization, distribution and transmembrane signaling. *Ann. NY Acad. Sci.* **763,** 22–42.

36. Allen, A. C., Gammon, C. M., Ousley, A. H., McCarthy, K. D., and Morell, P. (1992) Bradykinin stimulates arachidonic acid release through the sequential actions of an sn-1 diacylglycerol lipase and a monoacylglycerol lipase. *J. Neurochem.* **58,** 1130–1139.

37. Farooqui, A. A., Anderson, D. K., and Horrocks, L. A. (1993) Effect of glutamate and its analogs on diacylglycerol and monoacylglycerol lipase activities of neuron-enriched cultures. *Brain Res.* **604,** 180–184.

26

Patch-Clamp Recording Methods for Examining Adrenergic Regulation of Potassium Currents in Ocular Epithelial Cells

Jennifer S. Ryan, Chanjuan Shi, and Melanie E. M. Kelly

1. Introduction

Neurotransmitters act on cell-surface receptors to produce a wide range of effects on target cells *(1)*. These effects include modulation of ion channels and changes in cellular electrical properties. Deciphering the intracellular molecular pathways by which receptor activation is transduced into alterations in cell function is often difficult owing to the accessibility to the cells to be studied *in situ* and the lack of regulation over both the extracellular and intracellular environment. The use of viable in vitro isolated cell models and patch-clamp recording methodology *(2)* to assay ion channel activity allows the identification of receptors and coupled intracellular signaling molecules, which regulate the response of interest. We have used whole-cell, patch-clamp recording techniques to study adrenergic receptor (AR) modulation of ion channels in ciliary epithelial cells *(3)*. This approach has allowed us to measure current flowing via identified ion channels, and to identify the G-protein-coupled signaling pathway(s), which transduces AR activation to ion channel modulation.

The ciliary epithelium of the eye is a secreting epithelium comprised of two different epithelial layers, a nonpigmented ciliary epithelial (NPCE) cell layer whose basolateral membrane faces the posterior and vitreal spaces of the vertebrate eye, and a pigmented ciliary epithelial (PCE) cell layer whose basal surface faces the stromal side (**Fig. 1**). These two cell layers are coupled via gap junctions at their apical membranes *(4,5)*. The formation of aqueous humor in

From: *Methods in Molecular Biology, vol. 126: Adrenergic Receptor Protocols*
Edited by: C. A. Machida © Humana Press Inc., Totowa, NJ

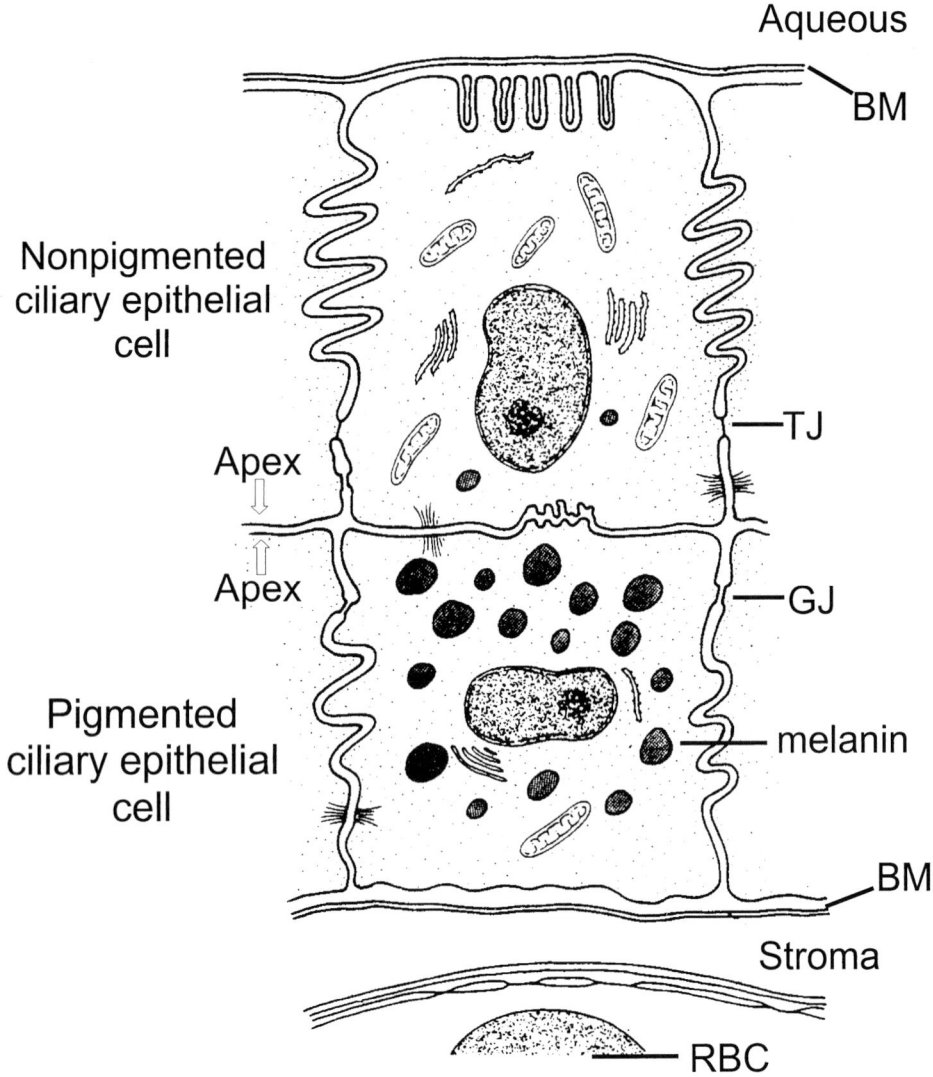

Fig. 1. Schematic diagram of pigmented and nonpigmented ciliary epithelial cells. The NPCE cell layer lies adjacent to the aqueous side, and the PCE cell layer faces the stroma. BM, basement membrane; GJ, gap junction; RBC, red blood cell; TJ, tight junction. Reproduced with permission from Caprioli *(25)*.

the eye occurs primarily by electrolyte secretion across the bilayered ciliary epithelium and requires the coordinated interaction of membrane-localized transporters and ion channels *(6)*.

The rate and quantity of aqueous humor production by the ciliary epithelium are an important determinant of intraocular pressure and are subject to modulation by the sympathetic neural activity *(4)*. Although the mechanisms of AR regulation in the ciliary epithelium are not entirely clear, α_1-, α_2- and β_2-ARs have been identified in the mammalian ciliary process *(7–10)* and, in general, stimulation of α_1- and β-ARs has been found to enhance aqueous production, whereas activation of α_2-ARs inhibits aqueous formation *(11,12)*. In the intact ciliary epithelium, AR stimulation results in alterations in cytosolic Ca^{2+} *(13)*, changes in transepithelial current, and modulation of ion channels *(14)*. Alterations in either K or anion channels affects salt secretion by the epithelium, thus leading to alterations in intraocular pressure in the eye. In this chapter, we will focus on the use of whole-cell, patch-clamp recording techniques to isolate whole-cell Ca^{2+}-activated K^+ (K_{Ca}) current, and to identify receptor subtypes and coupled signaling molecules in cultured pigmented ciliary epithelial cells. In PCE cells, K_{Ca} channel modulation by α_1-ARs involves a pertussistoxin (PTX)-insensitive G-protein(s) and a phospholipase C (PLC)/D-myoinositol (IP_3) signaling pathway *(3)*.

2. Materials

2.1. Tissue Dissection

1. Anesthesia, sodium pentobarbital (MTC Pharmaceuticals, Ontario, Canada).
2. Cooled (4°C) Dulbecco's phosphate-buffered saline (D-PBS; #14040-026; Gibco-BRL, Burlington, Ontario, Canada).
3. Glass dish with silastic-coated bottom for pinning out tissue and fine dissection.
4. Dissecting microscope with zoom lens.
5. Fiber optic illuminator (Fiber Lite; #9745-00; Cole-Palmer Instrument Co., Chicago, IL).
6. Iridectomy scissors, watchmaker's forceps, blunt-ended scissors (Fine Science Tools, Vancouver, BC, Canada).

2.2. Cell Culture

1. Collagenase D (#1088 858; Boehringer Mannheim, Laval, Quebec, Canada).
2. Pronase (#165921; Boehringer Manneheim).
3. Dulbecco's Modified Eagle's Media (DMEM; #11965-012; Canadian Life Technologies, Burlington, Ontario, Canada).
4. Fetal calf serum (FCS; #16000-028; Canadian Life Technologies).
5. Penicillin/streptomycin (#600-5140 PT; Canadian Life Technologies).
6. 12-mm Glass cover slips (#12545-80; Fisher Scientific, Nepean, Ontario, Canada).
7. 15-mL plastic centrifuge tubes (Sarstedt, Inc., Newton, NC).
8. Sterile heat-polished Pasteur pipets.
9. Nuncalon Multidish 4 (cat. #176740; Canadian Life Technologies).

2.3. Solutions and Drugs for Electrophysiology

All drugs and chemicals were obtained from Sigma (St. Louis, MO), except where noted otherwise.

1. Standard external solution: 100 mM Na$^+$-aspartate, 30 mM NaCl, 5 mM KCl, 1 mM CaCl$_2$, 1 mM MgCl$_2$, 10 mM Na$_2$HCO$_3$, 10 mM HEPES, and 10 mM glucose. Low Ca^{2+} external solutions: 100 mM Na$^+$-aspartate, 30 mM NaCl, 5 mM KCl, 0.2 mM CaCl$_2$, 0.25 mM EGTA, 1 mM MgCl$_2$, 10 mM Na$_2$HCO$_3$, 10 mM HEPES, and 10 mM glucose.

   The extracellular [Ca^{2+}] with this solution is estimated to be <10 nM. Calcium concentrations are calculated using software provided by A. French (Department of Physiology, Dalhousie University, Halifax, NS, Canada). All external solutions are continuously bubbled with 5% CO$_2$/95% air and adjusted to pH 7.4 with NaOH.

2. Standard electrode filling solution for whole-cell recordings: 110 mM K$^+$-aspartate, 30 mM KCl, 1 mM MgCl$_2$, 20 mM HEPES, 1 mM (EGTA), 0.4 mM CaCl$_2$, 1 mM adenosine triphosphate (ATP) (Mg), and 0.1 mM guanosine triphosphate (GTP) (Na$_2$) adjusted to pH 7.3. Free internal [Ca^{2+}] is estimated to be <100 nM. In some experiments, intracellular Ca^{2+} is buffered to <10 nM by inclusion of 10 mM BAPTA (#A9801; Sigma) in the electrode solution. Solution osmolarity is measured by freezing point depression (Osmette A, Fisher Scientific). The osmolarity of external solutions is 330 mos$M$ and the osmolarity of internal solutions is 320 mos$M$.

3. The K$_{(Ca)}$ channel blocker is iberitoxin (IbTX; Peninsula Laboratories Inc., Belmont, CA). IbTX is first dissolved in Millipore water to give a 10-mM stock. Aliquots of 10 µL stock solution are then lyophilized and stored at –80°C. For experimental use, stock aliquots are reconstituted and diluted into extracellular solution to a final concentration of 10 nM.

4. The Ca^{2+} ionophore, ionomycin is dissolved in dimethylsulfoxide (DMSO) to give a stock of 10 mM and stored at –80°C (*see* **Note 1**). Stock solutions are diluted with extracellular solution and used at a concentration of 10 µM.

5. Adrenergic agonists, epinephrine (EPI), the α$_1$-AR selective agonist, phenylephrine (PHe), and the α-AR antagonist prazosin (PZ) are prepared from stocks to give final concentrations cited in **Subheading 3.4.** For storage conditions and other cautionary measures regarding these drugs, refer to **Note 1**. For PTX-pretreatment, cells are incubated with 500 ng/mL of PTX in the culture medium 24 h before recording (*see* **Note 2**).

6. The phospholipase Cβ inhibitor U-73122 is dissolved in DMSO to give a stock of 5 mg/mL and frozen at –80°C until use. Prior to recording, U-73122 is included in the culture media at a concentration of 10 µM for a pretreatment time of 20 min. U-73122 is also included in the standard external solution and superfused at the same concentration as used in the pretreatment during current recording.

7. We use thapsigargin (TG) for depletion of cytosolic Ca^{2+} stores. TG is dissolved in DMSO and then made up as a 10-mM stock in Millipore water and stored at

–80°C. Before recording, cells are pretreated for 15 min in the dark with TG dissolved in the culture medium at a concentration of 5 μ*M*. Other signaling molecules, such as D-*myo*-inositol (1,4,5)-triphosphate (IP$_3$), are dissolved in internal pipet solution and included in the recording pipet at concentrations cited in **Subheading 3.5.** GDPβS, TG, IP$_3$, and U-73122 are purchased from Calbiochem (San Diego, CA).

8. For drug application: IbTX, TG, U-73122, and PZ are all added to the extracellular solution and superfused during electrophysiological recording at concentrations cited in **Subheadings 3.3.–3.5.**; EPI, PHe, and ionomycin are applied by pneumatic pressure injection; IP$_3$ and GDPβS are included in the intracellular recording solution.

2.4. Equipment Required for Electrophysiology

1. Three-dimensional hydraulic manipulator (Narishige Scientific Instrument Lab., WR-88, Ku, Tokyo 157, Japan).
2. Picospritzer II (General Valve Corp., Fairfield, NJ).
3. Borosilicate glass (#B150-110-10, Sutter Instruments, Novato, CA).
4. Two-stage vertical microelectrode puller (Narishige model PP83, Tokyo, Japan).
5. Electrode-salt bridge combination (Dri-ref-2, World Precision Instruments, Sarasota, FL).
6. Axopatch ID amplifier (Axon Instruments, Foster City, CA).

3. Methods
3.1. Cell Preparation and Identification

1. All experiments are carried out on pigmented epithelial cells isolated from rabbit ciliary epithelium using a method modified from **refs. *15,16.* Figure 2** provides a schematic outline of the ciliary epithelial cell isolation, as described below.
2. Five- to 10-wk-old pigmented rabbits (Reimens, St. Agatha, Ontario, Canada) are anesthetized with 8 mg/100 g of sodium pentobarbital, and eyes are enucleated in accordance with the Association for Research in Vision and Ophthalmology (ARVO) statement for the Use of Animals in Ophthalmic and Vision Research.
3. Enucleated eyes are placed in sterile D-PBS, and a small incision is made at the corneo-scleral junction using a razor bade. The anterior segment of the eye is removed by cutting around the globe, starting at the incision, with blunt-ended scissors.
4. The ciliary body is then freed from the iris by making a small incision between the iridal and ciliary processes using iridectomy scissors.
5. The epithelium is then dissected from the underlying stroma under a dissecting microscope at 200× magnification, and treated for 30–40 min with 5 mL D-PBS containing collagenase (1.5 mg/mL) and pronase (1 mg/mL).
6. The enzyme reaction is terminated by the addition of 1.5 mL of 10% fetal calf serum (FCS), and the cell suspension centrifuged at 2000*g* for 5 min, washed with fresh sterile DMEM, and recentrifuged for 5 min.

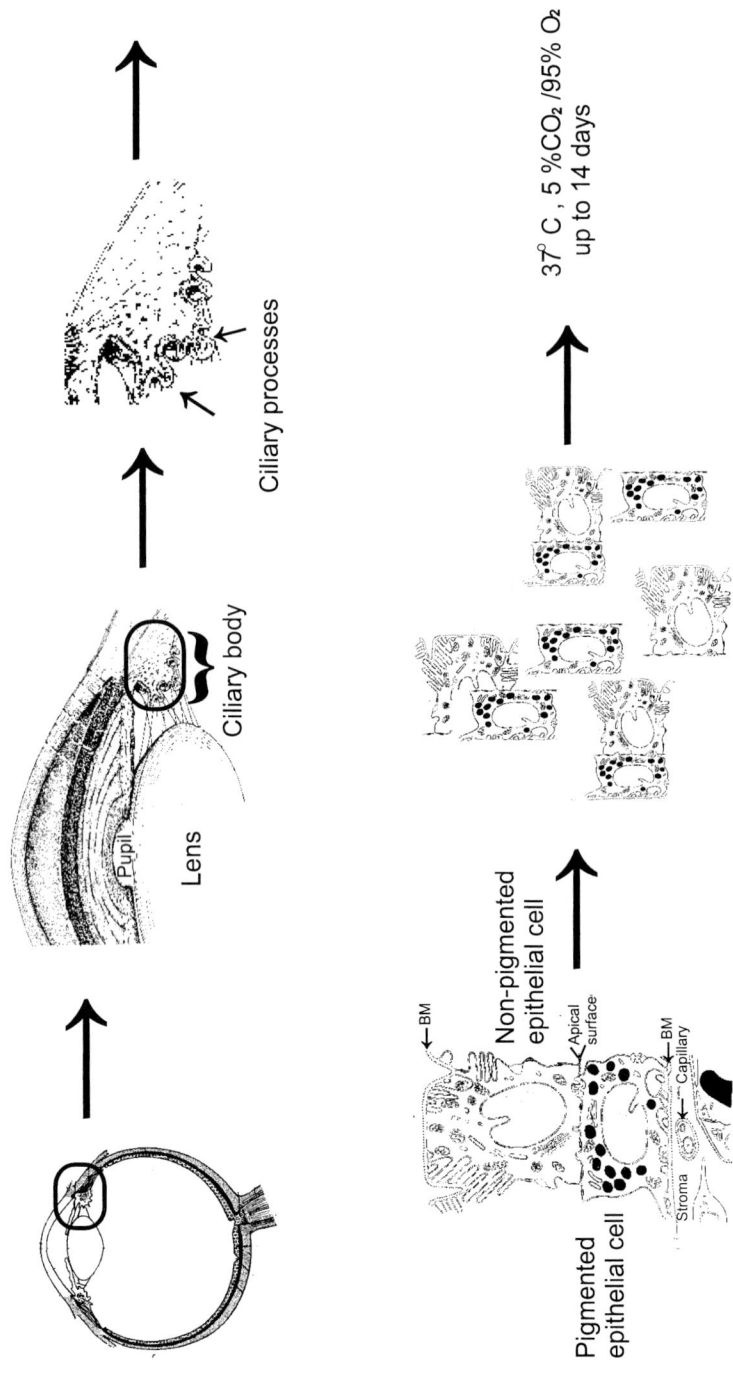

Ciliary processes

Ciliary body

Lens

Pupil

37° C , 5 % CO_2 /95% O_2
up to 14 days

Non-pigmented
epithelial cell

Apical
surface

BM

BM

Capillary

Stroma

Pigmented
epithelial cell

Fig. 2. Schematic diagram of rabbit PCE isolation and culture. *See text* for details.

7. The epithelial pieces are triturated gently with a sterile Pasteur pipet to yield single cells and small tissue explants.
8. Once isolated, cells are seeded onto glass cover slips at a density of 10^5 cells/mL in DMEM containing 5% FCS and antibiotics, and incubated in an atmosphere of 5% CO_2/95% O_2 at 37°C until use (up to a maximum of 14 d). PCE cells are clearly identified from NPCE cells by the presence of pigment granules. Furthermore, NPCE cells tend to have limited survival under the culture conditions used and, after 5, d ciliary epithelial cultures consisted primarily of PCE cells.

3.2. Electrophysiological Recording Techniques

1. Ionic currents in isolated cultured epithelial cells are studied using whole-cell, tight-seal, patch-clamp recording methods *(2)*.
2. For electrophysiological recording, cells attached to glass cover slips are placed in a shallow chamber and positioned on the stage of a Nikon inverted microscope (**Fig. 3**).
3. The chamber is superfused with standard external solution from a series of reservoirs and valves designed to provide a regulated gravity-fed flow rate of 1–2 mL/ min. Most drugs and ligands are also applied by bath superfusion from designated reservoirs. Test solutions are applied for a minimum of 5 and usually for 10 complete (1 mL) bath exchanges.
4. For application of test substances by pressure ejection, micropipets (≥2 mm in diameter) are positioned 50–100 mm from the cell using a three-dimensional hydraulic manipulator (Narishige Scientific Instrument Lab.) and 2–5 lb/in.^{2} pressure applied to the back of the micropipet using a Picospritzer II (General Valve Corp.).
5. Patch electrodes are pulled from borosilicate glass with an external diameter of 1.5 mm and an internal diameter of 1.1 mm (#B150-110-10, Sutter Instruments, Novato, CA), using a two stage vertical microelectrode puller (Narishige model PP83). Electrodes have resistances of 3–5 MΩ when filled with internal solution and are coated with beeswax to reduce capacitance.
6. The reference electrode is a sealed electrode-salt bridge combination (Dri-ref-2; World Precision Instruments). Offset potentials are nulled using the amplifier circuitry before seals are made on the cells. Liquid junction potentials (LJPs) arising between the bath and the electrode are measured experimentally and defined as the potential of the bath solution with respect to the pipet solution *(17)*. For whole-cell recording, all the data and current–voltage relationships are routinely corrected for LJPs which are calculated as:

$$V_m = V_p - LJP \qquad (1)$$

where V_m = the membrane potential of the cell and V_p is the pipet potential. To confirm experimentally generated measurements, LJPs are also calculated using a software program (JPCalc, version 2.00; P.H. Barry, Sydney, Australia). For the data shown, LJPs are 9.7 mV for standard low-Cl external Ringers and

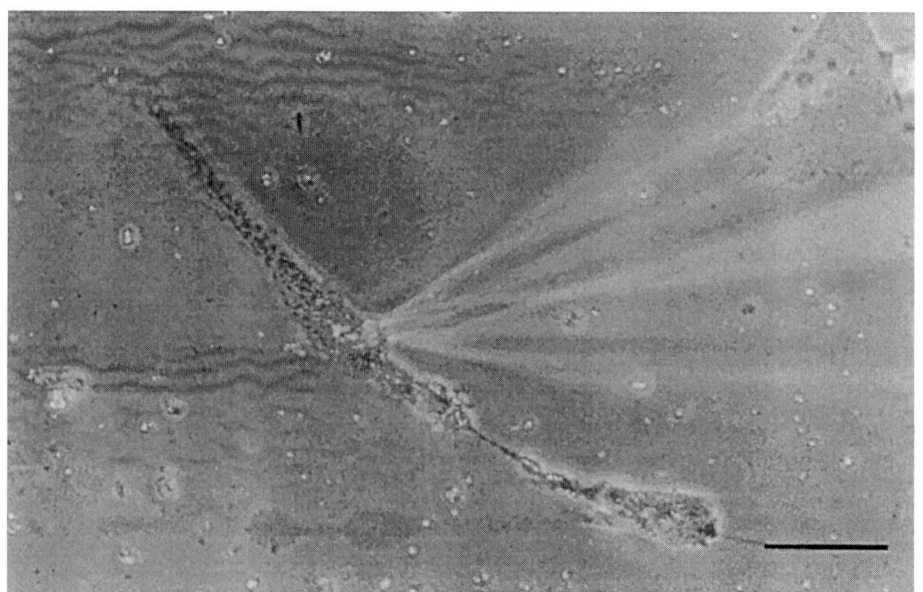

Fig. 3. Photomicrograph of a representative cultured PCE cell used for recording. The cell has been growing for 5 d in primary culture. The recording microelectrode is shown to the left of the cell. Scale bar is equal to 40 μm.

       K-aspartate pipet solution, and 5.4 mV for low-Cl/low-Ca external and BAPTA intracellular solution.

7. Membrane potential and ionic currents are recorded with an Axopatch 1D amplifier (Axon Instruments) at a temperature of 22–24°C. Currents are filtered with a four-pole low-pass Bessel filter and digitized at a sampling frequency of 5–10 kHz using pCLAMP software, version 6.0 (Axon Instruments). Current and voltage are displayed on a Kikuzui 5040 oscilloscope and on a Gould TA240 chart recorder, and stored on computer disk.

8. Currents are generally elicited in voltage-clamp mode using 500-ms long pulses from a holding potential of –62 mV with steps from –122 to +58 mV. Current amplitude is determined from the steady-state current at the end of the voltage step, and these values are then used to construct current-voltage plots and histograms.

9. In some cases, currents are normalized for cell capacitance and expressed as pA/pF. The significance of results is determined using Student's t-tests and data are considered significantly different at $p < 0.05$. Values are expressed as standard error of mean (SEM) where n = number of cells.

3.3. Recording Whole-Cell K Currents in PCE Cells

Voltage-dependent K^+ currents are a ubiquitous feature of PCE cells **(Fig. 4A)** *(3,6,15)*. Using standard extracellular and intracelluar recording

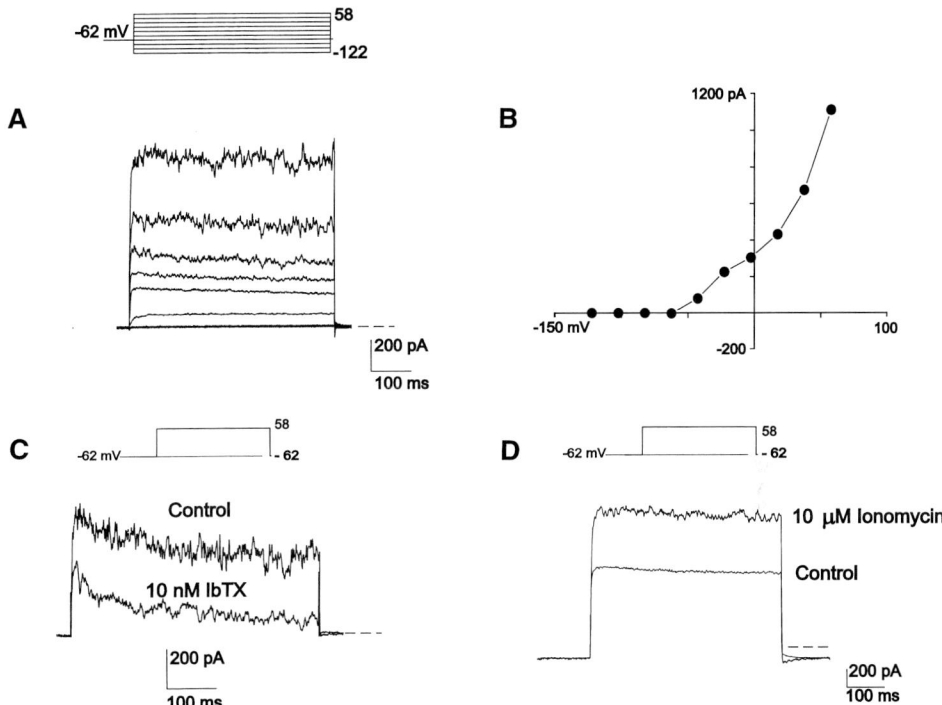

Fig. 4. Whole-cell K$^+$ current in cultured PCE cells. **(A)** Representative whole-cell currents elicited by a series of step depolarizations from −122 mV to +58 mV from a holding potential of −62 mV. Depolarization-induced outwardly rectifying K$^+$ current was observed in all PCE cells examined. **(B)** Current–voltage relationship measured from the traces shown in (A). **(C, D)**. Whole-cell current traces recorded from two representative PCE cells elicited by a voltage step to +58 mV from a holding potential of −62 mV. Outward current was reduced in the presence of 10 n*M* of the maxi-K ($I_{K(Ca)}$) channel blocker IbTX (C) and increased in the presence of 10 μ*M* of the Ca^{2+} ionophore ionomycin (D). Dashed line represents zero current in this and subsequent figures.

solutions, outward K$^+$ currents can be recorded in >95% of isolated rabbit PCE cells using whole-cell, patch-clamp techniques.

1. Under voltage-clamp, PCE cells are held at −62 mV, and the membrane potential is stepped in 20-mV increments to potentials negative and positive to the holding potential (*see* voltage protocol, top **Fig. 4**). An outwardly rectifying K$^+$ current is recorded in all cells examined. In 66% of the cells, the outward K$^+$ current appears "noisy" at more depolarized potentials, suggesting the presence of large-conductance channels (**Fig. 4A**). The current–voltage plot (**Fig. 4B**) shows that

the outward current activates at around −60 mV and increases with increasing depolarizing potentials.

2. IbTX is a selective blocker of large conductance Ca-activated K channels (K_{Ca}) *(18,19)*. Whole-cell outward current is first recorded in PCE cells in control extracellular solution, after which the cell is superfused with 10 nM IbTX solution for 5 min at a flow rate of 1–2 mL/min. Outward current measured at +58 mV is reduced 69 ± 5% by IbTX, demonstrating the presence of K_{Ca} channels in rabbit PCE cells (**Fig. 4C**). Further confirmation of the presence of $I_{K(Ca)}$ in PCE cells is obtained by pneumatic pressure application of 1 µM of the Ca^{2+} ionophore ionomycin (**Fig. 4D**), to increase cytosolic [Ca^{2+}]. Whole-cell outward current is recorded before and during a 40-s application of ionomycin via an application pipet. Consistent with the presence of K_{Ca} channels in PCE cells, ionomycin increases the outward K^+ current measured at +58 mV by a mean of 69%.

3.4. Identification of Adrenergic Receptor Modulation of I_{KCa} in PCE Cells

$K_{(Ca)}$ channels are targets for modulation by a number of agonists *(20)*. Adrenergic regulation of $I_{K(Ca)}$ in PCE cells is examined by recording whole-cell currents in standard recording solutions before and after pneumatic pressure application of the nonselective adrenergic receptor agonist EPI or the selective α_1-AR agonist, PHe.

1. Application of 100 µM EPI for 40 s from a puffer pipet increases the whole-cell outward current measured at +58 mV by 117% (**Fig. 5A**). To confirm that EPI is increasing the Ca^{2+}-sensitive component of the outward K^+ current, EPI is applied in the presence of 10 nM IbTX, and the outward K^+ current recorded. IbTX blocks the increase in the outward current induced by EPI, thus confirming that EPI is activating $I_{K(Ca)}$ in PCE cells (**Fig. 5B**).
2. Pneumatic pressure application of 100 µM of the selective α_1-AR agonist, PHe, for 40 s, also increases macroscopic $I_{K(Ca)}$ by 82% (**Fig. 5C**), suggesting that adrenergic modulation of I_{KCa} in PCE cells occurs via activation of α_1-ARs.
3. To confirm further the involvement of α-ARs, the PHe-mediated increase in $I_{K(Ca)}$ is measured in the presence of 100 µM of the α-AR antagonist, prazosin. The effect of 40 s of pneumatic pressure application of PHe on whole-cell outward current is recorded, and then the cell is superfused with prazosin at a flow rate of 1–2 mL/min for 5 min. PHe is applied again, and its effect on whole-cell outward current in the presence of PZ is recorded. Consistent with the involvement of α_1-ARs, 100 µM prazosin blocks PHe-mediated increase in $I_{K(Ca)}$ by 73% (**Fig. 5D**).

3.5. Signaling Pathways Involved in α_1-Adrenergic Receptor Modulation of I_{KCa}

In most cell types, α_1-ARs couple to PTX-insensitive G_q proteins linked to activation of PLC, the generation of IP_3, and subsequent Ca^{2+} stores release

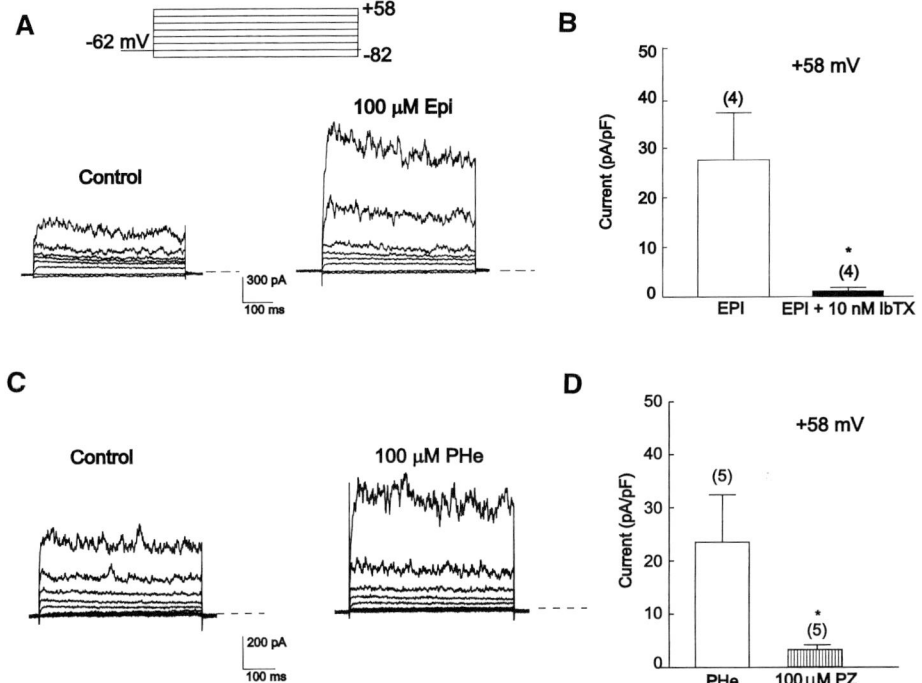

Fig. 5. Activation of $I_{K(Ca)}$ by AR activation. **(A, C)** Whole-cell currents elicited from voltage steps from –82 to +58 mV from a holding potential of –62 mV, recorded in the absence (control) and presence of 100 µM epinephrine (EPI) or phenylephrine (PHe). **(B)** Mean (SEM) increase in I_K measured at +58 mV after exposure of 4 cells to 100 µM EPI in the absence (EPI) and presence of IbTX (EPI + 10 nM IbTX). **(D)** Histogram shows mean (SEM) current measured at +58 mV for the same cells before and after 40-s application of 100 µM PHe in the absence of (PHe) or presence of 100 µM prazosin (PHe + PZ). Data have been normalized for cell capacitance.

(21,22). To investigate the involvement of this signaling pathway in PHe-mediated increases in $I_{K(Ca)}$, we employ a number of drugs that mimic or inhibit signaling molecules in this pathway.

1. Activation of $I_{K(Ca)}$ by phenylephrine could be mimicked by dialyzing PCE cells with 10 µM of IP$_3$. Dialysis of PCE cells with 10 µM IP$_3$ for 10 min increases the whole-cell outward current measured at +58 mV by a mean of 30 pA/pF **(Fig. 6A)**.
2. To test the involvement of IP$_3$-stimulated Ca^{2+} stores release in PHe-mediated $I_{K(Ca)}$ activation, we pretreat the cells with 5 µM of the membrane-permeant Ca^{2+}-ATPase blocker TG to empty sarcoplasmic reticular Ca^{2+} stores *(23)*. Before recording, PCE cells are incubated for 20 min in the dark in low-Ca^{2+} external

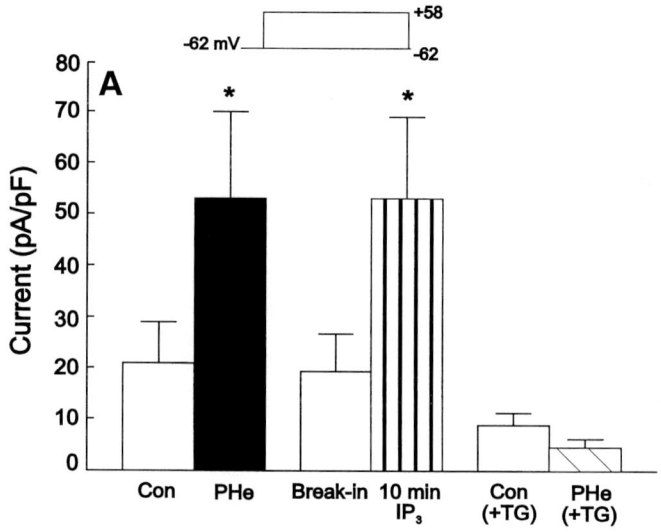

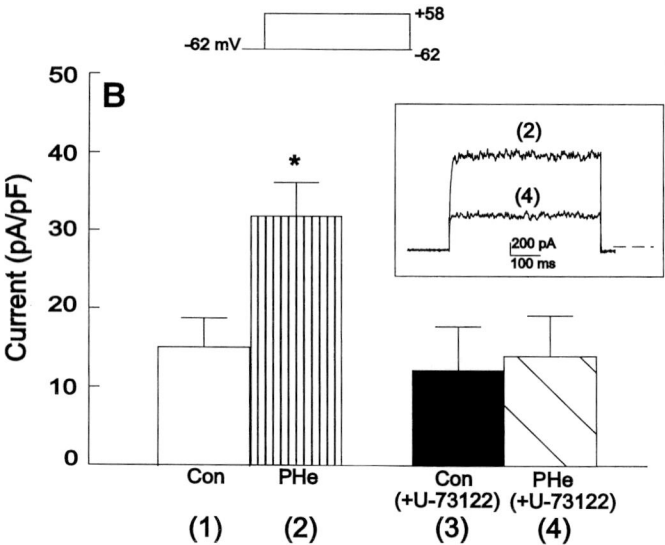

Fig. 6. Involvement of IP$_3$, PLC, and intracellular Ca^{2+} release in I$_{K(Ca)}$ activation. (**A**) Histogram shows mean (±SEM) current amplitude measured at +58 mV and normalized for cell capacitance in: 5 PCE cells before (Con) and after (PHe) 40 s of pressure application of 100 μM PHe; in 5 PCE cells treated for 30 min with 5 μM thapsigargin before (Con + TG) and after 40 s of pressure application of 100 μM PHe (PHe +TG); in 11 cells immediately after break-in to the whole-cell configuration (Break-in), and 10 min after intracellular dialysis with 10 μM IP$_3$ (10 min IP$_3$). (**B**) Inset shows representative whole-cell currents from a single PCE cell in response to a 500-ms step depolarization to +58 mV from a holding potential of −62 mV. Current is

solution containing 5 μM TG. Whole-cell current is then recorded in PCE cells, which are constantly superfused at a flow rate of 1 mL/min using the low Ca^{2+} extracellular solution containing 5 μM TG. The intracellular electrode solution in this experiment is the low-Ca^{2+} solution containing 10 mM BAPTA. Outward current is measured before and after pneumatic pressure application of 100 μM PHe dissolved in low-Ca^{2+} external solution. PCE cells from the same culture without TG pretreatment are used as control. In comparison to control PCE cells, TG pretreatment prevents PHe-mediated increases in $I_{K(Ca)}$ (**Fig. 6A**).

3. The PLC inhibitor U-73122 is employed to examine the effect of PLC inhibition on responses to PHe. Whole-cell currents are recorded before and after a 40-s pneumatic pressure application of 100 μM PHe in standard extracellular solution. The cell is then superfused at a flow rate of 1 mL/min for 5–10 min with standard extracellular solution containing 10 μM U-73122. The effect of PHe application on whole-cell outward current is then recorded in the presence of U-73122 and compared to control response in the absence of U-73122. PHe-mediated increases in $I_{K(Ca)}$ are blocked by U-73122 (**Fig. 6B**) confirming the involvement of PLC.

4. To examine the involvement of G-proteins in PHe-mediated increases in $I_{K(Ca)}$, whole-cell current is recorded in PCE cells dialyzed with 2 mM of the G-protein inhibitor GDPβS included in standard intracellular solution. The effect of GDPβS on the response of a cell to PHe is recorded 10 min after attaining the whole-cell configuration to ensure adequate exchange of the solution. The response of PCE cells from the same culture to PHe is also recorded following dialysis of the cells for 10 min with standard electrode solution containing 0.1 mM GTP. GDPβS significantly reduces the PHe-mediated increase in $I_{K(Ca)}$ compared to control cells from the same culture dialyzed with GTP (**Fig. 7A**).

5. The sensitivity of the PHe response to PTX is examined by treating PCE cells with PTX for 12–14 h (*see* **Note 2**). PTX blocks G-proteins of the $G_{i/o/z/t}$ class by ADP ribosylating cysteine residues at the GTP binding site on the α-subunit *(24)*.

Following PTX pretreatment, PCE cells are placed in the experimental chamber and superfused with standard external solution with K^+ aspartate internal solution. The whole-cell current response of control and PTX-treated cells is then recorded after a test application of 100 mM phenylephrine (by pneumatic pressure injection for 40 s). PTX treatment has no effect on the PHe-mediated increase in $I_{K(Ca)}$ measured at +58 mV (**Fig. 7B**), confirming the involvement of a PTX-insensitive G protein(s) in $α_1$-AR modulation of K_{Ca} channels.

Fig. 6. (*continued from opposite page*) shown after pressure application of 100 μM PHe in the same cell before (4) and after 20–30 min of incubation with 50 μM U-73122 (2). Histogram shows mean (±SEM) PHe-mediated increase in the current amplitude, normalized for cell capacitance, measured at +58 mV in 4 cells before (Con + U-73122) and after incubation with U-73122 (PHe + U73122).

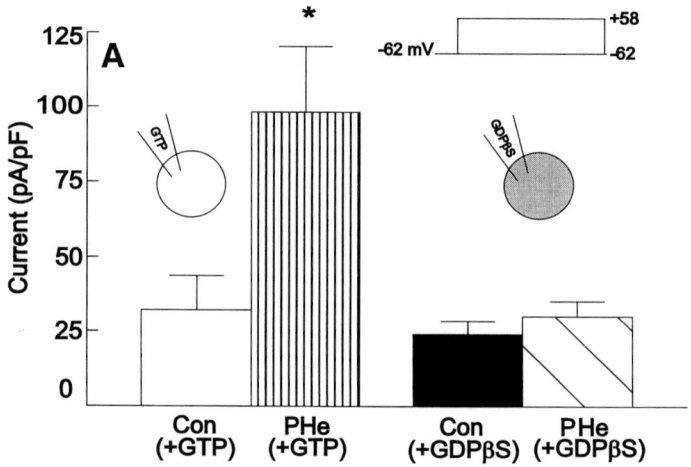

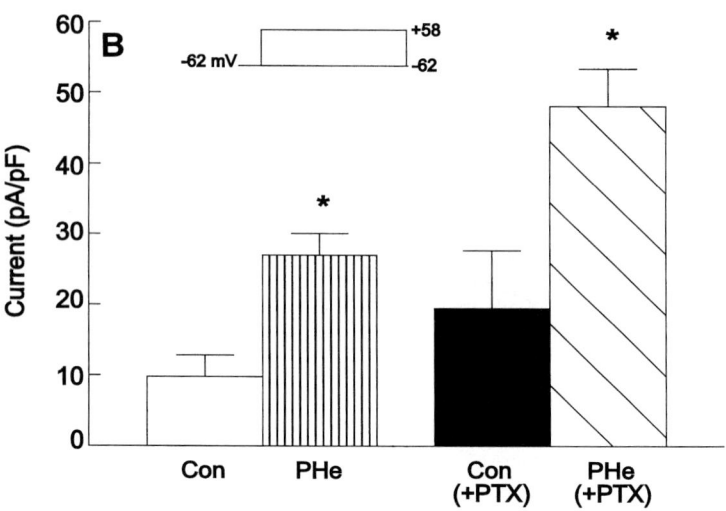

Fig. 7. G-proteins are involved in PHe-mediated $I_{K(Ca)}$ increases. (A) Histogram shows mean (±SEM) current amplitude measured at +58 mV with 0.1 mM GTP or 2 mM GDPβS in the pipet before (GTP; GDPβS) or after 40-s application of 100 μM PHe (PHe + GTP; PHe + GDPβS). (B) Mean (±SEM) current amplitude measured at +58 mV in 4 untreated cells before (Con) and after PHe application (+PHe), and in 4 cells pretreated with 500 ng/mL of PTX before (Con + PTX) and after PHe application (PHe + PTX). Data have been normalized for cell capacitance.

3.6. Summary

Using patch-clamp recording techniques, we have demonstrated that I_{KCa} in rabbit PCE cells can be activated by α_1-ARs coupled, via a PTX-insensitive G-protein to a PLC-dependent signaling pathway and intracellular IP_3-sensitive Ca^{2+} stores. In vivo, released transmitters and hormones could modulate K current and provide for both paracrine and endocrine regulation of aqueous humor secretion by the ciliary epithelium.

4. Notes

1. For light-sensitive drugs, such as EPI, ionomycin, TG, and PZ, experiments are performed in the absence of any external light source. For example, once the whole-cell configuration has been obtained, the light source for the microscope is turned off. In addition, all solutions and inlet reservoirs are made light-tight by wrapping in tin foil, and solutions are kept cool at $4°C$ in the dark until immediately before use.

2. In order to determine whether modulation of $I_{K(Ca)}$ by α_1-AR agonists, involved G-proteins of the PTX-sensitive $G_{i/o/z/t}$ class, PCE cells are incubated for 12–14 h with 500 ng/mL of PTX in culture medium at $37°C$ in an atmosphere of 95% O_2/5% CO_2. PCE cells from the same culture, but without PTX treatment, are used as the control.

Acknowledgments

The authors wish to thank Christine Jollimore for her technical assistance. This work was supported by the Medical Research Council of Canada, grant # MT-13484 and a National Science and Engineering Research Council of Canada PGSB studentship to J. S. R.

References

1. Wickman, K. and Clapham, D. E. (1995) Ion channel regulation by G proteins. *Physiol. Rev.* **75,** 865–885.
2. Hamill, O. P., Marty, A., Neher, E., Sakmann, B. and Sigworth, F. J. (1981) Improved patch-clamp techniques for high resolution current recording from cell and cell-free membrane patches. *Pflugers Arch.* **391,** 85–100.
3. Ryan, J. S., Tao, Q.-P., and Kelly, M. E. M. (1998) Adrenergic regulation of calcium-activated potassium current in cultured rabbit pigmented ciliary epithelial cells. *J. Physiol.* **511,** 145–157.
4. Cole, D. A. (1984) Ocular fluids, in *The Eye* (Davson, H., ed.) Academic, New York, pp. 269–390.
5. Raviola, G. and Raviola, E. (1978) Intercellular junctions in the ciliary epithelium. *Invest. Ophthalmol. Vis. Sci.* **17,** 958–981.
6. Jacob, T. J. C. and Civan, M. M. (1996) Role of ion channels in aqueous humor formation. *Am. J. Physiol.* **271,** C703–720.

7. Mittag, T. W. and Tormay, A. (1985) Adrenergic receptor subtypes in rabbit iris-ciliary body membranes: classification by radioligand studies. *Exp. Eye Res.* **40,** 239–249.

8. Mallorga, P., Buisson, S., and Sugrue, M. F. (1988) Alpha 1-adrenoceptors in the albino rabbit ciliary process. *J. Ocul. Pharmacol.* **4,** 203–214.

9. Jin, Y., Verstappen, A., and Yorio, T. (1994) Characterization of alpha 2-adrenoceptor binding sites in rabbit ciliary body membranes. *Invest. Ophthalmol. Vis. Sci.* **35,** 2500–2508.

10. Wax, M. B. and Molinoff, P. B. (1987) Distribution and properties of beta-adrenergic receptors in human iris-ciliary body. *Invest. Ophthalmol. Vis. Sci.* **28,** 420–430.

11. Ross, R. A. and Drance, S. M. (1970) Effects of topically applied isoproterenol on aqueous dynamics in man. *Arch. Ophthalmol.* **83,** 39–46.

12. Chiou, G. C. Y. (1983) Effects of α_1 and α_2 activation of adrenergic receptors on aqueous humor dynamics. *Life Sci.* **32,** 1699–1704.

13. Farahbakhsh, N. A. and Cilluffo, M. C. (1994) Synergistic effect of adrenergic and muscarinic receptor activation on $[Ca^{2+}]_i$ in rabbit ciliary body epithelium. *J. Physiol.* **77,** 215–221.

14. Krupin, T. E., Wax, M. B., Carre, D. A., Moolchandani, J., and Civan, M. M. (1991) Effects of adrenergic agents on transepithelial electrical measurements across the isolated iris-ciliary body. *Exp. Eye Res.* **53,** 709–716.

15. Fain, G. L. and Farahbakhsh, N. A. (1989) Voltage-activated currents recorded from rabbit pigmented ciliary body epithelial cells in culture. *J. Physiol.* **417,** 83–103.

16. Cilluffo, M. C., Fain, M. J., Fain, G. L., and Brecha, N. C. (1986) Culture of rabbit ciliary body epithelium. *Invest. Ophthalmol. Vis. Sci.* **27, Suppl.,** 322.

17. Barry, P. and Lynch, J. (1991) Liquid junction potentials and small cell effects in patch-clamp analysis. *J. Membr. Biol.* **121,** 107–117.

18. Giangiacomo, K. M., Garcia, M. L., and McManus, O. B. (1992) Mechanism of iberiotoxin block of the large-conductance calcium-activated potassium channel from bovine aortic smooth muscle. *Biochemistry* **31,** 6719–6727.

19. Giangicomo, K. M., Sugg, E. E., Garcia-Calco, M., Leonard, R. J., McManus, O. B., Kaczorowski, G. I., et al. (1993) Synthetic charybdotoxin chimeric peptides define toxin binding sites in calcium-activated and voltage dependent potassium channels. *Biochemistry* **32,** 2363–2370.

20. Hildebrandt, J. P., Plant, T. D., and Meves, H. (1997) The effects of bradykinin on K^+ currents in NG108-15 cells treated with U73122, a phospholipase C inhibitor, or neomycin. *Br. J. Pharmacol.* **120,** 841–850.

21. Berridge, M. J. (1993) Inositol triphosphate and calcium signaling. *Nature* **361,** 315–325.

22. Minneman, K. P. (1993) α_1-Adrenergic receptor subtypes, inositol phosphates and sources of cell Ca^{2+}. *Pharmacol. Rev.* **40,** 87–115.

23. Foskett, J. K., Roifman, C. M., and Wong, D. (1991) Activation of calcium oscillations by thapsigargin in parotid acinar cells. *J. Biol. Chem.* **266,** 2778–2782.

24. Ui, M. (1990) Pertussis toxin as a valuable probe for G-protein involvement in signal transduction, in *ADP-Ribosylating Toxins and G Proteins. Insights into Signal Transduction* (Moss, J. and Vaughan, M., eds.), American Society for Microbiology Press, Washington, DC, pp. 45–77.
25. Caprioli, J. (1987) The ciliary epithelia and aqueous humor, in *Adler's Physiology of the Eye: Clinical Application* (Moses, R. A. and Hart, W. M., eds.), Mosky, St. Louis, MO, pp. 204–222.

V

TRANSACTIVATOR ANALYSIS

27

Southwestern Blots for Detection of a DNA Binding Protein Recognizing the α_{1B}-Adrenergic Receptor Gene Promoter

Bin Gao and George Kunos

1. Introduction

The α_{1B}-adrenergic receptor (α_{1B}-AR) is a G-protein-coupled receptor that plays a key role in the sympathetic regulation of a variety of physiological processes, such as cardiac and smooth muscle contractility, contraction of the spleen, liver glycogenolysis, melatonin secretion by the pineal gland, and cell proliferation and hypertrophy *(1)*. Expression of the α_{1B}-AR gene is controlled by hormonal and developmental factors in a complex, tissue-specific manner. In order to clarify the molecular mechanisms of its transcription, we have cloned the rat α_{1B}-AR gene and identified its promoters and the *cis*-acting elements in its 5' regulatory domain *(2)*. Footprinting analyses using liver nuclear extracts have revealed a strongly protected region between –490 and –540 (oligo II) within the P2 promoter region *(3)*. Further experiments identified the nuclear protein bound to this region as nuclear factor-1 (NF1) or a closely related protein by using Southwestern blotting, DNA affinity chromatography, UV crosslinking of proteins to DNA, methylation interference analysis, and DNA gel-mobility shift assays *(4)*. All of these methods are based on the specific binding of protein transcription factors to DNA, but they differ in their respective roles in the characterization of transcription factors. For example, Southwestern blotting is used to estimate the molecular mass of a transcription factor. The successful detection of transcription factor binding in a Southwestern blot is also a prerequisite for using DNA affinity chromatography and screening of an expression library with DNA probes in the subsequent purification and isolation of the transcription factor. Therefore, before we

From: *Methods in Molecular Biology, vol. 126: Adrenergic Receptor Protocols*
Edited by: C. A. Machida © Humana Press Inc., Totowa, NJ

embarked on a DNA affinity purification protocol to identify the factor(s) that bind to footprint II within the P2 promoter, we determined their molecular mass as 32 and 34 kDa, using Southwestern blotting (**Fig. 1**).

As implied by its name, Southwestern blotting combines some features of the original Southern blotting by employing a radiolabeled DNA fragment as probe, and the Western blotting technique by having the protein(s) to be probed immobilized on a membrane *(5)*. The method is based on the specific binding of DNA to transcription factors. The first step is to select and radiolabel a suitable probe. Next, a nuclear extract or a partially purified transcription factor is fractionated by gel electrophoresis, followed by electrophoretic transfer of the proteins onto a nitrocellulose membrane. The transblotted proteins are then renatured by the withdrawal of protein denaturants, such as guanidinium chloride or urea. Finally, the membrane-bound protein is hybridized with a radiolabeled DNA probe corresponding to the binding site (footprint) for the transcription factor. After hybridization, the membranes are washed to remove unbound or loosely bound material, and exposed to X-ray film or subjected to phosphorimaging analysis. A detailed protocol for the detection of DNA binding protein(s) that recognizes the α_{1B}-AR gene P2 promoter by Southwestern blotting is given below.

2. Materials

2.1. Preparation of a Radiolabeled DNA Probe (Concatemers of Annealed Synthetic Oligo II)

1. 10X kinase buffer: 0.5 M Tris-HCl (pH 7.6), 0.1 M MgCl$_2$, 0.05 M dithiothreitol (DTT), and 0.01 M spermidine.
2. T4 polynucleotide kinase.
3. [γ-^{32}P] adenosine triphosphate (ATP) (3000 Ci/mmol, 10 mCi/mL).
4. Tris-EDTA (TE) buffer, pH 8.0: 10 mM Tris-HCl, pH 8.0, 1 mM ethylenediamine tetra-acetic acid (EDTA), pH 8.0.
5. Chroma spin column-10: Clontech (Palo Alto, CA).
6. 10X ligase buffer: 0.5 M Tris-HCl (pH 7.6), 0.1 M MgCl$_2$, 0.05 M DTT.
7. T4 polynucleotide ligase.
8. ATP.

2.2. Electrophoretic Blotting of Proteins

1. Apparatus, power supply, and reagents are similar to those used for sodium dodecyl sulfate-polyacrylamide gel electrophoresis (SDS-PAGE) of proteins.
2. Prestained protein standard.
3. Nitrocellulose membranes and Whatman 3MM paper.
4. Western transfer buffer: 48 mM Tris base, 39 mM glycine, 0.037% SDS, 20% methanol.

Extract (µg)

Liver   Brain

30 150 30 100   M(kDa)

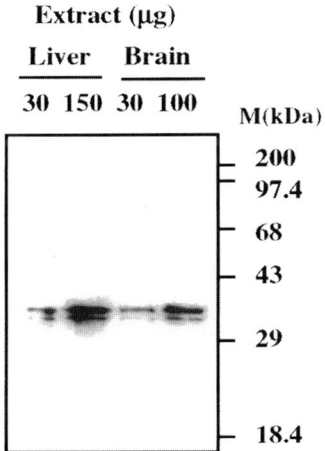

— 200
— 97.4

— 68

— 43

— 29

— 18.4

Fig. 1. Identification of the factors binding to footprint II on the P2 promoter of the α_{1B}-AR gene by Southwestern blotting. Nuclear extracts from rat brain or liver were electrophoretically fractionated and then transferred onto to a nitrocellulose membrane. After renaturing the proteins on the membrane by withdrawal of guanidinium chloride, the blots were hybridized with labeled concatemers of oligo II and analyzed with a PhosphorImager.

5. High-voltage (3000-V) power supply and apparatus for electrophoretic transfer of proteins or Bio-Rad semidry gel transfer apparatus.
6. Tris-buffered saline-Tween (TBST) buffer: 10 mM Tris-HCl, pH 7.9, 150 mM NaCl, 0.05% Tween 20.

2.3. Probing Protein Blots with Radiolabeled Concatemers of Oligo II

1. Z′ buffer: 25 mM HEPES-potassium hydroxide (KOH), pH 7.6, 100 mM KCl, 12.5 mM MgCl$_2$, 0.01 mM ZnSO$_4$, 20% glycerol, 0.1% Nonidet P-40, 1 mM DTT. Adjust the pH to 7.6 using 1 N KOH.
2. Non-fat dried milk.
3. Guanidinium chloride.
4. Poly(dI-dC).

3. Methods
3.1. Preparation of the Radiolabeled DNA Probe

1. Synthesize the sense and antisense strands of oligo II using an automated oligonucleotide synthesizer, and purify them by electrophoresis on a 15–19% polyacrylamide, 8 M urea gel (*see* **Note 1**).

2. End-label the purified sense and antisense strands of oligo II.

100 ng of purified sense and antisense of oligo II	4 µL
10X polynucleotide kinase buffer	2 µL
[γ-^{32}P]ATP (10 µCi/µL; 5000 Ci/mmol)	5 µL
T4 polynucleotide kinase (10 IU/µL)	2 µL
H$_2$O	to a final volume of 20 µL

   Incubate at 37°C for 60 min.

3. Add 80 µL of TE buffer, and separate the labeled oligonucleotide from the excess ATP by spin-column chromatography.

4. Quantify the incorporation of the radioisotope by liquid scintillation counting of a 1-µL aliquot of the spin-column eluate (the labeling reaction should yield ~2 × 10^8 total cpm incorporated).

5. Anneal the sense and antisense strands of oligo II by mixing them in equal amounts in a microcentrifuge tube, heating to 90°C for 5 min, and then allowing the mixture to cool to room temperature. Anneal at room temperature for at least 30 min.

6. After annealing, ligate the double-stranded oligo II to form concatemers.

^{32}P-labeled double-strand oligo II	100 µL
10X T4 DNA ligase buffer	12 µL
10 m*M* ATP	2 µL
T4 DNA ligase	2 µL
H$_2$O to final volume of 120 µL	

7. Ligate overnight at 16°C.

3.2. Electrophoretic Blotting of Proteins

1. Prepare SDS-polyacrylamide gel (10% resolving gel and 4% stacking gel, 1.5 mm thick) *(6)*.

2. Load about 50–150 µg of each crude nuclear extract into the bottom of the wells (*see* **Note 2**). Also load a prestained protein standard.

3. Electrophorese the gel at an initial voltage of 8 V/cm of gel. After the dye front has moved into the resolving gel, increase the voltage to 15 V/cm, and electrophorese the gel until the bromophenol dye reaches the bottom of the resolving gel (about 3 h).

4. Disconnect the power supply, and remove the gel from the glass plates. Mark the orientation of the gel by cutting its top left corner.

5. The gel is ready to be used for a Western transfer.

6. Wearing gloves, cut six pieces of Whatman 3MM paper and one piece of nitrocellulose membrane (Schleicher & Schuell) to the exact size of the gel. Mark the orientation of the nitrocellulose membrane by cutting the same corner as on the gel.

7. Float the nitrocellulose membrane on a water surface, and allow it to wet from beneath by capillary action. Then, soak the membrane in Western transfer buffer for 5 min.

8. Soak the six pieces of 3MM paper in Western transfer buffer.
9. Assemble a Western blot sandwich in the following order:
    Bottom plastic holder facing negative electrode.
    Fiber pad.
    Three pieces of 3MM paper.
    Gel.
    Nitrocellulose membrane.
    Three pieces of 3MM paper.
    Fiber pad.
    Top plastic holder facing positive electrode.
    Make sure that the layers are exactly aligned. Squeeze out any trapped air bubbles using a glass pipet.
10. Connect the electrical leads (positive or red lead to the top plastic holder). Apply a current of 0.65 mA/cm^2 of gel for a period of 1 h.
11. Disconnect the apparatus and gently peel off the top plastic holder, fiber pad, and three pieces of 3MM paper. Then check whether the prestained protein standard transferred completely from the gel. If transfer is complete, proceed to **step 12**. Otherwise, reassemble the sandwich and transfer for a longer time.
12. Wash the membrane with TBST buffer and proceed to **Subheading 3.3.**

3.3. Probing Protein Blots with Radiolabeled Concatemers of Oligo II

Perform all steps at 0–4°C.

1. After the electrophoretic blotting of the protein, place the nitrocellulose membrane in a dish.
2. Add enough volume of Z' buffer containing 6 M guanidinium chloride to cover the membrane, and incubate for 8 min with gentle shaking.
3. Discard the solution, replace it with Z' buffer (without guanidinium), and gently shake for another 8 min.
4. Decant one-half of the solution, and add the same volume of Z' buffer (without guanidinium chloride) into the dish. Incubate the membrane for 8 min with gentle shaking.
5. Repeat **step 4** five more times. The concentration of guanidinium chloride will be gradually reduced to 0.09 M.
6. Discard the solution, and wash the membrane twice for 5 min with Z' buffer.
7. Incubate the membrane with Z' buffer containing 3% nonfat dried milk for 30 min to block nonspecific binding sites on the membrane (*see* **Note 3**).
8. Discard the solution, and wash the membrane with Z' buffer containing 0.25% nonfat dried milk for 8 min.
9. Discard the solution, and place the membrane in a plastic bag. Add just enough Z' buffer to cover the membrane, containing 0.25% nonfat dried milk and 2 µg poly(dI-dC). Incubate for 30 min to block nonspecific DNA binding proteins.

10. Add ^{32}P-radiolabeled concatemers of oligo II to the contents of the bag, and incubate for another 30 min. (For some proteins, incubation at room temperature may increase the extent of DNA binding) (*see* **Note 4**).
11. Discard the radioactive probe solution, and wash the membrane three times with Z′ buffer (5 min each time).
12. Expose the membrane to X-ray film or analyze by phosphorimaging (*see* **Note 5**).

4. Notes

1. The DNA probe used in Southwestern blotting should contain a specific binding site for the transcription factor and should be first tested in gel-mobility shift assay. In order to obtain higher binding affinity, the oligonucleotide probe can be ligated to form concatemers.
2. If a crude nuclear extract is used, a maximum amount of extract that does not overlaod the gel should be used or varying amounts of extract should be loaded onto several lanes. The amount of extract should be determined empirically. For 0.75–1.5 mm thick 15 × 17 cm protein gels, 50–150 µg of crude nuclear extract can be loaded in each lane.
3. Since the crude nuclear extract contains nonspecific DNA binding proteins, such competitors as poly (dI-dC) should be used before and during probing with the specific DNA probe. It may be also advisable to include a control DNA probe that does not contain a binding site for the protein of interest.
4. If specific DNA binding requires more than one polypeptide species or if the DNA binding protein of interest has not been efficiently renatured, Southwestern blotting may fail to detect any bands. In such cases, UV crosslinking should be used. In this method, transcription factor(s) present in crude nuclear extracts is first allowed to bind to the DNA probe and then exposed to UV light to form a covalent bond with the probe. Excess unbound probe is removed by DNase I digestion. The molecular mass of the radiolabeled DNA/protein complex is then estimated by SDS-PAGE. Since the size of the DNA fragment is very small after DNase I treatment, the effect of the DNA on the electrophoretic mobility of the complex is minimal.
5. If several bands are present on the autoradiogram of the Southwestern blot, the membrane may be washed three times with Z′ buffer containing 0.2–0.3 *M* KCl. This higher salt content can remove probe that is bound nonspecifically to the membrane or the protein.

Acknowledgments

Work carried out in the authors' laboratories was supported by grants from the American Cancer Society (IN-105U), Thomas F. Jeffress and Kate Miller Jeffress Memorial Trust (J-379), and NIH (R29CA72681).

References

1. Graham, R. M., Perez, D. M., Hwa, J., and Piascik, M. (1996) α_1-Adrenergic receptor subtypes: molecular structure, function, and signaling. *Circ. Res.* **78,** 737–749.
2. Gao, B. and Kunos, G. (1994) Transcription of α_{1B}-adrenergic receptor gene is controlled by three promoters in rat liver. *J. Biol. Chem.* **269,** 15,762–15,767.
3. Gao, B., Spector, M., and Kunos, G. (1995) The rat α_{1B}-adrenergic receptor gene middle promoter contains multiple binding sites for sequence-specific proteins including a novel ubiquitous transcription factor. *J. Biol. Chem.* **270,** 5614–5619.
4. Gao, B., Jiang, L., and Kunos, G. (1996) Transcriptional regulation of rat α_{1B} (α_{1B}-AR) adrenergic receptor gene by nuclear factor 1 (NF1): a decline in the concentration of NF1 correlates with the downregulation of α_{1B}-AR gene expression in regenerating liver. *Mol. Cell. Biol.* **16,** 5997–6008.
5. Jackson, S. P. (1993) Detection and analysis of proteins expressed from cloned genes in *Gene Transcription, a Practical Approach.* (Hames, B. D. and Higgins, S. J., eds.), IRL at Oxford University Press, Oxford, UK, pp. 47–66.
6. Sambrook, J., Fritsch, E. F., and Maniatis, T. (1989) Identification and characterization of eukaryotic transcription factors, *Molecular Cloning: A Laboratory Manual*, 2nd ed. Cold Spring Harbor Laboratory, Cold Spring Harbor, NY, pp. 189–240.

28

DNase I Footprinting Analysis of Transcription Factors Recognizing Adrenergic Receptor Gene Promoter Sequences

Bin Gao and George Kunos

1. Introduction

Adrenergic receptors (ARs) are G-protein-coupled transmembrane glyco-proteins that play a key role in mediating the sympathoadrenal response to stress. Pharmacological studies have suggested the existence of multiple adrenergic receptor subtypes and sub-subtypes, and to date, nine distinct AR genes have been identified by molecular cloning, including three α_1-AR subtypes ($\alpha_{1A/D}$, α_{1B}, and α_{1C}), three α_2-AR (α_{2A}, α_{2B}, and α_{2C}) and three β-AR (β_1, β_2, and β_3) *(1,2)*. The expression of these receptors is regulated in a complex tissue-specific manner. Such regulation by hormones or developmental factors has been shown to occur at the transcriptional level under many of these conditions. To understand the molecular mechanisms involved in regulating the transcription of the AR under a variety of physiological and pathological conditions, many AR genes have been isolated. After the promoter sequences of receptor genes have been defined through the use of chloroamphenicol acetyltransferase (CAT) reporter assays, a computer-based search of the Sitedata database of previously identified *cis*-elements can be used to check whether these promoter sequences contain known sequence-specific protein binding sites. To determine whether, in a given tissue, a protein does in fact bind to a consensus recognition sequence on the DNA and/or to identify novel *cis*-elements, DNase I footprinting techniques can be used. Once protein binding to DNA is verified by using this technique, a number of other methods, including DNA-gel mobility shift and supershift assays and mutational

From: *Methods in Molecular Biology, vol. 126: Adrenergic Receptor Protocols*
Edited by: C. A. Machida © Humana Press Inc., Totowa, NJ

analyses can be used to characterize the *trans*-acting factors bound to any specific site.

DNase I footprinting has become the standard method for identifying specific protein binding sites on the promoter regions of genes *(3,4)*. The principle of this technique is illustrated in **Fig. 1.** A DNA probe radiolabeled at one end of a single strand (sense or antisense) is prepared and incubated with nuclear extracts. After binding has occurred, the remaining naked DNA probe as well as probe bound by protein is cut randomly once only or a few times with a limited concentration of DNase I. When the products are separated on a denaturing polyacrylamide gel, they form a "ladder" of DNA fragments of varying size. The DNA region to which the protein is bound is protected against cleavage by DNase I, which results in the loss of a specific subset of labeled DNA fragments. Thus, a gap in the DNA ladder (DNA "footprint") is observed when this sample is compared with the ladder generated from naked DNA.

By using the above techniques, we found that the α_{1B}-AR gene P2 promoter contains a number of *cis*-elements, including consensus binding sites for the transcription factors NF1, SP1, CP1, cAMP response element binding protein (CREB), AP2 *(5–7)*. In similar studies of the rat β_2-AR gene P2 promoter, we identified binding sites for the transcription factors Sp1, CREB, CP1, AP2, NF1, nuclear factor κB (NF-κB), and CCAAT enhancer binding protein (C/EBP) *(8)*. Here we describe in some detail the DNase I footprinting analysis of the rat α_{1B}-AR gene P2 promoter.

2. Materials

2.1. Preparation of End-Radiolabeled DNA Probe

1. α_{1B}-AR gene P2$_{(-813/-432)}$CAT plasmid DNA.
2. Restriction enzymes: *Xba*I, *Bam*HI.
3. Calf intestinal alkaline phosphatase (CIAP): Gibco-BRL (Rockville, MD).
4. Buffer-saturated phenol: The phenol was saturated with Tris-HCl (pH > 7.4).
5. Chlorform.
6. Phenol : chloroform : isoamyl alcohol (24:25:1, v/v): Gibco-BRL.
7. Ethanol.
8. Sodium acetate.
9. 10X kinase. buffer: 0.5 M Tris-Cl, pH 7.6, 0.01 M MgCl$_2$ 0.05 M DTT, and 0.01 M spermidine.
10. T4 polynucleotide kinase.
11. [γ-^{32}P]ATP (3000 Ci/mmol, 10 mCi/mL).
12. Agarose.
13. Ethidium bromide: 10 mg/mL in water.
14. Gene-clean.

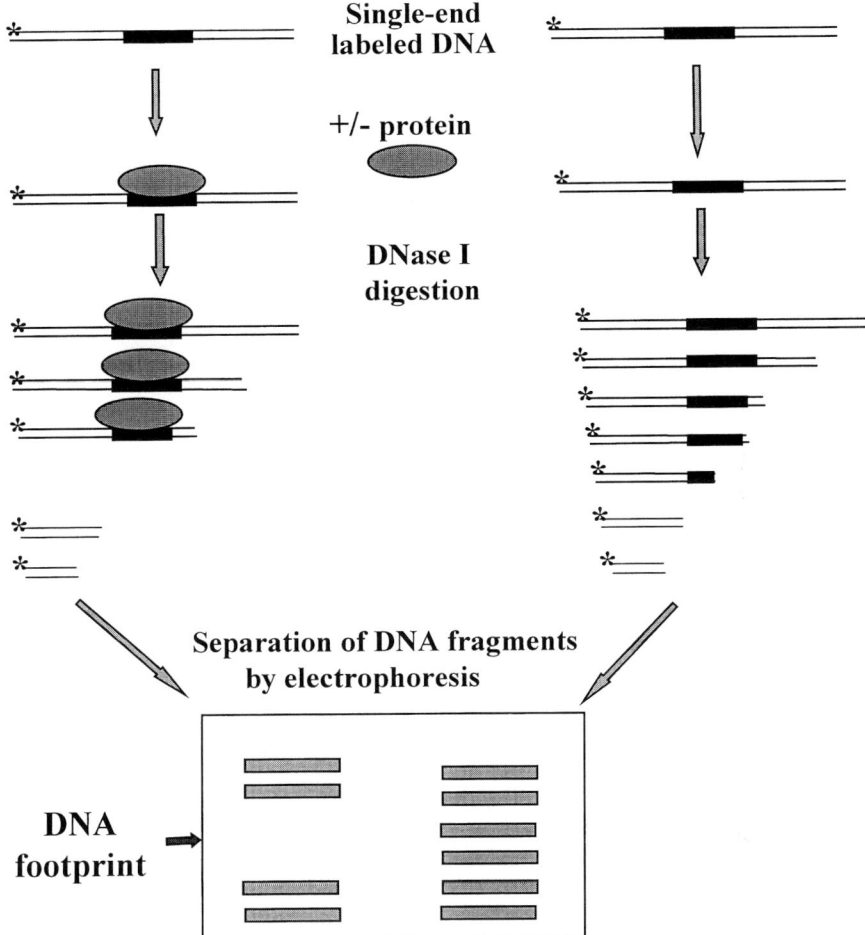

Fig. 1. The principle of DNase I footprinting. A DNA probe is labeled at a single end of one strand and is incubated with nuclear extract. Bound and naked probe is then digested with a limited concentration of DNase I to cut each DNA molecule once or twice only. This digestion will result in a ladder of DNA fragments of varying size when separated on a denaturing polyacrylamide gel. The nucleotides to which the protein is bound will be protected against DNase I cleavage. This will result in the loss of a specific subset of labeled DNA fragments in the gel. A gap in the DNA ladder (DNA footprint) is observed when this sample is compared with digests of the corresponding naked DNA.

2.2. DNase I Footprinting Assay

1. Binding buffer: 25 m*M* HEPES, pH 7.6, 40 m*M* KCl, 0.1 m*M* ethylenediamine tetra-acetic acid (EDTA), 1 m*M* dithiothreitol (DTT), 10% glycerol.
2. Poly (dI-dC).
3. Ca^{2+}/Mg^{2+} solution: 5 m*M* $CaCl_2$, 10 m*M* $MgCl_2$.
4. DNase I stop solution: 50 m*M* Tris-HCl, pH 8.0, 10 m*M* EDTA, pH 8.0, 2% sodium dodecyl sulfate (SDS), 0.1 mg/mL proteinase K (note: the proteinase K must be added just prior to use).
5. Proteinase K: Prepare stock solution (100 mg/mL), and store at –20°C.
6. RQ1 RNase-free DNase stock from Promega (Madison, WI): store at –20°C, and avoid freezing and thawing. Diluted DNase I should be discarded immediately after use.

2.3. Maxam and Gilbert Sequencing Reaction (G + A)

1. Sonicated DNA: Dissolve salmon sperm (ss) DNA or calf thymus DNA at a concentration of 1 mg/mL in water and then sonicate extensively to produce fragments about 500 nucleotides in length. Add NaCl to a final concentration of 0.15 *M*, purify the sonicated DNA by three sequential extractions with phenol:chloroform, precipitate DNA with ethanol, and dissolve in water at the concentration of 1 mg/mL.
2. 1 *M* piperidine formate (pH 2.0): Adjust a 4% solution of formic acid in water to pH 2.0 with 10 *M* piperidine.
3. 1 *M* piperidine: This solution should be freshly prepared by mixing 1 vol of piperidine (10 *M*; Fisher) with 9 vol of water in a glass tube.
4. Hydrazine stop solution: 0.3 *M* sodium acetate, pH 7.0, 0.1 m*M* EDTA, pH 8.0, 100 μg/mL yeast tRNA.
5. Sequencing gel-loading buffer: 98% deionized formamide, 10 m*M* EDTA, pH 8.0, 0.025% xylene cyanol FF, and 0.025% bromophenol blue.

2.4. Preparation of Denaturing Polyacrylamide Gel

1. 40% Acrylamide stock solution (39:1 acrylamide:*bis*-acrylamide): mix 39 g of acrylamide, 1 g of *N,N'*-methylene bisacrylamide and water to 100 mL final volume. Dissolve the reagents by heating the solution to 37°C with a constant stirring. Filter the solution through a 2-μm membrane filter, and store at 4°C in a foil-wrapped bottle. This solution is stable for several months.
2. 5X Tris-borate electrophoresis (TBE) buffer: 450 m*M* Tris-borate, 10 m*M* EDTA (to prepare this solution, mix 54.0 g Tris base, 27.5 g boric acid, 20 mL of 0.5 *M* EDTA, pH 8.0, and add water to 1-L final volume).
3. Tetramethylethylenediamine (TEMED): Store at 4°C.
4. 10% (w/v) Ammonium persulfate. This solution is stored at room temperature and is stable for several weeks.
5. Urea, ultra-pure.
6. Sequencing gel electrophoresis apparatus, power supply, and gel dryer.

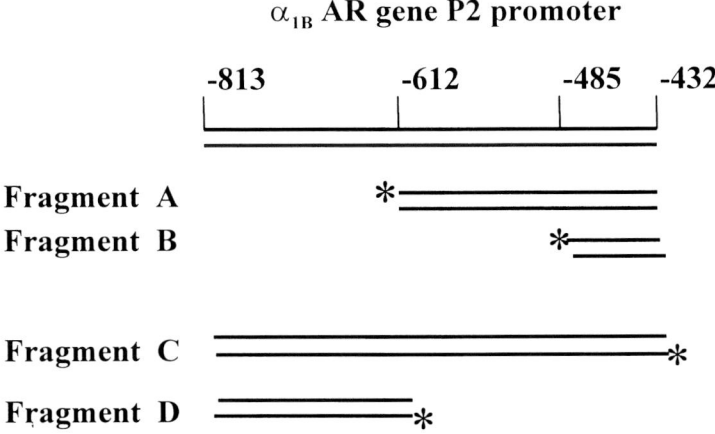

Fig. 2. The probes used in the footprinting analyses of α_{1B}-AR gene P2 promoter. The top line with negative numbers represents the rat α_{1B}-AR gene P2 promoter. Asterisks represent the ^{32}P end label. Fragments A and B were labeled at single end on sense strand. Fragments C and D were labeled at single end on antisense strand.

3. Methods

The first step in DNase I footprinting is the selection and radiolabeling of an appropriate probe. Since only the region within 25–100 nucleotides from the labeled end of the DNA fragment will be accurately resolved on a nondenaturing polyacrylamide gel, the end of the DNA probe should be approx 25–100 nucleotides upstream of the putative footprint region. If the location of the footprint region is unknown, it should first be determined by subjecting radiolabeled DNA fragments of different lengths to DNase I digestion. For example, in order to identify all possible footprints on the P2 promoter of the α_{1B}-AR gene, four probes described in **Fig. 2** were used. The second step of DNase I footprinting is to perform the binding reaction between DNA and protein (the latter usually contained in a nuclear or whole-cell protein extract). Third, a limited concentration of DNase I is added to the reaction mixture for a set period of time. Finally, the reaction is terminated and the DNA isolated and analyzed by denaturing polyacrylamide gel electrophoresis (PAGE).

Figures 2 and **3** illustrate DNase I footprinting of the P2 promoter of α_{1B}-AR gene. In order to generate suitable DNA probes, the P2 promoter region of the α_{1B}-AR gene was digested with various restriction enzymes to generate DNA fragments, which were then radiolabeled at a single end (**Fig. 2**). The following protocol describes the DNase I footprinting analysis of

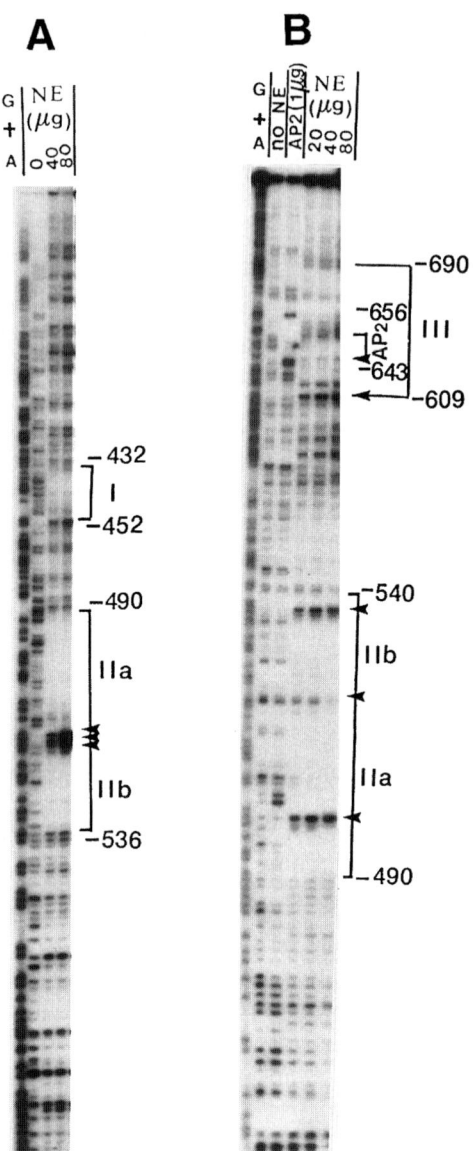

Fig. 3. DNase I footprinting analyses of α_{1B}-AR gene P2 promoter probes A (**A**) and C (**B**). (A) and (B) illustrate the footprints (marked by brackets and Roman numerals) obtained on DNA fragments A and C, respectively (fragments labeled as in **Fig. 2**). DNase I-hypersensitive sites are marked by filled arrows. G + A, Maxam-Gilbert G + A sequencing reaction; NE, rat liver nuclear extract.

the α_{1B}-AR gene region –432 to –612. **Figure 3A,B** illustrates the footprints obtained on both the sense and antisense strands of the P2 promoter.

3.1. Preparation of End-Labeled α_{1B}-AR Gene P2 Promoter Probe and β_2-AR Gene P2 Promoter Probe

End-labeling of a single strand of a DNA fragment can be achieved by using T4 polynucleotide kinase or by using the polymerase chain reaction (PCR) amplification method. Detailed protocols for end-labeling the α_{1B}-AR gene P2 promoter and the β_2-AR gene P2 promoter are given below.

3.1.1. End-Labeling the α_{1B}-AR Gene P2 Promoter Probe (Fragment A in *Fig. 2*) Using T4 Polynucleotide Kinase

1. Digest 10 µg of P2 promoter/pCAT construct with restriction enzyme *Bam*HI (at –612 upstream from ATG codon) and simultaneous phosphatase treatment (0.5 U CIAP/µg DNA) at 37°C for 1 h.
2. Extract the DNA once with phenol:chlorform (1:1) and once with chlorform before precipitating the DNA by adding sodium acetate, pH 5.2, to 0.3 *M* and 2 vol of 100% cold ethanol.
3. The DNA pellet is then dissolved in 10 µL of H_2O and subjected to 5' end-labeling reaction in a 1.5-mL microcentrifuge tube (*see* **Note 1**):

DNA	10 µL
10X kinase buffer	2 µL
T4 polynucleotide kinase (8 U/µL)	1 µL
[γ-^{32}P]ATP (3000 Ci/mmol, 10 mCi/mL)	2 µL
H_2O to final volume of	20 µL

4. Incubate at 37°C for 30 min. Add an additional 1 µL of T4 polynucleotide kinase and incubate for an additional 30 min at 37°C.
5. Phenol:chlorform extraction and ethanol precipitation of DNA as described in **step 2**. The DNA pellet is then dissolved in 20 µL H_2O and subjected to a second restriction enzyme digestion.

End-labeled DNA	20 µL
10X restriction buffer	3 µL
*Xba*I	1 µL
H_2O to final volume of	30 µL

6. Incubate at 37°C for 1 h.
7. The mixture is then electrophoresed on a 1.5% agarose gel and stained with ethidium bromide.
8. The single end-labeled DNA probe is purified from the gel by the glass powder method (Geneclean, BIO 101, CA).

3.1.2. End-Labeling of the β_2-AR Gene P2 Promoter Probe Using a PCR Method

1. Design reverse and forward primers.
2. Label one of these primers by using [γ-^{32}P]ATP and T4 polynucleotide kinase.

3. Purify this radiolabeled primer by using chroma spin-10 column (Clontech).
4. Using these two primers and β_2-AR gene P2 promoter/CAT construct as a template, PCR reaction is performed.
5. The PCR products are electrophoresed on a 1.5% agarose gel and stained with ethidium bromide.
6. The single end-labeled DNA probe is purified from the gel by the glass powder method.

3.2. DNase I Footprinting Analyses of the α_{1B}-AR and β_2-AR Gene P2 Promoters

1. Set up the following reaction in a microcentrifuge tube:

Nuclear extract (5–80 μg) (in binding buffer) (*see* **Note 2**)	10 μL
poly(dI-dC) (1 mg/mL)	3–2 μL
^{32}P-end-labeled α_{1B}-AR or β_2-AR gene promoter fragment (0.1–1 pmol) (*see* **Note 3**)	1 μL
Add binding buffer to final volume	100 μL

2. Gently mix each tube, and incubate for 25 min at 25°C (*see* **Note 4**).
3. During incubation, prepare an excess of DNase I stop solution (proteinase K solution must always be used fresh), and leave at room temperature.
4. Dilute 5 μL of stock RQ1 RNase-free DNase (1 U/μL) into 100 μL of cold Tris buffer, pH 8.0 immediately before use.
5. Add 100 μL of room temperature Ca^{2+}/Mg^{2+} solution into **step 1** reaction mixture, and incubate for 1 min.
6. Add 3 μL of the diluted RQ1 RNase-free DNase solution, mix gently but thoroughly, and incubate at room temperature for 1 min (*see* **Notes 4–7**).
7. Terminate the reaction by adding 40 μL of stop solution, vortex vigorously, and incubate for 30–45 min at 42°C (*see* **Note 8**).
8. Extract each reaction twice with 200 μL of phenol:chlorform.
9. Transfer the upper, aqueous phase to a fresh tube, and precipitate with ethanol.
10. Resuspend the pellet in 3 μL of loading buffer. Heat at 80°C for 2 min, and quickly chill on ice for 2 min.
11. Load and electrophorese in an 8% denaturing polyacrylamide sequencing gel (*see below*).

3.3. Maxam and Gilbert Sequencing Reaction (G + A)

To determine the precise DNA sequence protected from the DNase I digestion, Maxam and Gilbert sequencing reactions are electrophoresed in a denaturing polyacrylmide gel in parallel with the footprinting reaction.

1. Mix together in a 1.5-mL microcentrifuge tube:

Single-end-radiolabeled DNA probe	10 μL
Sonicated DNA (1 mg/mL)	4 μL
H_2O	10 μL

2. Chill to 0°C, add 4 μL of piperidine formate, pH 2.0, and incubate for 15 min at 37°C.
3. Add 240 μL of hydrazine stop solution (at 0°C) and 750 μL of ethanol (–20°C).
4. Store for 5 min at –70°C and centrifuge for 5 min at 12,000g at 4°C.
5. The pellet is dissolved in 300 μL of 0.3 M sodium acetate, pH 5.2, at 0°C; add 900 μL of ethanol (–20°C).
6. Store for 5 min at –70°C, and centrifuge for 5 min at 12,000g at 4°C.
7. Wash the pellet with 1 mL of ethanol (–20°C), remove the supernatant, and lyophilize the pellet.
8. Add 100 μL of 1 M piperidine, close the cap tightly, and mix.
9. Incubate for 30 min at 90°C and then lyophilize.
10. Dissolve the pellet in 20 μL of H_2O and lyophilize.
11. Repeat **step 10** once.
12. Add 10 μL of sequencing loading buffer. Vortex for 20 s.
13. Centrifuge and heat at 90°C for 1 min. Load 3 μL of each sample on separate lanes of a sequencing gel.

3.4. Electrophoresis of Samples

1. Prepare an 8% polyacrylamide sequencing gel. Mix the following components together:

40% Acrylamide	52 mL
5X TBE	20 mL
Urea (48 g) (takes up much volume)	
H_2O to final volume of	100 mL

Just before pouring the gel, add TEMED, 50 μL and 10% ammonium persulfate, 200 μL.
2. Clean the gel plates thoroughly. Siliconize one of the plates with Sigmacote, and wash with chloroform. Assemble the gel plates with spacers.
3. Pour the polyacrylamide gel, and insert the wide comb (not shark's teeth comb) into the gel solution (*see* **Note 9**).
4. After polymerization of the gel is complete, carefully remove the comb and the spacer from the bottom of the gel.
5. Attach the gel mold into the electrophoresis apparatus, and fill the reservoir with 0.5X Tris-borate EDTA buffer (TBE).
6. Prerun the gel for 30–60 min in 1X TBE at 35–40 W constant power.
7. Flush the loading wells, and load 3 μL of each predenatured sample into adjacent wells.
8. After all of the samples are loaded, connect the power, and electrophorese gel at constant power (35–40 W for a 20 × 40 cm gel, about 1700 V).
9. When the bromophenol blue reaches the bottom of the gel (about 2–3 h), disconnect the sequencing apparatus from the power supply.
10. Remove the glass plates from the apparatus, and remove and dry the gel.
11. Expose dried gel to X-ray film, or analyze by phosphorimaging (Phosphor-Imager).

4. Notes

1. Avoid using buffers containing ammonium salts for dissolving DNA prior to treatment with kinase, since ammonium ions are strong inhibitors of T4 polynucleotide kinase.
2. Nuclear extracts can be prepared by many methods, but high purity is required for quantitation of binding kinetics. The protocol described by Gorski et al. *(9)* is highly recommended. Nuclear extract should be dialyzed completely with binding buffer to remove $(NH_4)_2SO_4$ and NaCl, since these two salts inhibit DNase I activity.
3. The DNA probe used in the footprinting assay should be 0.1–1 pmol/assay (about 1000–10,000 cpm total).
4. DNA binding to protein is affected by a number of factors, such as temperature, time, pH, and ionic strength. For example, the concentration of K^+ should be 50–200 mM, since higher concentrations will reduce the specific binding affinity while increasing the specificity (stringency) of binding. DNase I digestion requires the presence of Mg^{2+} and Ca^{2+}, which may both alter DNA binding to protein. Therefore, incubation of DNA and protein in the presence of Mg^{2+} and Ca^{2+} should be limited to <1 minute.
5. The concentration of DNase I should be determined empirically. The amount of DNase I used with crude nuclear extracts should be higher than the amount used with purified protein, which in turn should be higher than the amount used in the control that lacks protein. Usually, DNase I should be added at a concentration that allows approx 1 or 2 random nicks/molecule.
6. The length of time for DNase I digestion is approx 1 min. The digestion time should be the same in all reactions and should be determined empirically. Since it is difficult to standardize the extent of DNase I digestion when working with a large number of samples, we recommend that no more than four samples be simultaneously digested and that the stop solution be close at hand for stopping the reaction.
7. DNase I should be handled carefully, since this enzyme is sensitive to denaturation by shaking. DNase I in reaction should be mixed manually by stirring or pipeting up and down.
8. After terminating the DNase I digestion reaction by adding the stop solution, the mixture should be extracted twice with phenol:chlorform to remove the protein completely. The DNA pellet after ethanol precipitation should be washed twice with 70% ethanol to remove residual EDTA and salts. If proteins and salts are not removed, bands in the sequencing gel may become compressed and narrow.
9. To obtain better resolution of DNase I footprint on the gel, a wide comb is preferred over a shark-toothed comb in preparing the DNA sequencing gel.

Acknowledgments

Work carried out in the authors' laboratories was supported by grants from the American Cancer Society (IN-105U), Thomas F. Jeffress and Kate Miller Jeffress Memorial Trust (J-379), and NIH (R29CA72681).

References

1. Hieble, J. P., Bylund, D. B., Clarke, D. E., Eikenburg, D. C., Langer, S. Z., Lefkowitz, R. J., et al. (1995) International Union of Pharmacology. X. Recommendation for nomenclature of alpha 1-adrenoceptors: consensus update. *Pharmacol. Rev.* **47,** 267–270.
2. Strosberg, A. D. (1995) Structure, function, and regulation of the three beta-adrenergic receptors. *Obes. Res.* **Suppl. 4,** 501S–505S
3. Brenowitz, M., Senear, D. F., and Kingston, R. E. (1988) DNase I footprint analysis of protein–DNA binding, in *Current Protocols in Molecular Biology*, vol. 2 (Ausubel, F. M., et al., eds.), John Wiley, New York, pp. 10–16.
4. Brenowitz, M., Senear, D. F., Shea, M. A., and Ackers, G. K. (1986) Quantitative DNase footprint titration: a method for studying protein-DNA interactions. *Methods Enzymol* **130,** 132–181.
5. Gao, B., Spector, M., and Kunos, G. (1995) The rat α_{1B}-adrenergic receptor gene middle promoter contains multiple binding sites for sequence-specific proteins including a novel ubiquitous transcription factor. *J. Biol. Chem.* **270,** 5614–5619.
6. Gao, B., Jiang, L., and Kunos, G. (1996) Transcriptional regulation of the rat α_{1B}- (α_{1B}-AR) adrenergic receptor gene by nuclear factor 1 (NF1): a decline in the concentration of NF1 correlates with the downregulation of α_{1B}-AR gene expression in regenerating liver. *Mol. Cell. Biol.* **16,** 5997–6008.
7. Chen, J., Spector, M. S., Kunos, G., and Gao, B. (1997) Sp1-mediated transcriptional activation from the dominant promoter of the rat alpha1B adrenergic receptor gene in DDT_1MF-2 cells. *J. Biol. Chem.* **272,** 23,144–23,150.
8. Jiang, L., Gao, B., and Kunos, G. (1996) DNA elements and protein factors involved in the transcription of the beta 2-adrenergic receptor gene in rat liver. The negative regulatory role of C/EBP alpha. *Biochemistry* **35,** 13,136–13,146.
9. Gorski, K., Carneiro, M., and Schibler, U. (1986) Tissue-specific in vitro transcription from the mouse albumin promoter. *Cell* **47,** 767–776.

29

Electrophoretic Mobility Shift Assay for Detection of DNA Binding Proteins Recognizing β-Adrenergic Receptor Gene Sequences

Philbert Kirigiti and Curtis A. Machida

1. Introduction

The β-adrenergic receptors (β-ARs) are important modulators in the sympathetic control of various metabolic processes in the central (CNS) and peripheral nervous system *(1–4)*. The β-ARs mediate the physiological effects of the catecholamines epinephrine and norepinephrine. These receptors are localized at multiple sites throughout the nervous system, and serve as important regulators of CNS-mediated behavior and several neural functions, including mood, memory, neuroendocrine control, and stimulation of autonomic function *(1–4)*. Abnormalities in the expression of the β-AR system have been implicated as playing potential roles in a variety of psychiatric and neurological diseases, including depression and disorders affecting autonomic nervous system activity. Although much is known about the physiological responses of β-ARs to catecholamines in the periphery, there is limited understanding of the molecular mechanisms that differentially regulate β-AR activation and expression in the CNS.

β_1-ARs are the primary β-AR subtype found in the CNS and are expressed at multiple neuroanatomical locations, including the cortex, striatum, cerebellum, and thalamus *(1–4)*. β_1-AR transcriptional start sites, promoter elements, and polyadenylation sites used for expression have been defined in the C6 glioma cell line and within primary myocytic cells *(5–7)*. Transfection analyses indicate that the primary rat β_1-AR promoter occurs within region –389 to –325 relative to the translational start site, with several other regions appearing either to enhance or repress basal β_1-AR expression in C6 cells *(5–7)*.

From: *Methods in Molecular Biology, vol. 126: Adrenergic Receptor Protocols*
Edited by: C. A. Machida © Humana Press Inc., Totowa, NJ

Comparison of the flanking sequences of the rat and primate β_1-AR genes reveal conserved regulatory sequences that may be important for β_1-AR gene expression and transcriptional modulation. These sequences include a cAMP response element (CRE), two T-cell enhancers (TCEs), an AP1 binding site, and a thyroid hormone receptor response element (TRE), all of which are clustered between −1330 and −1250 in the rat gene *(8)*.

1.1. Heterogeneity of Adrenergic Receptor Promoter Structure and Gene Transcription

Eukaryotic promoters typically contain TATA and CCAAT box sequences that interact with nuclear transcription factors *(9)*. Alternative promoters that lack the traditional TATA and CCAAT promoter elements have been described for some genes, including viral genes, housekeeping genes, and certain tissue-specific genes *(9)*. These TATA-less promoters include the GC-rich promoters that are usually constitutively expressed and initiate transcription from several potential start sites. The GC-rich promoters contain several GC boxes (GTCCGCCCT) that can serve as potential binding sites for transcription factor SP1 and other nuclear factors. The remaining TATA-less promoters do not contain GC-rich sequences or GC boxes, initiate transcription from limited start sites, and are regulated during cell differentiation and development.

The majority of the adrenergic receptor genes contain GC-rich promoters and utilize multiple transcriptional start sites *(10–13)*. The adrenergic receptor promoters also lack CCAAT sequences, and appear to regulate cell- or region-specific expression within the nervous system *(10–13)*. Thus, the adrenergic receptor promoters may contain other nonconventional promoter structures to regulate transcription. Interestingly, the human α_{2A}-AR gene contains a TATA box and GC box, but lacks the CCAAT element *(11)*.

1.2. DNA Binding Proteins and Electrophoretic Mobility Shift Assay (EMSA)

DNA binding proteins are involved in the regulation of a variety of molecular mechanisms, including transcriptional activation *(14)*. Tissue-specific transactivating proteins can bind to *cis*-acting DNA sequences or hormonal response elements (HREs) to regulate gene transcription.

The EMSA is commonly utilized to investigate the interactions between a target sequence (DNA or RNA) and cellular proteins (**Fig. 1**) *(15–17)*. EMSA is a simple and rapid technique used to detect nuclear or cytoplasmic proteins that can recognize a target sequence in a DNA or RNA fragment. This assay allows an investigator to verify functional interactions involving consensus DNA binding element(s) determined by sequence analysis, or to identify novel proteins that recognize and bind to a specific DNA sequence, isolated from a

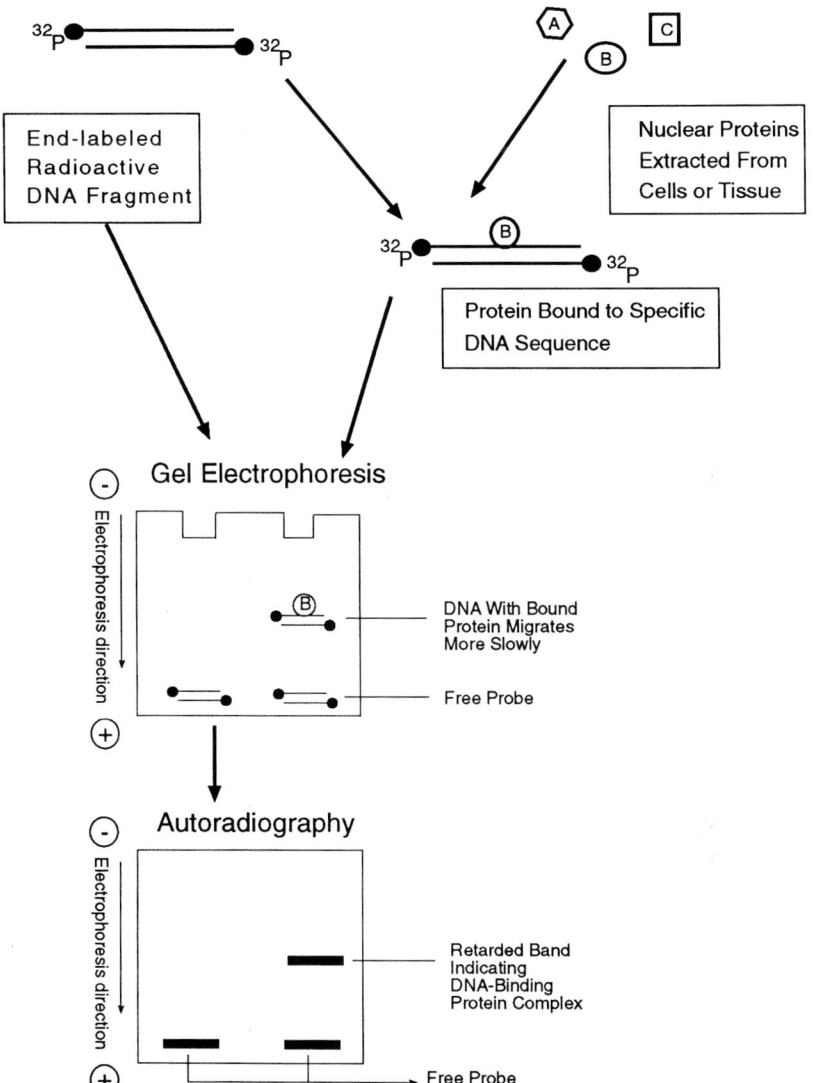

Fig. 1. Pictorial description of the EMSA. DNA fragment is end-labeled with [γ-^{32}P]ATP using polynucleotide kinase. Labeled DNA fragments are sequentlally purified and bound to nuclear protein extract derived from tissue or from cells. Samples are then electrophoresed in polyacrylamide gels and processed for autoradiography. Bands that have reduced electrophoretic mobilities indicate the presence of DNA binding protein complex.

complex pool of other proteins. This technique was originally developed for DNA binding proteins, but has also been extended to allow detection of RNA binding proteins.

In general, purified proteins or crude cell extracts are incubated with a $[^{32}]$P-radiolabeled DNA or RNA probe; the electrophoretic mobilities of the nucleic acid–protein complexes are shifted (slower migration) and can be distinguished from the free probe in a nondenaturing polyacrylamide gel (**Fig. 1**). For the detection of DNA binding proteins, such as transcription factors, either purified or partially purified proteins or nuclear cell extracts are generally utilized in binding reactions. For detection of RNA binding proteins, either purified or partially purified proteins, or nuclear or cytoplasmic cell extracts are utilized, depending on the location of the RNA binding protein of interest. The specificity of the DNA or RNA binding protein for the putative binding site is established by competition experiments using DNA or RNA fragments or oligonucleotides containing a binding site for the protein of interest (specific control), or other unrelated sequence (nonspecific control). The differences in the nature and intensity of the complex formed in the presence of specific and nonspecific competitor allow identification of specific interactions. Site-specific mutagenesis of the DNA binding site can then be conducted to determine the exact site of interaction. Altering the critical binding region may potentially abolish DNA–protein complex formation.

Another test of specificity is the antibody "supershift" assay. Antibodies to the bound protein, when added to the preformed DNA–protein complex, can further retard the electrophoretic mobility of the complex (supershift). Alternatively, antibodies to the putative DNA binding protein can be incubated with the purified protein or protein extract, prior to addition with labeled target DNA. In some cases, the antibody may prevent DNA–protein interactions by blocking the DNA binding domain and, as a result, eliminate its ability to induce mobility shifts. The antibody can furthermore be used to deplete the extract of a particular protein via immunoprecipitation.

1.3. EMSA Identification of CNS-Specific DNA Binding Proteins Recognizing β_1-AR Gene Sequences

To begin identifying potential tissue-specific transactivators of β_1-AR transcription, we systematically sub-divided the 5' flanking region (–3352 to –1) of the rat β_1-AR gene into smaller restriction or polymerase chain reaction- (PCR) generated fragments. DNA mobility shift analyses were conducted using these fragments as radioactive probes, binding to tissue extracts either abundant in (cortex, heart, lung) or devoid (liver) of β_1-ARs. Interestingly, we observed multiple unique DNA–protein complexes using CNS-derived extracts; a smaller number of complexes were also observed using peripheral tissues (heart and lung). One example demonstrating the specificity of interaction of β_1-AR upstream sequences to CNS-derived proteins is illustrated in **Fig. 2**.

1.4. Other Known Interactions of DNA Binding Proteins with β_1-AR Gene Sequences

In the case of negative transcriptional regulation, binding of nuclear proteins to an upstream DNA sequence might function by physically inhibiting transcription by RNA polymerase. Inducible cyclic AMP early repressor (ICER), a member of the cyclic AMP response element modulator (CREM) family of transcription factors, is a powerful repressor of cyclic AMP-mediated transactivation *(18,19)*. ICER is induced in C6 glioma cells by β_1-AR agonists, and interacts with the β_1-AR promoter at the CRE (position −1314 to −1307, relative to the translational start site) to downregulate β_1-AR mRNA transcription *(20,21)*. In a different case, the rat β_1-AR mRNAs appear to be negatively regulated with dexamethasone. The rat β_1-AR gene contains two potential GREs at position −2791 and −950, relative to the translational start; interestingly, the more proximal site at −950 appears to be functional and binds to the α-subunit of the human glucocorticoid receptor *(6)*.

2. Materials

2.1. Preparation of Nuclear and Cytoplasmic Extract from Tissues or from Cultured Cells

2.1.1. The Extraction of Nuclear and Cytoplasmic Proteins from Suspension and Adherent Cell Lines

1. Monolayer adherent mammalian cells (e.g., HeLa, CV-1) or suspension (e.g., Sup-T1) cultures (*see* **Note 5**).
2. Stock solutions:
   a. 1X Phosphate-buffered saline (PBS).
   b. $1M$ HEPES-KOH, pH 7.9.
   c. $1M$ $MgCl_2$.
   d. $4M$ KCl.
   e. $1M$ Dithiothreitol (DTT) stored at −20°C.
   f. 0.1 M Phenylmethylsulfonyl fluoride (PMSF): Solvent is 100% isopropanol. Store at 4°C.
   g. 1000X Protease inhibitor cocktail (1 mg/mL aprotinin, 0.5 mg/mL Leupeptin, 0.7 mg/mL pepstatin). Store at −20°C.
   h. 100% Sterile glycerol.
   i. $5M$ NaCl.
3. Buffer A: 10 mM HEPES-KOH, pH 7.9, 1.5 mM $MgCl_2$, 10 mM KCl, 0.5 mM

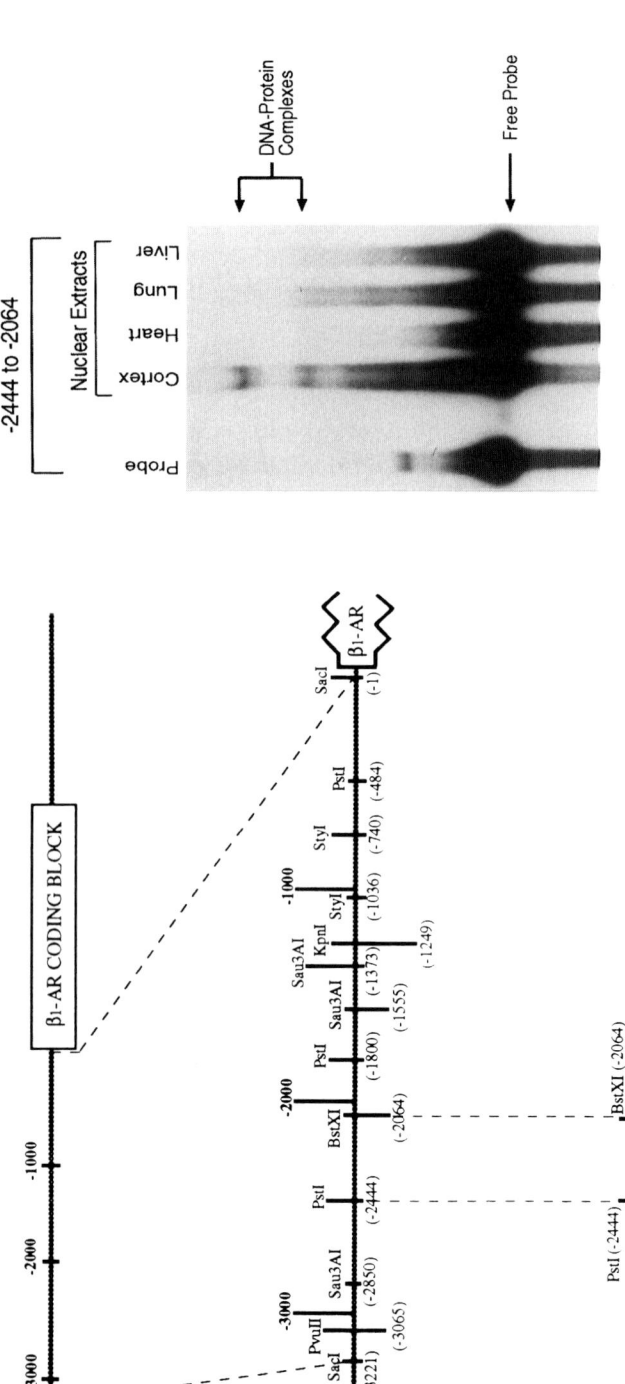

A

β₁-AR CODING BLOCK

B

-2444 to -2064

Nuclear Extracts

Probe | Cortex | Heart | Lung | Liver

DNA-Protein Complexes

Free Probe

Fig. 2. (**A**) Restriction map of the rat β₁-AR upstream region, extending from position −3221 to −1, relative to the translational start site. The restriction fragment probe used for EMSA displayed in panel B is the β₁-AR *PstI–BstXI* fragment encompassing sequence −2444 to −2064 (relative to the translational start site). (**B**) EMSA example demonstrating the specificity of interaction of β₁-AR upstream sequences to CNS-derived proteins: *PstI–BstXI* restriction fragment (approx 200 pmol) were end labeled with [γ-^{32}P]ATP (7000 Ci/mmol, 167 mCi) using T4 polynucleotide kinase (10 U, 90 min, 37°C). Labeled restriction fragments were sequentially purified through a Sephadex G-50 column and a Wizard purification resin (Promega Corp.) Labeled fragments (2.5 × 10⁴ cpm) were subsequently bound to nuclear protein extract (10 μg) derived from rat tissue (cortex, heart, lung, and liver) for 25 min at room temperature. Samples were electrophoresed in 4% polyacrylamide gel and exposed for autoradiography for 16–20 h.

DTT,* 0.2 mM PMSF,* 1X protease inhibitor cocktail (1 mg/mL aprotinin, 0.5 μg/mL leupeptin, 0.7 μg/mL pepstatin).*

4. 5X Buffer B: 0.15 M HEPES, pH 7.9, 0.7 M KCl, 0.015 M MgCl$_2$, 1X protease inhibitor cocktail.*

5. Buffer C: 20 mM HEPES-KOH, pH 7.9, 25% glycerol, 20 mM NaCl, 1.5 mM MgCl$_2$, 0.2 mM EDTA, 0.5 mM DTT, 0.2 mM PMSF, 1X protease inhibitor cocktail.*

6. Buffer D: 20 mM HEPES-KOH, pH 7.9, 25% glycerol, 420 mM NaCl (*see* **Note 24**), 1.5 mM MgCl$_2$, 0.2 mM ethylenediamine tetra-acetic acid (EDTA), 0.5 mM DTT,* 0.2 mM PMSF,* 1X protease inhibitor cocktail.*

7. Sterile cell scraper.

8. Bradford reagent: Available from Bio-Rad (Hercules, CA).

2.1.2. Extraction of Nuclear and Cytoplasmic Proteins from Mammalian Tissue

1. 0.5–2.0 g Mammalian tissue.

2. Dounce homogenizer with type B pestle: Available from Fisher Scientific (Santa Clara, CA).

3. All the reagents and stocks used in **Subheading 2.1.1.** and **items 1** and **2** above.

2.2. Preparation of Radiolabeled Synthetic Oligonucleotide, and PCR-Generated or Restriction Endonuclease-Generated DNA Fragment

1. 100 pmol DNA oligonucleotide, restriction endonuclease-generated DNA fragment, or PCR-generated DNA fragment (*see* **Note 27**).

2. T4 Polynuceotide kinase (PNK) with 10X PNK buffer or Klenow fragment of *Escherichia coli* DNA polymerase with 10X Klenow buffer. Enzymes are available from New England Biolabs (Beverly, MA).

3. ^{32}Phosphorus- (^{32}P)† labeled nucleotide, e.g. [γ-^{32}P]ATP (6000–7000 Ci/mmol): Isotope could be purchased from ICN Biomedicals (Irvine, CA).

4. Wizard® DNA Clean-Up System: This kit is available from Promega (Madison, WI).

5. Sephadex G-50 exclusion columns: This item is available from Pharmacia Biotech (Piscataway, NJ) (*see* **Note 11**).

6. 1X TE: 10 mM Tris-HCl, pH 8.0, and 1 mM EDTA, pH 8.0.

7. QIAquick gel extraction kit, a product of Qiagen (Santa Clara, CA).

8. DE-81 filter: Can be purchased from Fisher Scientific (Santa Clara, CA).

---

*Stock ingredient should be added to buffer immediately before usage.

†Handle ^{32}P radioisotope with care. If you have not periodically handled radioisotopes, consult a trained person who is aware of the danger of radioactive material and the use of appropriate shielding. Remember to wear hand and body dosimeters at all times when dealing with the isotope.

9. 70% Ethanol.
10. 5 mM dNTP mixture: This item is available from numerous vendors including Pharmacia Biotech.

2.3. Binding Assay and Electrophoretic Separation of Bound Complex or Complexes from Free Probe

2.3.1. Preparation of a Nondenaturing Gel

1. 5X Tris-glycine EDTA, pH 8.8 (TGE): 0.25 M Tris base, 1.18 M glycine, 0.01 M EDTA.
2. 46% (w/v) Acrylamide§: 45 : 1 (acrylamide : *bis*), filtered and degassed.
3. *N,N,N′,N,′*-Tetramethylethylenediamine (TEMED).
4. 10% Ammonium persulfate (AMPS) prepared fresh or <1 wk old: Store covered with aluminum foil at 4°C.
5. Vertical slab gel electrophoretic apparatus: Pharmacia Biotech Unit. The 16 × 18 cm Hoefer® apparatus is commonly used.
6. Gel Slick®: This reagent is available from J. T. Baker, Inc. (Phillipsburg, NJ).

2.3.2. Binding Reactions

1. 5X Binding buffer: 50 mM Tris-HCl, pH 7.5, 0.25 M NaCl, 5 mM EDTA, pH 8.0, 25% glycerol, 25 mM MgCl$_2$.
2. 1 µg/µL Poly-deoxy-inosinic-deoxy-cytidylic acid (Poly dI-dC).

2.3.3. Gel Electrophoresis and Analysis

1. 10X Tracking dye: 250 mM Tris-HCl, pH 7.5, 0.2% bromophenol blue, 0.2% xylene cyanol, 40% glycerol.
2. Autoradiographic film and film exposure cassette: Both items can be purchased from Kodak company (Rochester, NY).
3. Loading tips: This item is a available from numerous vendors, e.g., Fisher.
4. Plastic wrapper.
5. Blot paper, e.g., Whatmann 3MM, available from Island Scientific (Bainbridge Island, WA).
6. Green Glow®, Fluorescent dye: Available from Polymark (Cloxis, CA).

2.4. Competition Binding Assay

1. This will require all the materials listed in **Subheadings 2.2.** and **2.3.**
2. In addition, the assay will need an unlabeled nonspecific competitor (approximately the same length as labeled probe, but containing an unrelated sequence) and a specific nonradioactive competitor (same probe, but unlabeled).

---

§**Caution:** Acrylamide monomer is neurotoxic. Gloves should be worn while handling the solution, and face masks worn when weighing the powder.

2.5. Antibody Supershift Assay

1. This will require all items listed in **Subheadings 2.2.** and **2.3.**
2. Additional material required will consist of an antibody (monoclonal or polyclonal) specific for a DNA binding protein.
3. A nonspecific control antibody.

3. Methods

3.1. Preparation of Nuclear and Cytoplasmic Extract from Tissues or from Cultured Cells

The method utilized to isolate nuclear and cytoplasmic extracts employs the use of hypotonic lysis of the plasma membrane, followed by high-salt extraction of nuclei.

3.1.1. Extraction of Nuclear and Cytoplasmic Proteins from Adherent Cell Lines (see **Note 1**)

Note: All centrifugations should be performed at 4°C with samples kept on ice at all times. It is preferable to conduct protein isolations in a cold room.

1. Rinse the cells briefly with ice-cold 1X PBS. Using a cell scraper, scrape cells into 1 mL 1X PBS.
2. Transfer cell suspension to a microcentrifuge tube. If harvesting more than one dish of cells, pool scraped cells into a chilled 1.6-mL polypropylene tube, and place on ice.
3. Pellet cells at 300g for 5 min. Resuspend cells in 1.5 mL ice-cold 1X PBS, and transfer into 1.6-mL microcentrifuge tube.
4. Pellet cells at 300g for 5 min.
5. Add 100 µL/10^{7}cells of ice-cold buffer "A." Resuspend cells (by gently flicking the tube or very gentle trituration with a wide-bore pipet tip).
6. Allow cells to swell in buffer "A" for 10 min on ice (*see* **Note 2**).
7. Vortex for 10 s.
8. Centrifuge for 15 min at 500g. Transfer the supernatant to a fresh tube. Do not discard the pellet. Place on ice. Pellet will be used in **step 13** for nuclear extract.
9. To the supernatant, add the appropriate volume of cold 5X buffer "B." Mix gently by agitation at 4°C for 10 min.
10. Centrifuge lysate for 15 min at 14,000g.
11. Add sterile glycerol to lysate to attain a final concentration of 25%. Mix gently.
12. This would be the cytoplasmic extract portion. To quantitate the amount of cytoplasmic protein extract, continue to **step 20** of this section.
13. Resuspend pellet from **step 8** gently with 100X vol 1X PBS. This step rinses the extracted nuclei. Note the consistency of nuclei. Intact nuclei should have a sand-like appearance and consistency.
14. Centrifuge suspension for 1 min at 1000g to pellet nuclei (*see* **Note 9**).

15. Using a tube with graduations, approximate the packed nuclear volume (pnv). Resuspend nuclei gently in a volume of low-salt buffer "C" equal to 0.5 pnv.

16. In a dropwise fashion (*see* **Note 3**) with gentle stirring (you can use the pipet tip), add a volume of high-salt buffer "D" (*see* **Note 4**) equal to 0.5 pnv (from **step 15**).

17. Allow the nuclei to extract for 20–30 min with continuous gentle mixing (using a rotary shaker).

18. Pellet the nuclei by centrifuging for 30 min at 25,000*g*.

19. Draw off the supernatant, and place into a clean chilled tube on ice. This is the nuclear extract portion.

20. Perform Bradford protein assay on cytoplasmic extract portion from **step 12** and nuclear extract portion from **step 18**.

21. Aliquot and store extracts in –80°C at a final concentration of 1.0 µg/µL, diluting if necessary with buffer "D." Avoid multiple freeze–thaw of protein sample.

3.1.2. Extraction of Nuclear and Cytoplasmic Proteins from Suspension Cell Lines

1. Collect cells into a graduated conical tube at 300*g* for 10 min. Aspirate the supernatant and discard.

2. Using the graduations on the tube, approximate the packed cell volume (pcv). Resuspend the pelleted cells in five times the pcv with ice-cold 1X PBS.

3. Centrifuge the cells for 10 min at 300*g*. Aspirate the supernatant and discard.

4. Continue to **step 3** in **Subheading 3.1.1.**

3.1.3. Extraction of Nuclear and Cytoplasmic Proteins from Mammalian Tissue (see **Note 6**)

1. Using a sterile sharp blade, slice the tissue into small slices (*see* **Note 7**).

2. Place 0.5–2 g of sliced tissue into a sterilized glass Dounce homogenizer on ice. Approximate the tissue volume (*see* **Note 8**).

3. To the tissue, add 5X vol of hypotonic buffer "A." Swell tissue for 10 min on ice.

4. Homogenize sample with 15 gentle complete-cycle strokes using type B pestle. Transfer homogenate to a microcentrifuge tube and vortex for 10 s.

5. Continue to **step 8** in **Subheading 3.1.1.**

3.2. Preparation of Radiolabeled Synthetic Oligonucleotide, and PCR-Generated or Restriction Endonuclease-Generated DNA Fragment (see Note 10)

3.2.1. Preparation of a Radiolabeled Synthetic DNA Oligonucleotide Fragment

1. Synthesize both the sense and antisense DNA strands.

2. Adjust each oligonucleotide concentration so that both strands have an equal OD_{260}/mL value (e.g., 5 OD_{260}/mL).

Table 1
Preparation of End-Labeled Double-Stranded DNA Probe

100 pmol Double-stranded DNA	X μL[a]
10X PNK buffer	2 μL
170 mCi [γ-^{32}P]ATP (7000 Ci/mmol)	1 μL
T4 PNK enzyme	1 μL
Water (adjust volume to 20 μL)	Y μL[b]

[a]To be determined by the stock concentration of DNA.
[b]Volume of water to adjust the total volume to 20 μL.

3. Mix equal volumes of each oligonucleotide, e.g., 100 μL of sense strand plus 100 μL of antisense strand.
4. Heat for 5 min at 95°C. Cool slowly (overnight) at room temperature to anneal strands.
5. Quantitate annealed strands, and note the OD$_{260}$.
6. Annealed strands can be stored at –20°C indefinitely.
7. For 5′ end labeling, 100 pmol of double-stranded oligonucleotide are needed (*see* **Note 12**).
8. For the labeling reaction with PNK (*see* **Note 13**), use the ingredients described in **Table 1**.
9. Incubate reaction at 37°C for 90 min.
10. To purify the labeled oligonucleotide from free isotope, add 150 μL 1X TE to the labeling reaction, and apply the full volume to the Sephadex G-50 column (*see* **Note 11**).
11. Centrifuge column at 1500*g* for 5 min, and collect the eluate.
12. Subject the labeled probe for further purification using a Wizard® DNA Clean-Up Kit. Follow the protocol provided by manufacturer.
13. Spot 1 μL of the eluted probe on a DE-81 filter. Wash the filter twice with 70% ethanol, and count in liquid scintillation counter. Dilute an aliquot of the probe in water if necessary to acquire a count of 20,000 cpm/μL. The stock probe can be frozen approx 2 wk.

3.2.2. Preparation of a Radiolabeled Restriction Endonuclease-Generated DNA Fragment

1. Digest 1–4 μg of DNA (in a volume of 20 μL) with an appropriate restriction enzyme that will generate a 5′ overhang.
2. Add 20 μCi of the desired [α-^{32}P]dNTP (400–800 Ci/mmol), e.g., [α-^{32}P]UTP, 1 μL of an appropriate 5 m*M* dNTP solution, and 1–4 U of Klenow fragment to the reaction mixture.
3. Incubate the reaction for 15 min at 30°C. Stop the reaction by adding 1 μL of 0.5 *M* EDTA (20 m*M* final concentration) to the mixture or by heating the mixture for 10 min at 75°C.

Table 2
Preparation of a Nondenaturing Polyacrylamide Gel

Stock concentration	Final concentration	Volume, 30-mL gel
46% (w/v) Acrylamide	4%	2.6 mL
5X TGE (*see* **Note 16**)	1X	6 mL
Water	Adjust volume to	21.4 mL
10% AMPS[a]	1.00%	300 mL
TEMED[a]	0.10%	30 mL

[a]Add components last, just prior to pouring solution to the casting.

4. An alternative way of labeling is by 5′-end labeling using PNK (*see* **Note 10**). In this case, proceed as in **Subheading 3.2.1.**, from **step 8**. Instead of 100 pmol oligonucleotide, use 100 pmol of gel-purified restriction endonuclease-generated DNA fragment.
5. Continue through all the steps of purification and postlabeling as listed in **Subheading 3.2.1.**, **steps 9–13**.

3.2.3. Preparation of a Radiolabeled PCR-Generated DNA Fragment

1. Proceed by extracting fragment from agarose gel, using Qiaquick gel extraction kit or your favorite DNA extraction procedure.
2. Quantitate the eluted DNA fragment by measuring the OD_{260}. Calculate pmol amount (*see* **Note 12**).
3. For 5′-end labeling using PNK, *see* **Note 10**. In this case, proceed as indicated in **Subheading 3.2.1.**, from **step 8**. Instead of 100 pmol oligonucleotide, use 100 pmol of gel-purified DNA fragment from PCR.
4. Continue through all the steps of purification postlabeling as listed in **Subheading 3.2.1.**, **steps 9–13**.

3.3. Binding Assay and Electrophoretic Separation of Bound Complex or Complexes

3.3.1. Preparation of a Nondenaturing Gel

1. Assemble washed and gel-slicked glass plates following the gel apparatus assembly specifications (*see* **Note 14**).
2. Prepare a 4% nondenaturing polyacrylamide gel. The proportions described in **Table 2** are based on total gel volume of 30 mL.
3. Pour the gel mix between the plates. Do not create bubbles. Insert comb to a sufficient depth to accommodate approx 50 µL vol of sample (*see* **Note 15**).
4. Allow gel polymerization to occur for about 45–60 min.
5. Remove the comb gently. Wash the wells by gently dispensing 1X TGE buffer. Assemble the electrophoretic apparatus to its specified running position.

Table 3
Components for Binding Reaction

Content	Final concentration
Radiolabeled DNA probe	10,000–25,000 cpm
Poly dl-dC (1 mg/mL) (*see* **Note 26**)	0.1–2.0 µg
5X Binding buffer	1X
DNA binding protein[a]	5–15 µg crude extract
	5–25 ng purified protein
Water	Adjust to 20 µL

[a]Protein is added last to mixture.

6. Add 1X TGE to the lower and upper buffer chambers. Prerun the gel for 60 min at 100 V.

3.3.2. Binding Reactions

1. While the gel is prerunning, prepare the binding reactions in a 1.6-mL microcentrifuge tube. Follow the proportions indicated in **Table 3** (*see* **Note 17**).
2. When ready to perform binding reaction, thaw the DNA binding protein or extract on ice. Add the DNA binding protein to mixture, and mix gently by tapping or slow trituration.
3. Incubate the reaction mix at room temperature for 15–30 min (*see* **Note 18**).

3.3.3. Gel Electrophoresis and Analysis

1. Using a pipeter and fresh loading tips, load each binding reaction to the gel. Flank the samples in the gel by loading tracking dye alone at final dye concentration of 1X, diluting stock dye with water (*see* **Note 19**).
2. Electrophorese samples at 100–150 V. If power supply can deliver constant current, electrophorese at 30–35 mA, for the minimum required time to allow distanced separation of bound complex(es) with the free probe (time required is dependent on the size of the probe). Use tracking dye as a guide to determine the position of the free probe (*see* **Note 20**).
3. At completion of the electrophoresis, turn off the power, and disassemble the unit. Gently pry off the gel-slicked glass plate using a spatula. Remove spacers, but leave the gel on the nonslicked plate.
4. Cut blotting paper to match the size of the plates. Place blot paper centered on top of the gel, and allow all surfaces of paper to make contact to the gel.
5. Gently lift the blotting paper with the gel attached to the paper.
6. Mark orientation of the gel in order to identify sample order when viewing autoradiograph. Use a fluorescent Green Glow® dye to mark the orientation.
7. Cover the gel and paper with plastic wrap.
8. Expose to film with an intensifying screen at −80°C for 10–18 h (*see* **Note 21**).

Table 4
Components for Competition Binding Assay

Content	Final concentration
Radiolabeled DNA probe	10,000–25,000 cpm
Unlabeled specific competitor	1X, 5X, 10X, 50X
Unlabeled nonspecific competitor	1X, 5X, 10X, 50X,
Poly dl-dC (1 mg/mL)	0.1–2.0 µg
BSA (1 mg/mL)	300 µg/mL
5X Binding buffer	1X
DNA binding protein[a]	5–15 µg crude extract
	5–25 ng purified protein
Water	Adjust to 20 µL

[a]Protein is added last to mixture.

3.4. Competition Binding Assay

1. Follow the protocol for regular binding assay as described in **Subheadings 3.2.–3.3.1.** for the preparation of probe and gel.
2. Modify the binding reactions by adding a molar excess of unlabeled specific and nonspecific DNAs as indicated in **Table 4** (*see* **Note 22**).
3. Continue as indicated from **step 2** in **Subheading 3.3.2.**

3.5. Antibody Supershift Assay

1. Follow the basic protocol to prepare the probe and gel (**Subheadings 3.2.–3.3.1.**). Prepare binding reaction using **Table 6** as a guideline.
2. Antibody plus DNA–protein mixture should be incubated at room temperature for 20 min and transferred to 4°C overnight (*see* **Note 23**).
3. Perform two supershift assays, one with a specific antibody, and the other with an equivalent amount of a nonspecific antibody.
4. Continue as directed in **step 3** in **Subheading 3.3.2.**

4. Notes

See **Table 5** for a quick reference to troubleshooting.

1. The method described here is a quick and simple procedure for isolating nuclear and cytoplasmic extract from tissue or cultured cells. The core method discussed in this chapter has removed time-consuming ultracentrifugation and dialysis steps discussed by Dignam et al. *(22)*. We have observed minimal differences, if any, in the exclusion of these steps. The method discussed by Dignam et al. *(22)* requires a large amount of starting material and is a very laborious method. The method discussed here is relatively shorter and yields a sufficient amount of extract of similar quality.

Table 5
A Quick Guide to Troubleshooting Problems

Problem observed	Probable cause	Condition adjustment	Refer to
No shifted band observed	Insufficient amount of protein	Titrate the amount of protein	**Notes 16, 19, 20, 25,** and **26**
	Too much nonspecific competitor used	Titrate the amount of nonspecific competitor poly(dI-dC) or try different type	
	Specific activity of probe is too low	Use fresh ^{32}P label, and check incorporation percentage	
	Binding buffer is missing essential components	Add appropriate elements in buffer, e.g., zinc, magnesium	
	Unstable DNA-protein complex	Change buffer system and/or electrophorese at 4°C	
Bound complex does not enter the gel	Too much protein used	Titrate the amount of protein	**Note 28**
	Insufficient amount of nonspecific competitor added	Titrate the amount of nonspecific competitor poly(dI-dC) or try different type	
	Salt concentrations in binding buffer is not optimal	Titrate the amount of salt	
Multiple shifted bands produced	Protein degraded	Increase the concentration of protease inhibitor or use fresh extract	—
	Several proteins in extract recognize DNA sequence/s	Perform supershift, and competition assays to identify specific band	
Free probe not seen (in negative control lane)	Gel run was too long	Run gel for a shorter time	**Note 19**
Bands appear to smear	Incomplete gel polymerization	Allow longer polymerization time, and use fresh 10% AMPS	**Note 14, 19,** and **20**
	Incorrect voltage gradient	A voltage gradient of 10–15 V/cm is recommended, in some cases 30–35 V/cm	
	Plates have residual detergent	Clean and rinse plates thoroughly	

Table 6
Components for Antibody Supershift Assay

Content	Final concentration
Radiolabeled DNA probe	10,000–25,000 cpm
Poly dl-dC (1 mg/mL)	0.1–2.0 µg
BSA (1 mg/mL)	300 µg/mL
5X Binding buffer	1X
DNA binding protein[a]	5–15 µg crude extract
	5–25 ng purified protein
Antibody	1 µg
Water	Adjust to 20 µL

[a]Protein is added last to mixture followed immediately by addition of anti-

2. Cytoplasmic membrane disruption can be monitored at 5-min postinitiation of lysis by removing a small aliquot (5–10 µL) of suspended cells and adding 1 vol of 0.4% trypan blue. The dye is excluded from intact cells, but stains nuclei of lysed cells. Monitor until lysis is 80–90%.
3. It is important to add buffer "D" in a dropwise fashion with continuous mixing. This minimizes the formation of localized regions of high-salt concentration that can cause nuclei to lyse.
4. Add 0.5 pnv high-salt buffer "D" slowly until the nuclei pellet appears to shrink and aggregate ("ball up"). At this point, stop the addition of any extra buffer "D"; if shrinkage is not observed, add the complete 0.5 pnv of buffer "D."
5. An estimate of cell number is sufficient (e.g., confluent T80 flask of C6 cells contains approx 1×10^7 cells). It is important to note that this estimation varies from cell line to cell line. Larger cells utilize more surface area; therefore, total cell number per surface area is less compared to smaller cells. Also, some cell lines can utilize the growth surface in a different manner than others (e.g., aggregation of cells in small dense areas is observed for some cells, including HepG-2 cells). A suitable number of cells to achieve sufficient recovery of nuclear and cytoplasmic extracts is 5×10^5–10^7 cells. For animal tissue, 0.5–2 g can be used.
6. Owing to the fibrous nature of animal tissue, extraction of nuclear and cytoplasmic proteins from tissue, (e.g., brain, liver) will require a homogenization step. This step is not necessary for preparing extracts from cells in culture. To allow full access of the lysis buffer, a mechanical separation of these cells needs to be performed.
7. The slicing should be performed on a sterile clean surface (e.g, on tissue-culture dish). The dish should be placed on a dry ice block. We have found that freezing tissue makes slicing simpler. Also, frozen tissue minimizes proteolysis.
8. To approximate tissue volume, place the tissue into a graduated microcentrifuge tube prior to homogenization, and note volume. Also note the volume of liquid your Dounce homogenizer can accommodate. Calculating the approximate

postswelling volume is important, because it provides limitations on the quantity of tissue that can be processed with a single homogenization.

9. In **step 14** of **Subheading 3.1.1.**, nuclei can be further purified using a sucrose cushion and an ultracentrifugation procedure as described by Greenberg and Bender *(23)*.

10. Restriction endonuclease-generated DNA fragments are labeled by filling in a 5′ overhang with Klenow fragment of *Eschericia coli* DNA polymerase and a ^{32}P-labeled nucleotide. However, end-labeling could similarly be accomplished using polynucleotide kinase. Kinased probes should be avoided in experiments using crude protein preparations that might contain phosphatase activity.

11. This column can be prepared in the lab using 1X TE to swell the G-50 matrix, which is poured into a 1-cc syringe packed with an interfaced layer of glass wool. These columns can be stored at 4°C for 3 mo. Prepoured and equilibrated columns are also available commercially from vendors (e.g., Pharmacia Biotech).

12. To calculate pmol of double-stranded oligonucleotide, use the following conversion calculations. Use the example to calculate pmol for a double-stranded, 30-mer oligonucleotide with an $OD_{260} = 0.06$.
    a. $0.06 \times 150¶ \times 30$ μg/mL = 0.27 μg/μL
    b. 1 bp ≅ 660 g/mol
    c. ∴ 30-mer = 30 × 660 g/mol = 19,800 g/mol
    d. (0.27 g/L)/(19,800 g/mol) = 1.36×10^{-5} mol/L or 13.6 pmol/μL
    e. (13.6 pmol/μL) × 2** = 27.2 pmol/μL

13. Alternatively, the oligonucleotide can be designed to leave an overhang that can be filled in with labeled nucleotide and Klenow fragment.

14. Assure that plates do not retain any detergent residue from plate washing, since detergents will disrupt protein–DNA complex. This might cause a smearing effect on bands owing to dissociation of complex during the electrophoresis. An important step if using the TGE buffer system is to apply Gel Slick® to one of the plates and label the respective plate. TGE gels tend to stick to plates. This can cause extensive damage to the gel during the separation of gel from the plate, if the plate was not freshly slicked.

15. Very deep wells are more difficult to clean. It is important to assure that no bubbles have formed in the gel or between the individual teeth of the comb. This could cause uneven current flow and disrupt the uniformity of samples running in the gel. Comb selection and well width and number are also important factors. A well width of 5–7 mm is optimal.

16. Other commonly used buffers are Tris-acetate and Tris-borate. The Tris-glycine buffer system seems to be optimal for stability of most complexes. Concentrations of running buffers can also be varied from 0.25 to 2.0X final concentration. If using concentrations (≤0.25X), the buffer should be recirculated during elec-

---

¶Dilution factor used in quantitation, as determined with the spectrophotometer.
**There are two 5′ ends for each double-stranded oligonucleotide.

trophoresis. High-ionic-strength buffers can destabilize DNA–protein complexes. Although not a requirement, in some cases glycerol can be added to a final concentration of 2.5% (v/v) to help stabilize some interactions.

17. After adding all the components in the reaction with the exception of the protein extract, place the tube on ice, and allow the contents to chill. When protein is added, remove tube from ice to facilitate conditions for the binding. It is important to include nonspecific controls. One control would be the probe plus all contents, with the exception of the DNA binding protein being investigated.

18. Binding temperature should be optimized. For different proteins, optimal binding temperature can vary from room temperature to 37°C.

19. Bromophenol blue and/or xylene cyanol can cause some DNA–protein complexes to destabilize and dissociate. Adding tracking dye to sample should be avoided. For the tracking of the migration of free probe position, tracking dyes could be loaded in wells flanking the sample. Since bromophenol blue migrates at about the position of 70 bp and xylene cyanol at about 180 bp, it is simple to approximate the position of the free probe. Avoid introducing bubbles while loading.

20. Do not electrophorese at very high voltages that can cause excessive heat. In some cases, this may cause a smearing band effect instead of a discrete DNA–protein complex band. If performing the electrophoresis at room temperature, allow only slight warming of the plates to occur. The average time to electrophorese the samples should be 2 h. To accommodate higher voltages for faster run times, electrophorese at 4°C (cold room).

21. An alternative is to dry the gel under vacuum with heat, but this is lengthy and unnecessary. Exposing at –80°C provides equivalent film results. In most cases, it is necessary to re-expose gel for a longer or shorter time, depending on intensity of complex of interest.

22. It is important to note that the competitions are based on molar excess values. Therefore, you will need to calculate and approximate the number of fmol you are using for the binding reaction. This will then allow you to dilute the stock DNA probe to the required molar amount for use as a competitor.

23. Incubation temperatures with antibody and the duration at the specific temperature will need to be optimized for some reactions. In some cases, you will need to perform the binding reaction first for 20 min at room temperature without the antibody. The antibody is then added and (a) incubation is performed at 4°C overnight or (b) incubation is performed at room temperature for an additional 15–40 min postbinding. In cases where the DNA binding domain on the protein is located in close proximity with the epitope for antibody binding, inhibition of DNA–protein complex formation can occur. For these circumstances, incubation of the protein with increasing amounts of antibody will result in a gradual disappearance of the shifted band, therefore demonstrating specificity of the target protein to antibody.

24. If a full volume of buffer "D" is used, the final concentration of NaCl will be 210 mM. Most nuclear extractions can be attained by using this high-salt buffer. However, some nuclear extractions might need optimization by increasing or

decreasing the NaCl concentration in high-salt buffer "D." The range of salt used could be from 0.4 to 1.6 M in buffer "D." The range of the final NaCl concentration is 0.2–0.8 M. In some cases, when lysis does not occur, the optimal NaCl concentration within this stated range will need to be determined.

25. What parameters of the gel-shift assay require optimization? For novel DNA binding proteins, a variety of parameters may require optimization, including salt concentrations, incubation temperature, pH, and time of the binding reaction. Some proteins require heavy metals (such as zinc) for DNA binding. If cellular or nuclear extracts are used in the assay, care must be taken to inhibit nucleases, phosphatases, and proteases, which could interfere with the assay. Also, the DNA–protein complex may be unstable during electrophoresis; optimization of polyacrylamide concentration, buffer pH, ionic strength, and electrophoretic temperature may enhance stability.

26. Causes of high nonspecific background could be numerous. Increasing the amount of poly(dI-dC) can potentially resolve the problem. In some cases, when performing the binding assay for the first time, it is useful to titrate the amount of nonspecific competitor added, following a range of 0.2–2 μg. If a high background still persists, different types of nonspecific competitor (e.g., poly [dG-dC]) may be used. Some potential carriers, such as calf thymus DNA or salmon sperm DNA, should be avoided.

27. Several considerations are important in DNA probe choice. The size of the target DNA should be kept below 300 bp to facilitate electrophoretic separation of the unbound probe and DNA–protein complex. Double-stranded synthetic oligonucleotides as well as restriction fragments can be used as probes with the gel-shift assay. Short (approx 25-bp) oligonucleotides are preferred if the protein of interest has been previously identified, since the target binding site can be separated from potential binding sites for other factors. Larger restriction fragments can be utilized to map protein binding sites in putative promoter/enhancer regions. Subsequently, DNase I footprinting can be employed to determine the position of the binding site at the DNA sequence level.

28. Addition of protein causes probe to stick in the well. One major cause could be owing to excessive protein addition. This is easily resolved by titrating the amount of protein or nonspecific competitor (*see* **Note 26**). Another potential cause could be the salt concentration in the binding buffer. In addition, excessive salt or impurities that copurify with the DNA probe may often cause this problem. If probe size is too large, multiple interactions can occur and limit the ability of the complex(es) to enter the gel.

Acknowledgments

We acknowledge our laboratory colleagues Yong-feng Yang, Li Biao, Tarsem Moudgil, Gail Marracci, and Howard Nichols for their support in the development of this contribution. Special acknowledgments are extended to Howard Nichols for his technical assistance in conducting the experiment described in **Fig. 2**. We also thank Carol Houser for support in the preparation of the manuscript. C. A. M. is supported by NIH RR0163, HL42358, DK53462, and was a prior recipient of an American Heart Association Established Investigatorship.

References

1. Collins, S., Lohse, M. J., O'Dowd, B., Caron, M. G., and Lefkowitz, R. J. (1991) Structure and regulation of G protein-coupled receptors: The β_2-adrenergic receptor as a model. *Vitam. Horm.* **46,** 1–37.
2. Strosberg, A. D. (1995) Structural and functional diversity of β-adrenergic receptors. *Ann. NY Acad. Sci.* **757,** 253–260.
3. Hadcock, J. R. and Malbon, C. C.. (1991) Regulation of receptor expression by agonists: transcriptional and post-transcriptional controls. *Trends Neurosci.* **14,** 242–247.
4. Gudermann, F., Kalkbrenner, F., and Schultz, G. (1996) Diversity and selectivity of receptor-G protein interaction. *Ann. Rev. Pharmacol. Toxicol.* **36,** 429–459.
5. Searles, R., Midson, C. N., Nipper, V., and Machida, C. A. (1995) Transcription of the rat β_1-adrenergic receptor gene: Characterization of the transcript and identification of important sequence elements. *J. Biol. Chem.* **270,** 157–162.
6. Bahouth, S. W., Park, E. A., Beauchamp, M., Cui, X., and Malbon, C. C. (1996) Identification of a glucocorticoid repressor domain in the rat β_1-adrenergic receptor gene. *Receptors Transduc.* **6,** 141–149.
7. Tseng, Y. T., Waschek, J. A., and Padbury, J. F. (1995) Functional analyses of the 5′ flanking sequence in the ovine β_1-adrenergic receptor gene. *Biochem. Biophys. Res. Commun.* **215,** 606–612.
8. Searles, R. P., Nipper, V., and Machida, C. A. (1994). The rhesus macaque β_1-adrenergic receptor: Structure of the gene and comparison of the flanking sequences with the rat β_1-adrenergic receptor gene. *DNA Sequence* **4,** 231–241.
9. Leff, S. E., Rosenfeld, M. G., and Evans, R. M. (1986) Complex transcriptional units: Diversity in gene expression by alternative RNA processing. *Annu. Rev. Biochem.* **55,** 1091–1117.
10. Kobilka, B. K., Frielle, T., Dohlman, H. G., Bolanski, M. A., Dixon, R. A. F., Keller, P., et al. (1987) Delineation of the intronless nature of the genes for human and hamster β_2-adrenergic receptors and their putative promoter regions. *J. Biol. Chem.* **262,** 7321–7327.
11. Handy, D. E. and Gavras, H. (1992) Promoter region of the human α_{2a}-adrenergic receptor gene. *J. Biol. Chem.* **267,** 24,017–24,022.
12. Gao, B. and Kunos, G. (1994) Transcription of the rat α_{1b}-adrenergic receptor gene in liver is controlled by three promoters. *J. Biol. Chem.* **269,** 15,762–15,767.
13. Ramarao, C. S., Kincade-Denker, J. M., Perez, D. M., Gaivin, R. J., Riek, R. P., and Graham, R. M. (1992) Genomic organization and expression of the human α_{1b}-adrenergic receptor. *J. Biol. Chem.* **267,** 21,936–21,945.

14. Dignam, J. D., Lebovitz, R. M., and Roeder, R. G. (1983) Acute transcription initiation by RNA polymerase II in a soluble extract from isolated mammalian nuclei. *Nucleic Acids Res.* **11,** 1475–1489.

15. Kingston, R. E. (1993) DNA–protein interactions, in *Current Protocols in Molcular Biology*, vol. 1, (Ausebel, F., et. al., eds.), Wiley, Canada. pp. 12.0.3–12.2.11.

16. Garabedian, M. J., LaBaer, J., Liu, W.-H., and Thomas, J. R. (1993) Analysis of protein-DNA interaction, in *Gene Transcription: A Practical Approach* (Hames, B. D. and Higgins, S. J., eds.), IRL, Oxford, pp. 243–259.

17. Dent, C. L. and Latchman, C. L. (1993) The DNA mobility shift assay, in *Transcription Factors: A Practical Approach* (Latchman, D. S., ed.), IRL, Oxford, pp. 1–26.

18. de Groot, R. P. and Sassone-Corsi, P. (1993) Hormonal control of gene expression: multiplicity and versatility of cyclic adensosine $3',5'$-monophosphate-responsive nuclear regulators. *Mol. Endocrinol.* **7,** 145–153.

19. Molina, C. A., Foulkes, N. S., Lalli, E., and Sassone-Corsi, P. (1993) Inducibility and negative autoregulation of CREM: an alternative promoter directs the expression of ICER, an early response repressor. *Cell* **75,** 875–886.

20. Hosada, K., Feussner, G. K., Rydelek-Fitzgerald, L., Fishman, P., and Duman, R. S. (1994) Agonist and cyclic AMP-mediated regulation of β_1-adrenergic receptor mRNA and gene transcription in rat C6. *J. Neurochem.* **63,** 1635–1645.

21. Rydelek-Fitzgerald, L., Machida, C. A., Fishman, P. H., and Duman, R. S. (1996) Adrenergic regulation of ICER (inducible cyclic AMP early repressor) and β_1-adrenergic receptor gene expression in C6 glioma cells. *J. Neurochem.* **67,** 490–497.

22. Dignam, J. D., Lebovitz, R. M., and Roeder, R. G. (1983) Acute transcription inititaition by RNA polymerase II in a soluble extract from isolated mammalian nuclei. *Nucleic Acids Res.* **11,** 1475–1489.

23. Greenberg, M. E. and Bender, T. P. (1994) Identification of newly transcribed RNA, in *Current Protocols in Molcular Biology*, vol. 1 (Ausebel, F., et al., eds.), Wiley, Canada, pp. 4.10.6–4.10-7.

30

Determination of mRNA Stability and Characterization of Proteins Interacting with Adrenergic Receptor mRNAs

Burns C. Blaxall and J. David Port

1. Introduction

Modulation of gene expression at the level of mRNA stability has emerged as an important regulatory paradigm. Although several general models of eukaryotic mRNA decay have been described, the details of how mRNA decay occurs for G-protein-coupled receptors is minimally characterized. By contrast, considerable information is known about the regulation of proto-oncogene and cytokine/lymphokine mRNA stability (for review, *see* **ref.** *1*).

Shortly after the cloning of the hamster β_2-adrenergic receptor (AR) *(2)*, Hadcock and Malbon *(3)* were the first to demonstrate agonist-mediated downregulation of a G-protein-coupled receptor mRNA. In this particular case, the mechanism found to be responsible for decreased steady-state mRNA abundance was agonist-mediated destabilization of the mRNA *(4)*. Subsequently, a number of G-protein-coupled receptor mRNAs have been shown to be regulated at the level of mRNA stability *(5–10)*. However, it is abundantly clear for β-ARs that transcriptional as well as posttranscriptional regulatory mechanisms are important *(3,5,11–14)*.

Motifs common to proto-oncogenes and to cytokines/lymphokines that selectively target mRNA for rapid turnover or, conversely, for stabilization have been identified. These elements are most often localized to the 3′ untranslated region (UTR) and are typically composed of A + U-rich elements (AREs) that act as targets for protein binding (e.g., *15–17*). In general, it is assumed that binding of *trans*-acting factors to specific nucleotide sequences (AREs) modulates two processes: 3′ to 5′ exonucleolytic deadenylation

From: *Methods in Molecular Biology, vol. 126: Adrenergic Receptor Protocols*
Edited by: C. A. Machida © Humana Press Inc., Totowa, NJ

(i.e., removal of the poly(A) tail), and 5' decapping of the mRNA *(18,19)*. These events are typically sequential and lead to rapid degradation of many eukaryotic mRNAs. However, there are a number of exceptions to the above decay pathways. For example, several mRNAs are known to have coding region instability determinants *(20–22)*; however, there is no well-described consensus for these determinants, and information regarding specific proteins binding to these regions is just beginning to emerge. Conversely, there is significant information on non-ARE motifs that regulate mRNAs associated with iron metabolism *(1)*.

Following the work of Hadcock et al. *(4)*, we were able to describe proteins that bound to mRNAs encoding G-protein-coupled receptors *(23)*. By UV crosslinking, the $M_r \sim 35,000$ β-AR mRNA binding protein (β-ARB) was found to bind to an A + U-rich sequence within the hamster β_2-AR mRNA, but not to α_1-AR or β-globin mRNAs. Additional details regarding the interaction of β-ARB with mRNAs encoding G-protein-coupled receptors have been described by the Port and Malbon laboratories *(23–26)*. Interestingly, a 20 nt A + U-rich region of the hamster β_2-AR mRNA appears to be necessary (and is perhaps sufficient) for β-ARB binding and for agonist-mediated destabilization of the mRNA *(26)*. Unfortunately, the identity of β-ARB remains unknown *(24)*.

Noting an association between the presence of an AU-rich mRNA binding activity and the destabilization of target mRNAs, Brewer and colleagues purified and cloned a protein called "AUF1" for A + U-rich binding Factor 1 *(27)*. Further investigation of this protein indicates that there is a good correlation between the relative affinity of AUF1 for specific AU-rich mRNAs and their intrinsic stability *(28)*. That is, the "tighter" that AUF1 binds to an mRNA, the less intrinsically stable the mRNA. More recently, we have demonstrated that AUF1 binds to an A + U-rich region of human β_1-AR 3'UTR *(24)* as well as to AU-rich regions of other β-AR mRNAs *(29)*. However, the relationship between AUF1 binding, as well as that of other mRNA binding proteins, and stability of β-AR or other G-protein-coupled receptor mRNAs remains to be established. It is likely that other mRNA binding proteins, including members of the heterogenous nuclear ribonuclea protein (hnRNP) *(30–32)* and elav-like protein families *(33–36)* will be demonstrated to play a role in the regulation (stabilization/destabilization) of mRNAs encoding G-protein-coupled receptors.

2. Materials

A wide range of materials are needed to perform analysis of mRNA half-life and for analysis of mRNA/protein interaction. As with any experiment performed with RNA, care must be taken to prevent RNase contamination. Glassware and other solutions must be RNase-free and sterile. This often requires the use of diethylpyrocarbonate- (DEPC) treated water for making solutions. DEPC is a potent carcinogen.

2.1. Analysis of mRNA Half-Life

1. Cell-culture media, as needed.
2. RNA STAT 60 (Tel-Test B, Inc., Friendswood, TX).
3. Actinomycin D (Sigma, St. Louis, MO): This agent, like all DNA intercalating agents, is highly carcinogenic. Proper handling precautions and disposal are required.
4. $[\alpha\text{-}^{32}\text{P}]\text{UTP}$ (800 Ci/mmol, NEN, Life Science Products, Inc., Boston, MA).
5. Maxiscript Kit (Ambion Inc., Austin, TX) for production of riboprobe.
6. Megascript Kit and antisense probe template (Ambion Inc.) for production of low-specific activity 18S riboprobe.
7. RPA II Ribonuclease Protection Assay Kit (Ambion, Inc.).

2.2. In Vitro Transcription of Template RNA

1. Restriction endonucleases to produce linearized plasmid DNA template.
2. RNasin (Promega, Madison, WI).
3. 2.5 mM ATP, GTP, UTP, CTP (Promega).
4. $[\alpha\text{-}^{32}\text{P}]\text{UTP}$ (800 Ci/mmol, NEN).
5. m7Gppp(5′)G cap analog (New England BioLabs, Beverly, MA).
6. DEPC-treated water.
7. 5X transcription buffer (Promega).
8. RNase-free DNase (Promega).
9. Phenol/chloroform.
10. 100 and 70% Ethanol.
11. 3 M sodium acetate.

2.2.1. Preparation of S100 Cytosolic Extracts

1. Cell-culture media, as required.
2. Phosphate-buffered saline (PBS).
3. Aprotinin and leupeptin (10 mg/mL Sigma).

2.2.2. Preparation of S130 and Polysomes

1. Cell-culture media, as required (serum-free Dulbecco's Modified Eagle's Medium [DMEM]).
2. Buffer A: 10 mM Tris-HCl, pH 7.6, 1 mM potassium acetate, 1.5 mM magnesium acetate, 2 mM dithiothreitol (DTT).
3. Buffer B: 30% (w/v) sucrose, 10 mM Tris-HCl, pH 7.6, 1 mM potassium acetate, 1.5 mM magnesium acetate, 2 mM DTT.

2.3. UV Crosslinking

1. Sterile 96-well microtiter plate (Falcon, Becton Dickinson Labware, Bedford, MA).
2. Cytosolic protein, as required.

3. ^{32}P-labeled target RNA, as required.
4. Yeast tRNA.
5. DTT.
6. Sodium heparin (Sigma).
7. RNasin (Promega).

2.4. Nondenaturing Gel-Shift Analysis

1. Recombinant protein, as required.
2. ^{32}P-labeled RNA probe, as required.
3. Reaction buffer: 10 mM Tris-HCl, pH 7.5, 5 mM magnesium acetate, 100 mM potassium acetate, 2 mM DTT, 0.1 mM spermine, 0.1 mg/mL bovine serum albumin (BSA), 8 U rRNasin, 0.2 mg/mL yeast tRNA, 0.1 mg/mL poly-C, 5 mg/mL sodium heparin.

3. Methods

3.1. Methods for Measuring mRNA Half-Life

There are several ways to measure the half-life ($t_{1/2}$) or the rate of decay of target mRNAs. Each method has its relative strengths and weaknesses. All require isolation of highly purified, intact total cellular or poly(A)-selected RNA, accurate quantification of concentration, and an accurate measure of mRNA steady-state abundance of target and reference RNAs. Methods for isolation of RNA will not be discussed.

Prior to measuring half-life, pilot studies should be performed to determine relative mRNA abundance and to obtain a "ballpark" measure of rate of decay. For G-protein-coupled receptors, half-lives are commonly on the order of several hours *(4,5,10,37,38)*. However, depending on the specific mRNA in question, and on cell-culture and expression conditions, half-life may be as short as 30 min *(37,39)*.

Following quantification, equal amounts of total RNA from each time-point are used for determination of mRNA abundance by either Northern blot analysis or ribonuclease protection assay (RPA) using a sequence-specific antisense riboprobe. The relative abundance of target mRNA should be normalized to that of a stable, nonregulated control RNA, e.g., 18S rRNA or reduced glyceraldehyde-phosphate dehydrogenase (GAPDH). For each sample, the ratio of signal for target RNA/internal control RNA is converted to a percentage (at $t = 0$ to a value of 100%). The log percent ratio is plotted vs time, and mRNA half-life is based on the linear regression equation. The general formula for measurement of mRNA decay is: $\ln([mRNA]_t/[mRNA]0) = -k_{obs}t$ *(40)*. *See also* **Note 1**.

3.2. Chemical Inhibition of Transcription

Transcriptional inhibition to prevent *de novo* synthesis of mRNA is a commonly employed method used to determine mRNA half-life. Cells in culture are simply exposed to one of several inhibitors ($t = 0$) and RNA isolated from the cells at multiple time-points (0–24 h) following transcriptional inhibition. Readily available transcriptional inhibitors include: actinomycin D, 5,6-dichloro-1-β-D-ribofuranosylbenzimidazole (DRB), or α-amanitin. Each inhibitor has certain limitations, and all, over time, are cell-lethal. mRNA half-lives determined by means of chemical inhibition of transcription can be considerably different than those determined in the same system by other strategies, such as pulse-chase (whole-cell radioactive labeling followed by washout) or transient promoter activation. Thus, although chemical inhibition of transcription can be a rapid, efficient way to measure mRNA half-life, the above limitations must be considered prior to use of these agents. As an example, inhibition of transcription may affect the abundance of proteins that may directly or indirectly stabilize or destabilize the mRNA independent of the effect being investigated.

3.3. Transcriptional Pulse

Transient promoter activation (transcriptional pulse) provides a potentially less toxic means to measure mRNA half-life. One method involves placing the gene of interest downstream of the serum-inducible c-*fos* promoter *(41)*. Cells expressing the gene, either transiently or stably, are incubated in low serum (0.5%) to render them quiescent. Subsequently, the cells are exposed to 10–15% serum which rapidly and transiently (15–30 min) induces the c-*fos* promoter. Following withdrawal of serum stimulus, the half-life of the mRNA can be determined as described above *(41,42)*. Two caveats must be recognized. First, mRNAs with a very short half-life (<1 h) may be difficult to analyze accurately owing to the somewhat variable length of induction by the c-*fos* promoter. Second, serum starvation in and of itself may significantly alter mRNA metabolism, and is therefore only physiologically relevant to cells undergoing G_0 to G_1 transition.

A second method of transient transcriptional induction involves the use of one of several recently described tetracycline (Tet) regulatory promoter systems. Both "Tet-on" and "Tet-off" binary systems are commercially available (Clontech). Recently, Xu et al. *(18)* have demonstrated the ability to provide a transient burst of transcription of target mRNAs using this system. This gene-specific transcriptional pulse appears to provide accurate mRNA half-life measurements. Although difficult to develop, systems such as these may ultimately be the least toxic and most accurate method of studying mRNA stability.

3.3.1. Detection of mRNA Binding Proteins by UV Crosslinking

UV crosslinking of labeled, in vitro transcribed mRNAs to cytosolic proteins is a highly useful general method for determining the molecular weight and nucleotide specificity of proteins binding to mRNAs of interest (e.g., *17,20,23,24*). One limitation of UV-crosslinking is that detection of protein and mRNA interaction by this method does not address the biologically relevant issue of affinity. A number of protein sources from cultured cells or tissues may be used. Preparation of these protein sources is described below. (*See also* **Notes 2–4**).

3.3.2. Preparation of Cytosolic (S100) Extracts

1. Following drug treatment (if indicated), cells are washed in PBS, gently pelleted and resuspended in PBS, transferred to a sterile ultracentrifuge tube, and again gently pelleted. Excess buffer is removed, and 5 µL of each protease inhibitor aprotinin (10 mg/mL) and leupeptin (10 mg/mL) are added.
2. Cells are centrifuged at 100,000g for 90 min at 4°C. The resulting supernatants are transferred to Eppendorf tubes and placed on ice for immediate use, or stored at –80°C for future use (*23,24*). Protein concentration can be measured by any appropriate method.

An alternate preparation has been described by Dignam et al. (*43*).

1. Cells are gently pelleted (1000g for 10 min) and washed with cold PBS.
2. Cells are then lysed with a Dounce homogenizer in buffer A (10 m*M* HEPES NaOH ([pH 7.9 at 25°C], 1.5 m*M* MgCl$_2$, 10 m*M* KCl, 0.5 m*M* DTT, 0.5 m*M* phenylmethylsulfonyl fluoride [PMSF]), and nuclei are pelleted by centrifugation (1000g, 10 min). This step generates a nuclear extract that can be placed on ice for immediate use or stored at –80°C for future use.
3. The cytoplasmic fraction is mixed with 0.11 vol buffer B (0.3 m*M* HEPES, 1.4 *M* KCl, 30 m*M* MgCl$_2$) and centrifuged at 100,000g for 1 h. The supernatant is dialyzed against 20 vol of buffer C (20 m*M* HEPES, 25% [v/v] glycerol, 0.42 *M* KCl, 1.5 m*M* MgCl$_2$, 0.5 m*M* DTT, 0.5 m*M* PMSF). Following dialysis, the precipitate is pelleted (25,000g, 20 min), resuspended and aliquoted, and stored at –80°C for future use.

3.3.3. Preparation of Cytosolic (S130) and Polysome Fractions

These preparations were originally described by Brewer and Ross (*44*) and are a refinement beyond S100 preparations in that the concentration of RNA binding proteins is increased. However, this is coincident with increased RNase activities.

1. Cells are washed twice with cold, serum-free medium DMEM, and resuspended in 3.5 mL of buffer A (10 m*M* Tris-HCl, pH 7.6, 1 m*M* potassium acetate, 1.5 m*M* magnesium acetate, 2 m*M* DTT).
2. Cells (~2 × 10^8) are homogenized with a tightly fitting Dounce homogenizer (20 strokes) and centrifuged at 12,000g for 10 min at 4°C to pellet the nuclei.

3. The supernatant is removed and layered gently over a 1.5-mL cushion of buffer B (30% [w/v] sucrose, 10 mM Tris-HCl, pH 7.6, 1 mM potassium acetate, 1.5 mM magnesium acetate, 2 mM DTT) in a SW60 Ultraclear centrifuge tube (Beckman Coulter, Inc., Fullerton, CA).
4. The tube is centrifuged at 130,000g in a SW60 rotor for 2.5 h at 4°C.
5. The S130 supernatant is removed from above the S130/sucrose interface, aliquoted, and stored at –80°C.
6. To recover the polysome fraction, the sucrose cushion is aspirated off, and the tubes allowed to drain while kept at 4°C. The polysome fraction is then rinsed twice with buffer A, resuspended in 0.5 mL buffer A, aliquoted, and stored at –80°C. For purposes of quantification, a conversion factor of 11 A_{260} U/mg protein is used for the polysome fraction. The pellet is often difficult to resuspend. We typically use a pipet tip that has had its tip cut off sterilely.

3.4. In Vitro Transcription

1. Appropriately linearized, purified plasmid DNA encoding the RNA of interest is in vitro transcribed to produce capped, uniformly labeled mRNAs.
2. Messages are transcribed in the presence of RNasin (Promega), all four deoxyribonucleoside triposphates (dNTPs), and the radiolabel [α-^{32}P]-UTP (800 Ci/mmol, New Life Science Products). Cotranscriptional capping is performed by using the cap analog m^7(5′)Gppp(5′)G (New England BioLabs) at a concentration that is 10-fold in excess of the concentration of GTP.
3. RNase-free DNase is added to the mixture to remove template DNA, and the labeled transcript is extracted with phenol, then chloroform, and precipitated with 2.5 V ice-cold ethanol and 0.1 V 3 M sodium acetate.
4. Labeled transcript is reconstituted in RNase-free water and is used immediately (important for long RNAs [>500 nt]) or stored at –80°C until used within 24 h of synthesis. Size and integrity of the transcripts can be verified by agarose/formaldehyde gel electrophoresis.
5. The integrity and specific activity of the labeled RNA should be determined using standard methods.

3.5. UV Crosslinking

1. Radiolabeled mRNAs (1–4 × 10^6 cpm) are added to a mixture containing the S100 cytosolic fraction (30–100 µg total protein) or other protein source, 5 µg yeast tRNA, 4 mM DTT, 5 µg heparin, and 65 U of RNasin in a total volume of 50 µL.
2. Individual combinations of protein preparations and radiolabeled mRNAs are conveniently combined in wells of a 96-well microtiter plate and allowed to mix for 10 min at 22°C.
3. Samples are placed on ice and exposed briefly (~3 min) to short-wave (254 nm) UV radiation using a Stratalinker UV source (Stratagene, La Jolla, CA).
4. Noncrosslinked mRNA is digested away with a mixture of RNase A and RNase T1 at 37°C for 30 min.

5. Samples are solubilized in 50 μL (1:1) of Laemmli loading buffer for 10 min at 67°C and loaded onto a 10% sodium dodecyl sulfate- (SDS) polyacrylamide gel (5% stack) and electrophoresed for a total of 110 mA h. Gels are then stained, destained, and exposed for autoradiography for 1–3 d.

3.6. Nondenaturing Gel-Shift Analysis of mRNA/Protein Interaction

Nondenaturing gel-shift analysis of target RNAs with purified, recombinant proteins can be performed as described by DeMaria and Brewer *(28,45)* and based on the method of Carey *(46)*. This method is highly useful for determining the apparent affinity (K_d) of an mRNA binding protein for mRNAs. **Figure 1** is an example of a gel-shift experiment using the 3'UTR of granulocyte-macrophage colony-stimulating factor (GM-CSF) and purified recombinant His_6-p37AUF1 protein. *See also* **Notes 5** and **6**.

1. Increasing amounts of purified, recombinant protein (1–1000 n*M*; depending on ~K_d) are incubated with 1 fmol of ^{32}P-labeled RNA probe in a final volume of 10 μL. The final reaction contains 10 m*M* Tris-HCl, pH 7.5, 5 m*M* Mg acetate, 100 m*M* K acetate, 2 m*M* DTT, 0.1 m*M* spermine, 0.1 mg/mL BSA, 8 U RNasin, 0.2 mg/mL yeast tRNA, 0.1 mg/mL poly-C, and 5 mg/mL heparin.
2. Reaction mixtures are incubated on ice for 10 min and resolved on a nondenaturing 6% polyacrylamide gel (depending on the size of the target RNA). Gels are prerun for 30 min at 13 V/cm followed by a sample run of 2–3 h; gels are dried and subjected to autoradiography.
3. Percentage of remaining probe is plotted vs concentration of protein and nonlinear curve fitting is performed using a computer software, program such as GraphPad InPlot 4.1 (GraphPad Software, San Diego, CA). The apparent K_d of the protein for RNA is estimated as the concentration of protein at which ~50% of the RNA was bound (*see* **Subheading 4.**).

4. Notes

1. Measurement of mRNA half-life: Another means of nonchemical inhibition of transcription is the use of cells, such as tsAF8s, with a "ts" mutation in pol II activity. However, for short-lived mRNAs, the time lag of temperature shift and, thus, pol II inhibition can be substantial *(47)*.
2. UV crosslinking: Using increasing concentrations of nonlabeled competitor mRNAs in crosslinking assays may be used to help determine the specificity of mRNA binding proteins *(23)*.
3. UV crosslinking: Inclusion of thio-UTP along with [α-^{32}P]-CTP for radiolabeling may potentially increase crosslinking activity several-fold, and can be useful in studying mRNA binding proteins with low affinity or specificity for the target mRNA *(48)*.
4. UV crosslinking: Following UV crosslinking, one may identify or "rule out"

0 nM--750 nM→

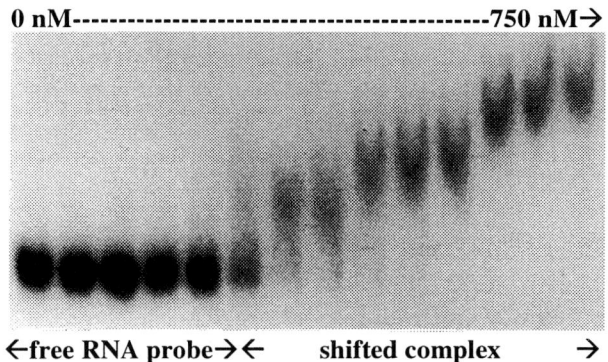

←free RNA probe→ ←   shifted complex   →

Fig. 1. Binding of purified, recombinant His_6-p37AUF1 protein to the 3'UTR of GM-CSF. Increasing amount of purified, recombinant AUF1 protein was added to a fixed amount of ^{32}P-labeled, in vitro transcribed mRNA encoding the 3'UTR of GM-CSF. Complex formation was visualized by nondenaturing PAGE followed by autoradiography.

specific mRNA binding proteins by immunodetection methods. For example, we wished to determine if β-ARB and AUF1 were related. By transferring the proteins of a UV-crosslinking reaction and performing a Western blot with an anti-AUF1 antibody, comparison of the crosslinking reaction autoradiogram to the Western blot revealed that AUF1 and β-ARB are unique proteins *(24)*.

5. Nondenaturing gel-shift analysis: Determination of an accurate K_d can be complicated by the fact that the affinity of mRNA binding protein can be affected by phosphorylation and by dimerization/multimerization. Thus, a protein may not bind with high affinity unless protein/protein interaction occurs *(45)*. Another issue is that the larger the segment of RNA, the more difficult it can be to resolve mRNA/protein complexes.

6. Nondenaturing gel-shift analysis: An alternative method for determining mRNA binding protein affinity for mRNA has been demonstrated using a filter assay *(49)*. The incubation of protein and probe is much the same as for gel-shift assays. However, the reactions are then placed on filters and washed. Scintillation analysis of remaining radioactivity on the filter is used to determine the K_d, where K_d is equal to the protein concentration at which 50% of maximal mRNA binding occurs. This method can be quite effective. However, it may mask some of the more complex mRNA/protein interactions (such as dimerization), which can be revealed by nondenaturing polyacrylamide gel electrophoresis (PAGE).

Acknowledgments

This work was supported in part by United States Public Health Services Grants HL51239 (J. D. P.), from the National Institutes of Health, and by Grant CWGS-46-97 from the Colorado Affiliate of the American Heart Association

(J. D. P.). B. C. B. was supported in part by a University of Colorado Cancer Center Seed Grant awarded to J. D. P., by a PhARMA Foundation predoctoral fellowship, and by Training Grant GM07635 from the National Institutes of Health.

References

1. Ross, J. (1995) mRNA stability in mammalian cells. *Microbiol. Rev.* **59,** 423–450.
2. Dixon, R., Kobilka, B., Strader, C., Benovic, J., Dohlman, H., Frielle, T., et al. (1986) Cloning of the gene and cDNA for mammalian β-adrenergic receptor and homology with rhodopsin. *Nature* **321,** 75–79.
3. Hadcock, J. R. and Malbon, C. C. (1988) Down-regulation of β-adrenergic receptors: agonist-induced reduction in receptor mRNA levels. *Proc. Natl. Acad. Sci. USA* **85,** 5021–5025.
4. Hadcock, J. R., Wang, H-Y., and Malbon, C. C. (1989) Agonist-induced destabilization of β-adrenergic receptor mRNA. Attenuation of glucocorticoid-induced up-regulation of β-adrenergic receptors. *J. Biol. Chem.* **264,** 19,928–19,933.
5. Danner, S., Frank, M., and Lohse, M. J. (1998) Agonist regulation of human β2-adrenergic receptor mRNA stability occurs via a specific AU-rich element. *J. Biol. Chem.* **273,** 3223–3229.
6. Devedjian, J-C., Fargues, M., Denis-Pouxiel, C., Daviaud, D., Prats, H., and Paris, H. (1991) Regulation of the α_{2a}-adrenergic receptor in HT29 cell line. *J. Biol. Chem.* **266,** 14359–14366.
7. Ferry, R., Unsworth, C., Artymyshyn, R., and Molinoff, P. (1994) Regulation of mRNA encoding 5-HT2a receptors in P11 cells through a post-translational mechanism requiring activation of protein kinase C. *J. Biol. Chem.* **269,** 31,850–31,857.
8. Izzo, N., Tulenko, T., and Colucci, W. (1994) Phorbol esters and norepinephrine destabilize α1b-adrenergic receptor mRNA in vascular smooth muscle cells. *J. Biol. Chem.* **269,** 1705–1710.
9. Lee, N., Earle-Hughes, J., and Frazer, C. (1994) Agonist-mediated destabilization of m1 muscarinic acetylcholine receptor mRNA. *J. Biol. Chem.* **269,** 4291–4298.
10. Mitchusson, K. D., Blaxall, B. C., Pende, A., and Port, J. D. (1998) Agonist-mediated destabilization of human β_1-adrenergic receptor mRNA; role of the 3′ untranslated region. *Biochem. Biophys. Res. Commun.* **252,** 357–362.
11. Collins, S., Caron, M., and Lefkowitz, R. (1991) Regulation of adrenergic receptor responsiveness through modulation of receptor gene expression. *Ann. Rev. Physiol.* **53,** 497–508.
12. Fitzgerald, L. R., Li, Z., Machida, C. A., Fishman, P. H., and Duman, R. S. (1996) Adrenergic regulation of ICER (inducible cyclic AMP early repressor) and β_1-adrenergic receptor gene expression in C6 glioma cells. *J. Neurochem.* **67,** 490–497.

13. Hosoda, K., Feussner, G., Rydelek-Fitzgerald, L., Fishman, P., and Duman, R. (1994) Agonist and cyclic AMP-mediated regulation of β_1-adrenergic receptor mRNA and gene transcription in rat C6 glioma cells. *J. Neurochem.* **63,** 1635–1645.

14. Hadcock, J. R. and Malbon, C. C. (1988) Regulation of β-adrenergic receptors by "permissive" hormones: glucocorticoids increase steady-state levels of receptor mRNA. *Proc. Natl. Acad. Sci. USA* **85,** 8415–8419.

15. Bohjanen, P., Petryniack, B., June, C., Thompson, C., and Lindsten, T. (1992) AU RNA-binding factors differ in their binding specificity's and affinities. *J. Biol. Chem.* **267,** 6302–6309.

16. Malter, J. (1989) Identification of an AUUUA-specific messenger RNA binding protein. *Science* **246,** 664–666.

17. Vakalopoulou, E., Schaack, J., and Shenk, T. (1991) A 32-kilodalton protein binds to AU-rich domains in the 3′ untranslated regions of rapidly degraded mRNAs. *Mol. Cell. Biol.* **11,** 3355–3363.

18. Xu, N., Chen, Y.-A., and Shyu, A-B. (1997) Modulation of the fate of cytoplasmic mRNA by AU-rich elements: key sequence features controlling mRNA deadenylation and decay. *Mol. Cell. Biol.* **17,** 4611–4621.

19. Couttet, P., Fromont-Racine, M., Steel, D., Pictet, R., and Grange, T. (1997) Messenger RNA deadenylation precedes decapping in mammalian cells. *Proc. Natl. Acad. Sci. USA* **94,** 5628–5633.

20. Prokipcak, R., Herrick, D., and Ross, J. (1994) Purification and properties of a protein that binds to the C-terminal coding region of human c-myc mRNA. *J. Biol. Chem.* **269,** 9261–9269.

21. Bernstein, P. L., Herrick, D. J., Prokipcak, R. D., and Ross, J. (1992) Control of c-myc mRNA half-life in vitro by a protein capable of binding to a coding region stability determinant. *Genes Dev.* **6,** 642–654.

22. Wisdom, R. and Lee, W. (1991) The protein-coding region of c-myc mRNA contains a sequence that specifies rapid mRNA turnover and induction by protein synthesis inhibitors. *Genes Dev.* **5,** 232–243.

23. Port, J. D., Huang, L-Y., and Malbon, C. C. (1992) β-adrenergic agonists that down-regulate receptor mRNA up-regulate a Mr 35,000 protein(s) that selectively binds to β-adrenergic receptor mRNAs. *J. Biol. Chem.* **267,** 24,103–24,108.

24. Pende, A., Tremmel, K. D., DeMaria, C. T., Blaxall, B. C., Minobe, W., Sherman, J. A., et al. (1996) Regulation of the mRNA-binding protein AUF1 by activation of the β-adrenergic receptor signal transduction pathway. *J. Biol. Chem.* **271,** 8493–8501.

25. Tholanikunnel, B. G., Granneman, J. G., and Malbon, C. C. (1995) The Mr 35,000 β-adrenergic receptor mRNA-binding protein binds transcripts of G-protein-linked receptors which undergo agonist-induced destabilization. *J. Biol. Chem.* **270,** 12,787–12,793.

26. Tholanikunnel, B. G. and Malbon, C. C. (1997) A 20-nucleotide (A+U)-rich element of the β_2-adrenergic receptor (β_2-AR) mRNA mediates binding to the

β_2-AR-binding protein and is obligate for agonist-induced destabilization of receptor mRNA. *J. Biol. Chem.* **272,** 11,471–11,478.

27. Zhang, W., Wagner, B., Ehrenman, K., Schaefer, A., DeMaria, C., Crater, D., et al. (1993) Purification, characterization, and cDNA cloning of an AU-rich element RNA-binding protein, AUF1. *Mol. Cell. Biol.* **13,** 7652–7665.

28. DeMaria, C. and Brewer, G. (1996) AUF1 binding affinity to A+U-rich elements correlates with rapid mRNA degradation. *J. Biol. Chem.* **271,** 12,179–12,184.

29. Blaxall, B. C., Pende, A., Brewer, G., and Port, J. D. (1996) Differential binding of p37AUF1 to beta-adrenergic receptor mRNAs. *Mol. Biol. Cell.* **7,** 300A.

30. Hamilton, B., Nagy, E., Malter, J., Arrick, B., and Rigby, W. (1993) Association of heterogeneous nuclear ribonucleoprotein A1 and C proteins with reiterated AUUUA sequences. *J. Biol. Chem.* **268,** 8881–8887.

31. Kiledjian, M., Novick, K., DeMaria, C. T., Brewer, G., and Xu, Q. (1997) Identification of AUF1/hnRNP-D as a component of the α-globin mRNA stability complex. *Mol. Cell. Biol.* **17,** 4870–4876.

32. Swanson, M. and Dreyfuss, G. (1988) Classification and purification of proteins of heterogeneous nuclear ribonucleoprotein particles by RNA-binding specificities. *Mol. Cell. Biol.* **8,** 2237–2241.

33. Levine, T., Gao, F., King, P., Andrews, L., and Keene, J. C. (1993) Hel-N1: an autoimmune RNA-binding protein with specificity of 3′ uridylate-rich untranslated regions of growth factor mRNAs. *Mol. Cell. Biol.* **13,** 3494–3504.

34. Ma, W.-J., Cheng, S., Campbell, C., Wright, A., and Furneaux, H. (1996) Cloning and characterization of HuR, a ubiquitously expressed elav-like protein. *J. Biol. Chem.* **271,** 8144–8151.

35. Jain, R. G., Andrews, L. G., McGowan, K. M., Pekala, P. H., and Keene, J. D. (1997) Ectopic expression of Hel-N1, an RNA-binding protein, increases glucose transporter (GLUT1) expression in 3T3-L1 adipocytes. *Mol. Cell. Biol.* **17,** 954–962.

36. Levy, N. S., Chung, S., Furneaux, H., and Levy, A. P. (1998) Hypoxic stabilization of vasular endothelial growth factor mRNA by the RNA-binding protein HuR. *J. Biol. Chem.* **273,** 6417–6423.

37. Danner, S. and Lohse, M. J. (1997) Cell type-specific regulation of β_2-adrenoceptor mRNA by agonists. *Eur. J. Pharmacol.* **331,** 73–78.

38. Nickenig, G. and Murphy, T. J. (1996) Enhanced angiotensin receptor type 1 mRNA degradation and induction of polyribosomal mRNA binding proteins by angiotensin II in vascular smooth muscle cells. *Mol. Pharmacol.* **50,** 743–751.

39. Li, Z., Vaidya, V. A., Alvaro, J. D., Iredale, P. A., Hsu, R., Hoffman, G., et al. (1998) Protein kinase C-mediated down-regulation of β_1-adrenergic receptor gene expression in rat C6 glioma cells. *Mol. Pharm.* **54,** 14–21.

40. Belasco, J. and Brawerman, G. (1993) *Control of Messenger RNA Stability*. Academic, New York.

41. Lagnado, C. A., Brown, C. Y., and Goodall, G. J. (1994) AUUUA is not sufficient to promote poly(A) shortening and degradation of a mRNA: the functional sequence within AU-rich elements may be UUAUUUA(U/A)(U/A). *Mol. Cell. Biol.* **14,** 7984–7995.

42. Shyu, A.-B., Greenberg, M., and Belasco, J. (1989) The c-*fos* transcript is targeted for rapid decay by two distinct mRNA degradation pathways. *Genes Dev.* **3,** 60–72.

43. Dignam, J. D., Lebovitz, R. M., and Roeder, R. G. (1983) Accurate transcription initiation by RNA polymerase II in a soluble extract from isolated mammalian nuclei. *Nucleic Acids Res.* **11,** 1475–1489.

44. Brewer, G. and Ross, J. (1990) Messenger RNA turnover in cell-free extracts. *Methods Enzymol.* **181,** 202–209.

45. DeMaria, C. T., Sun, Y., Long, L., Wagner, B. J., and Brewer, G. (1997) Structural determinants in AUF1 for high affinity binding to A+U-rich elements. *J. Biol. Chem.* **272,** 27,635–27,643.

46. Carey J. (1991) Gel retardation. *Methods Enzymol.* **208,** 102–107.

47. Savante-Bhonsale, S. and Cleveland, D. W. (1992) Evidence for instability of mRNAs containing AUUUA motifs mediated through translation-dependent assembly of a >20S degradation complex. *Genes Dev.* **6,** 1927–1939.

48. Ali, N. and Siddiqui, A. (1997) The La antigen binds 5′ noncoding region of the hepatitis C virus RNA in the context of the initiator AUG codon and stimulates internal ribosome entry site-mediated translation. *Proc. Natl. Acad. Sci. USA* **94,** 2249–2254.

49. Chung, S., Eckrich, M., Perrone-Bizzozero, N., Kohn, D. T., and Furneaux, H. (1997) The Elav-like proteins bind to a conserved regulatory element in the 3′-untranslated region of GAP-43 mRNA. *J. Biol. Chem.* **272,** 6593–6598.

VI

MICROSCOPY ANALYSIS

31

Use of Immunohistochemistry and Confocal Microscopy in the Detection of Adrenergic Receptors

Ruth L. Stornetta

1. Introduction

With the advent of recombinant protein systems to manufacture large volumes of cloned receptors, antibodies against the α-adrenergic receptors (α-ARs) have become available *(1)*. These antibodies are useful tools for the immunocytochemical detection of cells containing the adrenergic receptors and can be used with light-level microscopy. In order to determine whether the receptor protein thus detected is on the cell surface or is actually contained inside the cell, the technique of confocal microscopy can be applied (for example, *see* **ref.** *2*). This powerful technique finds a good application in this case, because double-labeling at the light level can also be done with relative ease (e.g, *see 3,4*) for the location of adrenergic receptors in catecholaminergic or serotonergic neurons) compared to asking the same question at the electron microscopic level. The use of fluorophore-tagged markers for the detection of bound primary antibodies is central to this method, and with the commercially available variety of these reagents, this becomes a relatively simple procedure. The procedure consists of a standard immunohistochemical protocol using the primary antibody against the adrenergic receptor followed by either fluorophore-tagged secondary antibody or indirect-tagged secondary antibody (i.e., biotin-tagged secondary antibody subsequently visualized with an avidin-tagged fluorophore). The bound complex including the fluorophore can then be visualized on laser excitation of the fluorophore at the appropriate wavelength and the subsequent emission of energy at a wavelength that can be detected by a confocal microscope. The labeled material is examined under a confocal microscope at different tissue depths, and digital images are taken

From: *Methods in Molecular Biology, vol. 126: Adrenergic Receptor Protocols*
Edited by: C. A. Machida © Humana Press Inc., Totowa, NJ

that can be used to reconstruct the appearance and disappearance of labeled material at different focal planes through the cell soma. This confocal method of optical "sectioning" allows the localization of immunocytochemical reaction product within the cell soma.

In order to detect the cell soma, either a general neuronal marker, such as microtubule-associated protein, may be used or a more specific marker for cells of interest (i.e., tyrosine hydroxylase) may be employed. The investigator might instead wish to use an antibody against a marker for an intracellular compartment. Whatever antibody is used as a second label should be raised in a different species from the α-adrenergic receptor antibody. The following method will describe a general double-labeling procedure for α_{2A}-AR combined with tyrosine hydroxylase (TH) examined using an argon-krypton confocal laser microscope.

2. Materials
2.1. Tissue Preparation

1. Heparin sodium USP injectable solution, 1000 USP U/mL.
2. 0.9% (w/v) sodium chloride in deionized water (physiological saline) with pH adjusted to 7.4 with the addition of a few drops of concentrated (0.5 M) sodium phosphate buffer.
3. 4% (w/v) Paraformaldehyde in 100 mM phosphate buffer. Make this solution by heating 200 mL of deionized water to 60°C. Add 20 g paraformaldehyde (*see* **Note 1**), and stir while adding concentrated sodium hydroxide by drops until the solution clears. Filter the solution. Add 100 mL of a prefiltered stock solution of 0.5 M sodium phosphate buffer, pH 7.4. and bring to a total volume of 500 mL with deionized water.
4. 20-mL scintillation vials are handy and inexpensive containers for storing the postfixing brain.
5. 100 mM sodium phosphate buffer, pH 7.4.

2.2. Tissue Processing

1. Tris-saline-azide (TSA) buffer: 50 mM Tris-HCl, pH 7.4, 0.9% (w/v) sodium chloride, 0.01%(w/v) sodium azide.
2. Tris-saline (TS) buffer: 50 mM Tris-HCl, pH 7.4, 0.9% (w/v) sodium chloride.
3. Blocking solution: 3% (w/v) normal goat serum in TSA.
4. Primary antibody against α adrenergic receptor.
5. Primary antibody against second marker in a different species (e.g., mouse TH, Chemicon, Temecula, CA).
6. Glutathione S-transferase (GST) (Santa Cruz Biotechnology, Santa Cruz, CA).
7. Biotin-tagged secondary antibody for rabbit or the species of the α adrenergic receptor antibody used (e.g., biotinylated antirabbit IgG [H + L]), Vector Laboratories, Burlingame, CA).

8. Direct-tagged secondary antibody, such as Alexa 488, for the species of the second marker (e.g., Alexa 488 antimouse IgG, Molecular Probes, Eugene, OR).
9. Avidin-"Neutralite"-Texas red (Molecular Probes).
10. Krystalon mounting medium (EMS Diagnostic Systems, Gibbstown, NJ).
11. Microscope slides and cover slips.

3. Methods

3.1. Tissue Preparation

1. Anesthetize animal, and prepare for transcardiac perfusion. Inject 500 U heparin directly into the heart. Perfuse animal first with a rinse of phosphate-buffered physiological saline immediately followed by 500 mL of freshly prepared 4% paraformaldehyde in 100 mM phosphate buffer.
2. Remove brain and postfix for 30–60 min (e.g., α_2-AR are sensitive to postfix time) in 4% paraformaldehyde at 4°C. If brains will be frozen and cut on a freezing microtome or cryostat, brains must subsequently be cryoprotected by placing at 4°C in a container (a scintillation vial may be used) of 20% sucrose (w/v) in 100 mM phosphate buffer until the brain sinks to the bottom of the container. This may take from 6 to 24 h.
3. Cut brain sections 30–50 μm on a vibratome, freezing microtome or cryostat into 100 mM phosphate buffer. Use a size 1, 0, or 00 paintbrush to pick up the brain section and place free-floating into the buffer solution. Multiwell trays are useful for keeping the tissue sections in order for further processing (*see* **Note 2**).

3.2. Tissue Processing for Immunocytochemical Detection of α_{2A}-Adrenergic Receptor and TH

1. Rinse sections free-floating 3 × 5 min in 50 mM TSA.
2. Incubate in blocking solution (e.g., 3% goat serum in TSA) for 30 min (*see* **Note 3**).
3. Rinse sections free-floating 3 × 5 min in TSA.
4. Incubate in primary antibody against α_{2A}-adrenergic receptor at 7.5 μg/mL prepared in 50 mM TSA and 1% goat (or other) serum (to block nonspecific binding). Time of incubation is 2 d at 4°C (*see* **Note 4**).
5. Rinse sections free-floating 3 × 5 min in 1% goat serum in 50 mM TS.
6. Incubate sections for 45 min in biotinylated antirabbit IgG 50 μL/10 mL (*see* **Note 5**).
7. Rinse sections free-floating 3 × 5 min in 50 mM TS.
8. Incubate sections for 30 min in avidin-Texas red at 1:150 (*see* **Note 5**).
9. Rinse sections free-floating 3 × 5 min in 50 mM TS.
10. Incubate sections for 3–4 h in primary antibody against TH (mouse TH, Chemicon) at 1:1000 concentration (10 μL/10 mL).
11. Rinse sections free-floating 3 × 5 min in 50 mM TS.
12. Incubate sections for 60 min in Alexa 488 antimouse IgG at 1:150 (*see* **Note 5**).
13. Rinse sections free-floating 3 × 5 min in 50 mM TS.

14. Mount sections onto microscope slides by sliding sections (with a size 1, 0, or 00 paintbrush) out of TS solution in a Petri dish. Let sections air-dry in a dark, dust-free place. Cover slip with Krystalon.

3.3. Examination and Documentation with Confocal Microscopy

1. Sections are examined with a high-power objective (×100) on Bio-Rad (Hercules, CA) argon-krypton UV confocal laser-scanning microscope. The laser excitation of 568 nm in combination with a Bio-Rad dichroic filter cube T2a was appropriate for detection of the Texas red fluorochrome. The laser excitation of 488 nm in combination with a Bio-Rad dichroic filter cube T1 was appropriate for detection of the Alexa 488 fluorochrome. A similar combination should be possible on other brands of confocal laser-scanning microscopes (*see* **Note 5**).
2. Most confocal systems allow capture of digital images. Images should be obtained throughout the depth of the section by scanning at appropriate intervals from the top of the section to the bottom (i.e., the *z* direction). Images may be reconstructed at appropriate intervals through the cell soma to demonstrate the appearance and disappearance of the immunoreaction product within the cell body. If a typical cell soma has a 10- to 20-μm diameter, this can be accomplished with sections at 0.4-μm intervals. Most images obtained by confocal microscopy can be imported into the Adobe PhotoShop software and can thus be further refined for presentation.

4. Notes

1. The "prill" form of paraformaldehyde (EMS, Fort Washington, PA) is easy and much safer to work with than standard reagent grades of this chemical.
2. For the processing of free-floating tissue sections, plastic mesh-bottom multiwell dishes (Nason Machine, Ft. Bragg, CA) or CoStar netwell cell-culture dishes (Corning, Acton, MA) are an invaluable tool. The sections may be placed into individual wells with up to 10 sections/well for processing. Solutions may be changed by simply lifting the wells or inserts out of the solution chamber. Otherwise, brush transferring of sections from one well of solution to another is tedious and destructive to the tissue. This process can also result in lost sections.

   The TSA buffer is made by adding 100 mg sodium azide/L of 50 m*M* Tris-saline. The azide serves to inhibit the growth of bacteria and other microorganisms. The Tris-saline buffer is kept at room temperature. Another option is to use freshly prepared 50 m*M* Tris-saline, without the addition of azide, for each day's solutions.
3. The blocking solution may be any animal serum at a concentration from 3 to 10% in TSA. Bovine serum albumin (BSA) can also be used.
4. When using antibodies directed against GST-fusion proteins, the antibody must be preincubated in GST at a threefold excess of GST to the antibody. The GST + antibody solution should be allowed to incubate undiluted for 3–4 h before use to allow preadsorption of any anti-GST antibodies. These anti-GST antibodies could

interfere with the specificity of the immunohistochemical assay. After this preincubation, the solution should be centrifuged to remove any precipitated particulate material. The supernatant can then be diluted to the desired concentration with 1% goat serum, 0.1% Triton in TSA.

5. The use of a biotinylated secondary antibody in combination with an avidin- or streptavidin-conjugated fluorochrome will result in a stronger signal, as a result of the amplification inherent in avidin–biotin systems, than the use of a direct fluorescent-tagged secondary antibody. In this case, the avidin–biotin system was used to detect the α_{2A}-AR owing to its lower expression than TH. TH is expressed in abundance, and the TH antibody can easily be detected with the direct-tagged secondary Alexa 488 antimouse IgG. The specific fluorochromes used will depend on the laser light excitation and detection wavelengths of the available confocal scanning laser microscope to be used. In the example used here, Texas red was used as the fluorochrome for detection of the α_{2A}-AR. Alexa 488 was used as the fluorochrome for detection of TH to visualize the cell soma of catecholaminergic cells. These fluorophores gave bright signals that did not fade appreciably over the time needed for confocal imaging.

References

1. Rosin, D. L., Zeng, D., Riley, T., Stornetta, R., Guyenet, P. G., and Lynch, K. R. (1991) Localization of α_{2A}-adrenergic receptors in cultured cells and rat brain using a subtype-specific polyclonal antibody [Abstract]. *Soc. Neurosci. Abst.* **18,** 98.

2. Lee, A., Talley, E., Rosin, D. L., and Lynch, K. R. (1995) Characterization of alpha $_{2A}$-adrenergic receptors in GT1 neurosecretory cells. *Neuroendocrinology* **62,** 215–225.

3. Rosin, D. L., Zeng, D., Stornetta, R. L., Norton, F. R., Riley, T., Okusa, M. D., et al. (1993) Immunohistochemical localization of α_{2A}-adrenergic receptors in catecholaminergic and other brainstem neurons in the rat. *Neuroscience* **56,** 139–155.

4. Guyenet, P. G., Stornetta, R. L., Riley, T., Norton, F. R., Rosin, D. L., and Lynch, K. R. (1994) Alpha 2A-adrenergic receptors are present in lower brainstem catecholaminergic and serotonergic neurons innervating spinal cord [published erratum in *Brain Res.* (1994) **646(2):** 356,357]. *Brain Res.* **638,** 285–294.

32

Distribution of α_{2A}- and α_{2C}-Adrenergic Receptor Immunoreactivity in the Central Nervous System

Diane L. Rosin

1. Introduction

α_2-Adrenergic receptors (α_2-ARs) are G-protein-coupled receptors that mediate the inhibitory effects of norepinephrine and epinephrine in the peripheral and central nervous system (CNS). In the CNS, these receptors have been implicated in the regulation of analgesia, memory, and central cardiovascular control *(1)*. Three subtypes of the receptor have been identified (α_{2A}-, α_{2B}-, and α_{2C}-ARs; for review, *see* **ref. 2**). Comparisons of the three subtypes show a high degree of sequence similarity and no apparent differences in their intracellular effectors. However, clues to the discrete functions of the subtypes have been provided by recent transgenic studies in which targeted mutation of the α_{2A}-AR or overexpression of the α_{2C}-AR in mice reveal specific effects on cardiovascular function, locomotor behavior, and thermoregulation *(3–6)* (for reviews, *see* **refs. 7** and **8**).

In situ hybridization *(9–11)* (for review, *see* **ref. 12** and immunohistochemical studies *(13–16)* have shown that the A and C subtypes of the α_2-ARs are widely distributed throughout the rat neuroaxis with the A subtype occupying a more prominent distribution than the C subtype. Discrete patterns of distribution of each subtype also suggest unique functional properties for the α_{2A}-AR and α_{2C}-AR subtypes, as discussed below.

Immunohistochemical localization of α_2-ARs, as with most G-protein-coupled receptors, requires the availability of highly selective antibodies that specifically recognize the subtype of interest. The basic structural motif of seven transmembrane-spanning domains linked by 3 intracellular and three extracellular loops is preserved throughout this family of receptors and conser-

From: *Methods in Molecular Biology, vol. 126: Adrenergic Receptor Protocols*
Edited by: C. A. Machida © Humana Press Inc., Totowa, NJ

vation of amino acid sequence across subtypes, and to some extent between groups of neurotransmitter receptors, is significant. An understanding of the molecular biology of neurotransmitter receptors, is therefore essential for the development of subtype-specific neurotransmitter receptor antibodies. Further, a high level of antibody sensitivity is necessary for detecting the low level of expression of this family of receptors in tissue.

1.1. Development of α_{2A}-AR and α_{2C}-AR Antibodies

For our studies on the distribution of α_{2A}- and α_{2C}-ARs in rat brain, we generated antibodies using recombinant fusion proteins as the antigens and carried the antibodies through a purification and rigorous characterization scheme (14–16). The details of this part of our work are beyond the scope of this chapter. However, we briefly summarize below the major steps involved in the development of these reagents. Readers may find additional details in Chapters 13 and 20.

Because of the high degree of sequence homology among adrenergic receptor subtypes, careful preparation of the antigen is central to successful generation of a specific antiserum. For this reason, we chose the third intracellular loops of the receptor subtypes, because they contain fairly long amino acid sequence unique to each subtype and regions of high predicted antigenicity. For each receptor subtype, a fragment (141–210 bp) of the α_{2A}-, α_{2B}-, or α_{2C}-AR gene was amplified by the polymerase chain reaction (PCR) using specific oligonucleotide primers. The amplified DNA fragments, which encode a 45–70 amino acid sequence of the third intracellular (i3) loop of each subtype, were subcloned into the pGEX.KG vector (17) downstream from the gene encoding glutathione-S-transferase (GST), a helminthic protein. Transcription of either parent protein (GST) or recombinant fusion protein (GST fused to the amino acid fragment of the α_2-AR i3 loop) was induced in cultures of *Escherichia coli* transformed with either the parent or recombinant plasmid, respectively, and the expressed proteins were purified as described previously (16,18). Purified fusion protein was used for immunizations, preparation of affinity columns, and antibody screening.

Antisera from animals immunized with the α_{2A}- or α_{2C}-AR fusion proteins were enriched for immunoglobulins by ammonium sulfate fractionation and further purified by affinity chromatography using the relevant GST/α_2-ARi3 fusion protein immobilized on a solid support. We found that immunohistochemical localization of receptors in brain tissue sections required affinity-purified antibodies; no specific immunoreactivity was detected otherwise.

Each of the antibodies that we developed was rigorously evaluated in a series of different assays to ensure the selectivity for the receptor and the subtype specificity (14–16). These include:

1. Immunoblot analysis of antibody titer and specificity by screening against the purified antigen or a nonrelevant subtype sequence.
2. Western blot analysis of antibody specificity by screening against membrane preparations of populations of transfected cells each expressing a different α_2-AR subtype (for additional details on this technique, the reader is referred to Chapter 21).
3. Immunohistochemical analysis of subtype specificity of the antibody using populations of transfected cells each expressing a different α_2-AR subtype.
4. Immunoprecipitation of photoaffinity-labeled α_2-ARs.

These assays demonstrate that the antibody specifically recognizes the appropriate recombinant receptor subtype either as a denatured protein, as a fixed protein in tissue, or as a functional receptor capable of binding the appropriate receptor ligand. In addition, specificity of the antibody is further established in each of these assays and, in immunohistochemistry of rat brain, by demonstrating a lack of specific response when the antibody is preadsorbed to the relevant fusion protein antigen prior to use. These criteria are essential to establishing confidence in the antibody and, therefore, for confidently interpreting immunohistochemical localization results using the antibody.

In addition to the antibodies that we developed against α_2-AR subtypes using the strategy described above, antibodies for the α_{2B}-AR subtype *(19)* and for all three subtypes of α_2-ARs *(20–22)* were developed independently in a similar manner by other groups. Furthermore, commercially produced α_2-AR antibodies, which have become available in recent years, specifically recognize the appropriate recombinant α_2-AR subtype as evaluated by a number of assays (e.g., Western blotting, immunoprecipitation). However, to our knowledge, the application of these commercially available antibodies to immunohistochemical localization of α_2-ARs in brain has not been achieved to date.

1.2. Immunohistochemical Localization of α_2-ARs in Rat Brain

Immunohistochemical distribution of the A subtype of α_2-ARs at the light microscopic level in rat brain has been reported by our group *(14,16,23)* and by Aoki and colleagues *(13)*. The A subtype shows an extensive distribution throughout the brain in perikarya and the neuropil. In contrast to the α_{2A}-AR immunoreactivity, only perikaryal α_{2C}-AR immunoreactivity is evident in the rat brain, though its distribution is also quite widespread *(15)*. Localization of the B subtype has been described in kidney *(19)*. However, we have not detected α_{2B}-AR immunoreactivity in brain (Rosin and Okusa, unpublished observations). Indeed, there is no evidence to date for expression of α_{2B}-AR protein in brain in adult animals—or at best very, very low expression. Despite the discrete localization of mRNA for the α_{2B}-AR in hypothalamus and thalamus *(9–11)*, it is possible that expression of this subtype is present only in early development *(24)*.

We have used our α_{2A}- and α_{2C}-AR antibodies for a number of applications that provide additional information on the cellular biology of the receptors and their function in brain. By employing double labeling immunohistochemical techniques, we have found a very specific distribution of α_{2A}- and α_{2C}-AR immunoreactivity in the catecholaminergic cell body regions of the midbrain, pons, and medulla *(23,25)*, and have identified labeling in a neuronal line of cells in culture derived from hypothalamic neuroendocrine cells *(26)*. Double-labeling immunofluorescence in combination with confocal microscopy has enabled the investigation of the distribution of receptor immunoreactivity in intracellular compartments *(15,23,27)* (for additional details on this technique, *see* Chapter 31).

1.3. Ultrastructural Analysis of α_2-ARs

Perhaps one of the greatest advantages in having antibodies for adrenergic receptor subtypes is that these reagents make it possible to examine the cellular and subcellular distribution of receptor subtype immunoreactivity at the electron microscopic level. Ultrastructural studies by our group and others have revealed the subcellular localization of the A *(13,28,29)* and C *(30)* subtypes in rat brain and have focused specifically on ultrastructural localization of the receptors in the cortex *(13,31,32)*, hippocampus *(28,33)*, locus ceruleus *(13,29,30,34)*, rostral ventrolateral medulla *(35)*, and gigantocellular reticular nucleus *(36)*. Using single- and double-label immunohistochemistry, these ultrastructural studies of α_2-ARs help to delineate the cellular substrates through which epinephrine and norepinephrine exert their inhibitory functions on specific neuronal populations. This approach can also provide some clues to the functional relevance of multiple receptor subtypes in the brain. For example, both the α_{2A}- and α_{2C}-AR subtypes are found in the locus ceruleus, but the presence of both receptors, conceivably within the same cell, may be explained by their segregated neuroanatomical localization. Whereas the activity of neurons in the locus ceruleus may be modulated by the action of norepinephrine on α_{2A}-ARs located both pre- and postsynaptically *(29)*, α_{2C}-ARs seem to have a predominantly postsynaptic localization where they may modulate excitatory input to the locus ceruleus *(30)*. Specific methodological details for immunoelectron microscopy are beyond the scope of this chapter, but will be referred to briefly. This method is discussed in more detail in Chapter 35.

2. Materials

2.1. Tissue Preparation for α_2-AR Immunohistochemistry

2.1.1. Fixation for Light Microscopy/Immunohistochemistry–Paraformaldehyde Perfusion (see **Note 1**)

1. Peristaltic pump with plastic tubing.
2. Surgical tools.
3. Single edge razor blades.
4. Syringes (1 mL) with 26-gage hypodermic needles.
5. Rongeurs.
6. Sodium phosphate buffer—0.5 M stock solution (dilute to 0.1 M as needed): solution 1: 0.5 M dibasic sodium phosphate—134 g $Na_2HPO_4 \cdot 7H_2O$ (heat to dissolve), dH_2O up to 1000 mL, ~pH 9.0. Solution 2: 0.5 M monobasic sodium phosphate—17.15 g NaH_2PO_4, dH_2O up to 250 mL, ~pH 4.5. Mix 1000 mL solution 1 with ~234.5 mL of solution 2 (or more, as needed) to bring pH to 7.4. Filter (qualitative filter paper).
7. Injectable heparin, 1000 U/mL.
8. Brain matrix (Ted Pella, Redding, PA)—for precise blocking of brain prior to sectioning.
9. Cannula: 13-gage blunt-end hypodermic needle.
10. Sodium pentobarbital (Nembutal; Abbott Laboratories, North Chicago, IL).
11. Phosphate- (0.02 M) buffered physiological saline (0.9% NaCl): 40 mL 0.5 M sodium phosphate buffer, pH 7.4, 9 g NaCl, dH_2O up to 1000 mL.
12. Paraformaldehyde fixative: 4% (w/v) paraformaldehyde in 0.1 M phosphate buffer, prepared fresh daily. In a well-ventilated fume hood, add 40 g paraformaldehyde (Electron Microscopy Services, Fort Washington, PA) to 300 mL distilled H_2O. Stir and heat to about 60°C. Add 1 M NaOH by drops until the solution clears and paraformaldehyde is completely dissolved. Cool to room temperature, add 100 mL 0.5 M phosphate buffer (*see* **item 6**), and filter (qualitative filter paper). Adjust pH to 7.4 and bring final volume to 500 mL with distilled H_2O (*see* **Note 3**).

2.1.2. Fixation for Light Microscopy/Immunohistochemistry–Periodate/Lysine/Paraformaldehyde (PLP) Perfusion

1. PLP fixative *(37)*: 0.1 M periodate, 0.075 M lysine, 2% paraformaldehyde, 0.0375 M phosphate buffer, pH 7.4. Solution A: 1.83 g lysine (Sigma, St. Louis) in 50 mL distilled H_2O. Adjust pH to 7.4 with 0.1 M NaH_2PO_4. Bring volume up to 100 mL with 0.1 M phosphate buffer (*see* **Subheading 2.1.1., item 11**). Solution B: 8% unbuffered paraformaldehyde (prepare as in **Subheading 2.1.1., item 12** excluding the addition of phosphate buffer). Just before use, mix three parts solution A

with one part solution B and dissolve 21.4 mg NaIO$_4$ (Sigma) for every 10 mL of fixative.

2. **Items 1–11** in **Subheading 2.1.1.**

2.1.3. Fixation for Preembedding Immunoelectron Microscopy–Acrolein/Paraformaldehyde Perfusion

1. **Items 1–6, 8–11** in **Subheading 2.1.1.**
2. Reservoirs for saline and fixatives connected with plastic tubing through a two-way tap to peristaltic pump and cannula.
3. Injectable heparin (10,000–20,000 U/mL). Dilute with buffered physiological saline (**item 11, Subheading 2.1.1.**) to 1000 U/mL.
4. 2% Paraformaldehyde (prepare as in **Subheading 2.1.1., item 12**).
5. Acrolein (Polysciences [Warrington, PA]; 10-mL aliquots)—store in well-ventilated cold room (4°C).
6. 3.75% Acrolein/2% paraformaldehyde—**acrolein is toxic**. All steps must be performed in a well-ventilated fume hood. Wear protective clothing and gloves. Have 2% paraformaldehyde in a container sealed with parafilm. Using a 10-mL syringe with small needle, withdraw acrolein from sealed vial, and inject through the parafilm into the 2% paraformaldehyde. Cover with more parafilm.

2.1.4. Preparation of Vibrotome Sections

1. Vibrotome.
2. Rapid bond glue.
3. 0.1 *M* sodium phosphate buffer, pH 7.4.
4. Stainless-steel injector razor blades (Ted Pella or Electron Microscopy Services).
5. Fine-tipped sable paintbrushes (nos. 0, 1, 2; *see* **Note 2**).

2.1.5. Preparation of Cryostat Sections

1. Cryostat, cryostat knife, and chucks.
2. TS buffer (*see* **Subheading 2.2.1., item 3**).
3. Dry ice.
4. Embedding medium (Tissue Embedding Medium or Tissue-Tek OCT compound; Ted Pella).
5. Fine-tipped sable paintbrushes (nos. 0, 1, 2; *see* **Note 2**).
6. Sucrose solution: 20% sucrose (w/v) in 0.1 *M* sodium phosphate buffer, pH 7.2. Store at 4°C.

2.1.6. Storage of Fixed Brain Sections

1. 24-Well plastic tissue-culture dishes.
2. Small glass vials with lids (e.g., 22-mL scintillation vials).
3. Cryoprotectant solution *(38)*: Combine 500 mL 0.1 *M* sodium phosphate buffer (500 mL dH$_2$O, 1.59 g NaH$_2$PO$_4$·H$_2$O, 5.47 g Na$_2$HPO$_4$), 300 mL ethylene glycol (Sigma), 300 g sucrose, 10 g polyvinylpyrolidone (PVP-40; Sigma). Stir to

dissolve (it is difficult to get PVP into solution). Adjust volume to 1000 mL with dH$_2$O. Store at 4°C.

2.2. Immunohistochemical Detection at the Light Microscopic Level of α_2-ARs in Brain Sections Using Peroxidase Methods

2.2.1. Detection Using the Avidin–Biotin Complex (ABC) Technique (39)

1. Tissue trays (*see* **Note 3**).
2. Orbital or rocking platform shaker.
3. Buffers: 50 m*M* Tris-buffered saline (TS)—7.02 g Tris-HCl, 0.67 g Tris base, 8.75 g NaCl, 0.375 g KCl, 1000 mL dH$_2$O; 50 m*M* Tris-buffered saline/0.01% sodium azide (TSA)—add 0.1 g NaN$_3$/L of TS. Prepare buffers and adjust to pH 7.4 with HCl.
4. 30% H$_2$O$_2$ (Sigma).
5. Blocking serum (goat serum, horse serum, and so forth; Gibco [Grand Island, NY] or Vector Laboratories [Burlingame, CA]) or bovine serum albumin (BSA; RIA grade, Fraction V powder, Sigma A7888; *see* **Note 4**).
6. Triton X-100 (Sigma).
7. GST—recombinant protein prepared from bacteria harboring the pGEX plasmid (*see* **refs. *16–18***); and α_2-adrenergic receptor i3 loop fusion proteins— recombinant protein antigens for blocking experiments (*see* **Note 5**).
8. Primary (α_2-AR antibodies preincubated with GST; *see* **Note 5**) and secondary antibodies (biotinylated goat antirabbit IgG and biotinylated horse antigoat IgG [Vector Laboratories] for detecting α_{2A}- and α_{2C}-AR antibodies, respectively).
9. Vectastain Elite ABC kit (Vector Laboratories)—prepare avidin–biotin reagents 30 min before use according to manufacturer's instructions, i.e., 2 drops A + 2 drops B in 10 mL TS (*see* **Note 6**).
10. Peroxidase substrates: 3,3′-diaminobenzidine (DAB, Sigma). Prepare 2% DAB stock solution: 2 g DAB, 100 mL dH$_2$O, filter. Store small aliquots at –20°C. Prepare working solution (0.02% DAB, 0.01% H$_2$O$_2$ in TS) just before use. Alternative substrate: VIP peroxidase substrate kit (Vector Laboratories; prepare according to manufacturer's directions) (*see* **Notes 7** and **8**).
11. Gelatin-coated microscope slides (*see* **Note 9**), cover slips, mounting media (DPX—Aldrich, Milwaukee, WI).
12. Fine-tipped sable paintbrushes (nos. 0, 1, 2) or fire-polished curved-tip glass rods (heat the tip of a 6-in. Pasteur pipet in the flame of a bunsen burner until the glass begins to melt and the tip fuses; turn slowly to form a gentle U-shaped tip) (*see* **Note 2**).

2.2.2. Detection Using the Peroxidase–Antiperoxidase (PAP) Technique (see **Note 6**)

1. **Items 1–7, 9–12** in **Subheading 2.2.1.**
2. Primary (α_2-AR antibodies preincubated with GST; *see* **Note 5**) and secondary

antibodies (goat antirabbit IgG and horse antigoat IgG (Sternberger Monoclonals, Baltimore, MD) for detecting α_{2A}- and α_{2C}-AR antibodies, respectively).

3. Rabbit and goat PAP (Sternberger Monoclonals) for detecting α_{2A}- and α_{2C}-AR antibody/secondary antibody complexes, respectively.

2.2.3. Detection of Two Antigens: Double-Labeling Technique Using PAP and ABC Techniques (see **Notes 10–17**)

Example: colocalization of α_2-ARs in catecholaminergic cells using tyrosine hydroxylase (TH) as a marker

1. All items in **Subheadings 2.2.1.** and **2.2.2.**
2. Primary antibodies for labeling specific cellular markers (e.g., TH, phenyl-*N*-methyl transferase [PNMT], 5-hydroxytryptamine [5HT]—*see* **Note 12**) and appropriate secondary antibodies (*see* **Note 13**).
3. Peroxidase substrates: working DAB solution (*see* **Subheading 2.2.1., item 10**), NiDAB (working DAB solution prepared as in **Subheading 2.2.1., item 10** with the addition of 5 mg/mL nickel ammonium sulfate) or benzidine dihydrochloride (BDHC, Sigma). Add 10 mg BDHC to 95 mL dH_2O, and stir 30 min to dissolve before filtering. Just before use, add 5 mg sodium nitroferricyanide and 5 mL 0.2 *M* sodium acetate buffer, pH 6.8) (*see* **Note 14**).
4. Acetate-buffered saline, pH 6.8.
5. α-Terpineol (Sigma): xylenes (1:1).

2.3. Detection of α_2-ARs in Brain Using Immunofluorescence Methods

2.3.1. Single-Label Immunofluorescence Using Avidin–Biotin Amplification System (see **Notes 6** and **18**)

1. **Items 1–9** and **12** in **Subheading 2.2.1.**
2. Fluorophore-tagged avidin or streptavidin:avidin-NeutraLite™ Texas red® conjugate (Molecular Probes, Eugene, OR), Alexa™ 488-streptavidin (Molecular Probes), Cy™-3 conjugated streptavidin (Jackson ImmunoResearch Laboratories, Inc., West Grove, PA), AMCA-avidin 7-amino-4-methylcoumarin-3 acetic acid AMCA); Jackson ImmunoResearch Laboratories, Inc.) (*see* **Note 19**).
3. Gelatin-coated microscope slides (*see* **Note 9**), cover slips, mounting media containing antibleaching compound (e.g., Vectashield, Vector Laboratories), fingernail polish (*see* **Note 20**).
4. Fluorescence microscope with filter cubes to discriminate fluorochromes with different excitation and emission spectra.

2.3.2. Single-Label Immunofluorescence Using Direct Fluorophore-Tagged Secondary Antibodies (see **Note 18**)

1. **Items 1, 3,** and **4** in **Subheading 2.3.1.**
2. Fluorophore-conjugated secondary antibodies (*see* **Subheading 2.3.1.**).

2.3.3. Double-Label Immunofluorescence Using Fluorophore-Tagged Secondary Antibodies and Avidin–Biotin Amplification System (see **Notes 18** and **19**)

1. **Items 1–4** in **Subheading 2.3.1.**
2. **Item 2** in **Subheading 2.3.2.**

2.3.4. Triple-Label Immunofluorescence for Detection of Two Antigens and Fluorescently Tagged Retrograde Marker (see **Note 21**)

1. Surgical and stereotaxic equipment.
2. **Items 1** and **2** in **Subheading 2.3.3.**
3. Fluoroscein isothiocyanate- (FITC) labeled microbeads (Luma-Fluor, Inc., New City, NY).

2.4. Pre-Embedding Immunohistochemical Labeling of Vibrotome Sections for Electron Microscopic Detection of α_2-ARs

A very brief outline of methods is given here. Additional details may be found in Chapter 35.

2.4.1. Single Labeling Using the Peroxidase Method

1. 10X stock solutions filtered (0.2 μm)—1.0 M TS (*see* **Subheading 2.2.1.**), 1.0 M PO$_4$ (*see* **Subheading 2.1.1.**); working solutions diluted to 0.1 M with glass distilled water and filtered (0.2 μm) on day of use.
2. 0.2 μm Cellulose nitrate filters (Fisher Scientific, Pittsburgh, PA).
3. NaBH$_4$ (Sigma).
4. Bovine serum albumin (BSA) (RIA grade, Fraction V powder, Sigma A7888).
5. Ceramic crucibles with perforated base.
6. **Items 1, 2,** and **6–12** in **Subheading 2.2.1.**

2.4.2. Double-Labeling Using the Peroxidase and Silver-Enhanced Immunogold Methods (see **Note 22**)

1. 1 nm Gold-tagged secondary antibody (Amersham [Piscataway, NJ], Nanoprobes [Stony Brook, NY], and others).
2. Amersham IntenSE silver intensification kit.
3. 10X stock solution 0.1 M phosphate-buffered saline (PBS; *see* **Subheading 2.1.1.**), filtered (0.2 mm)—do not use HCl to adjust pH!! Working solutions are diluted to 0.01 M with glass distilled water and filtered (0.2 μm) on day of use.
4. Washing buffer: 50 mL PBS, 0.25 mL gelatin stock (from Amersham immunogold kit; 0.1% final), 0.4 g BSA (0.8% final).
5. 0.2 M citrate buffer: 500 mL glass-distilled water, 29.45 g sodium citrate, adjust pH to 7.4 with citric acid (8.4 g citric acid in 200 mL glass-distilled water). Do not use HCl to adjust pH!!! Filter (0.2 μm) stock solution prior to storing, and refilter amount needed on day of use.

6. 0.1 *M* sodium phosphate buffer, filtered (0.2 μm; *see* **Subheading 2.4.1.**).
7. 25% Glutaraldehyde (Electron Microscopy Services).
8. Orange wood sticks (*see* **Note 23**).
9. 24-Well plastic tissue-culture dishes.
10. Ceramic crucibles with perforated base.
11. **Items 1–6** in **Subheading 2.4.1.**

2.4.3. Embedding Immunolabeled Tissue for Electron Microscopy

1. 4% Osmium tetroxide (Electron Microsopy Services).
2. 0.1 *M* sodium phosphate buffer, filtered (0.2 μm; *see* **Subheading 2.4.1.**).
3. Ceramic Coors multiwell dishes.
4. 30, 50, 70, 90, 100% EtOH (freshly opened bottle of absolute EtOH; solutions prepared fresh just before use).
5. Propylene oxide (Electron Microsopy Services).
6. EM-bed 812 resin kit (Electron Microsopy Services): Prepare fresh plastic resin on day of use, combining the four components of the kit as follows—22 mL EM-bed 812, 15 mL DDSA, 13.5 mL NMA, 525 μL DMP-30 in a 50-mL conical polypropylene tube. Mix by placing tube on a rotating mixer or rocking shaker for 30 min.
7. 22-mL glass scintillation vials.
8. Rotating mixer.
9. Aclar plastic sheets (Electron Microsopy Services).
10. Powder-free latex gloves.
11. 60°C oven.
12. Metal tray.
13. Lead brick.

2.5. Immunohistochemical Controls

1. Preimmune serum.
2. Antigen: purified α_{2A}- and α_{2C}-AR fusion proteins.

2.6. Data Analysis

1. Microscope.
2. Image analysis system.
3. Camera (35 mm or digital).
4. Software (Neurolucida®, Adobe Photoshop, Canvas).

3. Methods

3.1. Tissue Preparation for α_2-AR Immunohistochemistry

3.1.1. Fixation for Light Microscopy/Immunohistochemistry by Vascular Perfusion with Paraformaldehyde (see **Note 24**)

1. Deeply anesthetize rat with pentobarbital (100 mg/kg, ip; *see* **Note 25**).
2. Surgically expose the heart and aorta. Inject 500 U heparin into the tip of the heart (*see* **Note 26**).

3. Insert cannula (*see* **Subheading 2.1.1.**) into left ventricle, and gently push the cannula until the tip enters the ascending aorta. Clamp cannula in place with hemostat.
4. Make a small excision in the right atrium, and begin perfusing immediately with ~50 mL buffered saline at a flow rate of 50–75 mL/min.
5. Perfuse with 300–500 mL 4% paraformaldehyde fixative (*see* **Notes 1** and **24**).
6. Decapitate rat, and carefully remove brain from skull cavity using rongeurs.
7. Place brain in tissue matrix. Using single-edge razor blades, cut a 2- to 6-mm block of brain containing the areas of interest. Choose either **step 8** or **9** for postfixation.
8. To prepare tissue for vibrotome sectioning, immerse brain in 4% paraformaldehyde to postfix overnight at 4°C (*see* **Note 27**). Proceed to **Subheading 3.1.4.**
9. To prepare tissue for cryostat sectioning, immerse brain in 4% paraformaldehyde to postfix for 30–60 min (*see* **Note 27**). Proceed to **Subheading 3.1.5.**

3.1.2. Fixation for Light Microscopy/Immunohistochemistry by Vascular Perfusion with Periodate/Lysine/Paraformaldehyde (PLP) (see **Note 24**)

1. Follow **steps 1–4** in **Subheading 3.1.1.**
2. Perfuse with 300–500 mL PLP fixative.
3. Decapitate rat, and remove brain from skull cavity using rongeurs. Block brain to obtain desired regions.
4. To prepare tissue for cryostat sectioning, immerse brain in PLP fixative to postfix for 4 h (*see* **Note 28**).

3.1.3. Fixation by Vascular Perfusion for Pre-embedding Immunoelectron Microscopy

1. Deeply anesthetize rat with pentobarbital (100 mg/kg, ip; *see* **Note 25**).
2. Surgically expose the heart and aorta.
3. Insert cannula (*see* **Subheading 2.1.1.**) into left ventricle, and gently push the cannula until the tip enters the ascending aorta. Clamp cannula in place with hemostat.
4. Make a small excision in the right atrium, and begin perfusing immediately with 50 mL buffered saline containing heparin (1000 U/mL) followed immediately by 55 mL 2% paraformaldehyde/3.75% acrolein at a flow rate of 80–110 mL/min (*see* **Notes 1, 26,** and **29**).
5. Continue perfusing with 400 mL 2% paraformaldehyde (100–150 mL at 60–80 mL/min and then the remainder at 40–50 mL/min).
6. Remove the brain, cut into blocks containing the desired brain regions, immerse in 2% paraformaldehyde to postfix for 30 min (*see* **Note 27**). Process immediately for vibrotome sectioning (**Subheading 3.1.4.**).

3.1.4. Preparation of Vibrotome Sections of Rat Brain

1. Transfer brain to 0.1 *M* PO₄ buffer.
2. Using fine forceps, remove dura from the surface of the brain under a dissecting microscope.

3. Mount tissue on vibrotome cutting platform using rapid bond glue (such as Superglue®).
4. Immerse brain in filtered 0.1 *M* PO$_4$ and cut 30- to 50-μm sections (*see* **Note 30**).
5. Transfer sections free floating to buffer using soft, fine-tipped paintbrushes (*see* **Note 2**) and process immediately for immunohistochemistry, or transfer to cryogenic buffer and store sections at –20°C (*see* **Subheading 3.1.6.** and **Note 31**).

3.1.5. Preparation of Cryostat Sections of Rat Brain

1. Transfer brain to ice-cold sucrose solution for cryoprotection, and store overnight at 4°C (*see* **Note 32**).
2. To mount tissue on cryostat cutting chuck, embed tissue using frozen tissue embedding medium and freeze on dry ice. Transfer chuck to cryostat, and allow tissue to equilibrate to –20°C.
3. Cut 10- 20-μm sections (*see* **Note 30**).
4. Transfer sections free floating to buffer using soft, fine-tipped paintbrushes (*see* **Note 2**) and process immediately for immunohistochemistry, or transfer to cryogenic buffer and store sections at –20°C until needed (*see* **Subheading 3.1.6.** and **Note 31**).

3.1.6. Storage of Fixed Brain Sections (see **Note 31**)

1. Transfer freshly cut vibrotome or cryostat sections to ice-cold cryogenic buffer. Large numbers of sections can be stored in batch in scintillation vials, or serial sections may be organized in multiwell plastic tissue-culture plates.
2. Store at –20°C until needed for periods of up to 6 mo to a year.
3. Tissue sections stored in cryogenic buffer must be rinsed profusely in buffer prior to processing for immunohistochemistry.

3.2. Immunohistochemical Detection at the Light Microscopic Level of α$_2$-AR in Brain Sections Using Peroxidase Methods (see Notes 3 and 33)

3.2.1. Detection Using ABC Technique (see **Note 6**)

1. Rinse sections 4 × 5 min with TS (*see* **Notes 34–36**).
2. Block endogenous peroxidases by incubating in 1% H$_2$O$_2$/TS for 30 min (*see* **Notes 26** and **37**).
3. Rinse 3 × 5 min with TS.
4. Block tissue with 3% serum or 10% BSA in TS (*see* **Note 4**).
5. Rinse 3 × 5 min with 1% serum (or BSA) in TS.
6. Incubate tissue in primary antibody diluted with 1% serum (or BSA) and 0.1% TX-100 in TS for 24–48 h at 4°C (*see* **Notes 38** and **39**).
7. Rinse 4 × 5 min with TS/1% serum. (**Note:** if TSA was used for primary antibody incubation, begin using TS now to rinse out azide before getting to ABC step; *see* **Note 36**).

8. Incubate 45 min with biotinylated secondary antibody (diluted according manufacturer's specifications in TS/1% serum; *see* **Note 39**).
9. Rinse 3 × 5 min with TS/1% serum.
10. Incubate 30 min with ABC reagents (*see* **Notes 6** and **36** and **Subheading 2.2.1., item 9**).
11. Rinse 3 × 5 min with TS.
12. Prepare peroxidase substrate working solution just before use (DAB/TS/H_2O_2). Incubate sections in DAB solution for 2–6 min or until brown reaction product develops (*see* **Notes 8**, **40**, and **42**).
13. Rinse 4 × 5 min with TS (*see* **Notes 41** and **42**). Mount sections on gelatin-coated slides, allow to air-dry completely, and cover slip with DPX.

3.2.2. Detection Using PAP Technique (see **Note 6**)

1. Follow **steps 1–7** in **Subheading 3.2.1.**
2. Incubate in secondary antibody (1:150) prepared in TS/serum for 45 min (*see* **Note 6** regarding selection of secondary antibody and PAP complex reagent).
3. Rinse 3 × 5 min with TS/serum.
4. Incubate in PAP (1:100 in TS/serum) for 30 min.
5. Follow **steps 11–13** in previous section.

3.2.3. Double-Labeling Technique Using Peroxidase Substrates that Result in Reaction Products of Different Color (see **Notes 10–17**)

3.2.3.1. DOUBLE LABELING USING DAB AND NiDAB (E.G., LABELING α_{2C}-ARs AND TH IN CATECHOLAMINERGIC CELLS; *SEE* **NOTE 14**)

1. Follow **steps 1–5** in **Subheading 3.2.1.**
2. Follow **step 6** in **Subheading 3.2.1.** selecting for this incubation the primary antibody (#1) that will be detected using the ABC method (e.g., α_{2A}-AR antibody) (*see* **Note 15**).
3. Follow **steps 7–11** in **Subheading 3.2.1.** (select biotinylated secondary antibody directed against primary antibody used in **step 2**).
4. Prepare peroxidase substrate just before use (NiDAB/TS/H_2O_2; see **Subheading 2.2.3.**). Incubate sections in NiDAB solution for 2–6 min or until dark brown/black reaction product develops (*see* **Note 40**).
5. Rinse 4 × 5 min with TS (*see* **Note 41**).
6. Follow **step 6** in **Subheading 3.2.1.** selecting for this incubation the primary antibody (#2) that will be detected using the PAP method (e.g., TH antibody; *see* **Note 12**).
7. Rinse 4 × 5 min with TS.
8. Follow **steps 2–4** of **Subheading 3.2.2.** selecting a secondary antibody and PAP complex appropriate for recognition of the primary antibody (#2) used in **step 6** of this section.
9. Follow **steps 11–13** of **Subheading 3.2.1.**

3.2.3.2. DOUBLE LABELING USING DAB AND BDHC (E.G., LABELING α_{2C}-ARS AND TH IN CATECHOLAMINERGIC CELLS; *SEE* **NOTE 14**)

1. Follow **steps 1–5** in **Subheading 3.2.1.**
2. Follow **step 6** in **Subheading 3.2.1.** selecting for this incubation the primary antibody (#1; e.g., α_{2C}-AR antibody) that will be detected using the ABC method (*see* **Note 15**).
3. Follow **steps 7–12** in **Subheading 3.2.1.** (select biotinylated secondary antibody directed against primary antibody #1 that was used in **step 2**).
4. Rinse 4 × 5 min with TS.
5. Follow **step 6** in **Subheading 3.2.1.** selecting for this incubation the primary antibody (#2; e.g., TH antibody) that will be detected using the PAP method.
6. Follow **steps 2–4** in **Subheading 3.2.2.** selecting a secondary antibody and PAP complex appropriate for recognition of the primary antibody #2 used in **step 5** of this section.
7. Rinse 3 × 5 min with acetate buffered saline.
8. Prepare BDHC, and preincubate sections for 10 min.
9. Replace BDHC with BDHC containing 0.005% H_2O_2 (H_2O_2 added to BDHC just prior to use), and react sections for 30 s to 2 min (*see* **Note 40**).
10. Rinse 4 × 5 min with acetate-buffered saline.
11. Mount sections on gelatin-coated slides, and allow to air-dry completely. Dehydrate quickly (10 s each) through an ethanol series followed by 1 min in α-terpineol/xylene (1:1). Delipidate in xylenes and cover slip with DPX.

3.3. Detection of α_2-ARs Using Immunofluorescence Methods (see Note 33)

3.3.1. Single-Label Immunofluorescence (see **Note 18**)

1. Follow **steps 1** and **3–9** in **Subheading 3.2.1.**
2. Incubate with a fluorophore-tagged avidin (e.g., avidin-Texas red) diluted 1:150–1:200 in TS (*see* **Notes 19** and **20**).
3. Rinse 4 × 5 min with TS.
4. Mount sections on gelatin-coated slides, allow to air-dry completely, cover slip with Vectashield or other suitable mounting medium, and attach cover slips by sealing edges with nailpolish (*see* **Note 20**).
5. View slides on a fluorescence microscope (or by confocal microscopy—*see* **Note 43**) with filters appropriate for discrimination of excitation/emission spectra of different fluorophores.

3.3.2. Double-Label Immunofluorescence (see **Notes 6, 11–13, 16–20,** and **43**)

1. Follow **steps 1–3** in previous subheading to label sections with the primary antibody (#1) that has been selected for detection with the avidin–biotin method.
2. Incubate with primary antibody #2, which will be detected with a fluorophore-tagged secondary antibody prepared in TS/serum/TX-100.

3. Rinse 3 × 5 min with TS.
4. Incubate with fluorophore-tagged secondary antibody (1:200; e.g., FITC-tagged IgG; directed against primary antibody #2) for 1 h.
5. Follow **steps 3** and **4** of previous subheading.
6. View slides on a fluorescence microscope (or by confocal microscopy—*see* **Note 43**) with filters appropriate for discrimination of excitation/emission spectra of each of the two different fluorophores.

3.3.3. Triple Label (see **Note 21**)

1. Stereotaxically inject fluorophore-tagged microbeads into terminal field of interest.
2. Perfuse 7 d later (as in **Subheading 2.1.1.**), and cut sections as in **Subheading 2.1.4.**
3. Follow combined immunofluorescence protocol in **Subheading 3.3.2.**
4. View slides on a fluorescence microscope (or by confocal microscopy—*see* **Note 43**) with filters appropriate for discrimination of excitation/emission spectra of each of the three different fluorophores.

3.4. Pre-Embedding Immunohistochemical Labeling of Vibrotome Sections for Electron Microscopic Detection of α_2-ARs

A very brief outline of methods is given here. Additional details may be found in Chapter 35.

3.4.1. Single Labeling with Peroxidase Method

1. Perfuse as in **Subheading 3.1.3.**
2. Cut 30-μm vibrotome sections into 0.1 M phosphate buffer (PB) as in **Subheading 3.1.4.**
3. Rinse 4 × 5 min in PB.
4. Incubate 30 min in 1% NaBH$_4$/0.1 M PB (*see* **Note 37**).
5. Rinse copiously with 0.1 M PB until no more bubbles evolve from tissue.
6. Rinse 2 × 5 min with TSA (*see* **Note 36**).
7. Block with 10% BSA in TSA for 30 min.
8. Prepare primary antibody in TSA with 1% BSA. Incubate 24–48 h at 4°C (*see* **Notes 38** and **39**).
9. Rinse 3 × 5 min with 1% BSA/TS.
10. Incubate 60 min with biotinylated secondary antibody (diluted according to manufacturer's instructions in 1% BSA/TS; *see* **Note 39**).
11. Rinse 3 × 5 min with 1% BSA/TS.
12. Incubate 30 min with Vector Elite ABC reagents (*see* **Notes 6** and **36**).
13. Transfer sections to crucibles and rinse 3 × 5 min with TS (*see* **Notes 41** and **42**).
14. React with DAB substrate working solution (DAB/H$_2$O$_2$/TS) for 3–6 min (*see* **Notes 8** and **40**).
15. Rinse 3 × 5 min with TS and then 2 × 5 min with PBS (*see* **Note 41**).

16. Remove some sections to slide mount for light microscopy (*see* **Note 9**).
17. Prepare tissue for electron microscopy as in **Subheading 3.4.3.**

3.4.2. Double-Labeling Using Peroxidase Method and Silver-Enhanced Immunogold Methods (e.g., Dual ImmunoEM Detection of α_2-AR and TH) (see **Note 22**)

1. Follows **steps 1–7** in **Subheading 3.4.1.**
2. Prepare primary antibody cocktail of first primary antibody (e.g., α_2-AR antibody; detection will be with ABC method and DAB) and second primary antibody (cellular marker, such as TH; detection will be with immunogold-tagged secondary antibody), and incubate as in **step 8** of **Subheading 3.4.1.**
3. Follow **steps 9–15** of **Subheading 3.4.1.** selecting biotinylated secondary that recognizes first primary antibody.
4. Block in washing buffer (*see* **item 4, Subheading 2.4.2.**) for 10 min.
5. Incubate in gold-conjugated secondary antibody (1:50 in washing buffer) for 2 h. (Take IntenSE kit out of fridge at this point to allow 2 h for equilibration to room temperature before use.)
6. Rinse in washing buffer for 5 min.
7. Rinse 3 × 5 min with PBS.
8. Incubate covered for 10 min in 2% glutaraldehyde/PBS (note: shelf life of glutaraldehyde is 2 mo).
9. Rinse 2 × 5 min in PBS.
10. Rinse 2 × 5 min in 0.2 *M* citrate buffer.
11. Using orange wood sticks, transfer sections to fresh citrate buffer in sterile 24-well plastic tissue-culture dish (*see* **Note 23**).
12. Silver intensification—mix silver reagents A and B at a 1:1 ratio in empty wells (only a few wells at a time). Using wooden sticks, transfer sections from 0.2 *M* citrate buffer to silver-intensification reagents. Intensify for 6–9 min, and immediately transfer to fresh citrate buffer to stop reaction. Transfer sections again to crucibles (*see* **Note 42**) in a shallow bath of fresh citrate, and rinse 5 min in citrate followed by 2 × 5 min with 0.1 *M* PB.
13. Prepare tissue for electron microscopy as in **Subheading 3.4.3.**

3.4.3. Embedding Immunolabeled Tissue for Electron Microscopy

1. Place sections flat in Coors multiwell ceramic dishes containing 0.1 *M* PB.
2. Draw off PB, and replace with 1–2% osmium tetroxide /PB. Incubate 1 h (*see* **Note 44**).
3. Rinse 3 × 3 min with PB.
4. Dehydrate through a graded series of freshly prepared ethanol solutions (30, 50, 70, 95% EtOH, 5 min each).
5. Transfer sections (using orange wood stick) to glass scintillation vials containing 100% EtOH. Immerse sections in two changes of 100% ethyl alcohol (EtOH) (2 × 10 min), followed by two changes of propylene oxide (2 × 10 min; *see* **Note 45**).

6. Infiltrate sections with resin (propylene oxide:resin [1:1]; prepare fresh just before use) by incubating overnight at room temperature in capped vials on a rotating mixer. (Use gloves when working with resin, work in a fume hood, and dispose of waste properly.)

7. Replace solution with freshly prepared resin (no propylene oxide), and place on rotating mixer for 2 h.

8. Wearing powder-free gloves, wipe Aclar sheets and cover slips with freshly opened 100% EtOH, and place on clean metal tray (*see* **Note 46**).

9. Transfer sections carefully with an orange wood stick to clean Aclar sheet.

10. Cover sections with Aclar cover slip, and carefully press out extra resin and any air bubbles.

11. Cover plastic embedded sections with a lead brick.

12. Bake overnight at 60°C in drying oven.

13. Prepare sample for ultrathin sectioning (further details are beyond the scope of this chapter).

3.5. Immunohistochemical Controls

See **Note 47** for an in-depth description of immunohistochemical controls.

3.6. Data Analysis

This includes making permanent photographic records, drawing brain sections and plotting labeled cells, and generating images for publication (*see* **Note 48**).

4. Notes

1. Most of the fixatives used for tissue fixation are aldehydes that are toxic (acrolein, in particular, is highly reactive) and should be handled with caution; gloves and protective clothing are advisable. Use the prill form of paraformaldehyde, a dust-free product, for easier and safer transfer during weighing and solution preparation. Preparation of fixatives, perfusion of the animal, and transfer of perfusate to a waste container should be performed in a fume hood. All waste solutions containing fixative should be disposed of properly according to institutional guidelines.

2. Fine-tipped good-quality paintbrushes are used to pick up tissue sections after sectioning on the vibrotome or cryostat, to transfer sections to solutions during immunohistochemical processing and to aid in mounting sections on glass slides. Alternatively, a fine curved-tip glass rod (*see* **Subheading 2.2.1., item 12**) can be used for transferring sections between solutions.

3. During all immunohistochemical incubations, sections are shaken gently on an orbital shaker and are maintained at room temperature unless otherwise noted. We perform all immunohistochemical reactions using free-floating sections, though slide-mounted sections may also be used. Labeling is typically better with free-floating sections because of the greater surface area that is available to immunohistochemical reagents. Multiwell dishes with mesh bottoms in each well provide a convenient method for multiple solution changes and avoid excessive

handling or transferring of tissue sections (commercially available from Nasson Machine, Fort Bragg, CA). Individual mesh inserts for multiwell tissue-culture dishes, though not as convenient as the multiwell mesh dishes, serve the same purpose. Alternatively, slide-mounted sections may be advisable for tissue that is not heavily fixed or thinner sections that would not otherwise be durable enough to withstand the free-floating procedure (*see also* **Note 2**).

4. A pretreatment step using an excess of irrelevent blocking protein (generally animal serum or BSA) is usually essential for reducing nonspecific binding of antibodies to the tissue. If serum is used, it is usually selected from the same animal species that the secondary antibody is raised in because the secondary antibody is unlikely to recognize "self" proteins in the serum. It is especially important that blocking serum is **not** from the same species in which the primary antibody was raised, because secondary will bind to any immunoglobulin in the serum rather than specifically to the bound primary antibody.

5. Two recombinant proteins are important for immunohistochemical studies with the α_2-AR antibodies. Because the antibodies are directed against GST-α_2-AR fusion proteins that retain the GST sequence and because the antibodies are affinity-purified using a GST–fusion protein column, the affinity-purified antibody must be preadsorbed with GST prior to use. This precaution will remove any anti-GST antibodies that might otherwise contribute to background labeling. In addition, control sections in which antibody is preadsorbed with its companion fusion protein can be included in an experiment to demonstrate specificity of labeling.

   Preadsorption of antibody with GST or fusion protein can be performed in solution by mixing undiluted antibody with soluble recombinant protein, incubating at room temperature, following with a rapid centrifugation to remove particulate matter, and then diluting the supernatant to achieve the desired antibody concentration in the buffer of interest. This method works well for most applications. For tissue prepared for electron microscopy, cleaner results are found if the recombinant protein is coupled to a soluble support (e.g., sepharose or agarose beads). A convenient product is the AminoLink Coupling Gel (Pierce, Rockford, IL), which can be adapted to produce small batches of coupled beads in a manner similar to preparing an affinity column. A small aliquot of GST- or fusion protein-coupled beads is then incubated with antibody in a microcentrifuge tube and the supernatant is recovered after removing the beads by centrifugation.

6. Neurotransmitter receptors, such as the α_2-ARs are expressed in cells in relatively low abundance, and therefore immunohistochemical detection of these proteins generally requires the use of an amplified detection system to increase the signal intensity. We have had excellent results with the Vector Elite ABC kit (Vector Laboratories) in which the avidin–HRP (A) and biotin (B) reagents are provided separately and must be premixed to produce the avidin–horseradish peroxidase (HRP)/biotin complex (ABC) *(39)*. Vector also provides the biotinylated secondary antibody in the kit. In this set of reactions, binding of a biotinylated secondary antibody to the bound primary antibody is detected using an enzyme or fluorophore-tagged avidin or streptavidin, which binds with

extremely high affinity to biotin and provides sites for binding of additional AB complexes, thereby amplifying the single-copy primary antibody binding event. Because experiments may use primary antibodies made in different animal species, we use Vector's generic Elite ABC kit (containing only the A and B reagents) and purchase separately stocks of biotinylated secondary antibodies made in the appropriate matched species. Products that provide a premixed set of reagents are available from other companies and may incorporate streptavidin (e.g., Amersham) rather than avidin. Some investigators advise against using serum in the ABC/buffer reaction mixture, because serum may interfere with formation of the AB complex.

Amplification of signal can also be achieved using the PAP bridge method, although the PAP method is generally better for abundantly expressed proteins, such as TH. Unlabeled secondary antibody, selected on the basis of primary antibody being used, and the PAP complex reagents are available from Sternberger Monoclonals. Selection of the correct PAP complex is critical so that the secondary antibody bridges between the primary antibody and the antiperoxidase, i.e., the antiperoxidase must be made in the same species as the primary antibody, and the unlabeled secondary antibody must be directed against immunoglobulins from this same species.

7. The DAB substrate is available commercially in several different forms. We buy the crystalline form (Sigma), prepare a 2% (w/v) solution in water, and filter the solution to remove undissolved particulates. Small aliquots are stored at $-20°C$ and diluted in buffer as needed just before use. **Caution:** DAB is a carcinogen and must be handled using recommended chemical safety procedures (protective bench covering, disposable gloves, lab coat). DAB solutions must be collected separately and disposed of properly as chemical waste.

8. In addition to the most commonly used peroxidase substrate, DAB, there are other reagents available for single or double labeling that produce different colored products. For example, Vector offers several peroxidase substrates, including their VIP substrate, which produces a purple reaction product. This substrate in our hands produces labeling with a very clean background. Unlike the DAB product, which is extremely stable, the VIP product is labile in alcohol. It may be necessary to react the sections longer in VIP substrate to produce a darker reaction product than desired. Then, in order to dehydrate effectively the sections before cover slipping without losing the VIP label, a series of quick, short immersions through an ethanol series will dissolve only some of the excess VIP product and leave adequate labeling of the tissue.

9. Tissues that have been processed free-floating for immunohistochemistry are subsequently mounted on glass microscope slides for viewing and permanent storage. Slides are prepared for tissue mounting by gelatin coating (also called subbing) to provide an adherent surface to which sections will remain affixed. We prepare our own subbed slides in batches that can be stored for several months at room temperature (procedure described below). Precoated slides are also available commercially.

Subbing slides: Place glass slides in slide holders. Immerse slides for at least 20 min or overnight in Nochromix® or in soapy water mixed with bleach (the latter method works well and is a safer solution for handling). Rinse profusely with tap water, and follow with several dH$_2$O rinses. Keep slides immersed in dH$_2$O until ready to gelatin-coat. Prepare 0.5% gelatin by mixing 1.5 g gelatin/ 300 mL dH$_2$O (enough to fill a slide staining dish) with the addition of ~100–150 mg CrKSO$_4$ (to prevent bacterial growth on gelatin-coated slides). Heat to ~60–80°C to dissolve gelatin. Cool to <30°C. Filter through Whatman No. 2 paper into staining dish. Remove one slide holder from tray of water, and drain off excess water. Dip slides several times in gelatin, tapping gently to remove air bubbles, and immerse slides in gelatin for ~10 s. Remove slide holder from gelatin, allow liquid to drain, and blot slides on paper towel to remove remaining excess liquid. Dry slides in a drying oven for 2 h or overnight. Store in slide boxes at room temperature for up to 2–3 mo.

10. For dual labeling of two different antigens, primary antibodies generated in two different species are selected. In this way, species-specific secondary antibodies can be employed that will bind specifically only to the appropriate primary antibody and not crossreact with the other primary antibody.

11. Detection of two different primary antibodies in dual-labeled tissue—two fluorophores vs two chromogens. The choice of detection system depends partly on the nature of the sites being labeled. If it is expected that the two primary antibodies will colabel cells and that the intracellular pattern of labeling will be very similar, then immunofluorescence is the preferred method, because overlapping patterns of labeling can be discriminated optically by fluorescence microscopy. The two fluorophores must be chosen so that their excitation-emission spectra do not overlap and that there is no "bleed-through" from overlapping wavelengths of one fluorophore while viewing the other. It is also important to be sure that the selected fluorophores will be compatible with the properties of the available filters that are installed on the investigator's fluorescence microscope. Although dual immunofluorescent labeling is extremely valuable, if not essential, for some double-labeling studies, it suffers from the disadvantage that the label is not permanent (as chromogenic peroxidase products usually are) because the intensity of labeling fades with exposure to light (during both the processing of tissues and microscopic viewing of slide-mounted sections [*see also* **Note 20**]). By contrast a number of different reporter molecules (e.g., peroxidase [most common], alkaline phosphatase, glucose oxidase) are available for enzyme-based detection systems, the sensitivity of detection is greater than with fluorescent methods, and the chromogenic substrates generally yield a stable reaction product. Labeling on tissue sections is therefore permanent and can be viewed for unlimited periods of time without loss of intensity. However, the antigens must have a different spatial location to be able to discriminate two colorimetric products in the same cell.

12. Information on the cellular localization of α_2-ARs in specific brain areas can be enhanced by examining the receptor distribution with respect to specific cellular

markers. Some commonly used markers for which good antibodies are available include the catecholamine synthetic enzymes (TH, dopamine β-hydroxylase [DBH] and PNMT), the cholinergic marker choline acetyltransferase (ChAT), and serotonin (5-hydroxytryptamine, 5HT).

13. For the detection of two antigens, primary antibodies must be available for each antigen, and the antibodies must be raised in two different species. Secondary antibodies directed against immunoglobulins from each of the two different species must then be used. It is best if the secondary antibodies were raised in the same species, so that blocking serum from the same species can be used (*see also* **Note 4**) and so that there is no crossreactivity between secondary antibodies.

14. Several specific examples of dual-labeling approaches using colorimetric markers will serve as useful illustrations of general applications of the technique. A very common combination of chromogenic substrates is the standard DAB reaction that produces a brown product for one marker and the addition of nickel ammonium sulfate to the DAB substrate to produce a blue-black reaction product for a second marker. We have found this combination to be particularly useful for dual labeling of α_2-ARs with the ABC reaction and DAB (which reveals diffuse labeling throughout the cell) and of c-FOS with the PAP reaction and Ni-DAB (which reveals a dark discrete label that is confined to the cell nucleus). This approach is particularly well suited in this case because of the distinct cellular separation of the two antigens. Similarly, we have labeled TH, the catecholamine synthetic enzyme, with PAP and DAB (which reveals a diffuse brown label throughout the cell) in combination with the ABC reaction and Ni-DAB for the α_2-ARs (which appear as dark dots of intracellular reaction product). Alternatively, we have adapted a colorimetric method published previously *(40)* to localize antigens that may have overlapping spatial distributions within a cell. In this example, we used peroxidase reaction products that differ both in color and textural appearance and can easily be discriminated even with overlapping patterns of deposition within the same cell. Diffuse brown DAB reaction product resulting from the labeling of α_{2C}-ARs using the ABC method was distinct from the granular blue immunoreaction product resulting from labeling of TH with the PAP method and the peroxidase substrate, BDHC *(25)*. In all of the examples described above, it is important to optimize the labeling conditions to get an appropriate balance of reaction product intensities that can be discriminated light microscopically.

15. The choice of which primary antibody gets paired with each reaction method depends on which antigen requires the ABC method for amplification. The choice of chromogen is based on the relative distribution of the two antigens in the cell and how they can best be discriminated (*see* **Notes 10** and **11**). It is not important whether the ABC or PAP protocol is performed first, but the NiDAB reaction must precede the DAB reaction.

16. Occasionally, problems in detection may become limiting in designing double-labeling experiments when both antigens require amplification. If colorimetric reactions are feasible for double-labeling, it may be possible to use the ABC

system (which usually gives the best amplification) for the antigen requiring the most amplification, and the PAP reactions may provide sufficient amplification for the other antigen. Dual fluorescent labeling in these situations often does not provide adequate sensitivity. Although amplification for one antigen is feasible using a biotinylated secondary antibody followed by a fluorophore-tagged avidin, there is no amplification of the signal for the second antigen when a direct-tagged secondary is used.

17. In our hands, the best signal-to-noise ratio is achieved in double-labeling protocols when the antigens are visualized using the sequential series of reagents detailed in **Subheading 3.2.3.** rather than employing a cocktail of reagents. It is possible, however, to get consistent results if a cocktail of the two primary antibodies is incubated with the tissue followed by a sequential series of visualization steps for each bound primary antibody.

18. As discussed in **Note 6**, amplification of signal is usually essential for detection of antibody bound to receptors, particularly for immunofluorescence, because the sensitivity is lower than with peroxidase labeling. For single immunofluorescence, detection of antireceptor antibody is with a biotinylated secondary antibody followed by fluorophore-tagged avidin; detection with a fluorophore-tagged secondary antibody does not produce adequate α_2-AR immunoreactivity in our hands. Therefore, for dual immunofluorescent labeling, the primary antibody for the second antigen must be detected with a fluorophore-tagged secondary antibody. *See* **Note 19** regarding choice of fluorophores for double labeling.

19. Combinations of fluorophores for double (or triple) labeling are chosen so that there is no overlap in their excitation and emission spectra, thus making it feasible to view each one separately without bleed-through into the emission spectrum of the other label. For example, a combination of compatible blue and red fluorophores would be AMCA (350 nm ex, 445 nm em) and Texas red (590 nm ex, 610 nm em). A compatible green-red combination could be FITC (490 nm ex, 520 nm em) and Texas red. Many other possibilities exist as well with the commercial availability of numerous fluorophores. In addition to finding suitable wavelength combinations, the choice of fluorophores might also depend on the availability of some fluorophores that are more stable than others to bleaching.

20. Fluorophores used for immunofluorescence studies must be handled with some precautions, i.e., minimizing exposure to light, to prevent rapid fading of the signal. Tissue sections should be incubated in the dark after addition of fluorophore-tagged reagents. Sections mounted on slides can be cover slipped with mounting medium, such as Vector's Vectashield, that contains an antifade compound; slides should then be stored in the dark. Water-soluble mounting media do not form an adhesive bond with the slide; therefore, cover slips are affixed to the slide by sealing the edges of the cover slip with nailpolish. Within reason, it is best to keep microscopic analysis to a minimum amount of time, because UV excitation of the fluorophore can rapidly produce bleaching (fading) of the signal.

21. Triple labeling for three different antigens may be difficult because signal inten-

sity sometimes diminishes with increasing numbers of incubation and wash steps. However, we have had excellent results when the third label is not an immunohistochemical reaction. In this case, the direct detection of retrogradely transported fluorophore-tagged microbeads is combined with dual immuno-fluorescence (for example, *see* **ref. 23**).

22. Double-labeling techniques for electron microscopy (EM) are somewhat more limited than for light microscopy, particularly if restricted to pre-embedding immunoEM. One of the most commonly used methods for dual immunolabeling for EM combines the HRP-DAB reaction with immunogold labeling *(41)*. For preembedding labeling techniques, a small colloidal gold particle (usually 1 nm) bound to the secondary antibody will not pose problems with tissue penetration. However, silver intensification is required for visualization of the gold particle. Under the proper conditions, the immunoperoxidase and silver-intensified immunogold reactions products can be easily discriminated at the EM level.

23. It is essential that no metal ions be introduced to the reactions when labeling tissue with immunogold for electron microscopy, because the artifactual silver intensification of spurious metal ions will interfere with the specific immunogold labeling. Therefore, during immunogold procedures, do not use any metal tools, metal reaction vessels, or glass vessels (which adsorb metal ions). Because the bristles are bound with metal, paintbrushes cannot be used to transfer tissue sections to reaction solutions. Instead, orange wood sticks are broken to form a soft point, which can be used with care to lift sections from the solution during the immunogold-silver intensification procedure.

24. Choice of fixatives: Paraformaldehyde (4%) is our standard fixative for immuno-histochemistry at the light microscopic level and works well for most antigens. However, we have found that optimal labeling of α_{2A}-ARs, proteins that seem to be sensitive to overfixation *(14)*, results from perfusion with PLP, a fixative that contains 2% paraformaldehyde. We have not found fixation to be an issue for labeling of α_{2C}-ARs *(15)* since the antigenicity of these proteins is apparently uncompromised even with harsher fixatives, such as glutaraldehyde or acrolein. Other fixatives that are sometimes used for immunohistochemistry, such as Bouin's, formalin, and Zamboni's solutions, must be evaluated for each antibody that is tested. Not discussed in this chapter are other methods of preparing tissue for histology, such as paraffin embedding and sectioning, which are not used routinely for preparing brain tissue for immunohistochemistry.

25. This is a lethal dose of pentobarbital that keeps the animal deeply anesthetized during the surgical preparation for perfusion.

26. Vascular perfusion with saline (heparinized to prevent clotting) is used to flush blood from tissues. If blood is not adequately removed, endogenous peroxidases found in erythrocytes will contribute a high background labeling when peroxidase-based detection systems are used in immunohistochemical experiments.

27. The duration of postfixation must be determined by balancing adequate tissue fixation with preservation of antigenicity. Conditions also depend on whether the whole brain is immersed or if the brain is first blocked into smaller pieces, in

which case better penetration would be achieved in a shorter period of time. For parformaldehyde-perfused brains prepared for cryostat sectioning or acrolein-perfused brains that will be vibrotome sectioned, postfixation is generally 30 min to 2 h. Longer postfixation (overnight) is generally needed for paraformalde-hyde-fixed brain that will be sectioned on a vibrotome in order to increase the firmness of the tissue.

28. A compromise must be made in selecting postfixation time for brains that have been perfused with PLP in preparation for labeling α_{2A}-ARs. We postfix for 4 h to provide enough additional fixation to cut sections that will survive the labeling process while not overfixing and disrupting the antigenicity of the α_{2A}-ARs.

29. Fixative for electron microscopy: Preservation of ultrastructural morphology is excellent after acrolein perfusion and immunoreactivity for α_{2A}- and α_{2C}-ARs is maintained (although a two- to fivefold higher concentration of pri-mary antibody is often required to produce product intensity that is comparable with levels found in tissue prepared for light microscopy). Tissue fixation with acrolein may not be compatible, however, for all antibodies owing to the high degree of crosslinking that occurs and the potential for disruption of critical epitopes. If antigenicity is lost, fixation with glutaraldehyde (0.5% or higher in 4% paraformaldehyde) or picric acid/paraformaldehyde can be used as alterna-tives. However, conditions for optimal fixation and immunolabeling must be de-termined for each tissue/antibody combination.

30. The choice of tissue section thickness may depend on the purpose of the experi-ment, the location of the epitope, and other practical matters. Thinner sections afford better penetration of antibodies and other immunohistochemical reagents resulting in better labeling (perhaps through the entire thickness of the section) with lower background. These advantages may be diminished, however, because thinner sections are more difficult to handle, may not hold up as well through long procedures and multiple transfers, and may be more difficult to slide mount.

31. Cryostat- or vibrotome-cut tissue sections may be stored immediately after sec-tioning in a cryoprotectant solution at –20°C prior to immunohistochemical assay. Long-term storage (up to several years) is often very convenient for tissues of particular value (e.g., animals from specific treatment groups, genetically manipulated animals) and provides a ready source of material that typically shows no loss of morphological preservation or intensity of immunohistochemical labeling. Indeed, tissue penetration may be improved in some cases. Never-theless, the stability of the antigen of interest must be evaluated for each protein of interest.

32. Cryoprotection prior to cryostat sectioning is accomplished by infiltrating the tissue with sucrose to minimize cellular damage that could be produced by the formation of ice crystals during a quick freeze. Sucrose reduces the rate and size of ice crystal formation.

33. The most common detection systems for localizing antibody bound to tissue are indirect enzymatic and fluorescence methods. In the enzymatic methods, an enzyme, such as HRP or alkaline phosphatase, is an end component of a series of

reactions. The enzyme may be coupled to the secondary antibody or part of an amplification system, such as in the PAP (unlabeled secondary antibody followed by PAP complex and peroxidase substrate) or the ABC method (biotinylated secondary antibody followed by HRP-avidin/biotin complex and peroxidase substrate). A specific substrate that is provided for the enzymatic reaction is converted to a stable, chromogenic product that is deposited on the tissue. For example, in the commonly used HRP system, the substrate DAB produces a brown peroxidase reaction product.

Fluorescent labeling can be achieved by using either (a) a secondary antibody tagged directly with a fluorophore or (b) a biotinylated secondary antibody followed by fluorophore-tagged avidin. The latter choice, which is comparable to the ABC peroxidase method, is advisable when amplification of the signal is required.

34. Fixed tissue sections must be rinsed adequately prior to immunohistochemical assays to remove any unreacted aldehydes that could interfere by nonspecifically binding antibodies to the tissue.

35. Buffers typically used for immunohistochemical reactions include phosphate and Tris buffers, and although some laboratories may adhere strictly to one particular buffer, the choice of buffer is generally not critical. Some protocols advise against the use of Tris buffers, because the amine group in Tris buffers may potentially produce artifacts by crosslinking with residual aldehyde molecules. We use Tris buffers routinely and have not found this to be a problem.

36. Tissue sections processed through long immunohistochemical procedures at room temperature are subject to bacterial contamination. Bacterial growth can be prevented by the addition of a bacteriostatic agent to buffers (e.g., 0.01% [w/v] sodium azide [Sigma] or thimerosol. **Caution:** Both of these substances are toxic). However, sodium azide must not be used in buffers during the peroxidase reaction, because azide inhibits the HRP enzyme.

37. Tissue pretreatment is essential for maintaining a low level of background labeling. In immunohistochemical protocols using colorimetric reactions for light microscopy, endogenous peroxidases are quenched by incubating tissue in H_2O_2. Unreacted aldehydes in tissue prepared for electron microscopy are quenched by incubation in sodium borohydride (*see also* **Note 26**).

38. For light microscopy, a low concentration of detergent (e.g., 0.1% Triton X-100) is routinely added during incubation of tissue with primary antibody to improve tissue penetration. Owing to the loss of ultrastructural morphology, which is evident on electron microscopic examination, it is advisable to avoid such tissue pretreatment. If tissue penetration is a problem, however, lower concentrations of Triton X-100 (e.g., 0.01–0.03%) may be employed to enhance labeling while keeping loss of morphology to a minimum. Alternatively, a rapid freeze–thaw can be used for tissue prepared for EM in a step prior to incubation with primary antibody. Sections must first be cryoprotected by equilibration in sucrose and are then subjected to a rapid freeze–thaw cycle by brief immersion in liquid nitrogen. A freeze–thaw procedure enhances antibody penetration by disrupting cell

membranes and therefore must be evaluated carefully for suitability of ultrastructural morphology. By contrast, subtle morphological changes are generally not evident when viewed by light microscopy, and tissue prepared by cryostat sectioning is often very easy to immunolabel owing to the enhanced penetration produced by the freezing process prior to and during sectioning.

39. Incubation conditions for optimal labeling, including a dilution curve for the antibody and time-course, must be determined empirically for each primary antibody. Typically, the incubation time of primary antibody with tissue sections is 24–48 h at 4°C; the total time may be reduced by first incubating at room temperature for 2–6 h before completing the overnight incubation in the cold. Immunohistochemical labeling of monolayers of cells in culture may need as little as a 1-h incubation of primary antibody with cells. We use our affinity-purified α_2-AR antibodies at a concentration of 2.5 µg/mL for immunohistochemical labeling of sections that will be viewed by light microscopy; higher concentrations are needed for immunofluorescence and immunoelectron microscopy (5–10 µg/mL). In addition to optimizing primary antibody concentration, adjusting the concentration of secondary antibody may reduce nonspecific background labeling without compromising specific immunolabeling. For example, although Vector recommends a dilution of 1:200 of its biotinylated secondary antibodies, we have observed lower background without a loss of specific labeling using the secondary antibody at a dilution of 1:400.

40. Incubation time for DAB (or other peroxidase substrate) must be determined empirically for each reaction, but 2–10 min are typical. It is best to observe the reaction immediately after addition of the DAB/peroxide substrate solution, using a dissecting microscope if needed, to achieve optimal labeling of desired structures without excessive background labeling. Stop the reaction by removing the DAB solution and rinsing sections rapidly with buffer. **Caution:** DAB is a carcinogen and must be handled using recommended chemical safety procedures (protective bench covering, disposable gloves, lab coat). DAB solutions must be collected separately and disposed of properly as chemical waste.

41. After completing the DAB reaction, sections must be rinsed copiously to remove particulate DAB from the tissue and from the reaction vessel. DAB and its peroxidase product tend to stain plasticware, especially the plastic mesh on the bottom of our histology dishes, leaving a residue of particulate material that later erodes into the buffer of subsequent reactions. Therefore, after numerous rinses and before mounting on slides, sections should be transferred to buffer in a clean Petri dish to rinse away undissolved DAB particles that might adhere to the sections and obscure specific labeling.

42. To avoid the problem of residual particulate matter, ceramic crucibles with perforated bases are used for the DAB incubation and wash steps. The DAB does not stick as readily to the crucible as it does to plastic mesh, and the crucibles are easier to clean. It is essential in preparing sections for electron microscopy that a vessel, such as the crucibles, be used for the DAB steps that will yield meticulously clean sections.

43. Immunofluorescently labeled sections can be viewed with a fluorescence microscope or may be subjected to confocal microscopy. The preparation of the tissue is identical. However, fluorophores for confocal microscopy should be chosen that exhibit adequate stability during laser excitation (so that the signal is not bleached during the analysis) and whose emission spectra are compatible with the available detectors in the confocal microscopy equipment. In many cases, the same fluorophores are used for both applications.

44. Osmium tetroxide is a highly reactive fixative that must be handled carefully and disposed of properly. Wear disposable gloves, work in a fume hood, and store waste properly for subsequent disposal following institutional guidelines.

45. Propylene oxide is toxic and must be handled carefully and disposed of properly. Use glass pipets and containers (not plastic, which will dissolve in propylene oxide), wear double gloves, work in a fume hood, and store waste properly for subsequent disposal following institutional guidelines.

46. It is important to work as cleanly as possible throughout the immunohistochemical procedure and the preparation of tissue for electron microscopy to prevent tissue sections from becoming contaminated with particulate matter that could obscure labeling when viewed at the EM level. Among the many precautions, powder-free gloves should be worn during the embedding of labeled tissue for EM.

47. Control experiments for immunohistochemistry are somewhat limited in scope, but provide some level of confidence that the observed pattern of labeling is specific. Depending on the reagents available, the following may be suitable control experiments, and each provides different types of information:

    a. Preimmune serum used in place of antisera (from the same immunized animals) is an internal control for a polyclonal antibody and indicates the baseline level of background labeling resulting from incubation with serum derived from the animal prior to immunizing. It is not an option for monoclonal antibodies (MAbs) or purified polyclonals in which serum is no longer the baseline reagent.

    b. Omission of primary antibody from the series of immunohistochemical reaction will reveal whether the signal can be attributed to nonspecific binding of secondary antibody to the tissue.

    c. If the appropriate reagents are available, preadsorption control experiments of the antibody with the antigen can be used to demonstrate that labeling is specific for an epitope contained within the antigenic protein.

    As controls for both light (colorimetric or fluorescence detection) and electron microscopy experiments, we incubate adjacent sections with α_{2A}- or α_{2C}-AR antibody that has been preadsorbed with the relevant α_{2A}- or α_{2C}-AR fusion protein. If the labeled site contains an epitope that corresponds to the protein of interest, then comparison to the paired adjacent section (incubated with antibody alone) will show that specific labeling is eliminated by preadsorption of the antibody with its fusion protein. However, even with favorable results of preadsorption experiments, there is some possibility that the antibody recognizes an irrelevant epitope in the cell. Every effort must be made in the development and testing of new antibodies to establish with reasonable certainty the specificity of the antibody prior to its use in immunohistochemical applications.

48. Documentation of immunohistochemical results can take several forms. For publication purposes, permanent photographic records can be made either with a 35mm or digital camera configured in a light microscope. Final output can be black and white or color slides, photographic prints, or digital images. Software, such as Adobe Photoshop, can be used to make and digitally label composites of photos obtained either by capturing images directly from the camera or by scanning slides, negatives, or prints. For mapping studies, it is often useful to illustrate the distribution of immunohistochemical reaction product by drawing brain sections and plotting (and electronically counting, if needed) labeled cells. We have found the Neurolucida® software (MicroBrightfield, Colchester, VT) to be extremely useful for this purpose.

References

1. Ruffolo, R. R., Jr., Nichols, A. J., Stadel, J. M., and Hieble, J. P. (1993) Pharmacologic and therapeutic applications of alpha 2-adrenoceptor subtypes. *Annu. Rev. Pharmacol. Toxicol.* **33**, 243–279.

2. Bylund, D. B., Eikenberg, D. C., Hieble, J. P., Langer,S. Z., Lefkowitz, R. J., Minneman, K. P., et al. (1994) International Union of Pharmacology nomenclature of adrenoceptors. *Pharmacol. Rev.* **46**, 121–136.

3. Sallinen, J., Haapalinna, A., Viitamaa, T., Kobilka, B. K., and Scheinin, M. (1998) Adrenergic α_{2C}-receptors modulate the acoustic startle reflex, prepulse inhibition, and aggression in mice. *J. Neurosci.* **18**, 3035–3042.

4. Sallinen, J., Link, R. E., Haapalinna, A., Viitamaa, T., Kulatunga, M., Sjöholm, B., et al. (1997) Genetic alteration of α_{2C}-adrenoceptor expression in mice: Influence on locomotor, hypothermic, and neurochemical effects of dexmedetomidine, a subtype-nonselective α_2-adrenoceptor agonist. *Mol. Pharmacol.* **51**, 36–46.

5. Lakhlani, P. P., MacMillan, L. B., Guo, T. Z., McCool, B. A., Lovinger, D. M., Maze, M., et al. (1997) Substitution of a mutant α_{2a}-adrenergic receptor via "hit and run" gene targeting reveals the role of this subtype in sedative, analgesic, and anesthetic-sparing responses *in vivo*. *Proc. Natl. Acad. Sci. USA* **94**, 9950–9955.

6. MacMillan, L. B., Hein, L., Smith, M. S., Piascik, M. T., and Limbird, L. E. (1996) Central hypotensive effects of the α_{2a}-adrenergic receptor subtype. *Science* **273**, 801–803.

7. Rohrer, D. K. and Kobilka B. K. (1998) Insights from in vivo modification of adrenergic receptor gene expression. *Annu. Rev. Pharmacol. Toxicol.* **38**, 351–373.

8. MacDonald, E., Kobilka, B. K., and Scheinin, M. (1997) Gene targeting—homing in on α_2-adrenoceptor-subtype function. *Trends Pharmacol. Sci.* **18**, 211–219.

9. McCune, S. K., Voigt, M. M., and Hill, J. M. (1993) Expression of multiple alpha

adrenergic receptor subtype messenger RNAs in the adult rat brain. *Neuroscience* **57,** 143–151.

10. Nicholas, A. P., Pieribone, V., and Hokfelt, T. (1993) Distributions of mRNAs for alpha-2 adrenergic receptor subtypes in rat brain: an in situ hybridization study. *J. Comp. Neurol.* **328,** 575–594.

11. Scheinin, M., Lomasney, J. W., Hayden-Hixson, D. M., Schambra, U. B., Caron, M. G., Lefkowitz, R. J., et al. (1994) Distribution of alpha 2-adrenergic receptor subtype gene expression in rat brain. *Brain Res. Mol. Brain Res.* **21,** 133–149.

12. Nicholas, A. P., Hökfelt, T., and Pieribone, V. A. (1996) The distribution and significance of CNS adrenoceptors examined with in situ hybridization. *Trends Pharmacol. Sci.* **17,** 245–255.

13. Aoki, C., Go, C. G., Venkatesan, C., and Kurose, H. (1994) Perikaryal and synaptic localization of alpha 2A-adrenergic receptor-like immunoreactivity. *Brain Res.* **650,** 181–204.

14. Talley, E. M., Rosin, D. L., Lee, A., Guyenet, P. G., and Lynch, K. R. (1996) Distribution of α₂A-adrenergic receptor-like immunoreactivity in the rat central nervous system. *J. Comp. Neurol.* **372,** 111–134.

15. Rosin, D. L., Talley, E. M., Lee, A., Stornetta, R. L., Gaylinn, B. D., Guyenet, P. G., et al. (1996) Distribution of α₂C-adrenergic receptor-like immunoreactivity in the rat central nervous system. *J. Comp. Neurol.* **372,** 135–165.

16. Rosin, D. L., Zeng, D., Stornetta, R. L., Norton, F. R., Riley, T., Okusa, M. D., et al. (1993) Immunohistochemical localization of alpha 2A-adrenergic receptors in catecholaminergic and other brainstem neurons in the rat. *Neuroscience* **56,** 139–155.

17. Guan, K. L. and Dixon, J. E. (1991) Eukaryotic proteins expressed in *Escherichia coli*: an improved thrombin cleavage and purification procedure of fusion proteins with glutathione S-transferase. *Anal. Biochem.* **192,** 262–267.

18. Smith, D. B. and Johnson, K. S. (1988) Single-step purification of polypeptides expressed in Escherichia coli as fusions with glutathione *S*-transferase. *Gene* **67,** 31–40.

19. Huang, L. P., Wei, Y. Y., Momose-Hotokezaka, A., Dickey, J., and Okusa, M. D. (1996) α₂B-Adrenergic receptors: Immunolocalization and regulation by potassium depletion in rat kidney. *Am. J. Physiol. Renal, Fluid Electrolyte Physiol.* **270,** F1015–F1026.

20. Kurose, H., Arriza, J. L., and Lefkowitz, R. J. (1993) Characterization of alpha 2-adrenergic receptor subtype-specific antibodies. *Mol. Pharmacol.* **43,** 444–450.

21. Huang, Y., Gil, D. W., Vanscheeuwijck, P., Stamer, W. D., and Regan, J. W. (1995) Localization of alpha 2-adrenergic receptor subtypes in the anterior segment of the human eye with selective antibodies. *Invest. Ophthalmol. Vis. Sci.* **36,** 2729–2739.

22. Stone, L. S., Broberger, C., Vulchanova, L., Wilcox, G. L., Hokfelt, T., Riedl, M. S., et al. (1998) Differential distribution of alpha2A and alpha2C adrenergic receptor immunoreactivity in the rat spinal cord. *J. Neurosci.* **18,** 5928–5937.

23. Guyenet, P. G., Stornetta, R. L., Riley, T., Norton, F. R., Rosin, D. L., and

Lynch, K. R. (1994) Alpha 2A-adrenergic receptors are present in lower brainstem catecholaminergic and serotonergic neurons innervating spinal cord [published erratum appears in *Brain Res.* [1994] **646(2),** 356,357]. *Brain Res.* **638,** 285–294.

24. Winzer-Serhan, U. H. and Leslie, F. M. (1997) α_{2B}-adrenoceptor mRNA expression during rat brain development. *Dev. Brain Res.* **100,** 90–100.

25. Lee, A., Wissekerke, A. E., Rosin, D. L., and Lynch, K. R. (1998) Localization of α_{2C}-adrenergic receptor immunoreactivity in catecholaminergic neurons in the rat central nervous system. *Neuroscience* **84,** 1085–1096.

26. Lee, A., Talley, E., Rosin, D. L., and Lynch, K. R. (1995) Characterization of alpha 2A-adrenergic receptors in GT1 neurosecretory cells. *Neuroendocrinology* **62,** 215–225.

27. Lee, A. and Lynch, K. R. (1996) Intracellular α_{2A}-adrenergic receptors in neurons and GT1 neurosecretory cells, in *Alpha2-Adrenergic Receptors: Structure, Function and Therapeutic Implications* (Limbird, L. and Lanier, S., eds.), pp. 129–139.

28. Milner, T. A., Lee, A., Aicher, S. A., and Rosin, D. L. (1998) Hippocampal α_{2A}-adrenergic receptors are located predominantly presynaptically but are also found postsynaptically and in selective astrocytes. *J. Comp. Neurol.* **395,** 310–327.

29. Lee, A., Rosin, D. L., and Van Bockstaele, E. J. (1998) α_{2A}-adrenergic receptors in the rat nucleus locus coeruleus: subcellular localization in catecholaminergic dendrites, astrocytes, and presynaptic axon terminals. *Brain Res.* **795,** 157–169.

30. Lee, A., Rosin, D. L., and Van Bockstaele, E. J. (1998) Ultrastructural evidence for prominent postsynaptic localization of α_{2C}- adrenergic receptors in catecholaminergic dendrites in the rat nucleus locus coeruleus. *J. Comp. Neurol.* **394,** 218–229.

31. Venkatesan, C., Song, X. Z., Go, C. G., Kurose, H., and Aoki, C. (1996) Cellular and subcellular distribution of α_{2A}-adrenergic receptors in the visual cortex of neonatal and adult rats. *J. Comp. Neurol.* **365,** 79–95.

32. Aoki, C., Venkatesan, C., Go, C. G., Forman, R., and Kurose, H. (1998) Cellular and subcellular sites for noradrenergic action in the monkey dorsolateral prefrontal cortex as revealed by the immunocytochemical localization of noradrenergic receptors and axons. *Cereb. Cortex* **8,** 269–277.

33. Miettinen, R., Holmberg, M., Tapiola, T., Regan, J. W., Riekkinen, P. J., Sr., and Scheinin, M. (1998) α_{2A} and α_{2C}-adrenoceptors reside in segregated locations in the rat hippocampus. *Soc. Neurosci. Abstracts* **24,** 596.

34. Saunders, A., Rosin, D. L., and Van Bockstaele, E. J. (1998) Ultrastructural evidence for prominent postsynaptic co-localization of α_{2C}-adrenergic and mu (μ)-opiate receptors in the rat locus coeruleus. *Soc. Neurosci. Abstracts* **24,** 854.

35. Milner, T. A., Rosin, D. L., Lee, A., and Aicher, S. (1999) Alpha$_{2A}$-adrenergic receptors are primarily presynaptic heteroceptors in the C1 area of the rat rostral ventrolateral medulla. *Brain Res.* **821,** 200–211.

36. Aicher, S., Drake, C. T., Rosin, D. L., and Milner, T. A. (1999) α_{2A}-adrenergic receptors in the gigantocellular reticular formation mediate vasodepressor effects of clonidine and are located primarily at postsynaptic dendritic sites. *J. Comp. Neurol.* submitted.

37. McLean, I. W. and Nakane, P. K. (1974) Periodate-lysine-paraformaldehyde fixative. A new fixation for immunoelectron microscopy. *J. Histochem. Cytochem.* **22,** 1077–1083.
38. Watson, R. E., Jr., Wiegand, S. J., Clough, R. W., and Hoffman, G. E. (1986) Use of cryoprotectant to maintain long-term peptide immunoreactivity and tissue morphology [published erratum appears in *Peptides* [1986] **7(3),** 545]. *Peptides* **7,** 155–159.
39. Hsu, S. M., Raine, L., and Fanger, H. (1981) Use of avidin-biotin-peroxidase complex (ABC) in immunoperoxidase techniques: a comparison between ABC and unlabeled antibody (PAP) procedures. *J. Histochem. Cytochem.* **29,** 577–580.
40. Levey, A. I., Bolam, J. P., Rye, D. B., Hallanger, A. E., Demuth, R. M., Mesulam, M. M., et al. (1986) A light and electron microscopic procedure for sequential double antigen localization using diaminobenzidine and benzidine dihydrochloride. *J. Histochem. Cytochem.* **34,** 1449–1457.
41. Chan, J., Aoki, C., and Pickel, V. M. (1990) Optimization of differential immunogold-silver and peroxidase labeling with maintenance of ultrastructure in brain sections before plastic embedding. *J. Neurosci. Methods* **33,** 113–127.

33

Quantitative Light Microscopic Autoradiography of α_2-Adrenergic Radioligand Binding Sites

Ursula H. Winzer-Serhan and Frances M. Leslie

1. Introduction

Adrenergic receptors (ARs) belong to the family of G-protein-coupled receptors and can be classified into three subfamilies, with each one having at least three distinct subtypes *(1)*. Although all three subfamilies have similar affinity for the endogenous ligands norepinephrine and epinephrine, they can be characterized by their different pharmacological profiles, which was the first evidence for receptor heterogeneity, prior to the molecular cloning of nine different adrenergic receptor genes *(2)*.

Membrane binding studies contributed significantly to the characterization of adrenergic receptors, which allowed the detection of subtle pharmacological differences and resulted in the classification into α_1-, α_2-, and β-ARs. However, the analysis of anatomical distribution and characteristics of radioligand binding sites within brain and peripheral tissues is best studied in cryostat-cut tissue sections and visualized autoradiographically *(3,4)*. Receptor autoradiography can achieve high anatomical resolution, approaching the limits of light microscopy, which allows the qualitative analysis of the localization of receptor binding sites within discrete subnuclei and/or laminae. Furthermore, quantitative analysis permits the pharmacological characterization of receptors. K_d, K_i, or B_{max} values can be determined, even in small and discrete areas and, thus, can be used to study changes in receptor properties in response to treatment.

The principle of receptor autoradiography is that radiolabeled drugs, in low concentrations, selectively bind to a receptor, for which they have highest affinity. The receptors retain their pharmacological properties *in situ*, which

From: *Methods in Molecular Biology, vol. 126: Adrenergic Receptor Protocols*
Edited by: C. A. Machida © Humana Press Inc., Totowa, NJ

permits the direct measurement of drug–receptor interactions under steady-state conditions. However, binding conditions are optimized for maximal specific binding, with a high ratio of signal to noise (total to nonspecific binding), and do not necessarily represent functional receptors. In this chapter, we will describe quantitative receptor autoradiography in rat brain sections. This method has been used successfully to study α_2-AR binding sites in adult and developing rat brain *(5–9)*. Although we concentrate on the detection of α_2-AR binding sites, this method is also suitable for detecting other adrenergic or nonadrenergic receptor sites.

2. Materials

1. Glass slides.
2. Isopentane.
3. Tissue Tek OCT compound (VWR, South Plainfield, NJ).
4. Black slide box (VWR).
5. Tupperware container with desiccant.
6. [^{3}H]Rauwolscine (41–60 Ci/mol, Amersham Pharmacia Biotech, Piscataway, NJ), stored at –80°C.
7. [^{3}H]RX781094 ([^{3}H]Idazoxan) (41–60 Ci/mmol, Amersham Pharmacia Biotech), stored at –80°C.
8. [^{125}I]*para*-Clonidine (2200 Ci/mmol, Dupont NEN, Boston, MA), stored at 4°C.
9. Phentolamine (Sigma, St. Louis, MO).
10. Hyperfilm ^{3}H or Hyperfilm β-max (Amersham Pharmacia Biotech).
11. Film cassette.
12. D-19 developer (Kodak, Rochester, NY).
13. Rapid fix (Kodak).
14. Cryostat.
15. [^{3}H]-microscale standards (Amersham Pharmacia Biotech).
16. [^{14}C]-microscale standards (Amersham Pharmacia Biotech).
17. Computer-based image analysis system M3 (MCID, Imaging Res. Inc., St. Catharines, Canada).
18. Sprague-Dawley adult male rats.
19. Sodium-potassium-phosphate (Na_2KHPO_4) buffer, pH 7.7: 1.16 g KH_2PO_4, 8.52 mM, and 0.26 g Na_2HPO_4, 58.2 mM in 1 L double-distilled H_2O, made fresh on day of experiment.
20. Gelatin solution: 1% w/v gelatin, 0.05% (w/v) chromium potassium sulfate.
21. 10% HCl, 90% ethanol solution.

3. Methods
3.1. Slide Preparation

Glass slides are coated with gelatin to assure the adhesion of the tissue to the slide and to minimize nonspecific binding of the radioligand off the sections.

Gelatin has been found to do a good job, although other coating materials might work as well. For neonatal tissue, which tends to float off the slide more easily than adult sections, the gelatin concentration is doubled. Gelatin-coated slides can be stored at room temperature for a long time. However, we prefer to use the slides within 6 mo after coating.

1. Place glass slides into a metal rack.
2. Wash slides overnight in diluted detergent solution (Liquinox 1:1000).
3. Rinse slides in hot running water for 15 min.
4. Rinse slides in distilled water for 15 min.
5. Soak slides in 10% HCl/90% ethanol for 20 min at room temperature.
6. Rinse slides thoroughly in running distilled water for 15 min.
7. Dip slides in 37°C gelatin solution for 15 s.
8. Dry slides in a dust-free environment overnight, and store until use.
9. Before cutting, place gelatin-coated slides in a black slide box on ice for at least 30 min.

3.2. Tissue Preparation

The preservation of tissue morphology for light microscopy is important for the accurate anatomical analysis of receptor binding sites, which also need to be preserved for pharmacological analysis. We use fresh unfixed frozen tissue, stored at −80°C, which allows the comparison of different ligand binding sites in adjacent sections. This might not be possible in prefixed tissue, because the fixation might differentially affect different receptor populations. In addition, mRNA expression can be detected in adjacent sections (as described elsewhere, *10*) and compared to the distribution of binding sites.

1. Decapitate rats, and quickly remove tissue.
2. Freeze tissue in −20°C isopentane for 30 s.
3. Remove tissue, and place on dry ice for 5 min.
4. Store tissue at −70°C in air-tight labeled plastic bags for up to 3 mo.
5. Remove tissue from freezer, and place on dry ice.
6. Mount tissue onto cryostat chuck with Tissue Tek.
7. Place chuck inside the cryostat for 30 min, so the tissue can adjust to cutting temperature (−18°C).
8. Cut 20-μm thick sections and thaw-mount onto ice-cold gelatin-coated slides.
9. Leave the mounted sections on ice while cutting the rest of the tissue.
10. Place several sections onto the same slide.
11. Cut adjacent sections for total and nonspecific binding, and a few extra sections for scrapes and early development.
12. After sectioning, place slides in black slide box inside a larger Tupperware container filled with desiccant, close and seal Tupperware container with parafilm, and dry slides for 4 h at 4°C.
13. Store sections in air-tight boxes together with desiccant at −70°C for up to 1 mo.

3.3. Radioligand Binding

For any radioligand binding experiment, it is important to use fresh incubation solutions prepared in clean plastic or glass bottles. Any contamination with detergents or ions could compromise the experiment. It is also important that the slides are dried quickly and thoroughly after washing to avoid dissociation of the ligand from the binding sites. The binding conditions for the different α_2-AR ligands are summarized in **Table 1** (*see* **Notes 1–6**).

1. Remove slide boxes from the –70°C freezer, and maintain at room temperature for 30 min before opening the boxes.
2. Prepare fresh incubation and washout solutions, and chill on ice where necessary.
3. Transfer slides into slide racks (assays with more than eight slides) or glass coplin jars (assays with eight or fewer slides).
4. Preincubate in incubation buffer at 22°C.
5. Prepare ligand solution for total and nonspecific binding; add radiolabeled ligand (total binding) and the nonspecific inhibitor, Phentolamine (nonspecific) to incubation buffer.
6. Transfer slides to incubation solution with the radioactive ligand.
7. Transfer slides to washout buffer, and rinse in two or three changes of fresh buffer.
8. Dry slides quickly under the cold air stream of hair dryers for 1 h.
9. Store slides in open slide box placed in closed Tupperware container with desiccant overnight.
10. Do scrapes on sections that were not dried. Wipe off sections for total and nonspecific binding with a small filter paper. Place filter paper in a scintillation vial with scintillation fluid, and count the next day for an initial evaluation of the binding experiment.
11. Tape slides onto cardboard together with standards of known radioactivity.
    a. For [^{3}H]Rauwolscine or [^{3}H]Idazoxan binding, expose sections and ^{3}H-standards to tritium sensitive Hyperfilm 3H for 2–3 mo. Expose extra slides for total binding on separate film for early development to determine the optimal exposure time.
    b. For [^{125}I]*para*-Clonidine binding, expose sections and ^{14}C-standards to β-max film for 1–3 d.
12. Develop film at 22°C in D19 developer for 4 min.
13. Wash film briefly in ddH$_2$O.
14. Fix in Rapid Fix for 5 min.
15. Rinse film in running water for 30 min and air-dry.

3.4. Quantification of Receptor Binding Sites

The basic problem in quantitative receptor autoradiography is to establish reliably the relationship between relative optical density and tissue radioactivity. For tritiated ligands, [^{3}H]-labeled standards generate a good and reliable

Table 1
Radioligand Binding Conditions for Three Different α_2-AR Specific
Ligands Used in Receptor Autoradiography Assays

Radioligand	Inhibitor	Incubation buffer	Preincubation	Incubation	Washout	Film exposure
[³H]Rauw. (0.6 nM)	Phentolamine (1 μM)	Na₂KHPO₄, pH 7.7	15 Min at 22°C in Na₂KHPO₄ buffer	2 h, 4°C in Na₂KHPO₄	1 × 8 min in 50 mM Tris, pH 7.4, 0°C	8–12 wk
[³H]Idaz. (0.8 nM)	Phentolamine (1 μM)	Na₂KHPO₄, pH 7.7	15 Min at 22°C in Na₂KHPO₄	2 h, 4°C in Na₂KHPO₄	2 × 1 min in 50 mM Tris, pH 7.4, 0°C	8–12 wk
[¹²⁵I]*para*-Clonidine (0.2 nM)	Phentolamine (1 μM)	170 mM Tris, 20 mM MgCl₂, pH 7.6	20 Min at 22°C in Tris buffer	90 Min at 22°C in Tris buffer	2 × 5 min in 170 mM Tris 20 mM MgCl₂, pH 7.6, at room temperature	2 d

estimate of the bound ligand per milligram of tissue. For [¹²⁵I]-labeled ligands, [¹⁴C]-labeled brain paste standards (generated as described, *11*) are used that can be calibrated against [¹²⁵I] standards (Amersham Pharmacia Biotech) of known radioactivity. A calibration curve of relative optical density of [¹⁴C] standard against [¹²⁵I] radioactivity is then generated for each exposure time, i.e., 24, 48 or 72 h *(12,13)*. After that, [¹⁴C]-standards provide a good estimate of [¹²⁵I]-ligand bound/mg tissue (*see* **Note 7**).

1. Analyze images using a computer-based image analysis system (MCID).
2. Measure the relative optical density (ROD) of the standards, and generate a standard curve of ROD vs concentration of radioactivity in dpm/mg tissue or nCi/mg tissue (1 nCi = 2.2×10^3 dpm). Construct a fourth degree polynomial curve for optimal fit of the measurement, which is generated by the program.

   To account for the decay of [¹²⁵I], correct the standard values by dividing them by the decay factor for [¹²⁵I]*para*-Clonidine at the time of film exposure. This is usually not necessary for [³H]-labeled ligands because of the long half-live of tritium.
3. Measure optical density of discrete regions for total and in corresponding areas for nonspecific binding.
4. Subtract the values of nonspecific binding from total binding to calculated specific binding.

5. Calculate corresponding values in nCi/mg tissue by interpolation from the standard curve (dpm/mg wt/2200 = nCi/mg wet wt; for example: 1100 dpm/mg wet wt = 0.5 nCi/mg wet wt).

6. Calculate fmol/mg tissue using the given specific activity (SA) of the radioligand (nCi/mg wet wt/SA Ci/mmol = mmol/mg wet wt; for example: at an SA of 50 Ci/mmol, 0.5 nCi/mg wet wt = 10 fmol/mg wet weight).

4. Notes

1. The assay conditions described in this chapter have been optimized with regard to incubation buffer, time, and temperature to achieve maximal specific binding to α_2-ARs in unfixed tissue sections *(4,5)*. They have successfully been used to characterize α_2-AR binding sites in adult and developing rat brain *(6,7,8,14)*. Other investigators have used different assay conditions with regard to tissue preparation, incubation buffer, temperature, ionic strength, or radiolabeled ligand, all factors that can alter the binding characteristics of the receptors. Although the qualitative characterization of α_2-AR binding sites is similar *(15–18)*, quantitative analysis and, in particular, the comparisons of pharmacological values, such as K_d or K_i, are difficult to compare if assay conditions vary among studies *(4,19)*. Thus, if quantitative results from different studies are compared, the assay conditions used in these studies need to be carefully evaluated.

2. For the characterization of α_2-AR binding sites with the [³H]-labeled ligands Idazoxan and Rauwolscine, we use the isotonic buffer Na_2KHPO_4, which results in a high signal-to-noise ratio and, therefore, high specific binding. Other investigators rely on nonisotonic buffers, such as Tris, which we also use for [¹²⁵I]*para*-Clonidine or glycylglycine buffer. Autoradiographic distributions of [³H]Idazoxan or [³H]Rauwolscine obtained in either Na_2KHPO_4 or glycylglycine buffers are almost identical in distribution and intensity *(14)*, whereas specific labeling to α_2-AR binding sites is reduced in Tris buffer *(4)*.

3. The concentration of the ligands used in receptor autoradiography is very important to characterize accurately the number of binding sites and their distribution. In general, for optimal autoradiographic images, ligand concentrations are in the range of the equilibrium dissociation constant (K_d). Using these conditions, it is possible in rodents to differentiate between α_2-AR subtypes owing to the lower affinity of [³H]Rauwolscine to α_{2A}-AR *(20)*. In rat brain, where the majority of α_2-AR are either of the α_{2A}- or α_{2C}-AR subtype, the high affinity [³H]Rauwolscine binding sites correspond to α_{2C}-AR *(5,6,8)*. Thus, using the two different α_2-AR-specific ligands [³H]Rauwolscine and [³H]Idazoxan or [¹²⁵I]*para*-Clonidine, it is possible to label differentially α_{2A}- and α_{2C}-AR in rat brain (**Fig. 1**).

4. In recent years, the specificity of imidazoline-based ligands, such as [³H]Idazoxan, for α_2-AR binding sites has been questioned *(21)*. Under different experimental conditions, they also bind to imidazoline receptors with high affinity, which could result in incomplete displacement of the specific radiolabeling and an overestimation of α_2-AR-specific binding sites. However, the conditions

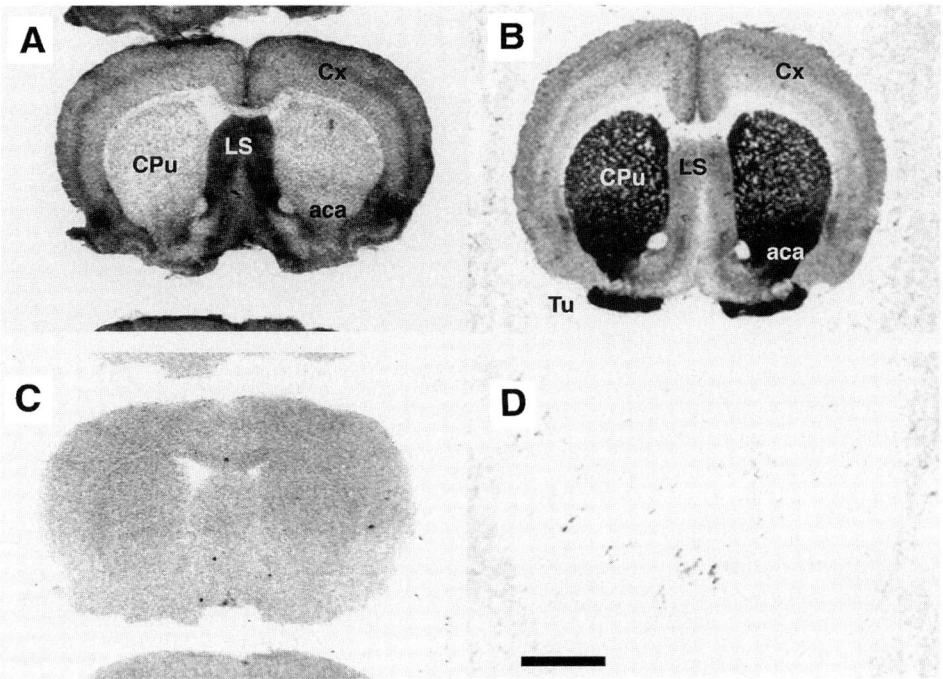

Fig. 1. Autoradiographic localization of α_2-AR high-affinity binding sites in rat brain at the level of septum. Comparable sections were labeled using 0.8 nM [^{125}I]*para*-Clonidine (**A,C**) and 0.5 nM [^{3}H]Rauwolscine (**B,D**) in the absence (total A, B) or presence (nonspecific C, D) of 1 µM phentolamine. The distribution of high-affinity [^{125}I]*para*-Clonidine binding sites closely resembles the adult distribution pattern of α_{2A}-AR mRNA *(2,7)*, whereas the distribution of high-affinity [^{3}H]Rauwolscine binding sites resembles the adult distribution of α_{2C}-AR mRNA *(3,8)*. Scale bar-300 µm. Abbreviations: aca, anterior commissure anterior, CPu, caudate-putamen; Cx, cortex; LS, lateral septum; Tu, olfactory tubercle.

described in this chapter favor binding to α_2-AR sites *(14)*. Displacement studies with nonlabeled adrenergic ligands, such as epinephrine, norepinephrine, or other α_2-AR-specific ligands, such as yohimbine, UK14304, or Clonidine, can be used to characterize the binding to α_2-AR sites and to determine the K_i values for different drugs *(5,6,14)*. However, to circumvent the problem of non-α_2-AR labeling, a number of other radiolabeled α_2-AR-specific ligands, such as [^{3}H]RX821002 or [^{3}H]MK912, are commercially available and have been used successfully to characterize α_2-AR-specific binding sites *(22,23)*.

5. One possible reason for blurry images is that the sections were not dried fast enough after washing so that the ligand diffused into surrounding tissue. This is especially a problem for fast dissociating ligands or when a large number of slides

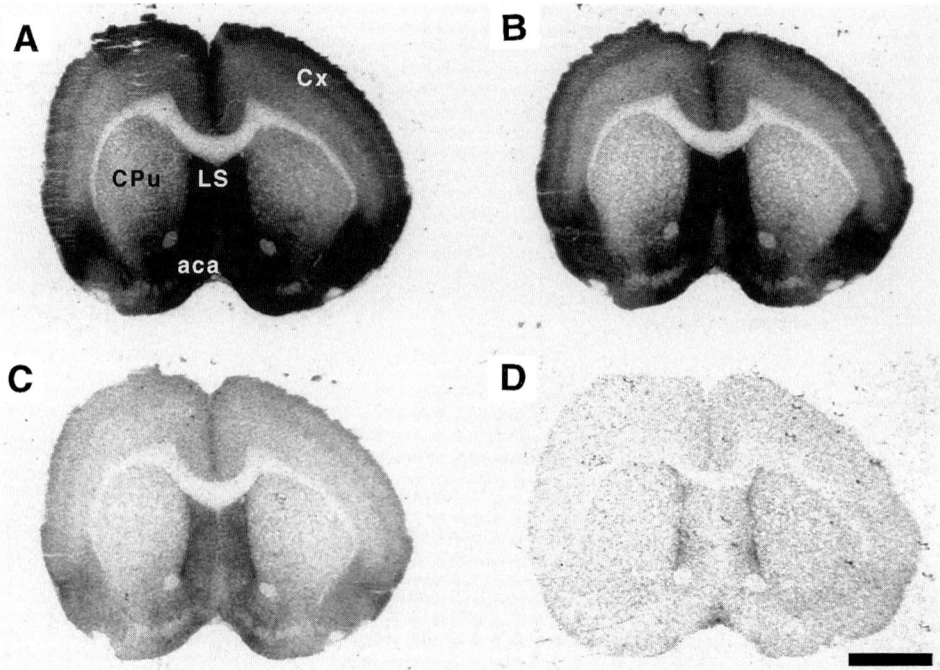

Fig. 2. Autoradiographic images of [³H]Idazoxan binding in the presence of different concentrations of Rauwolscine. Sections were incubated with 0.8 n*M* [³H]Idazoxan
plus **A**: 0.0 n*M*, **B**: 12 n*M*, **C**: 111 n*M* of rauwolscine. Nonspecific binding was determined in the presence of 10 μ*M* phentolamine (**D**). Scale bar-300 μm. Abbreviations:
aca, anterior commissure anterior, CPu, caudate-putamen; Cx, cortex; LS, lateral
septum.

   are processed at the same time. To improve the quality of images, the slides need
   to be dried as rapidly as possible.
 6. The anatomical identification of α₂-AR binding sites in small brain nuclei or
    layers using only autoradiograms can be very difficult. To facilitate the qualita-
    tive analysis, sections are counterstained after film exposure. The tissue is fixed
    in 4% paraformaldehyde and stained with cresyl violet.
 7. Quantitative analysis of autoradiographic images is possible, because the gray
    levels of the film are proportional to the concentration of radioactivity, as
    detected with [³H]- or [¹⁴C]-labeled standards. Using the specific activity of the
    radioligand, the values of radioactivity are converted into fmol/mg wet wt. This
    allows the comparison of pharmacological parameters between different experi-
    ments and treatment groups. For reliable measurements, it is important to obtain
    optical density readings in the linear portion of the calibration curve. Very light
    levels of optical density may overestimate the actual number of binding sites,

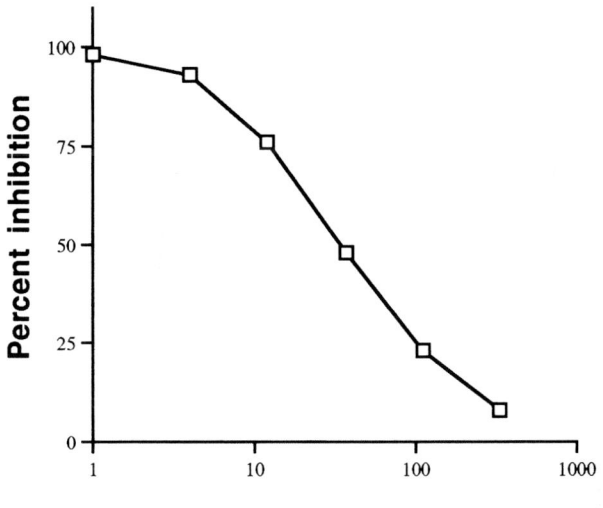

Rauwolscine concentration [nM]

Fig. 3. Inhibition of [³H] Idazoxan binding. Quantitative analysis of 0.8 n*M* [³H]Idazoxan binding in the lateral septum (LS) in the presence of increasing concentrations of Rauwolscine (0.14–333 n*M*). Nonspecific binding determined in the presence of 10 μ*M* phentolamine was subtracted from total and a dose–response curve for Rauwolscine inhibition of [³H]Idazoxan specific binding constructed. Under the binding conditions described in **Table 1**, the K_i value for Rauwoslcine displacing [³H]Idazoxan was 13 n*M* *(5)*.

whereas readings in very dark areas underestimate the number of binding sites owing to film saturation.

For analysis of B_{max} and K_d values, tissue sections are incubated with increasing concentrations of radiolabel (i.e., 0.2–20 n*M*) and for K_i values at a constant concentration of radioligand (i.e., 8 n*M* [³H]Idazoxan) with increasing concentrations of a competing nonradioactive ligand (i.e., 0.1–333 n*M* Rauwolscine) (**Fig. 2**). By subtracting corresponding values for nonspecific binding, a saturation or inhibition curve can be obtained (**Fig. 3**), and the B_{max}, K_d, or K_i values extrapolated or calculated by programs, such as LIGAND.

Acknowledgment

This work was supported by an NIDA grant T32DA 07027.

References

1. Bylund, D. B., Wikenberg, D. C., Hieble, J. P., Langer, S. Z., Lefkowitz, R. J., Minneman, L. P., et al. (1994) IVth International union of pharmacology nomencalature of adrenoceptors. *Pharmacol. Rev.* **46,** 121–136.

2. Nicholas, A. P., Hökfelt, T., and Pieribone, V. (1996) The distribution and significance of CNS adrenoceptors examined with in situ hybridization. *TIPS* **17,** 245–255.

3. Leslie, F. M. (1987) Methods used for the study of opioid receptors. *Pharmacol. Rev.* **39,** 197–249.

4. Leslie, F. M., Kornblum, H. I., and Boyajian, C. L. (1988) Pharmacological characterization of ligand binding sites in *Receptor Localization: Ligand Autoradiography* (Leslie, F. M. and Altar, C. A., eds.), Liss, New York, pp. 49–66.

5. Boyajian, C. R. and Leslie, F. M. (1987) Pharmacological evidence for alpha-2 adrenoceptor heterogeneity: Differential binding properties of [^{3}H]Rauwolscine and [^{3}H]Idazoxan in rat brain. *J. Pharmacol. Exp. Ther.* **241,** 1092–1098.

6. Boyajian, C. R., Loughlin, S. E., and Leslie, F.M. (1987) Anatomical evidence for alpha-2 adrenoceptor heterogeneity: Differential autoradiographic distributions of [^{3}H]Rauwolscine and [^{3}H]Idazoxan in rat brain. *J. Pharmacol. Exp. Ther.* **241,** 1079–1091.

7. Winzer-Serhan, U. H., Raymon, H. K., Chen, Y., Broide, R. S., and Leslie, F. M. (1997) Expression of α_2 adrenoceptors during rat brain development: I. α_{2A} mRNA expression. *Neuroscience* **76,** 241–259.

8. Winzer-Serhan, U. H., Raymon, H. K., Chen, Y., Broide, R. S., and Leslie, F. M. (1997) Expression of α_2-adrenoceptors during rat brain development: II. α_{2C} mRNA expression and [^{3}H]Rauwolscine binding. *Neuroscience* **76,** 261–272.

9. Winzer-Serhan, U. H. and Leslie, F. M. (1999) Alpha$_{2A}$ adrenoceptor expression during neocortical development. *J. Neurobiol.* **38,** 259–269.

10. Winzer-Serhan, U. H., Broide, R. S., Chen, Y., and Leslie, F. M. (1999) Highly sensitive radioactive in situ hybridization to detect α_2-adrenoceptor subtype mRNA in adult and developing rat brain. *Brain Res. Protocols* **3,** 229–241.

11. Miller, J. A. (1991) The calibration of ^{35}S or ^{32}P with ^{14}C-labeled brain paste standards for quantitative autoradiography using LKB ultrofilm or Amersham hyperfilm. *Neurosci. Lett.* **121,** 211–214.

12. Miller, J. A. and Zahniser, N. R. (1987) The use of ^{14}C-labeled tissue paste standards for the calibration of ^{125}I-labeled ligands in quantitative autoradiography. *Neurosci. Lett.* **81,** 345–350.

13. Baskin, D. B. and Wimpy, T. H. (1989) Calibration of [^{14}C]plastic standards for the quantitative autoradiography of [^{125}I]labeled ligands with Amersham hyperfilm βmax. *Neurosci. Lett.* **104,** 171–177.

14. Raymon, K. H., Smith, T. D., and Leslie, F. M. (1992) Further pharmacological charaterization of [^{3}H]Idazoxan binding sites in rat brain: evidence for predominant labeling of α_2-adrenergic receptors. *Brain Res.* **582,** 261–267.

15. Jones, C. R. and Palacios, J. M. (1991) Autoradiography of adrenoceptors in rat and human brain: α-adrenoceptors and idazoxan binding sites. *Prog. Brain Res.* **88,** 271–291.

16. Hudson, A. L., Mallard, N. J., Tyacke, R., and Nutt, D. J. (1992) [^{3}H]-RX821002: a highly selective ligand for the identification of α_2-adrenoceptors in the rat brain. *Mol. Neuropharmacol.* **1,** 219–229.

17. Alburges, M. E., Bylund, D. B., Pundt, L. L., and Wamsley, J. K. (1993) Alpha2-agonist binding sites in brain: (^{125}I)para-iodoclonidine versus (^{3}H)para-aminoclonidine. *Brain Res. Bull.* **32**, 97–102.

18. Wamsely, J. K., Alburges, M. E., Hunt, M. A. E., and Bylund, D. B. (1992) Differential localization of α_2-adrenergic receptor subtypes in brain. *Pharmacol. Biochem. Behav.* **41**, 267–273.

19. Deupree, J. D., Hinton, K. A., Cerutis, D. R., and Bylund, D. B. (1996) Buffers differentially alter the binding of [^{3}H]Rauwolscine and [^{3}H]RX821002 to the alpha2- adrenergic receptor subtypes. *J. Pharmacol. Exp. Ther.* **278**, 1215–1227.

20. Link, R., Daunt, D., Barsh, G., Chruscinski, A., and Kobilka, B. (1992) Cloning of two mouse genes encoding alpha-2 adrenergic receptor subtypes and identification of a single amino acid in the mouse alpha-2C10 homolog responsible for an interspecies variation in antagonist binding. *Mol. Pharmacol.* **42**, 16–27.

21. Michel, M. C., Brodde, O. E., Schnepel, B., Behrendt, J., Tschada, R., Motulsky, H.J., et al. (1989) [^{3}H]Idazoxan and some other alpha 2-adrenergic drugs also bind with high affinity to a nonadrenergic site. *Mol. Pharmacol.* **35**, 324–330.

22. Wallace, D. R., Muskardin, E. T., and Zahniser, N. R. (1994) Pharmacological characterization of [^{3}H]idazoxan, [^{3}H]RX821002 and p-[^{125}I]iodoclonidine binding to α_2-adrenoceptors in rat cerebral cortical membranes. *Eur. J. Pharmacol.* **258**, 67–76.

23. Uhlen, S., Lindblom, J., Johnson, A., and Wikberg, J. E. (1997) Autoradiographic studies of central alpha 2A- and alpha 2C-adrenoceptors in the rat using [^{3}H]MK912 and subtype-selective drugs. *Brain Res.* **770**, 261–266.

34

In Situ Hybridization of Adrenergic Receptor mRNA in Brain

Mark Stafford Smith and Debra A. Schwinn

1. Introduction

The advent of transgenic mice has ushered in an explosion of new knowledge regarding key roles of adrenergic receptors (ARs) in central nervous system- (CNS) mediated blood pressure control, states of alertness/sedation, analgesic pathways, and temperature control; these findings have renewed interest in examining human CNS tissue for expression of adrenergic receptor subtype mRNA in specific brain regions. *In situ* hybridization enables detection of rare mRNA species in specific cells within the human CNS. However, identification of adrenergic receptor mRNA in human tissues is difficult owing to restricted access to tissue for harvest immediately after death. Therefore, *in situ* hybridization methods need to be carefully optimized to preserve remaining mRNA signal in human studies. Although [α-^{35}S]uridine triphosphate (UTP) isotopic *in situ* hybridization is our preferred method, a summary of nonisotopic *in situ* hybridization is also included (in **Subheading 4.**). [α-^{35}S]-UTP isotopic *in situ* hybridization is highly sensitive, enabling identification of rare RNA species in human tissues *(1)*. We have had best results in human tissues when the actions of tissue RNases on mRNA signal can be arrested within 2 h of death. We have published identification of α$_2$-AR mRNA in human spinal cord using this technique *(2)*. This chapter is designed to facilitate the experienced researcher switching from nonhuman models to human tissues, as well as providing the investigator unfamiliar with *in situ* hybridization with a step-by-step approach.

In addition to radiolabeled probes, nonisotopic methods for *in situ* hybridization are increasingly employed. The reader is referred to *Current Protocols*

From: *Methods in Molecular Biology, vol. 126: Adrenergic Receptor Protocols*
Edited by: C. A. Machida © Humana Press Inc., Totowa, NJ

in Molecular Biology (3,4) for more details and updates of these particular methods. Overall principles and concepts of nonisotopic *in situ* hybridization are similar to isotopic methods presented in detail above. Several different methods of cRNA probe labeling can be utilized, including biotin and digoxigenin incorporated directly into DNA probes (these are the most frequently used methods). Biotin and digoxigenin are then detected colorimetrically using any number of fluorochromes, as well as alkaline phosphatase and horseradish peroxidose-conjugated antibodies (which produce color precipitates). The example method presented in this chapter utilizes digoxigenin-labeled probes and alkaline phosphatase.

2. Materials
2.1. Reagents

1. Ethanol (EtOH)—100% (Baxter, McGaw Park, IL).
2. Mounting medium (Pro-Texx) (Baxter).
3. Optimal Cutting Temperature (OCT) Compound (Tissue Tek) (Baxter).
4. Filter paper (Whatman) (Baxter).
5. Slides—silylated (CEL Associates, Houston, TX).
6. Cover slips (Fisher Scientific, Pittsburgh, PA).
7. Parafilm (Fisher Scientific).
8. Paraformaldehyde (4%) (Fisher Scientific).
9. Hematoxylin (Gill Biological, Lerner Laboratories, Pittsburgh, PA).
10. Developer—D19 (Kodak/IBI, New Haven, CT).
11. Fixer (Kodak/IBI).
12. Nuclear Track Emulsion (NTB2 or NTB3) D19 (Kodak/IBI).
13. Disposable Spin Columns (NuClean R50) (Kodak/IBI).
14. Ammonium acetate (Mallinckrodt AR, Paris, KY).
15. Xylenes (Mallinckrodt AR).
16. Radioactivity ([α-^{35}S]UTP) (NEN DuPont, Wilmington, DE).
17. Eosin Phloxine Stain Working (Poly Scientific, Bay Shore, NY).
18. 100 mM dithiothreitol (DTT) (Promega).
19. ATP, CTP, and UTP (Promega).
20. Riboprobe System II Kit (Promega).
21. RNasin ribonuclease inhibitor (Promega).
22. SP6 or T7 polymerase (Promega).
23. Transcription buffer (5X) (Promega).
24. Scintillation cocktail (Research Products International, Mount Prospect, IL).
25. Plastic hybridization boxes (Rubbermaid, Wooster, OH).
26. β-ME (Sigma Chemicals, St. Louis, MO).
27. Chloroform, molecular biology grade (Sigma Chemicals).
28. Denhardt's Solution (50X), molecular biology grade (Sigma Chemicals).
29. Dextran sulfate, (50%) molecular biology grade (Sigma Chemicals).

30. Diethyl pyrocarbonate (DEPC) (Sigma Chemicals).
31. 1 *M* DTT, molecular biology grade (Sigma Chemicals).
32. Formamide (high and low grade) (Sigma Chemicals).
33. Phenol:chloroform:isoamyl alcohol 25:24:1, saturated with 10 m*M* Tris-HCl, pH 8.0. 1 m*M* ethylenediamine tetra-acetic acid (EDTA) (TE), molecular biology grade (Sigma Chemicals).
34. Proteinase K (Sigma Chemicals).
35. RNase A (Sigma Chemicals).
36. Salmon sperm DNA (Sigma Chemicals).
37. 20X Saline-sodium citrate (SSC) (Sigma Chemicals).
38. Sucrose, molecular biology grade (Sigma Chemicals).
39. tRNA (Sigma Chemicals).
40. Desiccant pellets (United Desiccants, Voorhees, NJ).
41. Agarose (USB, Cleveland, OH).
42. Eppendorf tubes (1.2 mL) (VWR, Raleigh, NC).

2.2. Equipment

1. Black electrical tape (any hardware store).
2. Chemical hood for work with stains/xylenes.
3. Coplin jar for emulsion (Fisher Scientific).
4. Darkroom with #2 Kodak safe light.
5. Leitz Kryostat 1720 digital with knives, or equivalent cryostat.
6. Oven/incubator, 50°C.
7. Refrigerated water bath, 15°C (Fisher Scientific).
8. Slide dryer (EM Corp, Chestnut Hill, MA).
9. Slide grip, plastic, five slides (VWR).
10. Slide holders and staining dish set, Tissue-Tek slide staining set with 12 solution wells (Baxter, Stone Mountain, PA).
11. Speed Vac (Baxter, Stone Mountain, PA).
12. Water bath, 40°C, in a darkroom.

2.3. Solutions

1. Ammonium acetate (10 *M*), DEPC-treated and autoclaved.
2. SSC buffer:
   20X SSC: 0.3 *M* sodium citrate, pH approx 7.0, containing 3 *M* NaCl.
   2X SSC: 1:10 dilution (v/v) of 20X SSC concentrate in DEPC-treated water.
   1X SSC: 1:20 dilution (v/v) of 20X SSC concentrate in DEPC-treated water.
3. "Box buffer": Volume required depends on the size of container used.
   Made by mixing equal volumes of:
   a. Formamide (full strength).
   b. 2X SSC.
4. DEPC-treated water: 5 L. Prepare 0.1% solution (1.0 mL DEPC/L) with highest-

quality filtered water, stirring overnight vigorously under a hood at room temperature. Autoclave (to inactivate DEPC) for 40 min.

5. Ethanol: 100, 95, 70, 50%. Dilute with DEPC-treated water.
6. Hybridization buffer. **Note**: The recipe below is for 1.5 mL total. Prepare enough hybridization buffer to allow for at least 600–700 μL/slide. (*See* detailed protocol for scale-up calculations.)

"Best frozen" formamide	750 μL
Denhardt's solution (50X)	30 μL
Dextran sulfate (50%)	240 μL
DTT (1 *M*)	30 μL
Salmon sperm DNA (10 mg/mL)*	150 μL
SSC (20X stock)	150 μL
tRNA (10 mg/mL)	150 μL
Total	1500 μL

7. Paraformaldehyde (4%) for tissue fixation
   a. Prepare 1 L phosphate-buffered saline (PBS) (1X)—use quality filtered water in baked flask (2 L).
   b. Treat solution with 0.1% DEPC (1.0 mL DEPC/L stirring overnight vigorously under a hood at room temperature). Autoclave (to inactivate DEPC) for 30 min.
   c. Approximately 30 min after autoclaving has been completed (when solution cools to about 70–80°C), slowly add 40 g paraformaldehyde under a hood (also use a mask). While cooling, stir for approx 1 h. (Note: Solution will be cloudy after stirring.)
   d. Filter-sterilize (0.2 μm filter unit, 1 L).
   e. Store in a light-protected bottle at 4°C. Shelf life approx 2 wk.
8. PBS (1X) Dry powder packages are available to dissolve in DEPC-treated water:
   Potassium chloride 2.7 m*M*
   Phosphate buffer salts 10 m*M*
   Sodium chloride 120 m*M*
   pH 7.4 at 25°C.
9. Sodium phosphate (0.5 *M*), pH 6.5, in DEPC-treated water and filtered (0.2 μm).
10. 20% Sucrose in PBS.
    a. Prepare 1 L of PBS (1X) using quality filtered water in baked flask (2 L).
    b. DEPC-treat solution at 0.1% final concentration (=1.0 mL DEPC/L) stirring overnight vigorously under a hood at room temperature. Autoclave (to inactivate DEPC) for 30 min.
    c. Cool until just warm to touch.
    d. Add 20 g sucrose/100 mL DEPC-treated PBS (Sigma, molecular biology reagent grade), stirring until sucrose dissolves completely.
    e. Store at 4°C.
11. Triton X-100 (0.3%) in 1X PBS.

---

*Heat shear salmon sperm DNA at 80°C for 10 min prior to adding.

2.4. Wash Solutions

1. Formamide/2X SSC (1:1).
2. Formamide/2X SSC and β–ME (1:1000 v/v) (1:1)
3. 2X SSC and 300 μL/L β–ME (50°C).
4. Formamide/2X SSC and 300 μL/L β-ME (1:1, 50°C).

2.5. Dehydration Solutions

1. 50% EtOH/0.3 M ammonium acetate.
2. 70% EtOH/0.3 M ammonium acetate.
3. 95% EtOH/0.3 M ammonium acetate.
4. 100% EtOH/0.3 M ammonium acetate.

2.6. Solutions and Chemicals for Nonisotopic In Situ *Hybridization*

1. Buffer A: 100 mM Tris-HCl, pH 7.5, 150 mM NaCl.
2. Buffer B: 3% normal goat serum, 3% Triton X-100.
3. Antidigoxigenin solution: 1:1000 to 1:5000 dilution of anti-dig-Fab fragments (Boehringer Mannheim, Indianapolis, IN) and 3% normal goat serum in buffer A.
4. Buffer C: 100 mM Tris-HCl, pH 9.5, 100 mM NaCl, 50 mM MgCl$_2$.
5. Nitroblue tetrazolium chloride.
6. Dimethylformamide (Boehringer Mannheim).
7. 5-Bromo-4-chloro-3-indolyl phosphate toluidine salt.

3. Methods

3.1. Preparation of Human Tissues (see Notes 1 and 2)

During tissue harvest (ideally collected <2 h after death), cut tissue blocks into manageable pieces (i.e., section size should permit slide-mounting without distortion).

1. To provide initial tissue fixation, immediately immerse tissue in ice-cold 4% paraformaldehyde solution (4°C) in 50-mL disposable conical tubes (4–6 h, no more than 24 h).
2. To eliminate air and freezing artifact, rinse tissue in ice-cold 20% sucrose in PBS, and then immerse in fresh ice-cold 20% sucrose in PBS (~5X tissue volume). Keep at 4°C, and do not remove until the tissue sinks (usually 24–72 h, up to 10 d for lung tissue).
3. Embed tissue in OCT compound using a square-based (2- to 3-cm sides) aluminum foil mold. Place tissue in mold with attention to orientation (the flat base will become the surface to be sectioned); the tissue block should not be higher than 2 cm. Add OCT compound slowly around the base, finally covering tissue; make sure to avoid bubbles (makes sectioning difficult).
4. Freeze the tissue-OCT block slowly (over 1–5 min) by lowering a corner of the mold into liquid nitrogen. The rate of freezing can be judged by the opaque white

appearance of frozen OCT. Tissue frozen too quickly will crack. Tissue blocks are stored at −70°C until use.

5. Note: An alternate tissue preparation method involves immediate freezing of samples in liquid nitrogen at harvest; tissues are then stored at −70°C. OCT is added to the tissue just prior to sectioning (**steps 3** and **4**). If this method is employed, then a tissue fixation step must be added just prior to hybridization (*see below*).

6. Slide preparation and cryosectioning: Prior to tissue sectioning, assure that the cryostat (Leitz Kryostat 1720 digital) and tissue block are at −20°C. Place tissue block in cryostat 30 min before sectioning to allow temperature equilibration.

7. Label silyated slides using lead pencil (most durable in our experience). Keep slides at room temperature. With the many slide immersions and washes during this protocol, tissue loss is a concern; silyated slides (and gentle handling) reduce separations of sections from slides.

8. To mount for cutting, adhere the tissue block to the −20°C chuck (within the cryostat) using several drops of fresh OCT; the flat base should be oriented for sectioning. Add more OCT around the block base for bracing.

9. Cut 5- to 10-μm tissue sections and thaw-mount (touch labeled side of room temperature silyated slide gently to tissue section at approx 45°; tissue will transfer to slide). Consider placing several tissue samples on each slide.

10. Air-dry slides at room temperature for 1–5 min, and then transfer to a cold (−20°C) slide box, which may be temporarily stored in the cryostat, or a nearby freezer.

11. Add desiccant pellets (to avoid section damage, face blank side of slide to pellets); store at −70°C until use.

3.2. Generation of Radiolabeled Riboprobes (see **Note 3**)

1. Linearization of DNA template: A DNA template is used to generate RNA probe using RNA polymerases (SP6, T3, or T7). A plasmid containing the desired DNA and two polymerases are necessary (to generate sense and/or antisense probes); pGEM vectors (Promega) are convenient in this regard.

2. Additional positive control probes, such as antisense β-actin, may be useful; β-actin is abundant in many human tissues. This probe can be used to confirm *in situ* hybridization (e.g., with a negative test probe result) and can also help define tissue architecture.

3. Linearize probe DNA using a restriction enzyme located at one end of the fragment in the following mixture (including 20 μg DNA):

    __ μL   (20 mg DNA total)
    __ μL   desired restriction enzyme (5 U/μg DNA, max 20 μL)
    20 μL   10X enzyme buffer
    __ μL   DEPC-treated water (to bring volume to 200 μL)
  = 200 μL   total volume

4. Mix well and incubate at 37°C for 1 h.

5. Confirm that plasmid linearization has occurred; use 5 µL of the mix to run on a 1% agarose gel.

6. Remove nucleases; add 10 µL proteinase K to the mixture (10 µg proteinase K/1 µg DNA; for 30 min at 37°C).

7. Extract DNA; add 100 µL TE-saturated phenol and 100 µL chloroform to the 200 µL DNA mixture. Vortex and then spin in microfuge at maximum speed (5 min). Pipet top 200 µL (water layer containing DNA) into new Eppendorf tube.

8. Precipitate DNA; add 100 µL DEPC-treated 7.5 M NH$_4$-acetate, and 600–900 µL 100% ethanol (precipitation is maximized at –20°C overnight).

9. Create DNA pellet; centrifuge (12,000g) for 30 min at 4°C in refrigerated centrifuge; pour off solution, leaving pellet. Wash the pellet with 500 µL cold 70% ethanol, centrifuge for 10 min at 4°C, and pour off solution, leaving pellet. Repeat this step using 500 µL cold 100% ethanol.

10. Lyophilize pellet for 10 min at room temperature.

11. Resuspend pellet in DEPC-treated H$_2$O.

12. Transcription of Riboprobe: All transcription reagents (except the radioactivity) may be purchased in a kit; we have used the Riboprobe system II (Promega). Prepare radioactive nucleotides; for each riboprobe to be generated, lyophilize 0.25 mCi [α-^{35}S]UTP (NEN, 1200 Ci/mmol) in a 1.5-mL Eppendorf tube. Be sure radioactivity is not limiting compared with amount of probe added.

13. Generate riboprobe: add the following ingredients, in order (top to bottom), keeping at room temperature.

    8.5 µL DEPC-treated H$_2$O

    4 µL transcription optimized 5X buffer

    2 µL 100 mM DTT.

    3 µL (1 µL each of ATP, CTP, and GTP).

    0.5 µL RNasin ribonuclease inhibitor (40 U/µL stock).

    1 µL linearized DNA template (1 µg/µL, *see* **Step 1** *above*). Note: Heat DNA to 80°C (5 min) immediately before adding.

    1 µL SP6, T3, or T7 RNA polymerase, 15 U/µL.

    20 µL total volume.

14. Incubate at 37°C for 60 min.

15. Stop the reaction; add 30 µL 1X transcription buffer (dilute 5X transcription buffer with DEPC-treated H$_2$O). Total volume is now 50 µL.

16. Use radioactivity incorporation to confirm successful RNA transcription: Place 1 µL probe mixture on each half of a DE-81 filter (Whatman); wash one of the two halves in 25 mL 0.5 M sodium phosphate (pH 6.5); gently rotate for 15 min. Count each half filter (washed and unwashed) in 10 mL scintillation cocktail. Divide washed (incorporated radioactivity) by unwashed (total radioactivity) counts, and then multiply by 100 to determine % incorporation. Repeat transcription if <20% incorporation (60–80% is considered good).

17. If efficient incorporation is confirmed, purify the probe using extractions: Add 25 µL cold (4°C) chloroform and 25 µL cold (4°C) TE-saturated phenol; vortex;

spin at maximum speed in a microfuge for 5 min; remove top (water) layer (50 μL) containing labeled RNA probe into a clean 1.5-mL Eppendorf tube. Repeat for two further extractions, using 50 μL cold (4°C) chloroform. Add 1 μL tRNA (10 mg/mL) to each probe suspension to facilitate later precipitation.

18. For further probe purification: Use Nuclean R50 Disposable Spun Column; centrifuge for 4 min at 3000*g* (the columns adsorb unincorporated nucleotides, and labeled probe is contained in eluate).

19. Add 1–2 μL 1 *M* DTT to the collected sample. At this point, the probe may be precipitated overnight by adding 1/5 vol (5 μL) 10 *M* ammonium acetate and 2.5 vol (75 μL) cold 100% ethanol (store at –20°C or –70°C).

20. Centrifuge sample for 45 min (12,000*g*, 4°C). Carefully remove solution leaving pellet, and then invert tube on filter paper to drain.

21. Wash pellet with 200 μL 70% ethanol (prepared with DEPC-treated H_2O), and centrifuge for 10 min (12,000*g*, 4°C).

22. Wash pellet with 200 μL 100% ethanol, and centrifuge for 10 min (12,000*g*, 4°C).

23. Lyophilize for 10 min at 42°C, and then resuspend pellet in 100 μL DEPC-treated H_2O and 1 μL 1 *M* DTT.

24. Spot 1 μL of sample on filter paper (do not wash), and count as before.

3.3. Hybridization Conditions (see Notes 4 and 5)

1. In preparation for hybridization:
   a. Turn on heat block to 80°C.
   b. Turn on hybridization oven to 50°C.
   c. Prepare hybridization containers (Rubbermaid, plastic), and place box buffer-soaked filter paper on floor of the container (high humidity avoids tissues drying during hybridization). Place parallel pieces of pipets in containers, on which slides are placed to avoid contact with the filter paper.

2. Prehybridization is not usually necessary in human tissues for best results, unless background signal is expected to be high. Prehybridization involves placing 600–700 μL/slide hybridization buffer (*see* recipe *below*) without radiolabeled probe over tissue samples (2–6 h, 50°C) before the hybridization step.

3. Preparation of hybridization solution: add [^{35}S]-labeled linearized probe to hybridization buffer to achieve a count of 5000–7000 cpm/μL. Calculate the amount of probe required assuming 600–700 μL/slide (i.e., 2 x 10^6 cpm/slide) as follows:

For 12 slides, total amount [^{35}S]-labelled linearized probe (mL) required =

$$(600 \ \mu L/slide \times 12) \times 6000 \ cpm/\mu L \qquad (1)$$

Total cpm/μL of [^{35}S]-labeled linearized probe counted
(typically ~400,000 cpm/mL)

$$= \frac{7200\ \mu L \times 6,000\ \text{cpm}/\mu L}{400,000\ \text{cpm}/\mu L} \qquad = 108\ \mu L\ \text{radiolabeled RNA probe} \qquad (2)$$
$$\text{(in 7.2 mL hybridization buffer)}$$

4. To confirm count of 5000–7000 cpm/μl, warm hybridization solution to 50°C (10 min in the hybridization oven, until thin), then pipet 5 μL into a vial containing 5 mL scintillation cocktail, and count. If count is outside the limits, use this information to calculate the required addition of radiolabeled probe (if count is too low) or hybridization buffer (if count is too high) to achieve the correct count. Then recheck the count. Keep hybridization solution at 50°C until ready to use. Each experiment requires two tubes (sense and antisense probes).

5. Remove slide boxes from –70°C freezer; without removing the slides, warm slowly with hair dryer (10 min).

6. Fixation: a balance must be found between inadequate (which will not allow good preservation of architecture) and excessive (which will obstruct probe access to the tissue) tissue fixation.

   a. If tissue sections were snap-frozen in liquid nitrogen, then fix tissue section on slides with cold 4% paraformaldehyde (10 min).

   b. If tissue is fixed with cold 4% paraformaldehyde immediately following harvest, then either bypass **step a** completely, or fix tissue section on slides with cold 4% paraformaldehyde for only 5 min. **Note**: Depending on your decision, RNase concentration in the washes the following day will vary.

   c. Management of slides for cell controls (i.e., cells with known mRNA expression). Note: Separation of cells from slides may occur if care is not observed during washes.

   First, fix tissue section on slides with cold 4% paraformaldehyde (5–10 min), and then wash briefly in 1X PBS (~30 s). Incubate in 0.3% Triton X-100 (a detergent to permeabilize cells), diluted with 1X PBS (5 min). Wash again briefly in 1X PBS, and then treat as all other slides.

7. Immerse all slides twice in 2X SSC (3–5 min/rinse).

8. Remove slides individually, use blotting paper to absorb solution from back and nontissue areas of front of slide (will dictate region of probe/hybridization buffer spread).

9. Add 600–700 μL/slide of hybridization buffer/probe solution (cover the tissue). If the tissue dries during the overnight incubation, the experiment is wasted; therefore, it is critical to use sufficient solution.

10. Place slides in prepared hybridization boxes, suspended on pipets over box buffer-soaked filter paper. Ensure that no slides are touching filter paper, since the hybridization solution will then be wicked off and the tissue will dry. When a box of slides is complete, seal the lid with parafilm, and incubate in the hybridization oven overnight (50°C).

11. Preparation for washes: two water baths, one outside a hood (35°C) and the second inside a hood (50°C), and slide racks for gentle dipping (we use 25 slide holders, which fit into 250-mL dipping containers [Tissue-Tek]. Calculation of

required volumes of wash solutions depend on the slide number, container, and rack size. The following amounts reflect volumes required for one rack (250 mL/container).

2X SSC	1000 mL	room temperature
2X SSC	250 mL	50°C
2X SSC and RNase[†]	250 mL	35°C
50% Formamide/50% 2X SSC	250 mL	50°C
50% Formamide/50% 2X SSC + β–ME[†]	250 mL	50°C

Concentrations of RNase in the wash solutions are dictated by the method used to fix the slides:

10 µg/mL—if the tissue was frozen after fixation in 4% paraformaldehyde, but was not fixed again prior to hybridization.

5 µg/mL—if the tissue was (a) frozen after fixation in 4% paraformaldehyde and fixed for 5 min with 4% paraformaldehyde prior to hybridization, or (b) frozen in liquid N_2 after harvest and fixed for 10 min with 4% paraformaldehyde prior to hybridization).

12. When wash solutions are ready, containers are removed from the hybridization oven and slides are lifted out individually; hybridization solution is allowed to run off each slide (onto a fresh bench pad or into a container for disposal with radioactive precautions), which is then placed in a slide rack for washes. Although solution is poured off, it remains critical that tissue samples do not dry; the first slide must still be moist when the last is processed!

13. Washes—immerse racks stacked with slides in the following sequence. Note: For optimal tissue preservation, sections on end slides (at each end of rack) should be turned inward, facing the center of the rack (i.e., not be directly exposed to turbulence).
    a. 2X SSC + 300 µL/L β–ME (50°C, 10 s).
    b. 50% Formamide/50% 2X SSC + 300 µL/L β–ME (50°C, 10 min).
    c. 50% Formamide/50% 2X SSC + 300 µL/L β–ME (50°C, 20 min).
    d. 2X SSC (room temperature, 10 min).
    e. 2X SSC + RNase (*see* concentrations *above*) (35°C, 30 min).
    f. 2X SSC (room temperature, 5 min).
    g. 50% Formamide/2X SSC (50°C, 10 min).
    h. 2X SSC (room temperature, 5 min).
    i. 2X SSC (room temperature, 10 min).

14. Dehydrate slides by placing slides into the following solutions:
    a. 50% EtOH/0.3 *M* ammonium acetate (2 min).
    b. 70% EtOH/0.3 *M* ammonium acetate (2 min).
    c. 95% EtOH/0.3 *M* ammonium acetate (2 min).
    d. 100% EtOH (5–10 s, no longer!).

[†]β–ME and RNase are added just before the solutions are utilized.

15. Air-dry slides in hood for 30 min. Sections containing vessels require up to 2 h to prevent depression at the vessel center and edge. Depression can cause changes in refraction once emulsion is added.

3.4. Sample Visualization (see Note 6)

1. Preparation for emulsion application—autoradiography emulsion should be warmed in a coplin jar placed in a 40°C water bath (in a darkroom using a #2 Kodak safelight). Do not remove the emulsion from its light-protected container until you are ready to use it. Prepare slides in slide holders for air-drying (prior to entering the darkroom), making sure the frosted labeled portion of each slide is inserted into the slide holder (this part will not be immersed in emulsion). We use Oncor slide holders; each slide dryer holds four racks of five slides. Bring slides into darkroom, wait several minutes for your eyes to adjust; avoid all light sources except the Kodak #2 safelight until the emulsion application is completed and the slide boxes sealed to avoid light leaks.
2. Pour autoradiography emulsion into warmed glass slide holder.
3. Dip slides in emulsion for 2 s. Hold slides up to examine visually that tissue sections have been completely covered; this can be a major source of lost data.
4. Place slide holders into the dryers. Oncor dryers use cool circulating air.
5. Allow slides to dry for about 2 h.
6. Under the Kodak #2 safelight, remove slides from slide holders and place in black slide boxes. Leave room for two to three desiccant pellets (to minimize moisture during storage at 4(C) at each end of boxes, and ensure sections on outermost slides are facing inward to avoid tissue damage.
7. Apply black electrical tape around the slide box lids to keep them closed and avoid light leak. We wrap a complete batch of slide boxes in tin foil before placing them in the darkroom refrigerator. Slides should be stored anywhere from 25 d (for stronger signals) to 42 d (for weaker signals). Four to 5 wk are a good starting point for most signals in our experience. Background signal increases with longer exposures.
8. Preparation for emulsion development: The Kodak safelight #2 should be used throughout the development process. Mix the developer (D19 Kodak developer mixed 1:1 with distilled H_2O, cooled to exactly 15°C in a refrigerated water bath). Prepare undiluted Kodak fixer.
   Note: A fresh lot of developer mix and fixer should be used for each rack of slides. This is usually best achieved by preparing as many containers of developer mix and fixer as required for a whole batch prior to starting the developing process.
9. At a prescribed time, several weeks after emulsion application, remove slides from the darkroom refrigerator (using safelight precautions). Warm slide boxes for either 30 min with a hair dryer (cool setting) or 90 min at room temperature (this will avoid condensation).

10. Place slides in dipping rack (similar to those used for washes). Develop slides by immersing slides into the following solutions:
Kodak developer: distilled H_2O (1:1 mix) (15°C, 4 min).
Distilled, filtered H_2O rinse (room temperature, 20 s).
Kodak fixer (room temperature, 5 min) (lights may be turned on when all slides have passed this step).
Distilled, filtered H_2O rinses (three times, room temperature, 5 min each)
11. Slides may remain briefly in final H_2O rinse for transfer to staining area.
12. Preparation for staining and counterstaining: Perform procedure in a hood for proper ventilation. Staining is performed to facilitate observation of tissue morphology. One container of hematoxylin solution and one container of eosin-phloxin stain should be filtered (these can be used for a whole slide batch). Set up sequential dipping containers for stain, counterstain, H_2O rinses, EtOH, and xylenes. Slides should remain in the slide racks used for developing. All solutions are at room temperature. Fresh filtered H_2O should be used for each rinse; all other solutions may be reused for a complete slide batch.

Graded ethanol immersions decrease lipid tissue content, leaving mainly fibroskeleton; they also decreasing background. The ascending ethanol and xylenes series also dehydrates the tissue and clears the stain. A "milky" appearance of xylenes indicates water saturation, and the solution should be discarded (in accordance with safety standards). Xylenes help the tissue to have the same index of refraction as Pro-Texx mounting media and also provide a medium in which Pro-Texx is soluble.
13. Counterstain slides by immersing slides into **steps a–e**; if counterstaining is not desired, use only **steps "f"** and onward:
    a. Filtered hematoxylin (1 min).
    b. Filtered H_2O (few seconds).
    c. Filtered H_2O (2 min).
    d. Filtered H_2O (5 min).
    e. Filtered eosin-phloxine stain (brief dip).
    f. 70% EtOH (3 min).
    g. 95% EtOH (3 min).
    h. 100% EtOH (3 min).
    i. 100% EtOH (3 min).
    j. Xylenes (5 min).
    k. Xylenes (5 min).
    l. Xylenes (5 min).
14. Remove individual slides from the last xylenes wash. Lay the slide flat on a paper towel, and adhere a cover slip to cover tissue section with Pro-Texx mounting medium. Note: Avoid even small bubbles under the cover slip, since they will interfere with analysis of results.
15. Allow slides time to dry (at least 24 h) before moving slides, or attempting to clean excess remaining emulsion from the back of the slide and around cover

slip. Glass cleaner will soften the emulsion, which can be scraped off with a razor blade. Avoid scratching slides, since this may impair darkfield imaging.

16. Enhance labeling blurred through the hybridization process. Store slides in dust-free slide folders.

17. Positive signal (radiolabeled mRNA) will show on slides as silver granules. Silver granules appear as black speckles using light-field microscopy and as white granules using dark-field microscopy.

3.5. Nonisotopic In Situ *Hybridization*

1. The labeling reaction is identical to isotopic methods, except for incorporation of digoxigenin-labeled UTP, taking advantage of RNA polymerases present in the DNA template to make labeled RNA probes.

2. After precipitation of digoxigenin-labeled probes, the pellet should be resuspended in 50 μL TE and 1 μL RNasin (vigorous vortexing and incubation at 37°C for 10 min may be necessary to dissolve the pellet).

3. During hybridization, approx 1 μL digoxigenin-tagged probe solution is used/ 100 μL hybridization solution.

4. Digoxigenin-tagged RNA visualization (using alkaline phosphatase-linked antibodies): slides are rinsed in buffer A for 5 min.

5. Rinse slides in buffer B (blocking buffer to block nonspecific binding) for 30 min.

6. Rinse slides in buffer two to three times, and excess buffer is blotted.

7. Add 200–400 μL of antidigoxigenin solution/slide, and incubate for 5 h at room temperature with gentle shaking.

8. Rinse slides with buffer A twice for 10 min.

9. Equilibrate slides in buffer C for 5 min, while the chromophor substrate for the alkaline phosphatase reaction is created (340 μg nitroblue tetrazolium chloride (34 μL of 100 mg/mL in 70% dimethylformamide; Boehringer Mannheim) and 175 μg 5-bromo-4-chloro-3-indolyl phosphate toluidine salt (35 μL of 50 mg/mL in 100% dimethylformamide)/10 mL buffer C).

10. Incubate slide overnight in an airtight container at room temperature.

11. Wash slides in buffer C twice ($\approx$5 min each), and then briefly in distilled H_2O and 70% EtOH prior to air-drying.

12. Store slides protected from light.

4. Notes

1. *In situ* experiments require considerable commitment of resources and time. A schematic overview including time estimates is presented in **Fig. 1**. Critical elements of protocol steps are highlighted below.

2. Preparation of human tissues: Optimal tissue fixation and sectioning are essential for penetration of probe, tissue architecture preservation, RNA retention, and overall success.

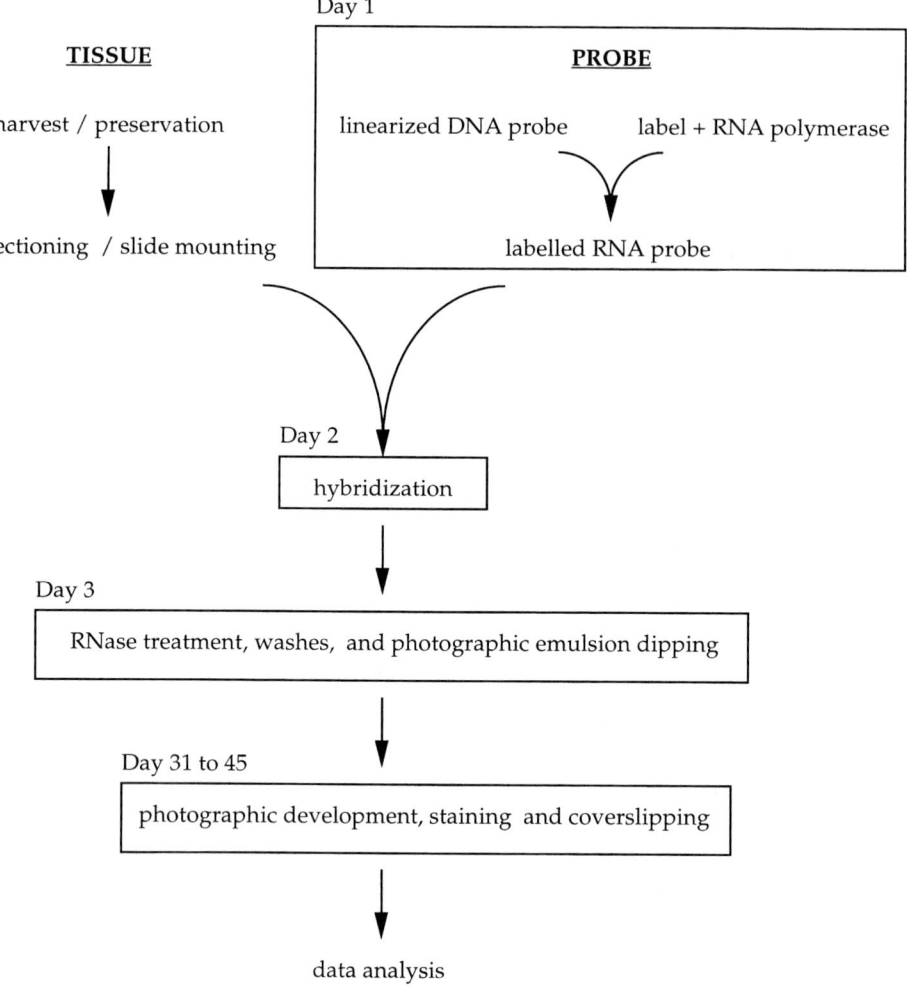

Fig. 1. Overall schematic of *in situ* hybridization.

3. Generation of radiolabeled riboprobes: Maximizing radionucleotide incorporation is important to increase sensitivity; ^{35}S provides a balance between cellular resolution, sensitivity, and acceptable exposure times (compared with ^{32}P or ^{3}H).
4. Hybridization conditions: Includes stringency, temperature, washes, and tissue dehydration.
5. Since RNase contamination prevents any chance of successful *in situ* hybridization, it is critical to assure that all solutions and hybridization work areas are RNase-free. RNase precautions should **always** be observed, including:
   a. Washing work areas and reusable items with bleach.
   b. Using disposable items whenever possible (e.g., autoclaved pipet tips).

c. Dedicated equipment items (e.g., hybridization ovens).

d. Using DEPC-treated water to make solutions.

e. Dedicating chemicals for RNA work, using bleach-rinsed, autoclaved spatulas to weigh chemicals.

f. Autoclaving suitable solutions after mixing, except sucrose, paraformaldehyde, and SSC.

g. Wearing clean disposable gloves in the RNase-free area.

h. Note: Ideally, the work area for RNase treatments after hybridization should be geographically separate from the main RNase-free work area.

6. Sample visualization: Including application of photographic emulsion, developing, fixing, staining, and counterstaining of individual slides.

References

1. Wilson, K. H., Schambra, U. B., Smith, M. S., Page, S. O., Richardson, C. D., Fremeau, R. T., et al. (1997) *In situ* hybridization: Identification of rare mRNA's in human tissues *Brain Res. Protocols* **1,** 175–185.

2. Smith, M. S., Schambra, U. B., Wilson, K. H., Page, S. O., Hulette, C., Light, A. R., et al. (1995) α_2-adrenergic receptors in human spinal cord: specific localized expression of mRNA encoding α_2-adrenergic receptor subtypes at four distinct levels. *Mol. Brain Res.* **34,** 109–117.

3. Perry-O'Keefe, H. and Kisinger, C. M. (1992) Chemiluminescent detection of nonisotopic probes, in *Current Protocols in Molecular Biology* (Ausubel, F. M., Brent, R., Kingston, R. E., Moore, D. D., Seidman, J. G., Smith, J. A., et al., eds.), John Wiley, New York, pp. 3.19.1–3.18.8.

4. Boyle, A. and Perry-O'Keefe, H. (1992) Specialized applications. Labeling and colorimetric detection of nonisotopic probes. In *Current Protocols in Molecular Biology* (Ausubel, F. M., Brent, R., Kingston, R. E, Moore, D. D., Seidman, J. G., Smith, J. A., et al., eds.), John Wiley, New York, pp. 3.18.1–3.18.19.

35

Use of Electron Microscopy in the Detection of Adrenergic Receptors

Chiye Aoki, Sarina Rodrigues, and Hitoshi Kurose

1. Introduction

1.1. Physiological Function of Adrenergic Receptors

Adrenergic receptors (ARs) belong to a superfamily of the G-protein-coupled receptors and are categorized by their binding to endogenously occurring catecholamines, i.e., norepinephrine and epinephrine. Adrenergic receptors are classified into three groups (α_1-, α_2-, and β-ARs), each of which is further divided into three subtypes. The α_1- (α_{1A}-, α_{1B}-, and α_{1D}-ARs) couple with G_q family of G-proteins (G_{11}, G_{14}, G_{15}, and G_{16}) and result in activation of phospholipase C-βs that liberate two second messengers, diacylglycerol and inositol-1,4,5-trisphosphate. The three subtypes of α_2-ARs are designated α_{2A}-, α_{2B}-, and α_{2C}-AR. On binding with agonists, α_2-AR inhibit adenylyl cyclase and calcium channels, but activate potassium channels through coupling to the G_i family of G-proteins (G_{i1}, G_{i2}, G_{i3}, and G_o). Finally, the three groups of β-AR are designated β_1-, β_2-, and β_3-AR: these increase the intracellular cAMP content by activating G_s, which is coupled to the enzyme, adenylyl cyclase *(1)*.

Functions of the adrenergic receptors in vitro and in vivo have been analyzed mostly by administrating subtype-selective agonists or antagonists *(2)*. However, although ligands specific for the three major adrenergic receptor types are available and have yielded much useful information, most of the ligands currently available do not exhibit sufficient specificity for discriminating among the subtypes (e.g., the A-subtype of α_2-AR). Thus, an alternative approach for identifying the function of the subtypes has been to knock out the

From: *Methods in Molecular Biology, vol. 126: Adrenergic Receptor Protocols*
Edited by: C. A. Machida © Humana Press Inc., Totowa, NJ

gene encoding for the particular receptor subtype. This approach has not always been met with success either, probably because other subtypes of catecholaminergic receptors compensate for the knockout. This is one reason many of the advances in our knowledge about the catecholaminergic receptor subtypes are derived from immunocytochemistry. For brain research, in particular, the immunocytochemical approach has been useful. This is because brain function depends critically on the connectivity formed among neurons. Thus, the effect of catecholamines within brain could differ greatly depending on the site of action of the neurotransmitter and on the receptor subtype located near the release of the neurotransmitter. The site of action of catecholamines can vary by region (e.g., visual vs auditory vs multimodal pathways), cell type within the region (e.g., neurons using excitatory transmitter for projecting long distances vs those using an inhibitory transmitter for local circuits or nonneuronal cells, such as astrocytes), and by the subcellular compartment (dendritic shafts, where primarily inhibitory inputs from other neurons are received, vs dendritic spines, where primarily excitatory inputs are received, or in axons, where outputs to other neurons are propagated and transmitted via release of neurotransmitters). For example, our study using an antiserum capable of selectively recognizing the A subtype of α_2-ARs revealed that these occur presynaptically (**Fig. 1A, B**), some of which were positively identified as noradrenergic axon terminals *(3)*. This result was expected, since earlier physiological studies had shown that α_2-ARs operate as autoreceptors, inhibiting release of norepinephrine or epinephrine *(4)*. However, we also observed that these receptors occur in noncatecholaminergic axon terminals, indicating that these may also operate as heteroreceptors regulating the release of transmitters other than norepinephrine and epinephrine. Furthermore, this receptor has been observed postsynaptically within the cerebral cortex *(5,6*; **Fig. 1***)* and the hippocampus *(7)*, even though electrophysiological studies have indicated a lack of α_2-AR-mediated postsynaptic effects in these forebrain structures *(8)*. Differences in findings such as these indicate that α_2-AR in these structures, unlike those in the brainstem, may activate intracellular second messenger cascades without activating potassium channels.

Similarly, a series of studies using antisera directed against distinct domains of β-ARs have revealed interesting differences in the receptor's conformation across developmental states and cell types within intact cerebral cortical tissue. The first polyclonal antiserum that became available for ultrastructural studies was raised by Joh, using the antigen harvested from frog erythrocyte membranes by Strader and colleagues *(9)*. This antiserum yielded immunolabeling of various portions of neurons, including perikarya and axons, but primarily distal dendrites. Astrocytic processes also were immunolabeled using this antiserum *(9,10)*. In sharp contrast to this result, it was observed that

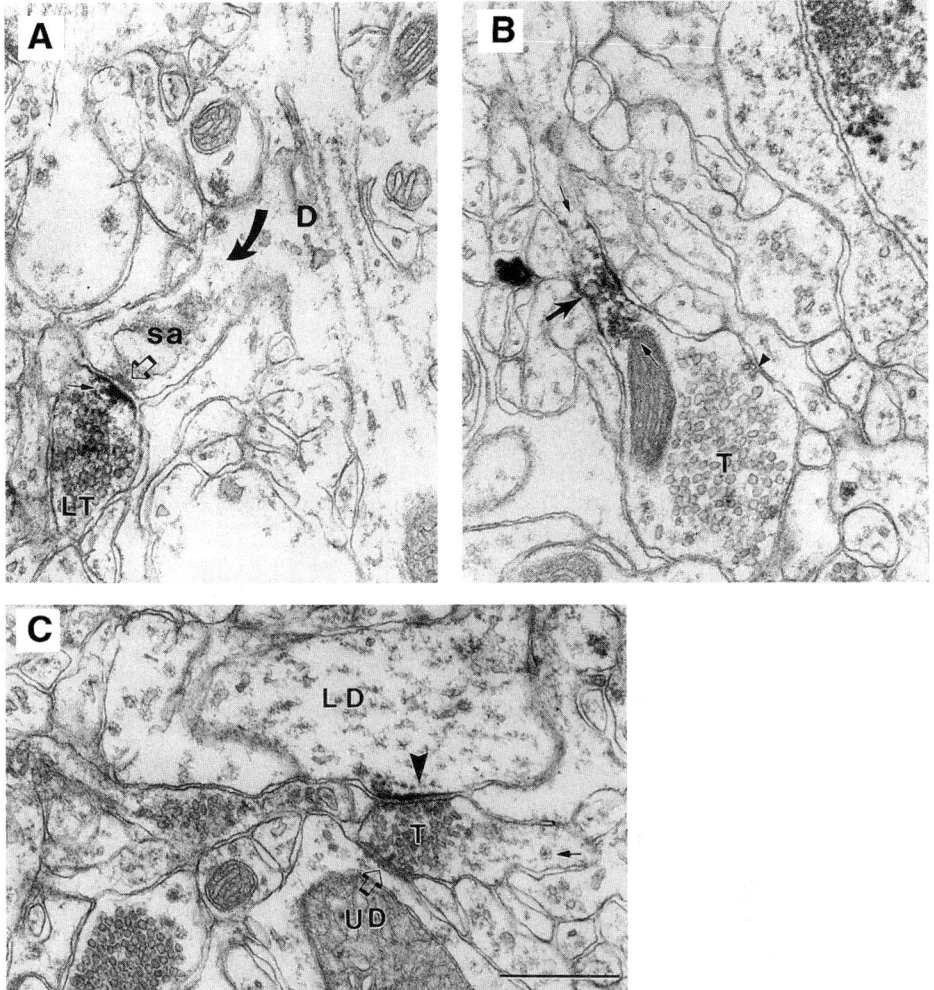

Fig. 1. EMs obtained from the monkey prefrontal cortex immunolabeled using the α_{2A}-AR antiserum and HRP-DAB as the immunolabel. **(A)** The receptor occurs directly over the presynaptic plasma membrane of a labeled terminal (small arrow in LT) forming a synaptic junction (open arrow) with a spine emanating from a dendritic shaft (arrow and D; sa = spine apparatus). **(B)** The receptor also occurs away from the synaptic region of a terminal, T, and instead at an intervaricose portion (large arrow and in between the two small arrows). The arrowhead points to a patch of plasma membrane undergoing endocytosis. **(C)** A terminal, T, forming a synaptic junction simultaneously with two dendrites, one of which exhibits immunoreactivity over the postsynaptic membrane (LD, arrowhead) and another that is unlabeled (UD). The tissue was fixed by transcardial perfusion with a mixture of acrolein and paraformaldehyde, and then postfixed following the ICC procedure with osmium tetroxide. Calibration bar = 500 nm for all panels (from **ref. 6**, reproduced with permission from Oxford University Press).

polyclonal antisera and monoclonal antibodies (MAbs) directed against the third intracellular loop region yielded immunolabeling primarily of perikaryal regions of neurons *(11)*, although distal dendrites, spines *(11)*, and axons, including presynaptic portions of axons *(12)*, also were immunoreactive. Finally, another polyclonal antiserum directed against the C-terminus of β-ARs recognized primarily astrocytic processes in adulthood *(13–16*; **Fig. 2***)* but also immunolabeled the earliest-formed synapses within neonatal cortices *(17*; **Fig. 3***)*. Each of these immunolabeling patterns was confirmed to be specific by showing abolishment of antigenicity following preadsorption of the primary antibodies (*see* **Notes 1** and **2**). Future studies that examine the relationship of β-ARs to the molecules known to interact with them, such as β-arrestin, β-AR kinase, and G_s proteins, under physiologically specified conditions promise to provide detailed knowledge required for understanding the dynamic regulation of cell physiology by epinephrine and norepinephrine.

Immunocytochemistry has also been useful for studying the subcellular compartment of different receptor subtypes that are coexpressed within single cells. For example, adipocytes and heart cells express both the $β_1$- and $β_3$-AR-subtypes of β-ARs. Interestingly, however, such coexpression does not allow for functional compensation, even though both subtypes are coupled to G_s and adenylyl cyclase *(18,19)*. It is possible that differential localization of the receptors within each cell, i.e., compartmentation, influences the response, since signaling molecules and second messengers are not expected to diffuse freely within cells. In another example, Jurevicus and Fischmeister *(20)* reported the functional compartment of the $β_1$-AR-mediated cAMP accumulation that is important for increase of calcium current through L-type calcium channel in heart, indicating that the $β_1$-AR, but not the $β_3$-AR, was closely associated with the effector molecule. In another example, it has been reported that the $α_{2C}$- and $α_{1A}$-ARs mainly localize intracellularly *(21,22)*. The compartmentation can facilitate efficient signal transduction from the receptor to cellular response and avoid unfavorable responses. In cells as polarized as neurons and glia, precise knowledge about receptor localization becomes ever more important, as diffusion of second messengers become restricted to single dendritic spines, dendritic shafts, or axon varicosities.

The cellular mechanism by which receptors, G-proteins, and effector molecules become properly localized is yet unknown. Clearly, elucidation of such cellular mechanism requires precise knowledge about the subcellular localization of the receptor and related elements, for which specific antibodies that recognize the receptors are required.

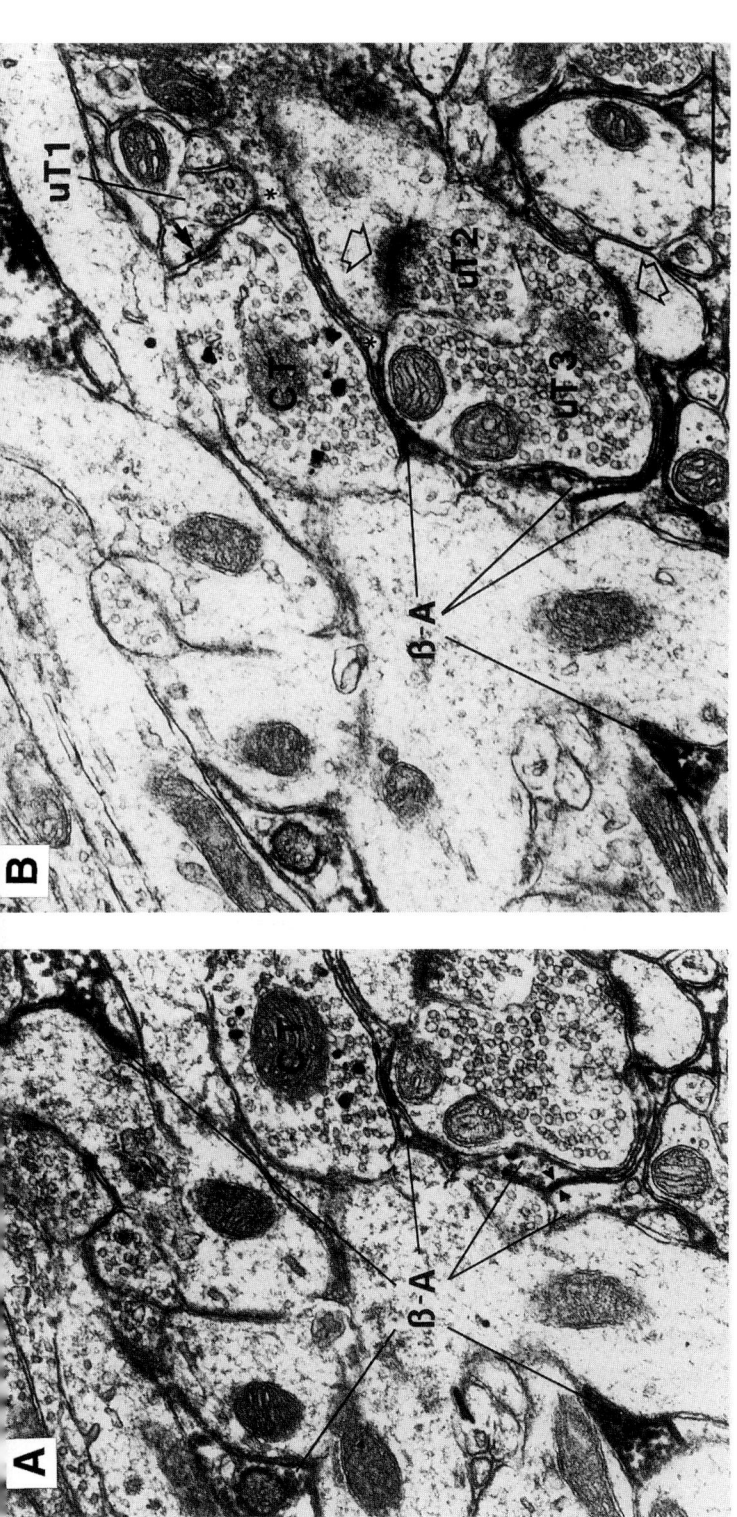

Fig. 2. EMs obtained from serially collected ultrathin sections of tissue immunolabeled dually for the catecholaminergic terminal (CT) marker, tyrosine hydroxylase, and the C-terminus of β-ARs. CT are identified by the SIG label, whereas β-ARs are identified by the HRP-DAB label. β-ARs occur in fine astrocytic processes (β-A) that surround neurites, including one catecholaminergic terminal (CT) and three unlabeled terminals (uT1, uT2, and uT3). uT2 and uT3 are forming asymmetric synaptic junctions with dendritic spines (open arrows point to the thick postsynaptic densities), indicating that they utilize glutamate for excitatory synaptic transmission. An arrangement of this sort, whereby β-A occur inserted between CTs and excitatory synapses, is supportive of the idea that activated β-A enhance excitatory synaptic transmission by reducing astrocytic uptake of L-glutamate. Arrowhead pairs in panel A point to a gap junction formed between two β-As. Calibration bar = 500 nm. (From **ref. 13**, reproduced with permission from the Society for Neuroscience).

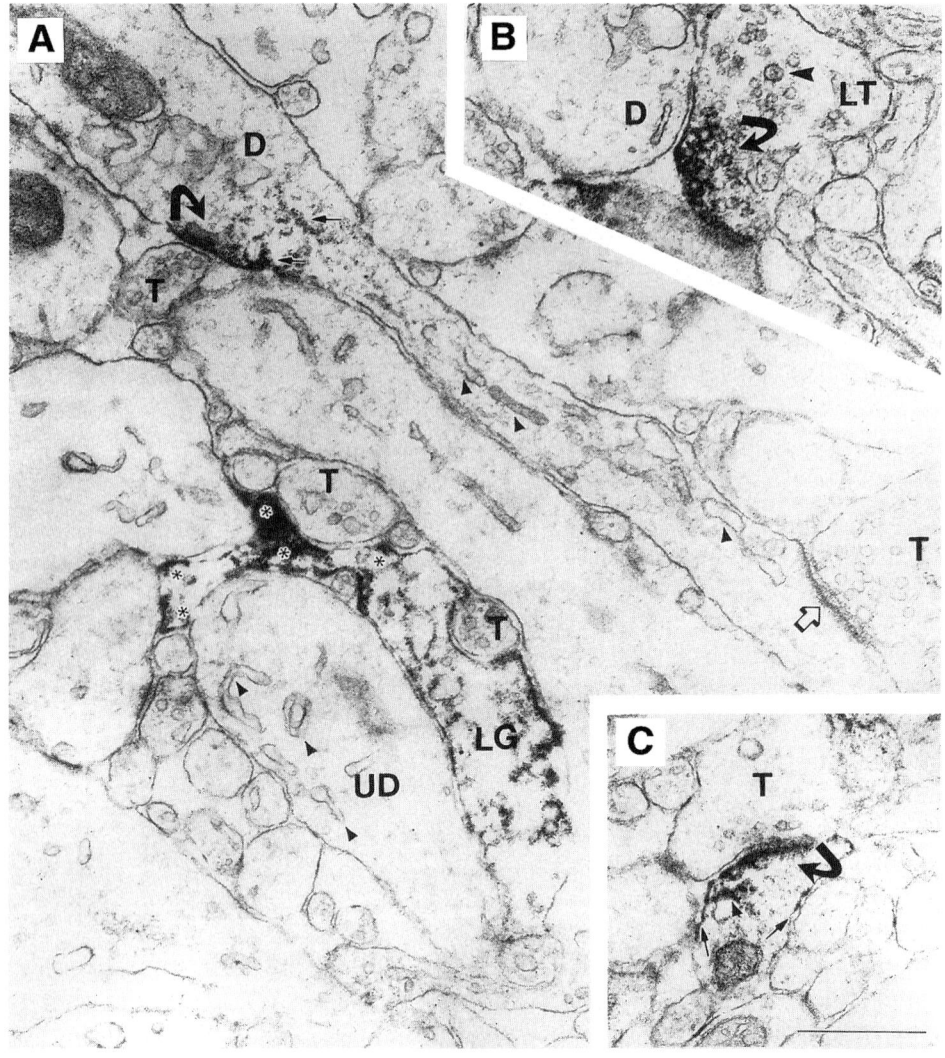

Fig. 3. EMs obtained from postnatal d 10 rat visual cortex, showing β-AR immu-noreactivity using HRP-DAB label and an antiserum directed against the C-terminus of the receptor. (**A**) A dendritic shaft receiving two synaptic inputs from unlabeled axon terminals (T), one of which is immunolabeled over the postsynaptic density (curved arrow) and another that is unlabeled (open arrow points to the postsynaptic density). Arrowheads point to the unlabeled smooth saccules, indicating robust mem-brane turnover, that accompany the process of maturation. Another profile exhibits immunoreactivity along the plasma membrane: judging from its irregular contour (asterisks), it is most likely a glial process (LG). (**B**) The same antiserum labels β-ARs in presynaptic terminals (LT), identified by the clustering of small clear vesicles (curved arrow). The same terminal contains a dense-cored vesicle (arrowhead). (A, C) β-AR immunoreactivity at postsynaptic sites occur not only over postsynaptic

1.2. Overview of Methods Used for Detecting Catecholaminergic Receptors

Much about the signal transduction mechanism has been learned also by expressing the receptors in physiologically "irrelevant" cell lines, i.e., those that are derived from cell types that, when within intact tissue, are devoid of the particular receptor. However, the level of expression of the exogenously transfected receptors tends to be high (above pmol/mg protein) compared to the level for endogenously expressed receptors (typically 10–200 fmol/mg protein). Although this difference serves as an advantage for measuring signal transduction, one still needs to be prudent about checking the behavior of the receptors in native and intact cells, where fine-tuning of receptor-mediated effects could depend critically on the exact location of the receptors in relation to other modulating and competing biochemical pathways.

Antibodies that are capable of recognizing the posttranslationally modified receptors promise to be powerful tools for analyzing the physiological conditions and consequences of posttranslational modification. The types of posttranslational modification include phosphorylation of specific intracellular domains in association with receptor desensitization, glycosylation of extracellular domains, and the addition of palmitoyl and myristoyl groups during intracellular trafficking.

Electron microscopic immunocytochemistry (EM-ICC) is very useful for determining existence and coexistence within fine processes of particular receptors, receptor subunits, neurotransmitters, or enzymes involved in generation of second messengers. The visualization of fine organelles is especially useful for distinguishing fine processes as astrocytic, axonal, or spinous, many of which often are less than a micrometer in diameter. Most importantly, electron microscopy (EM) is essential for identifying the morphological characteristics of synaptic junctions. Conversely, EM has been useful for identifying the presence at catecholaminergic receptors at sites lacking morphological characteristics of synaptic junctions, thereby providing strong support for the idea that catecholaminergic neuromodulation can occur by volume transmission, in addition to the more conventional transmission whereby transmitters released from axon terminals remain within the junctional cleft *(23)*. For this reason,

Fig. 3. *(continued from opposite page)* densities, but adjoining intracellular membranes and plasma membranes (small arrows). Calibration bar = 500 nm (from **ref. 17**, reproduced with permission from Cambridge University Press).

the methods chosen for EM-ICC aim for optimal tissue preservation while also avoiding loss of antigenicity brought about by excessive tissue preservation.

1.3. Overview of the EM-ICC Techniques Described in This Chapter

In this chapter, we will describe various techniques available for visualizing antibody–antigen complexes for EM-ICC, as well as their advantages and limitations. This discussion will be limited to immunolabeling intact brain tissue, for which the authors have direct experience and have obtained useful results. Problems encountered with producing polyclonal antisera and possible solutions to these problems also are included under **Subheading 4.** (*see* **Notes 1–7**).

2. Materials

The sources listed here have been used by the authors and shown to yield useful data. On the other hand, other sources are also likely to provide reagents of sufficient purity or specificity.

2.1. Tissue Source

Brain tissue can be obtained from a variety of animals. In our hands, vertebrates ranging from amphibians (e.g., frogs) to nonhuman primates have yielded useful data when using polyclonal antisera. Ultrastructural analysis is facilitated when tissue is preserved by rapid transcardial perfusion of fixatives, as detailed under **Subheading 3.**

2.2. Fixatives

All fixatives used for EM are volatile and highly reactive with tissue. Thus, these materials and particularly the solutions must be handled under a well-working hood. Some fixatives may also need to be collected after use. One should check with the local administrator regarding hazardous waste disposal.

The following sources have been used and yielded good preservation of the ultrastructure:

1. Paraformaldehyde (granular) and glutaraldehyde, both EM grade, from Electron Microscopic Sciences (Port Washington, PA), and acrolein, from Polysciences (Warrington, PA) or EM Corp. (Chestnut Hill, MA).
2. Acrolein has a more limited shelf life than glutaraldehyde and is highly volatile. For this reason, one should plan to store the vials in an explosion-proof refrigerator (which helps contain gas leakage, should the glass bottle break). Polysciences charges delivery and "poison charges" for this chemical owing to special handling required during delivery. These fixatives can be delivered transcardially, using 0.1 M phosphate buffer, pH 7.4, as a solvent.

2.3. EM Reagents

Heavy metals, osmium tetroxide, and Lowicryl can be purchased from EM Sciences (Fort Washington, PA), EM Corp., or from Ted Pella (Redding, CA).

1. Heavy metals used for EM, such as osmium tetroxide, uranyl acetate, and lead citrate, are biohazards. These should be handled with gloves and disposed as directed by the local administrator handling hazardous wastes.
2. Osmium tetroxide is also volatile and, thus, must be handled only under the hood.
3. Embedding resins can be allergenic: these also should be handled using gloves. Lowicryl, an embedding resin for postembedding immunocytochemistry, must be handled with nonlatex gloves (vinyl is recommended).
4. Ethyl alcohol, 200 proof, from Quantum Chemicals (Tuscola, IL).
5. Aclar sheets from Allied Signal Plastics (Pottsville, PA).
6. *Para*-phenylenediamine from Sigma (St. Louis, MO).
7. Iridium tetrabromide from Pfaltz and Bauer (Saterbury, CT), used at 0.5%, stirred the day before use in a maleate buffer (MB).
8. Tannic acid, 1%, dissolved immediately before use in MB.
9. Uranyl acetate can be used at a concentration of 1%, dissolved overnight in MB for osmium-free tissue processing, or 4% in 70% ethanol for osmium-fixed tissue.

2.4. Glassware and Plasticware

1. Porcelain crucibles (13-, 18-, 24-, and 40-mL capacity).
2. Glass spot dishes.
3. 0.22-μm Pore syringe "Acrodisc," made of nylon or Tuffryn, can be obtained from Gelman Sciences or from Fisher Scientific.
4. Beem capsules, used for embedding vibratome-sectioned, immunolabeled sections, can be obtained from EM Scientific.

2.5. Secondary Antibodies (Antirabbit IgG or Antimouse IgG)

1. Antibodies conjugated to colloidal gold, of sizes ranging from 1 up to 30 nm from Amersham (Arlington, IL) or Goldmark Biologicals (Phillipsburg, NJ): These come in two grades—those for histology and others for Western blots. We have tried using the histology-grade variety only.
2. Biotinylated antibodies from Vector Labs (Burlingame, CA) or from Jackson ImmunoResearch (West Grove, PA).
3. Unlabeled secondary antibodies (antirabbit IgG, antimouse IgG) from Jackson ImmunoResearch. These are available with or without affinity purification. The former requires to be used at a higher concentration, but would be expected to yield greater specificity. Biotin-labeled secondary antibodies of superb quality also can be purchased from Vector Labs.

2.6. Stocks Solutions

1. 0.2 *M* Phosphate buffer (PB), pH 7.4.
2. 25% Glutaraldehyde, packaged in ampules (EM Scientific).

3. 100% Acrolein, packaged in rubber-stopped bottles (Polysciences).
4. 4% Osmium tetroxide, packaged in ampules (EM Scientific).

2.7. Buffers and Their Uses

1. 0.1 M PB, pH 7.4: isotonic, strong buffer.
2. Phosphate-buffered saline (PBS): 0.01 M PB, 0.9% NaCl, pH 7.4: the buffer most widely used for rinsing and short-term storage of brain sections (*see* **Subheading 3.3.**).
3. PBS-bovine serum albumin (BSA): PBS, 1% BSA (from Sigma): the buffer used for minimizing nonspecific immunolabeling (*see* **Subheading 3.3.**).
4. PBS azide: PBS, 0.05% sodium azide: useful buffer for long-term storage of sections in the cold room (*see* **Subheading 3.1.7.**).
5. 0.1 M Citrate buffer, pH 6.5 (5.4 g of trisodium salt, using monohydrate salt of citric acid to adjust the pH to 6.5). This buffer is compatible with silver intensification of immunogold labels (*see* **Subheading 3.4.**).
6. TBST: 0.05 M Tris-buffered saline, 0.1% Triton X-100: this buffer is used for postembedding immunogold labeling (*see* **Subheading 3.5.**).
7. 0.1 M MB, pH 6.0: this buffer is used for processing sections for EM without the use of osmium tetroxide (*see* **Subheading 3.6.**).

2.8. Miscellaneous Solutions

1. Heparin, anticoagulant, from Elkins Sinn (Cherry Hill, NJ): Used during transcardial perfusions (*see* **Subheading 3.1.3.**, **step 4**).
2. Cryoprotectant: 0.05 M PB, 25% sucrose, 10% glycerol. Used for freeze–thaw permeabilization of tissue to enhance immunolabeling (*see* **Subheading 3.1.6.**, **step 3**).
3. Silver-intensification reagent, for enlarging 1-nm colloidal gold particles used for immunolabeling: a light-insensitive variety can be purchased from Amersham (*see* **Subheading 3.4.**).
4. 50% Sodium diethyl dithiocarbamate, dissolved in saline: a zinc chelator that minimizes background associated with silver-intensification of colloidal gold particles (*see* **Subheading 3.4.**).

3. Methods
3.1. Tissue Preparation and Storage
3.1.1. Choice of Methods for Fixation of Tissue: Transcardial Perfusion vs Immersion

For optimal preservation of cellular morphology and of the distribution of molecules within cells, transcardial perfusion with fixatives is required. However, there sometimes are needs to analyze the ultrastructure of tissue that has not undergone transcardial perfusion. One example is the need to analyze the ultrastructure of biopsy samples or blocks of tissue obtained postmortem. Even brains that have undergone transcardial perfusion with fixative may need to be postfixed by immersion for further improvement of structure. Under such circumstances, tissue may be fixed by immersion. Since penetration of fixatives through tissue is a slow process, relative to the rate of ultrastructural deterioration owing to anoxia, immersion fixation necessarily results in suboptimal conditions for ultrastructural analysis, particularly within portions of tissue removed from tissue surface. On the other hand, the surface-most portions of such tissue may be usable, since fixatives reach these portions with minimal delay. It is not advisable to use tissue that has undergone freezing prior to fixation: such tissue exhibits gross destruction of membranes, resulting from expansion of water during ice formation. Even when freezing follows fixation, destruction of the plasma membrane is not entirely avoidable during the freeze–thaw process. The problem with damaged plasma membranes is that identification of the boundaries of individual cellular processes by EM becomes difficult. This limitation, in turn, prevents analysis of the subcellular distribution of antigens.

3.1.2. Choice of Fixatives

Paraformaldehyde, used most widely for light microscopy, and acetone, used more for cultured cells, are not sufficient for ultrastructural preservation, since these fixatives do not preserve membranes of intracellular organelles or of the plasma membrane adequately for analysis. The most widely used fixative for EM is glutaraldehyde. This aldehyde has been used by electron microscopists at concentrations ranging from 0.05 up to 5%. Although the preservation of the ultrastructure is improved with increasing concentrations of glutaraldehyde, concentrations >0.1% have led to marked reduction of immunoreactivity for catecholaminergic receptors (unpublished observations). On the other hand, a brief (<7 min) exposure of tissue to another highly reactive, small aldehyde, i.e., acrolein *(24)*, at concentrations ranging from 3.0 to 3.75% has permitted good preservation of the ultrastructure as well antigenicity of catecholaminergic receptors *(3,5–7,10–17,* **Figs. 1–3***)*. Thus, authors of this chapter and others using antisera directed against catecholaminergic receptors have often used the following combination of fixatives: 0.05 or 0.1% glutaraldehyde in combination with 4% paraformaldehyde or 3–3.75% acrolein in combination with 2–4% paraformaldehyde.

3.1.3. Detailed Description of Transcardial Perfusion for EM

Transcardial perfusion with fixatives is one of the most critical steps for successful ultrastructural preservation of tissue. The aim of transcardial perfusion is to achieve ultrastructural preservation before morphological (and presumably chemical) alterations of tissue are triggered by anoxia. In order to minimize artifactual alterations of tissue, anoxia may be minimized by maintaining artificial ventilation during transcardial perfusion. The other key to success is speed, i.e., minimizing the number of seconds that lapses from the onset of anoxia (which begins the moment the diaphragm is cut for gaining access to the heart) up to tissue fixation (i.e., which must be preceded by steps whereby fixatives diffuse out of the blood vessel lumen and into the surrounding neuropil). A number of factors determine the speed. These include the rate of diffusion of the fixative, the efficiency with which one gains entry to the heart by dissection, and the rate of flow of the fixative through the cardiovascular system. For maximizing the rate of diffusion of the fixative within tissue, we recommend the use of small, highly reactive aldehydes, such as acrolein. Regarding swift entry into heart, one simply needs to practice the dissection procedure to gain expertise. The rate of flow of the fixative is best controlled using a peristaltic pump. This assures that the rate is maintained at a high level, but not overly high to cause rupture of blood vessels. For adult brains of most mammals, a flow rate setting of 70 mL/min is recommended. In order to avoid blockage of blood vessels by coagulated and aldehyde-fixed blood cells, one should flush the cardiovascular heart with heparinized saline (100–1000 U/mL of Heparin, added to 0.9% NaCl) prior to perfusion with fixatives. This saline flush, however, should be kept to a minimum in order to minimize delay of perfusion with fixatives.

For preparation of tissue for the immunocytochemical detection of catecholamine receptors, we and others have found the following aldehydes to be suitable, both for retention of antigenicity and ultrastructural preservation: a mixture of 3% acrolein and 4% paraformaldehyde, buffered with 0.1 M phosphate buffer (pH 7.4), perfused over a period of 3–7 min, followed by perfusion with 4% paraformaldehyde in phosphate buffer without acrolein *(3,5–7,10–17)*. Alternatively, a mixture of 0.1% glutaraldehyde with 4% paraformaldehyde perfused over a period of 30 min has been useful *(10)*.

Specifically, the following steps are recommended for transcardial perfusion:

1. Anesthetize the animal deeply (for experiments requiring SIG as immunolabels, inject diethyl dithiocarbamate [1 g/kg, ip] 15 min prior to **step 2**: *see* **Subheading 3.4.** for further details).
2. Open the chest cavity. Note the time the diaphragm has been cut.
3. Use a metal cannula connected to the peristaltic pump tubing to gain entry into the left ventricle. Snake the metal cannula tip into the ascending aorta.
4. Initiate perfusion using heparinized saline.
5. Continue perfusion with the aldehyde mixture.

3.1.4. Sectioning of Tissue for Pre-Embedding vs Postembedding EM-ICC

By far the most favorable sectioning procedure for EM-ICC detection of antigens is to use a vibratome. This procedure avoids freeze–thawing of tissue, which can, in turn, cause morphological damage owing to formation of large ice crystals. Vibratomes can readily generate sections as thin as 30 µm from moderately fixed brains. The stronger the fixation, the thinner the sections can be. Conversely, weakly fixed tissue, such as early postnatal tissue or those fixed with low concentrations or minimal volume of fixatives (e.g., 2% paraformaldehyde), need to be sectioned at greater thicknesses, e.g., 100 µm for postnatal d 3 rat brain sections, with greater vibration amplitude and with slower blade strokes.

Alternative choices for tissue sectioning include using the freezing microtome or a cryostat. However, these alternatives are less desirable because of unavoidable tissue damage caused by the freeze–thaw steps, even when precaution is taken to cryoprotect the tissue. When sectioning in a frozen state is unavoidable, one must take every precaution to avoid formation of large ice crystals that damage membranes: this is best managed by immersing the smallest possible block in a cryoprotectant, such as a mixture of sucrose (25%) and glycerol (10%), buffered with 0.05 M phosphate buffer, and then freezing rapidly using Freon or isobutyl alcohol chilled to a temperature colder than –70°C by using liquid nitrogen. Further details of cryoultramicrotomy can be found in manuscripts by Tokuyasu et al., Liou et al., and Sitte *(25,26)* since discussion of this technique is beyond the scope of this chapter.

3.1.5. Termination of the Fixation

For obtaining specific immunolabeling, it is desirable to control the termination as well as the initiation of fixation. Tissue fixed using highly reactive aldehydes, such as glutaraldehyde and acrolein, continue to form covalent bonds with primary amine groups of proteins, even after tissue has been sectioned and all excess aldehydes have been removed by rinsing. In order to terminate the aldehydes' crosslinking actions, one needs to treat sections with reducing agents, such as sodium borohydride *(27)*, that render the aldehyde groups nonreactive by converting them to alcohol groups or by treating with excess of primary amine groups. Acrolein- and glutaraldehyde-fixed sections of about 40-µm thickness require immersion for 30 min in a solution of 1% sodium borohydride, buffered with 0.1 M phosphate buffer. This solution must be made immediately prior to use. Following the 30-min incubation period, sections should be rinsed in 0.1 M phosphate buffer until bubbles cease to emerge.

3.1.6. Tissue Permeabilization

Tissue permeabilization is a step taken to increase penetration of immunoreagents, particularly antibodies, into tissue. For antigenic sites embedded within organelles, such as within vesicles, this step appears to be essential. For antigens that are soluble, such as those in the cytosol and for intracellular domains of membranous proteins, including the adrenergic receptors, permeabilization may be kept to a minimum. For EM-ICC, the permeabilization methods involving extraction of lipids from the plasma membrane, such as incubation in nonionic detergents (e.g., Triton X-100), interferes with ultrastructural analysis. Thus, detergent-treatment should be avoided. In cases where tissue penetration is required, methods compatible with EM include the following three:

1. Add low concentrations of Triton X-100 (<0.06%) or Photo-flo (0.1–0.3%) to primary antibody solution (refer to EMs shown in **ref. 28** to see the extent of membrane damage).
2. Incubate sections briefly (<30 min) in buffer containing low concentrations of Triton X-100 (0.1% or less), prior to incubation in primary antibody solution *(29)*.
3. Rapidly freeze–thaw, following cryoprotection. This is the most preferred method for ultrastructural analysis, since the destruction of membranes is minimized. There are many methods for this treatment. One that works is to cryoprotect by incubating sections for at least 1 h in 25% sucrose and 10% glycerol in 0.05 M phosphate buffer and to subject to rapid freezing using liquid Freon, followed by liquid nitrogen *(30)*. Such sections are thawed by pouring warm cryoprotectant over them. An alternative cryoprotectant is dimethyl sulfoxide (DMSO) *(31)*. Its infiltration can be achieved by incubating sections in increasing concentrations of DMSO (5, 10, 20%, 10 min for each concentration), each buffered with 0.1 M phosphate buffer. Cracking of sections is minimized by laying them flat on nylon mesh during the freezing and thawing steps.

3.1.7. Section Storage

Sections can be stored at 4–6°C for several months with minimal loss of ultrastructural details or antigenicity. The recommended storage buffer is PBS-azide.

3.2. Choice of Labels for EM-ICC

3.2.1. Horse Radish Peroxidase-3,3'-Diaminobenzidine (HRP-DAB)

The synaptic molecules are most readily detected using the enzymatically amplified method, i.e., the avidin–biotin–horseradish peroxidase complex, with DAB as substrate (HRP-DAB). Our previous experience with this label indicates that when used judiciously (i.e., with minimal peroxidase reaction period), HRP-DAB provides subcellular localization of antigens precise enough to differentiate labeled from unlabeled portions of dendrites (*see* **Figs. 1C** and **3A**). For example, we *(3,5,6,10–17)* and others *(7,32)* have used this label to distinguish immunolabeled from unlabeled synaptic membranes that are positioned immediately adjacent to one another.

The advantage of using HRP-DAB is that this label is compatible with conventional resins and strong fixative for membranes, such as osmium tetroxide. These reagents provide excellent preservation of tissue, and this factor is useful, although not absolutely necessary, for identifying various types of synapses, such as nascent synapses within developing tissue and symmetric synaptic junctions. We have noted that catecholaminergic synaptic junctions, identified by the presence of catecholaminergic receptors, differ from glutamatergic synapses in that the former frequently lack the conventional morphological features of synapses. This conclusion could not have been made if the analyzed specimens were fraught with difficulty caused by suboptimal preservation of the ultrastructure, such as the loss (rather than absence) of conventional morphological features of synapses.

HRP-DAB is compatible with pre-embedding silver-intensified gold immunolabeling (**Fig. 2**, detailed below), thereby allowing for identification of two antigens within single fine processes. Moreover, HRP-DAB allows for light microscopic inspection prior to EM processing, thereby allowing for assessment about the specificity of immunolabeling by comparing with expected (previously reported) distribution patterns of immunoreactivity across cell types and brain regions.

However, the HRP-DAB label is not free of limitations. Although the enzymatic amplification afforded by this method allows for excellent detection of antigens, diffusion of HRP-DAB labels precludes identification of antigenic sites as membranous vs cytosolic, nor are quantitative measurements, such as the concentration of antigens within single immunolabeled profiles, possible. For questions requiring this level of resolution and quantification, the pre- and postembedding gold methods, respectively, are recommended (*see below*). Additionally, the HRP-DAB label, when weak, is difficult to distinguish from the conventional counterstain, lead citrate. Thus, one may wish to omit the lead citrate counterstain to optimize detection of HRP-DAB labels.

3.2.2. Silver-Intensified Pre-Embedding Colloidal Gold (SIG)

An alternative method with which receptors can be labeled is the SIG method (*see* **Note 8**). Penetration of the secondary antibody is optimized by using small sizes of conjugated colloidal gold. One-nanometer colloidal gold particles are commercially available. In our experience, this smallest size colloidal gold is necessary for detection of antigens within fine processes, such as dendritic spines and axons. Since 1 nm is below the limit of resolution of electron microscopes, these colloidal gold particles will need to be silver-intensified for EM, as well as light microscopic detection of immunolabels. For detection of antigens in larger profiles, such as cell bodies and dendrites, colloidal gold of larger sizes, such as the 5- and 10-nm variety, can be used without silver intensification for EM. However, these larger sizes of colloidal gold will still require silver intensification for light microscopy.

These colloidal gold labels will offer greater subcellular localization than HRP-DAB for questions, such as the membranous vs cytosolic localization, since the label is not diffusible (**Fig. 4**). As with HRP-DAB, sections immunolabeled with SIG can be examined by both light and EM, thus allowing for assessment of specific immunolabeling based on the cellular and areal distribution pattern. Moreover, the SIG label is compatible with osmium fixation of membranes and thus can be combined with HRP-DAB (for which osmium treatment is required to render the DAB reaction product electron-dense).

The shortcomings of the SIG method are that the labels are not enzymatically amplified. For this reason, the procedure using SIG label is less sensitive than those using HRP-DAB labels, judging from the antibody concentration required to attain equivalent immunolabeling (estimated to be one-tenth). Others have also noted that immunogold labeling rarely occurs directly over postsynaptic densities, even for the presumed synaptic molecules, such as receptors. Instead, immunolabeling tends to occur at the edges of synaptic specializations *(33)* owing, possibly, to steric hindrance caused by colloidal gold particles even for cases where the smallest available size (1 nm) is used. Nevertheless, the pre-embedding SIG procedure continues to be an excellent label for combining with HRP-DAB and for discriminating localization of antigens to cytosol or over membranes.

3.2.3. Postembedding Colloidal Gold (PEG)

Finally, the PEG procedure would be useful for questions requiring the most precise localization of antigens, such as the potential coexistence of two molecules within single PSDs or of their coexistence along the same patch of nonsynaptic plasma membrane or cytoplasmically. For example, it has been demonstrated that β-ARs regulate N-methyl-D-aspartate (NMDA) receptors, suggesting that the two receptors coexist with single dendrites, thereby allowing for their interaction following near-synchronous depolarization of noradrenergic and glutamatergic fibers *(34)*. The PEG method, combined with a pre-embed label, could readily determine whether the two receptors, indeed, coexist within single fine processes (**Fig. 5**). Yet another useful application of PEG is for comparing the concentration of antigens across PSDs of different

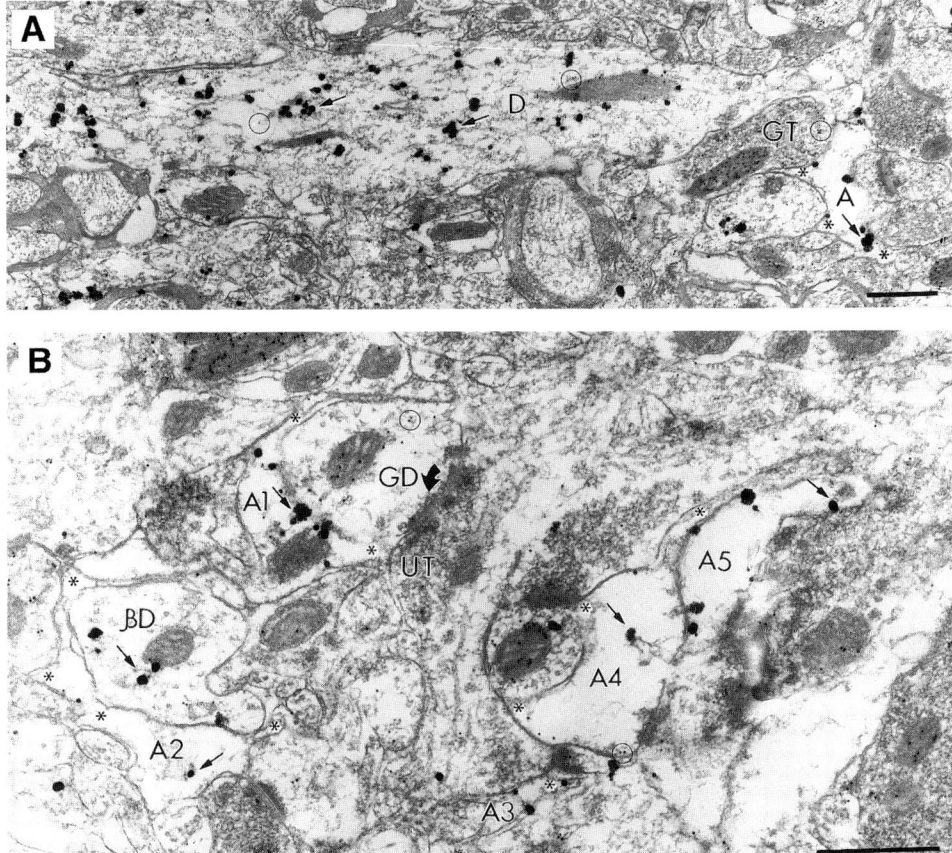

Fig. 4. EMs revealing the differential localization of β-ARs in astrocytes and neurons of adult visual cortex by SIG and their relation to GABA-immunoreactivity. **(A)** The dendrite from layer 5/6a of adult rat visual cortex exhibits numerous SIG particles (e.g., arrows), reflecting the presence of cytosolic β-ARs. Within a neighboring astrocytic process (A), the SIG particles are close to the plasma membrane. Asterisks point to the irregular contours of the astrocyte. The same dendrite exhibits numerous 10-nm colloidal gold particles, resulting from EM-ICC detection of GABA, using PEG as label (e.g., circled particles). A terminal, GT, contacting the dendrite exhibits high density of PEG labeling, indicating that it is a GABAergic terminal. **(B)** At a higher magnification, the neuropil from layer 2/3 of rat visual cortex exhibits five astrocytic processes (A1–A5). A1–A4 are immediately adjacent to asymmetric synaptic junctions (probably excitatory). A1, A3, and A5 exhibit robust β-AR immunoreactivity (small arrows) primarily along the plasma membrane, but A2 and A4 exhibit much lower levels of β-AR immunoreactivity and at sites away from the plasma membrane. Note that A4 is GABA-immunoreactive. In contrast, a dendritic process exhibits robust immunoreactivity for β-ARs at sites away from the plasma membrane (e.g., arrow in βD). A1 contacts an unlabeled terminal, UT, and also envelopes a dendrite, GD, identified to be GABAergic by the prevalence of PEG labels (circle), and synaptically associated with UT (curved arrow points to the postsynaptic density). Calibration bar = 500 nm.

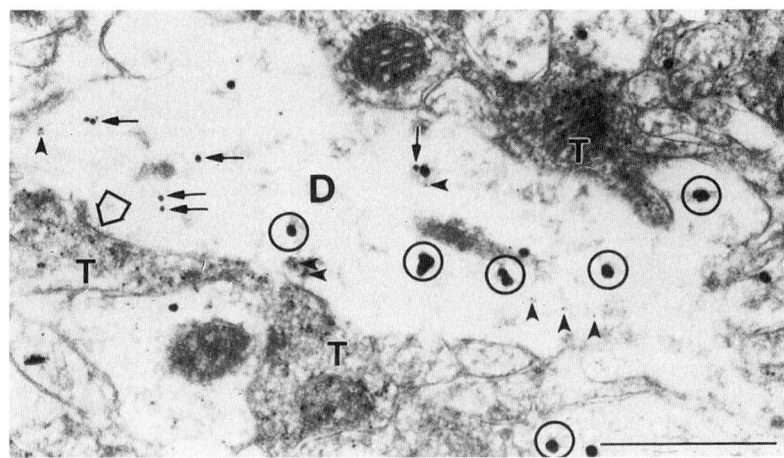

Fig. 5. An EM of the neuropil of an adult visual cortex, showing the coexistence of three antigens within a single dendrite. Triple EM-ICC was achieved by combining SIG to immunolabel NR2A subunits of NMDA-type glutamate receptors (circles), 30 nm colloidal gold-PEG for β-ARs (small arrows), and 10 nm colloidal gold-PEG for the inhibitoryneurotransmitter, GABA (arrowheads). This result indicates that a GABAergic inhibitory interneuron is receptive to noradrenaline as well as L-glutamate. The dendrite is also receptive to GABA, since it is forming a contact with two GABAergic terminals (T), one of which is associated with a morphologically identifiable synapse (open arrow points to the postsynaptic membrane). The localization of the two receptors away from the plasma membrane may be an indication that the receptors are in a desensitized state because of synaptic transmission that occurred during or prior to tissue fixation. Calibration bar = 500 nm.

synapse types (e.g., noradrenergic synapses formed on pyramidal neurons vs those formed on inhibitory interneurons).

Successful dual localization of the two receptor subunits to single PSDs and at sites away from synapses has been achieved by using two sizes of colloidal gold for PEG labels and/or by combining PEGs with SIG *(35)*. These studies indicate that PEG is applicable for the precise localization of neurotransmitter receptors. Moreover, the PEG procedure can, in some cases, be applied to osmium-fixed tissue for combining with HRP-DAB (HRP-DAB requires osmium treatment for rendering the label electron-dense for EM) (Erisir et al., unpublished observations). Results from these studies indicate that not only double, but also triple, EM-ICC is a possibility by combining SIG, HRP-DAB, and one or two PEG labels.

Should the molecule of interest not withstand the pre-embed osmium fixation, then PEG will need to be performed on tissue in which the osmium fixation of membranes is substituted by a protocol using tannic acid in combination with uranyl acetate and iridium tetrabromide *(35–39)*. The preservation of membranes is not as complete as with osmium tetroxide, but is still useful for yielding information regarding the subcellular distribution of receptors. This point is evident by comparing **Figs. 1–3**, which used osmium tetroxide, with **Figs. 4** and **5**, which were preserved without the use of osmium tetroxide.

Should the antigen of interest not withstand the heat needed to polymerize conventional resins (60°C for Epon and Epon-Spurr), an alternative resin, Lowicryl, can be used, since this resin can be polymerized at temperatures below –40°C. In recent years, results obtained using this resin have revealed exquisite, highly localized distribution of glutamate receptors *(38,39)*. Tissue to be embedded using Lowicryl cannot be fixed by osmium, since osmium interferes with UV irradiation required for Lowicryl polymerization.

In short, by combining multiple EM-ICC techniques, one can maximally analyze the cellular and molecular details of synapses while also compensating for the well-known short-comings of each method.

3.3. The Procedure of Using HRP-DAB as the Immunolabel

As noted above, the most sensitive method currently available for EM-ICC uses DAB as substrate for HRP, which, in turn, is attached to antibody–antigen complexes via avidin–biotin links (ABC). The alternative peroxidase-based EM-ICC procedure, peroxidase–antiperoxidase (PAP), has been described in detail and is also applicable for the detection of adrenergic receptors *(10)*. However, this procedure will not be described here, since it is less sensitive than the ABC-DAB procedure.

The development of HRP-DAB immunolabels by the ABC method involves linking of biotinylated secondary antibodies (antirabbit IgG) to biotinylated HRP, using the four binding sites on avidin as bridges. Specifically, the following procedure is recommended, as described by the manufacturer (Vector Labs):

1. Incubate sections for a minimum duration of 30 min in blocking buffer (blocks nonspecific immunolabeling), consisting of PBS, pH 7.4, containing 1% BSA.
2. Incubate overnight at room temperature (or up to 3 d at 4°C) in blocking buffer containing an empirically determined dilution of the primary antibody.
3. Rinse sections three times at 5-min intervals (3 × 5 min) in PBS.
4. Incubate in blocking buffer containing an empirically determined dilution of biotinylated antirabbit IgG. The biotinylated secondary antibodies from Vector usually works well at dilutions of 1:100 to 1:200 and require 30-min incubation, whereas the affinity-purified biotinylated secondary antibodies from Jackson ImmunoResearch need a dilution of approx 1:50 and a 1-h incubation period.
5. Prepare the ABC solution (two drops of solution A and two drops of solution B from Vector's Elite kit and PBS as diluent), and allow solution to stand for 30 min prior to use.

6. Rinse sections 3 × 5 min in PBS.
7. Incubate for 30 min in the ABC solution.
8. Rinse sections 3 × 5 min in PBS.
9. Immerse sections in the HRP substrate, consisting of 11 mg of DAB hydrochloride and 5 µl of 30% H_2O_2 in 50 mL of PBS. The reaction time can vary from a few minutes to 10 min or more, depending on the condition of the primary antibody incubation and the antibody's titer.
10. Examine the sections by light microscopy by mounting them temporarily on clean slides to confirm that the expected staining pattern has been achieved.
11. Mount sections for light microscopy, or else, follow the procedure outlined under **Subheading 3.5.** for processing sections for EM. Alternatively, sections can be stored without loss of DAB immunolabels for at least 1 wk, if maintained at 4°C in PBS.
12. Follow the procedure outlined under **Subheading 3.6.** for processing sections for EM.

3.4. The Procedure of Using SIG as the Immunolabel

Secondary antibodies are available conjugated to varying sizes of colloidal gold. For our purposes, which were to visualize immunoreactivity within fine processes (i.e., <1 µm in diameter), we have opted to use colloidal gold of 1–1.4 nm in diameter. Although this size of colloidal gold requires silver intensification for EM visualization, the extra steps required for silver intensification are well worth the trouble, since these allow for detection of antigens within dendritic spines as well as axons. In contrast, secondary antibodies conjugated to larger sizes (>5 nm) of colloidal gold do not require silver intensification for EM visualization, but preclude analysis of small profiles, such as axons, spines, or astrocytic processes.

Tissue to be used for silver-intensified colloidal gold labeling should have the endogenous zinc chelated, since zinc, together with colloidal gold, becomes silver-intensified, yielding particles indistinguishable from silver-intensified colloidal gold particles *(40)*. This is achieved by injecting sodium diethyl dithiocarbamate, ip (1 g/kg), 15 min prior to transcardial perfusion of the animal.

The protocol that has yielded useful data is as follows, which applies osmium tetroxide as the fixative of membranes. Alternatively, sections can be processed for EM without treatment with osmium tetroxide, in order to avoid any loss of silver-intensified gold particles. Procedure for the osmium-free treatment of sections for EM is outlined in **Subheading 3.6., step 2**.

1. Follow steps described in **Subheading 3.3.**, except that the dilution of the primary antibody is prepared at a concentration that is 4–10 times more concentrated than that used for HRP-DAB (the higher concentration of antibodies compensates for the relatively weaker sensitivity of the SIG method).

2. Incubate sections for 3–4 h at room temperature in a 1:50 dilution of 1 nm gold-conjugated antirabbit IgG. The diluent should be a blocking buffer, consisting of PBS containing 1% BSA.
3. Rinse 3 × 5 min in PBS.
4. Postfix sections using PBS containing 2% glutaraldehyde.
5. Rinse 3 × 5 min in PBS.
6. Rinse for 1 min in 0.2 M citric acid buffer (trisodium salt citrate adjusted to pH 6.5, using monohydrate salt of citric acid, prepared fresh using ultrapure or double-distilled water). The reason for switching buffers at this step is that chloride ions interfere with the silver-intensification step. The Silver IntensEM kit from Amersham recommends that sections be immersed in water prior to silver intensification. However, water immersion causes deterioration of the ultrastructure. Thus, we have preferred using isotonic citrate buffer, described above.
7. Silver-intense the colloidal gold particles by immersing sections for 3–15 min in Silver IntensEM kit (equal volumes of solution A and solution B, as directed by Amersham). Use a nonmetal instrument for transferring sections across different buffers, since metals interfere with the silver intensification.
8. Terminate the silver intensification by rinsing sections in the citrate buffer, and then in PBS.
9. Follow the procedure outlined under **Subheading 3.6.**

3.5. The Procedure of Using PEG as the Immunolabel

PEG is achieved by applying primary and secondary antisera directly on ultrathin sections prepared from resin-embedded tissue. Thus, considerations must be made for retaining antigenicity of the molecules throughout the procedure required for embedding sections in resins (which typically involve dehydration and irradiation with heat [ca. 60°C] or UV for a few days). Furthermore, one must expect further loss of antigenicity during the steps taken to incubate ultrathin sections in buffers for PEG labeling. For this reason, the PEG procedure should be performed soon after preparing ultrathin sectioning. Moreover, it is advantageous to use the strongest fixative possible for transcardial perfusion, such as a high concentration of glutaraldehyde, in order to minimize leaching of antigens out of ultrathin sections. Of course, the choice of fixatives is likely to be constrained further by the potential loss of antigenicity owing to denaturation of the molecule caused by strong fixations.

A few choices for embedding resins are available. For studies involving epitopes of molecules or antigens that are not heat-sensitive, EMBED 812 (Epon) or Epon-Spurr would be the resin of choice. Of the two, Epon-Spurr is more hydrophilic, thereby allowing greater penetration of ultrathin sections by immunoreagents. Should the antigen be heat-sensitive, then one will need to resort to using embedding resins that do not require heat for polymerization. Lowicryl is one such resin, for it can be polymerized at freezing temperatures by UV irradiation. One additional advantage of Lowicryl is that it is a hydrophilic resin, allowing for excellent penetration of immunoreagents through the thickness of ultrathin sections. However, UV radiation also may cause loss of antigenicity.

In general, all solutions listed below, except antisera, should be filtered using 0.22-μm Millipore filters. Small-size filters that fit on the tip of syringes are useful for this purpose. Grids are incubated by submerging them, face-up, in droplets. It is best to submerge grids by sliding them sideways into droplets, formed on the surface of parafilm or silicone mats for grids. The specific steps are as follows, based on a procedure optimized by Phend et al. *(37)*.

1. Wash grids in 0.05 *M* Tris buffer, made isotonic with 0.9% sodium chloride, pH adjusted to 7.6, and with 0.1% Triton X-100 added to enhance penetration of immunoreagents (TBST, pH 7.6).
2. Incubate in primary antiserum solution, diluted using TBST, pH 7.6. For most antisera, overnight incubation at room temperature should be sufficient. The dilution of the antiserum should be determined empirically to yield immunolabeling with this duration of incubation. For most antisera, the concentration needs to be approx 10-fold of what is needed for HRP-DAB labels. BSA need not be added to TBST.
3. Wash 2 × 5 min in TBST, pH 7.6.
4. Wash 30 min in TBST, pH 7.6.
5. Condition grids for 5 min in TBST at pH 8.2.
6. Incubate for 1 h in secondary antiserum (e.g., antirabbit IgG) conjugated to colloidal gold (5- to 25-nm sizes available commercially) at a dilution of 1 : 25 to 1 : 35, diluted using TBST, pH 8.2.
7. Wash 2 × 5 min in TBST, pH 7.6.
8. Wash 2 × 5 min in distilled water.
9. Air-dry
10. Counterstain for 5–10 min, using 5% uranyl acetate dissolved in 100% methyl alcohol (optional—step may be skipped to prevent obscuring weak HRP-DAB labels).
11. Rinse quickly by immersing vertically held grids repeatedly in a small beaker filled to the rim with 100% methyl alcohol.
12. Air-dry.
13. Rinse in distilled water.
14. Counterstain with lead citrate, 2%, for 0.5–2 min (optional).
15. Rinse in distilled water.
16. Air-dry.

3.6. Processing of Vibratome Sections for EM

The procedure described below involves postfixation, flat-embedding, capsule-embedding, and then preparation of ultrathin sections. Two procedures are outlined: one for HRP-DAB and SIG (**steps 1, 3–7**), and another—osmium-free *(2–7)*—that may need to be followed for SIG and PEG immunolabels, depending on the susceptibility of the antigen to denaturation caused by chemical and heat treatments. The osmium-free method is as outlined previously by Phend et al. *(37)*.

1. Postfix with osmium tetroxide. Return the sections to 0.1 M phosphate buffer. Postfix with osmium tetroxide, for 1–2 h at room temperature. This step is important for membrane preservation. If transcardial fixation is suspected to have achieved only weak fixation of tissue, then the osmium tetroxide fixation can be preceded by another postfixation, consisting of 2% glutaraldehyde diluted with PBS, to be administered for 10 min at room temperature.

    The recommended concentration of osmium tetroxide is 1–2%, diluted with 0.1 M phosphate buffer. It is advisable that the wells in which sections are placed for this postfixation condition be as flat as possible in order to avoid inducing curvature on the sections. Also, the wells should be covered and placed under a good working hood, in order to contain the highly volatile heavy metal solution. The fixation step is terminated by rinsing in 0.1 M PB.

2. Post-fixation without osmium tetroxide: All steps are performed on ice until the acetone step. Sections should be maintained flat at all times.
    a. Rinse 2 × 5 min in 0.1 M maleate buffer, pH 6.0 (MB).
    b. Incubate for 40 min in 1% tannic acid, dissolved immediately before use in MB.
    c. Rinse 2 × 3 min in MB.
    d. Incubate for 40 min in 1% uranyl acetate, dissolved overnight in MB while protected from light (one can use aluminum foil to shield the solution from light).
    e. Rinse 2 × 3 min in MB.
    f. Incubate for 20 min in 0.5% iridium tetrabromide, stirred from the day before in MB.
    g. Rinse 2 × 3 min in MB.

3. Dehydrate using increasing series of ethanol, beginning with 30% and up to 70%. For osmium-free tissue, follow this step by a 15-min incubation in 1% *para*-phenylenediamine hydrochloride, made fresh and protected from light using 70% ethanol. For osmium-fixed and osmium-free tissue, follow this by a 1–4 h *en bloc* staining with 1–4% uranyl acetate dissolved in 70% ethanol. This step not only counterstains, but also helps with ultrastructural preservation. Follow this step with further dehydration, up to 100% ethanol. From this point on, sections should be held in a tightly sealed vial to avoid humidity. Follow the ethanol dehydration with immersion in 100% acetone or 100% propylene oxide, followed by immersion for 4–20 h in a solution consisting of 50% acetone (or propylene oxide) and 50% resin (e.g., EMBED 812 or Epon-Spurr).

4. Immerse for 4–20 h in 1:3 ratio of acetone and resin and then in 100% resin.

5. Flat-embed: This consists of first placing sections flat on the surface of Aclar

plastic sheets that have been cleaned by scrubbing with 100% ethanol. These sections are cover slipped using another smaller sheet of Aclar plastic. Squeeze out the excess resin, as one would in cover slipping for light microscopy. Cure Aclar-sandwiched sections. For EMBED 812 and Epon-Spurr, this requires that the Aclar-sandwiched sections be placed in an oven for 12–20 h at a setting of 60°C. For Epon-Spurr, 37°C will suffice. Resins designed for PEG, such as Lowicryl, require that the Aclar- sandwiched sections be placed under UV light within a chamber made free of oxygen (since oxygen inhibits polymerization of the resin). Such a chamber can be prepared easily by using dry ice to displace air and using Saran wrap to seal the chamber. Such a chamber should be lined with aluminum foil to maximize the use of reflected UV irradiation for polymerization of the resin. Additionally, the chamber needs to be placed in a freezer set at a temperature ranging from –20 to –30°C to avoid denaturation of antigen during polymerization. Once cured, such flat-embedded sections can be stored indefinitely at room temperature, and also inspected and photographed with light microscopes.

6. Capsule-embed desired portions of sections. This is achieved by first peeling off one of the two pieces of Aclar plastic sheets, while keeping the section adhered to the remaining one sheet of Aclar. Portions of the sections, still adhered to a single sheet of Aclar, can be cut using a scissor or razor blade to a size small enough to fit into EM Beem capsules. Cut off the conical tip of Beem capsules using a razor blade. Place the flat-embedded section on the flat, inside surface of the cap, with the Aclar sheet facing the bottom. Fit the Beem capsule (opened at its other end) into the cap. Fill the capsule up to the razor blade-cut edge with fresh embedding resin. Cure in the oven for another 12–20 h.

7. Prepare ultrathin sections using an ultramicrotome. Counterstain the ultrathin sections with lead citrate (optional). Tissue fixed with osmium tetroxide and uranyl acetate often exhibit sufficient contrast for viewing under EMs, and the lead counterstain sometimes obscures weak HRP-DAB immunolabeling.

4. Notes

1. Polyclonal antibodies show several advantages over MAbs. Polyclonal antibodies usually:
   a. Can recognize denatured antigens.
   b. Show good specificity on Western blots.
   c. Are excellent reagents for immunoprecipitation owing to multivalent interaction of antibodies that recognize the same antigen.
   d. Yield strong signals for cell staining. However, major problems are the nonspecific binding and limited supply.

2. When raising rabbit antibodies directed against the human adrenergic receptors, we found that the quality of antibodies varied among rabbits and among antigens. Moreover, one should be aware that even rabbits yielding good antisera cannot be expected to yield unlimited volumes of good antisera, since the titer can drop with aging.

3. As the length of antigen increases, solubility decreases. When glutathione-*S*-transferase- (GST) fusion proteins cannot be recovered in supernatants after lysis of *Escherichia coli*, we recommend making new fusion protein constructs instead of trying to increase the solubility. In general, production of new fusion proteins is not very time-consuming (i.e., polymerase chain reaction [PCR] and sequencing). There are reports demonstrating increases of fusion proteins' solubility. One is cotransfection with or fusing with the protein of the thioredoxin gene *(41,42)*. Another is the addition of 2.0% sarkosyl to solubilize the fusion proteins from inclusion bodies *(43)*. In our experiments, cotransfection of thioredoxin gene did not help much to increase solubility of the fusion protein, and the sodium sarcosine–solubilized fusion proteins from inclusion bodies did not bind to glutathione–agarose beads well even after Triton X-100 was added to scavenge the sarcosine. In our opinion, these alternative methods should be followed only when left with no choice about the portions of the protein needed for GST fusion.

4. When using Affi-gels from Bio-Rad, we recommend that the fusion proteins be fused both to Affi-gels 10 and 15 because of the uncertainty of the quality of these columns. Regarding the column elution conditions, buffers of acidic and alkaline pH are commonly used. Beware that not all nonspecific binding can be eliminated, even when using different fusion protein constructs. In some cases, crude sera are more effective than affinity-purified antisera for immunoprecipitation.

5. Background staining or nonspecific binding is often encountered on Western blots. To reduce background, we have found the following methods to be helpful alone or in combination:
   a. Reduce the concentration of primary and secondary antibodies, and increase the incubation time with antibodies from 1 h to overnight. This will increase the sensitivity and reduce the nonspecific staining.
   b. Block and wash the membrane with and incubate the antibodies in RIPA buffer. RIPA buffer consists of 150 m*M* NaCl, 1.0% Nonidet P-40, 0.5% deoxycholate, 0.1% sodium dodecyl sulfate (SDS), 50 m*M* Tris (pH 8.0).
   c. Use different blocking buffers, such as 5% dry milk, in Tween-PBS or dry milk in RIPA buffer. Inclusion of serum obtained from the animal species used for the production of secondary antibody may also be helpful.
   d. Preadsorb the antiserum with total proteins obtained from tissue or cells known not to express the antigen.
   e. Affinity-purify the polyclonal antibodies using a peptide fragment from the antigen, rather than the entire fusion protein.

6. The detection limit on Western blots is approx 10 fmol/sample for glycosylated receptors. The limitation is owing, in part, to heterogenous glycosylation of the receptor, causing varied molecular weight of the molecule and, consequently, diffuse bands on Western blots. Thus, deglycosylation may help to increase the sensitivity of detection by sharpening the band.

   The other source of limitation is the amount of protein that can be applied to SDS-polyacrylamide gel electrophoresis (PAGE) gel slots. For most SDS-PAGE gels, the upper limit is about 200 µg/lane. Should the receptor expression level be above 100 fmol/mg protein, detection becomes feasible by SDS-PAGE. However, detection becomes nearly impossible when the expression level is 10–20 fmol/mg protein, as found in most intact cells.

Immunoreactive bands on Western blots must always be tested for specificity. This can be achieved by including a peptide fragment with the antiserum during incubation of the blot. Alternatively, the antiserum may be preadsorbed prior to incubating with the antiserum. Tissues or cells prepared from knockout mice are excellent sources for obtaining negative controls.

Enhanced chemiluminescence (ECL) is a widely used detection method for Western blot owing to its superb sensitivity. However, the enhanced sensitivity may also bring about smeared staining. This problem may be overcome by washing the membrane more rigorously and repeatedly and by using more stringent conditions.

7. G-protein-linked receptors have a tendency to aggregate. This tendency is increased when the protein suspension containing the receptor is boiled, as is done customarily prior to loading polyacrylamide gels. Such aggregated receptor molecules tend to stay in the stacking gel, rather than entering into the separation gel. Since the receptor protein can be electrophoresed successfully without boiling, the boiling step should be omitted.

8. Further details and examples of the application of the SIG procedure alone and in combination with HRP-DAB can be found in **ref. *44***.

References

1. Strader, C. D., Fong, T. M., Graziano, M. P., and Tota, M. R. (1995) The family of G-protein-coupled receptors. *FASEB J.* **9,** 745–754.
2. Rohrer, D. K. and Kobilka, B. K. (1998) Insights from in vivo modification of adrenergic receptor gene expression. *Annu. Rev. Pharmacol. Toxicol.* **38,** 351–373.
3. Aoki, C., Go, C. G. Venkatesan, C., and Kurose, H. (1994) Perikaryal and synaptic localization of α_{2A}-adrenergic receptor immunoreactivity in brain as revealed by light and electron microscopic immunocytochemistry. *Brain Res.* **650,** 181–204.
4. Kalsner, S. and Westfall, T. C. (1990) Presynaptic receptors and the question of autoregulation of neurotransmitter release, *Ann. NY Acad. Sci.* **604,** 652.
5. Venkatesan, C., Kurose, H., and Aoki, C. (1996) Cellular and subcellullar distribution of α2A- adrenergic receptor in the visual cortex of neonatal and adult rats. *J. Comp. Neurol.* **365,** 79–95.
6. Aoki, C. (1998) Cellular and subcellular sites for noradrenergic action in the monkey dorsolateral prefrontal cortex as revealed by the immunocytochemical localization of noradrenergic receptors and axons. *Cereb. Cortex* **8,** 269–277.
7. Milner, T. A., Lee, A., Aicher, S., and Rosin, D. L. (1998) Hippocampal α_{2A}-adrenergic receptors are located predominantly presynaptically but are also found postsynaptically and in selective astrocytes. *J. Comp. Neurol.* **395,** 310–327.
8. McCormick, D. A., Pape, H. C., and Williamson, A. (1991) Actions of norepi-

nephrine in the cerebral cortex and thalamus: implications for function of the central noradrenergic system. *Prog. Brain Res.* **88,** 293–305.

9. Strader, C. D., Picke, V. M., Joh, T. H., Strohsacker, M. W., Shorr, Lefkowitz, R. J., and et al. (1983) Antibodies to the beta-adrenergic receptor: attenuation of catecholamine-sensitive adenylate cyclase and demonstration of postsynaptic receptor localization in brain, *Proc. Natl. Acad. Sci USA* **80,** 1840–1844.

10. Aoki, T. H., Joh, and Pickel, V. M. (1987) Ultrastructural localization of immunoreactivity for β-adrenergic receptors in the cortex and neostriatum of rat brain. *Brain Res.* **437,** 264–282.

11. Aoki, C., Zemcik. Z. A., Strader, C. D., and Pickel, V. M. (1989) Cytoplasmic loop of β-adrenergic receptors: synaptic and intracellular localization and relation to catecholaminergic neurons in the nuclei of the solitary tracts. *Brain Res.* **493,** 331–347.

12. Aoki, C. and Pickel, V. M. (1990) Ultrastructural immunocytochemical evidence for presynaptic localization of beta-adrenergic receptors in the striatum and cerebral cortex of rat brain. *Ann NY Acad Sci.* **604,** 582–585.

13. Aoki, C. (1992). C-terminal fragment of β-adrenergic receptors: astrocytic localization in the visual cortex and their relation to catecholamine axon terminals, *J. Neurosci.* **12,** 781–792.

14. Aoki, C. and Pickel, V. M. (1992) Ultrastructural relations between β-adrenergic receptors and catecholaminergic neurons, *Brain Res. Bull.* **29,** 257–264.

15. Aoki, C. and Pickel, V. M. (1992) C-terminal tail of beta-adrenergic receptors: immunocytochemical localization within astrocytes and their relation to catecholaminergic neurons in the N. tractus solitarii and area postrema, *Brain Res.* **571,** 35–49.

16. Aoki, C., Lubin, M., and Fenstemaker, F. (1994) Columnar activity regulates astrocytic β-adrenergic receptor-like immunoreactivity in V1 of adult monkeys. *Vis. Neurosci.* **11,** 179–187.

17. Aoki, C. (1997) Differential timing for the appearance of neuronal and astrocytic beta-adrenergic receptors in the developing rat visual cortex as revealed by light and electron-microscopic immunocytochemistry. *Vis. Neurosci.* **14,** 1129–1142.

18. Susulic, V. S., Frederich, R. C., Lawitts, J., Tozzo, E., Kahn, B. B., Harper, M.-E., et al. (1995) Targeted disruption of the β_3-adrenergic receptor gene. *J. Biol. Chem.* **270,** 29,483–29,492.

19. Rohrer, D. K., Desai, K. H., Jasper, J. R., Stevens, M. E., Regula, D. P., Jr., Barsh, G. S., et al. (1996) Targeted disruption of the mouse beta1-adrenergic receptor gene: developmental and cardiovascular effects. *Proc. Natl. Acad. Sci. USA* **93,** 7375–7380.

20. Jurevicus, J. and Fischmeister, R. (1996) cAMP compartmentation is responsible for a local activation of cardiac Ca^{2+} channel by β-adrenergic agonists. *Proc. Natl. Acad. Sci. USA* **93,** 295–299.

21. von Zastrow, M., Link, R., Daunt, D., Barsh, G., and Kobilka, B. K. (1993) Subtype-specific differences in the intracellular sorting of G-protein-coupled receptors. *J. Biol. Chem.* **268,** 763–766.

22. Hirasawa, A., Sugawara, T., Awaji, T., Tsumaya, K., Ito, H., and Tsujimoto, G. (1998) Subtype-specific differences in subcellular localization and chlorethylclonidine (CEC) inactivation of α_1- adrenoceptors (ARs): CEC alkylates only the accessible cell surface α_1-ARs irrespective of the subtypes. *Mol. Pharmacol.* **52,** 764–770.
23. Fuxe, K. and Agnati, L. F. (eds.) *Volume Transmission in the Brain: Novel Mechanisms for Neural Transmission.* Raven, New York.
24. King, L. C., Lechan, R. M., Kugel, G., and Anthony, E. L. P. (1983) Acrolein: a fixative for immunocytochemical localization of peptides in the central nervous system, *J. Histochem. Cytochem.* **31,** 62–68.
25. Sitte, H. (1996) Advanced instrumentation and methodology related to cryoultramicrotomy: a review. *Scanning Microsc.* **10 (Suppl.),** 387–463.
26. Liou, W., Geuze, H. J., and Slot, J. W. (1996) Improving structural integrity of cryosections for immunogold labeling. *Histochem. Cell Biol.* **106,** 41–58.
27. Eldred, W. D., Zucker, C., Karten, H. J., and Yazula, S. (1983) Comparison of fixation and penetration enhancement techniques for use in ultrastructural immunocytochemistry, *J. Histochem. Cytochem.* **31,** 285–292.
28. Aoki, C., Starr, A., Kaneko, T., and Pickel, V. M. (1991) Identification of mitochondrial and non-mitochondrial glutaminase within select neurons and glia of rat forebrain by electron microscopic immunocytochemistry. *J. Neurosci. Res.* **28,** 531–548.
29. Erisir, A. and Aoki, C. (1998) Combined use of biocytin with avidin–biotin peroxidase for dual pre-embedding electron microscopy. *J. Neurosci. Methods* **81,** 189–197.
30. Sesack, S. R., Aoki, C., and Pickel, V. M. (1994) Ultrastructural localization of D2 receptor-like immunoreactivity in midbrain dopamine neurons and their striatal targets. *J. Neurosci.* **14,** 88–106.
31. Wouterlood, F. G. and Jorritsma-Byham, B. (1993) The anterograde neuroanatomical tracer biotinylated dextran-amine: comparison with the tracer Phaseolus vulgaris-leucoagglutinin in preparations for electron microscopy. *J. Neurosci. Methods* **48,** 75–87.
32. Baude, A., Molnar, E., Latawiec, D., McIlhinney, R. A. J., and Somogyi, P. (1994) Synaptic and nonsynaptic localization of the GluR1 subunit of the AMPA-type excitatory amino acid receptor in the rat cerebellum. *J. Neurosci.* **14,** 2830–2843.
33. Bernard, V., Somogyi, P., and Bolam, J. P. (1997) Cellular, subcellular and subsynaptic distribution of AMPA-type glutamate receptor subunits in the neostriatum of the rat. *J. Neurosci.* **17,** 819–833.
34. Raman I. M., Tong, G., and Jahr, C. E. (1996) Beta-adrenergic regulation of synaptic NMDA receptors by cAMP-dependent protein kinase. *Neuron* **16,** 415–421.
35. He, Y., Janssen, W. G. M., Vissavajjhala, P., and Morrison, J. H. (1998) Synaptic distribution of GluR2 in hippocampal GABAergic interneurons and pyramidal cells: a double-label immunogold analysis. *Exp. Neurol.* **150,** 1–13.

36. Kharazia, V. N., Phend, K. D., Rustioni, A., and Weinberg, R. J. (1996) EM localization of AMPA and NMDA receptor subunits at synapses in rat cerebral cortex. *Neurosci Lett.* **210,** 37–40.

37. Phend, K. D., Rustioni, A., and Weinberg, R. J. (1995) An osmium-free method of Epon embedment that preserves both ultrastructure and antigenicity for post-embedding immunocytochemistry. *J. Histochem. Cytochem.* **43,** 283–292.

38. Rubio, M. E. and Wenthold, R. J. (1997) Glutamate receptors are selectively targeted to postsynaptic sites in neurons. *Neuron* **18,** 939–950.

39. Wang, Y.-X., Wenthold, R. J., Ottersen, O. P., and Petralia, R. S. (1998) Endbulb synapses in the anteroventral cochlear nucleus express a specific subset of AMPA-type glutamate receptor subunits. *J. Neurosci.* **18,** 1148–1160.

40. Veznedaroglu, E. and Milner, T. A. (1992) Elimination of artifactual labeling of hippocampal mossy fibers seen following preembedding immunogold-silver technique by pretreatment with zinc chelator. *J. Microsc. Res. Tech.* **23,** 100, 101.

41. Yasukawa, T., Kanei-Ishii, C., Maekawa, T., Fujimoto, J., Yamamoto, T., and Ishii, S. (1995) Increase of solubility of foreign proteins in *Escherichia coli* by coproduction of the bacterial thioredoxin. *J. Biol. Chem.* **270,** 25,328–25,331.

42. LaVallie, E. R., DiBlasio, E. A., Kovacic, S., Grant, K. L., Schendel, P. F., and McCoy, J. M. (1993) A thioredoxin gene fusion expression system that circumvents inclusion body formation in the *E. coli* cytoplasm. *Biotechnology* **11,** 187–193.

43. Frangioni, J. V. and Neel, B. G. (1993) Solubilization and purification of enzymatically active glutathione *S*-transferase (pGEX) fusion proteins. *Anal. Biochem.* **210,** 179–187.

44. Pickel, V. M., Chan, J., and Aoki, C. (1993) Electron microscopic immunocytochemical labeling of endogenous and/or transported antigens in rat brain using silver-intensified one-nanometre colloidal gold, in *Immunohistochemistry II* (Cuello, A. C., ed.) John Wiley, New York, pp. 265–280.

Index